The Earth System

The Earth System

SECOND EDITION

Lee R. Kump
The Pennsylvania State University
Department of Geosciences
and Earth System Science Center

James F. Kasting
The Pennsylvania State University
Department of Geosciences and Meteorology
and Earth System Science Center

Robert G. Crane
The Pennsylvania State University
Department of Geography
and Earth System Science Center

PEARSON

Prentice
Hall

Upper Saddle River, New Jersey 07458

Library of Congress Cataloging-in-Publication Data

Kump, Lee R.
    The earth system / Lee R. Kump, James F. Kastin, Robert G. Crane.—
2nd ed.
        p. cm.
    Includes bibliographical references.
    ISBN 0-13-142059-3
    1. Gaia hypothesis.   I. Kasting, James F.   II. Crane, Robert G.   III.
Title.
    QH331.K798 2004
    577'.1—dc21   2003012099

Executive Editor: Patrick Lynch
Assistant Editor: Melanie Van Benthuysen
Editor-in-Chief: Sheri L. Snavely
Vice President of Production and Menufacturing: David W. Riccardi
Executive Managing Editor: Kathleen Schiaparelli
Managing Editor: Beth Sweeten
Production Editor/Composition: Pine Tree Composition, Inc.
Development Editor: Erin Mulligan
Manufacturing Buyer: Alan Fischer
Manufacturing Manager: Trudy Pisciotti
Marketing Manager: Christine Henry
Managing Editor, Audio/Visual Assets: Patty Burns
AV Editor: Jessica Einsig
Art Studio: Progressive Publishing Alternatives
Art Director: Jayne Conte
Cover Designer: Bruce Kenselaar
Photo Researcher: Kathy Ringrose
Cover Illustration: From "Sunlight Harvesting" by Glynn Gorick (113 Hemingford Road, Cambridge UK/www.gorick.u-net.com)
About the Cover: The cover illustrates a freshwater ecosystem with four trophic levels (i.e., a food chain) represented by four
    common species: *Chlamydomonas* (algal cells, part of the phytoplankton) are eaten by *Daphnia* (water fleas, part of the zoo-
    plankton), which are eaten by roach fish, which in turn are eaten by herons. This relationship, while familiar to most of us, may
    appear unfamiliar here because the image is scaled to reflect each species' biomass in the Earth system. Thus, the algae appear
    the largest while the heron is the smallest. The algae use sunlight, depicted in the lower left corner as photons being absorbed
    by chlorophyll molecules, to produce chemical energy (food) from carbon dioxide and water. That chemical energy is then uti-
    lized at each of the other higher trophic levels. The hydrologic cycle, represented by the clouds and the stream, is also driven
    by solar energy. In this one image, then, we see an example of the interconnectedness inherent in the study of the Earth
    system.

Any uncredited photos were supplied by the authors.

 © 2004, 1999 by Pearson Education, Inc.
Pearson Education, Inc.
Upper Saddle River, New Jersey 07458

Printed in the United States of America
10 9 8

ISBN 0-13-142059-3

Pearson Education Ltd., *London*
Pearson Education Australia Pty. Limited, *Sydney*
Pearson Education Singapore, Pte. Ltd.
Pearson Education North Asia Ltd., *Hong Kong*
Pearson Education Canada, Ltd., *Toronto*
Pearson Educación de Mexico, S.A de C.V.
Pearson Education—Japan, *Tokyo*
Pearson Education Malaysia, Pte. Ltd.

Contents

About the Authors

Lee R. Kump received his AB degree in geophysical sciences from the University of Chicago in 1981 and his PhD in marine sciences from the University of South Florida in 1986. He has been on the faculty of the Department of Geosciences at Penn State since 1986, where he now serves as Professor of Geosciences and affiliate of the NASA Astrobiology Institute and Penn State's Earth System Science Center (ESSC). Dr. Kump is the former coeditor of the preeminent Earth sciences journal *Geology* and is now editor of the *Virtual Journal of Geobiology* and associate editor of *Geochimica et Cosmochimica Acta.* He is a fellow of the Geological Society of America, and received the Distinguished Service Medal from the Geological Society of America in 2000. Dr. Kump's research interests include the behavior of nutrient and trace elements in natural environments, the evolution of ocean and atmosphere composition on geologic time scales, biogeochemical cycling in aquatic environments, and environmental change during extreme events (mass extinctions, extreme warm periods, glaciations) in Earth history.

James F. Kasting is a Professor at Penn State University, where he holds joint appointments in the Departments of Geosciences and Meteorology and is an affiliate of the NASA Astrobiology Institute and Penn State's ESSC. He received his undergraduate degree from Harvard University in Chemistry and Physics and did his PhD in Atmospheric Sciences at the University of Michigan. Prior to coming to Penn State in 1988, he spent 7 years in the Space Science Division at NASA Ames Research Center. Dr. Kasting is a Fellow of the American Association for the Advancement of Science and of the International Society for the Study of the Origin of Life. His research focuses on the evolution of planetary atmospheres, particularly the question of why the atmospheres of Mars and Venus are so different from that of Earth. Dr. Kasting is also interested in the question of whether habitable planets exist around other stars and how we might look for signatures of life by doing spectroscopy on their atmospheres.

Robert G. Crane received his PhD in Geography from the University of Colorado, Boulder. After working as a Research Associate in the National Snow and Ice Data Center and the World Data Center-A for Glaciology in Boulder, he spent a year teaching at the University of Saskatchewan before moving to Penn State in 1985. Dr. Crane's research has been on microwave remote sensing of sea ice, ice–climate interactions, and, more recently, regional-scale climate change, climate downscaling techniques, and climate change and variability in southern Africa. He is coeditor of a text on the applications of artificial neural networks in geography. Currently Dr. Crane holds the position of Professor in the Department of Geography and an affiliate of the ESSC. He also serves as the Associate Dean for Education in the College of Earth and Mineral Sciences at Penn State.

Preface

This is not a traditional Earth science textbook. Such books treat individual components of the Earth system—the solid Earth, atmosphere, and oceans—separately, with little consideration of the interplay among them or the important interactions with living organisms (the stuff of ecology texts). And, although they are the focus of this book, the modern environmental problems of global warming, ozone depletion, and loss of biodiversity are treated in a fundamentally different way here than in most texts. Here we recognize that these problems have analogues from Earth history: The geological past is the key to the present and to the future.

Content

Chapter 1, "Global Change," is an overview of these important issues—the observational data that convince us that serious problems exist and the events in Earth's history that illuminate how the Earth system responds under stress. The rest of the book is organized into three major sections. Chapters 2 through 9 are devoted to an exploration of how Earth "works." They develop the notion that processes active on Earth's surface are functioning together to regulate climate, the circulation of the ocean and atmosphere, and the recycling of the elements. The biota play an important role in all of these processes. Chapters 10 through 15 take the reader through the history of Earth, highlighting those events that provide lessons for the future. The final four chapters focus on the future of the Earth system, addressing the modern problems of global change and the prospect of life on other planets in the context of what was presented in the first two sections.

Revisions to the First Edition

In the four years since the first edition of this book came out, a lot has changed. Atmospheric CO_2 has increased by about 7 parts per million, freon-11 concentrations have decreased by 6 parts per trillion, and global surface temperatures have continued their inexorable but ragged rise. For this reason alone—just to keep up with the new data on global change—a book like this one needs to be regularly updated. However, it is not just the data that are changing. Ideas have been evolving as well during the past four years. New geologic evidence indicates that "Snow-ball Earth" episodes actually occurred not just once but several times during Earth's history. The case has been made that CH_4, rather than (or in addition to) CO_2, was the main greenhouse gas that helped to keep the early Earth warm despite reduced solar luminosity. The IPCC (Intergovernmental Panel on Climate Change) released a new report that for the first time states unambiguously that human activities are responsible for at least part of the observed surface temperature increase. And NASA's generous support for the new discipline of "astrobiology" has made us even more aware of the tight connections between the evolving Earth and its biota.

We have tried to reflect these and other changes in the revised edition of our book. We have added two new chapters: Chapter 6 (on global climate models) and Chapter 9 (on the biota, ecosystems, and biodiversity). We've also expanded our discussion of early Earth, now devoting two chapters to the topic: Chapter 10, on the origin of Earth and of life, and Chapter 11, on the effect life has had on the development of the atmosphere. Some of this involved simply reorganizing material that had previously been included in other chapters; however, a significant amount of new material has been added. Chapter 9 recognizes the importance of numerical modeling in the establishment of policy for a changing world. Chapter 9 highlights the role that the biota plays in the Earth system. Chapters 10 and 11 draw on the "universal" tree of life derived from sequencing of ribosomal RNA that places humans and fungi as closer relatives than different forms of bacteria. The order of Chapters 11 and 12 (Chapters 8 and 9 in the first edition) has been switched to reflect the increased importance of the O_2/CH_4 story for Precambrian paleoclimates.

Organization and Pedagogy

As in the first edition, we have employed a number of pedagogical features to assist in the learning process. Each chapter begins with "Key Questions" (objective questions students should be able to answer after they have read the chapter) and a "Chapter Overview" (a broad preview of the chapter to come). Within each chapter are boxed essays that provide interesting asides, more detailed or quantitative treatments of material in the text, or recent advances in scientific understanding. A "Chapter Summary" is provided in outline form at the end of each chapter to aid in reviewing the most important concepts.

"Suggested Readings" include both general readings and advanced readings for students (and instructors) interested in further information about the subject matter. These are followed by "Key Terms" lists, which consist of **boldfaced** terms that are introduced in the chapter and that appear in the "Glossary" in the back of the book. "Review Questions" focus students' review on important concepts and require only brief answers, whereas "Critical-Thinking Questions" are thought questions or analytical exercises that require students to synthesize concepts presented in the chapter. In designing this second edition, we have tried to organize our topics more logically and categorize special topics into "boxes" of different types, with the designations **A Closer Look**, which offers a closer examination of topics discussed in the book; **Important Concepts**, with in-depth presentations of fundamental concepts from the natural sciences essential to our understanding of the Earth system; and **Thinking Quantitatively**, which emphasizes how mathematics is used to better understand the workings of the Earth system. Instructors may choose whether to make any or all of these boxes assigned reading. We have also corrected errors pointed out to us by our students and by other faculty using the book, and we've brought the data graphs up to date. We hope that these changes will help make the book easier to use in a variety of different courses, as well as being more accessible and informative to students.

Chapter Sequencing

We anticipate that this book will be used in a variety of ways. We teach a general education class at The Pennsylvania State University that covers approximately three-quarters of the book during one semester. Several instructors teach this course, but not all of us choose to cover the same chapters. An instructor who is most interested in climate issues, for example, might use Chapters 1–6, 8 12, 14–16, and 19. One who is most interested in biodiversity might choose Chapters 1, 2, 8–11, 13, and 18. The course can also be tailored to emphasize either Earth history (Chapters 1, 2, 3, and 9–15) or modern global environmental problems (Chapters 1–6, 8, 9, and 16–19). By providing more material than can easily be covered in a one-semester course, we provide the flexibility to emphasize topics or topic areas that are of interest to different instructors and different groups of students.

Acknowledgments

In addition to the many people who helped with the first edition, we are especially grateful to Assistant Editor Melanie Van Benthuysen, Executive Editor Patrick Lynch, Developmental Editor Erin Mulligan, and Production Editor Patty Donovan. The following colleagues provided reviews of the first edition that were invaluable in our preparation of this edition:

Eric Barron, *Pennsylvania State University*
Chris Duncan, *University of Massachusetts, Amherst*
Jim Evans, *Utah State University*
Jonathan K. Filer, *Towson University*
Woody Hickcox, *Emory University*
Jean L. Hoff, *St. Cloud State University*
Hobart M. King, *Mansfield University of Pennsylvania*
Michael F. Rosenmeier, *University of Pittsburgh*
Cameron P. Wake, *University of New Hampshire*
Dean Wilder, *University of Wisconsin–La Crosse*

Lee R. Kump
James F. Kasting
Robert G. Crane
The Pennsylvania State University

Global Change

Key Questions

- What is meant by a "systems approach" to Earth science?
- How does global warming differ from the greenhouse effect, and is global warming actually occurring today?
- What is the Antarctic ozone hole, and what is its significance?
- Should we be concerned about tropical deforestation?
- What can understanding Earth's past tell us about Earth's future?

Chapter Overview

Earth is currently being altered at an unprecedented rate by human activity. The buildup of greenhouse gases in the atmosphere is expected to warm Earth's climate in the future—and may have done so already. The accumulation of chlorine-containing compounds in the atmosphere has damaged the ozone layer over part of the globe. Deforestation of the tropics may be causing large decreases in biodiversity. How serious are these problems, and how do they compare with past changes in the Earth system? Chapter 1 lays out the evidence of these changes and explains why an integrated, systems approach is useful in analyzing them.

Introduction

Our world is changing. In fact, Earth has always been changing and will continue to do so for ages to come. Yet, there is a difference between the changes occurring now and those that occurred previously. Earth is changing faster today than it has throughout most of its 4.6 billion-year history. Indeed, it may be changing faster than it ever has, except perhaps in the aftermath of giant meteorite impacts. The cause of this accelerated pace of change is simple: human activity. Human populations have expanded in numbers and in their technological abilities to the point at which we are now exerting a significant influence on our planet. The effects of our actions are seen most clearly in the thin envelope of gases that supports our existence, the *atmosphere,* but they are observable elsewhere as well. Forests, mountains, lakes, rivers, and even the oceans exhibit the telltale signs of human activity.

To what extent are these **anthropogenic** (human-induced) changes a cause for concern? All of us can think of situations in which human influence has clearly been detrimental to the environment, for example, cities plagued with polluted air and water. But these are local problems, and they are hardly new. Humans have generated local pollution ever since they first developed agricultural societies around 10,000 years ago. Human inhabitants of Easter Island (which lies off the southwest coast of South America) may have set the stage for the demise of their culture about 700 years ago through **deforestation**—that is, by the clearing of all the trees—of the island. Advanced technology is not needed to damage one's immediate surroundings.

Today, however, because technological advances abound and because there are simply more people on Earth than ever before, human influence extends to the global environment. For example, *global climate,* the prevailing weather patterns of a planet or region over time, is being altered by the addition of greenhouse gases to the atmosphere. **Greenhouse gases** are gases that warm a planet's surface by absorbing outgoing *infrared radiation*—radiant heat—and reradiating some of it back toward the surface. This process is called the **greenhouse effect.** (The analogy is not perfect, however, because the glass walls of a greenhouse keep the air warm by inhibiting heat loss by upward air motions rather than by absorbing infrared radiation.) The greenhouse effect is a natural physical process that operates in all planetary atmospheres. For example, the greenhouse effect, and not solely proximity to the Sun, is thought to account for the high surface temperature of Venus—460°C, compared with about 15°C at Earth's surface. On Earth, some greenhouse gases (such as water vapor) are entirely natural, but others are partly or wholly anthropogenic. The most abundant anthropogenic greenhouse gas on Earth is **carbon dioxide,** CO_2, which is produced by the burning of **fossil fuels** (fuels such as coal, oil, and natural gas that are composed of the fossilized remains of organisms) and by deforestation. When trees are cut down, they decay, and the carbon in their trunks, branches, and leaves is released as CO_2. Carbon dioxide is also a component of volcanic emissions, and it is cycled rapidly back and forth by living plants and animals. Thus, its abundance is controlled by a combination of natural and human-controlled processes.

Humankind is also capable of damaging Earth's fragile ozone layer. The **ozone layer** is a chemically distinct region within the *stratosphere,* part of the atmosphere. The ozone layer protects Earth's surface from the Sun's harmful *ultraviolet radiation.* Ultraviolet radiation is what gives us suntans but also sunburns. **Ozone** (O_3) is a form of oxygen that is much less abundant than, and chemically unlike, the oxygen that we breathe (O_2). As we shall see, the **ozone hole** over Antarctica, a patch of extremely low ozone concentration in the ozone layer, is known to be anthropogenic in origin.

We are also now deforesting parts of the planet—mainly the tropics—at a rate that was not possible until the 19th century. As we cut down the forests, we kill off many species of plants and animals that live there. Hence, we are now causing substantial decreases in **biodiversity,** or the number of species present in a given area.

The effects of these global environmental problems on humans are more difficult to assess than are the effects of local air and water pollution. Depletion of the ozone layer is a worrisome prospect, but serious losses of

ozone have so far been confined to the region near the South Pole, where few people live. Small decreases in ozone have been observed at mid-latitudes, but these are not yet thought to pose a serious hazard to health. Loss of biodiversity in the tropics has thus far only indirectly affected people who live at temperate latitudes. Tropical deforestation and fossil fuel burning could affect everyone by causing **global warming,** a warming of Earth's atmosphere due to an anthropogenic enhancement of the greenhouse effect. However, some people (those living in Alaska or Siberia, for instance) might see global warming as less of a threat than would others. How can we decide which global environmental problems are truly urgent and which may simply deserve careful, long-term study?

Three Major Themes

One major theme of ours will be global environmental issues such as these. We should be able to make our own decisions as to which modern environmental problems are worth worrying about and which, if any, are not. Making such decisions intelligently requires at least some knowledge of the scientific questions involved. Some of the issues, global warming in particular, are also politically contentious, because the scientific questions surrounding them are not entirely answered and because the actions needed to address them are potentially very costly. In such cases, it is important that both policymakers and citizens understand the problem at a reasonably detailed level.

To understand how humankind is changing the environment today, we need also to understand how the environment was changing before humans came on the scene. Otherwise, it is difficult to distinguish short-term, anthropogenic trends from longer-term, natural trends. So, a second major theme of ours is global change in the past. Climate is a good example of the overlap of short and long time scales of global change, and one to which we will return frequently. Earth's climate is predicted to warm over the next few decades to centuries as a consequence of the buildup of CO_2 and other greenhouse gases in its atmosphere. Evidence of past climates has come from cores drilled into sediments on the ocean floor. (**Sediments** are layers of unconsolidated material that is transported by water or air.) This evidence indicates that we are in the midst of a relatively short *interglacial period* (a warm interval marked by the retreat of Northern Hemisphere ice sheets) in between *glacial periods* (cold intervals marked by the buildup of these ice sheets). Hence, in the absence of anthropogenic influence, the planet would be destined over the next few thousand years to slip slowly into the next Ice Age. Which of these tendencies—global warming or the transition to a glacial period—will win out? We argue later that warming is likely to win out in the short term, because the rate of increase of atmospheric CO_2 and other greenhouse gases is faster than the historical rate of interglacial-to-glacial climate change. Thus, the question of time scales is important. Understanding how and why climate has changed in the past can help us understand how it may change in the future.

We are introduced to these two major themes in this chapter. A third major theme of ours is *systems*—in particular, the *Earth system*. We examine this theme more thoroughly in Chapter 2. For now, let us say just that a **system** is a group of components that interact. The **Earth system** is composed of four parts: the atmosphere, the hydrosphere, the biota, and the solid Earth (Figure 1-1). As we have seen, the **atmosphere** is a thin envelope of gases that surrounds Earth. The **hydrosphere** is composed of the various reservoirs of water, including ice. The **biota** include all living organisms. (Some ecologists define the *biosphere* as the entire region in which life exists, but we will avoid that term here because it overlaps our other system components.) The **solid Earth** includes all **rocks,** or consolidated mixtures of crystalline materials called *minerals,* and all unconsolidated rock fragments. It is divided into three parts: the core, mantle, and crust. The **core** of any planet or of the Sun is the central part. Earth's core is a dense mixture of metallic iron and nickel and is part solid, part liquid. The **mantle** is a thick, rocky layer between the core and crust that represents the largest fraction of Earth's mass. The **crust** is the thin, outer layer, which consists of light, rocky matter in contact with the atmosphere and hydrosphere.

One of our goals is to show how the different components of the Earth system interact in response to various internal and external influences, or *forcings*. A well-known example of a forcing is the variation in the amount of sunlight received in each hemisphere during the course of a year. The response to this forcing, which is governed by the interaction between the atmosphere and the hydrosphere, is the seasonal cycle of summer and winter. But there are other, more subtle forcings at work as well that may engage all four components of the Earth system. Some examples are given later in this chapter.

Chapters 3 through 9 describe the various components of the Earth system in some detail. These chapters are not particularly distinctive; many Earth science texts do much the same thing. However, Chapter 1 and all the later chapters are devoted to problems, such as global climate history and modern global change, that cut across traditional disciplinary boundaries and that involve interactions among different parts of the Earth system. It is here that this book differs from most other introductory textbooks. The systems approach adopted here can lead to a more in-depth understanding of such problems by providing a convenient way of analyzing complex interactions and predicting their overall effect.

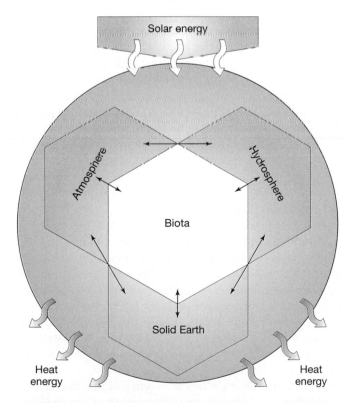

FIGURE 1-1

Schematic diagram of the Earth system, showing interactions among its four components. (From R.W. Christopherson, *Geosystems: An Introduction to Physical Geography, 3/e,* 1997. Reprinted by permission of Prentice Hall, Upper Saddle River, N.J.)

Global Change on Short Time Scales

We start our discussion of the Earth system by introducing three major global environmental changes that are occurring today: global warming, ozone depletion, and tropical deforestation. Afterward, we will backtrack to discuss how the Earth system operated in the past and how that may help us predict what will happen to it in the future.

Evidence of Global Warming

The most pervasive, and at the same time controversial, environmental change that is occurring today is global warming. This issue is extremely complex because it involves many different parts of the Earth system. It is controversial because it is difficult to separate anthropogenic influences from natural ones and because its causes are deeply rooted in our global industrial infrastructure; hence, these causes would be difficult to eliminate. A major goal of this book, therefore, is to help the reader understand global warming and to put it in the context of past climatic change.

Although the terms "greenhouse effect" and "global warming" are sometimes used interchangeably, the two phenomena are very different. The greenhouse effect is an indisputably real, natural process that keeps the surfaces of Earth and the other terrestrial planets warmer than they would be in the absence of an atmosphere. Global warming is an increase in Earth's surface temperature brought about by a combination of industrial and agricultural activities. These activities release gases that bolster the greenhouse effect. At present, not all scientists are convinced that global warming has begun. Most researchers agree that the climate has warmed over the past century, but not all of them believe that this warming is a result of human activities. However, the number of global warming skeptics has dwindled over the past several years. The influential Intergovernmental Panel on Climate Change released a report in 2001 that says that the evidence that humans have warmed the climate is now very strong. The debate is shifting to the question of how big that warming will become in the future.

Measurements of Atmospheric CO_2: *The Keeling Curve.*
The data that have aroused much of the current concern about global warming are shown in Figure 1-2. The graph shows the atmospheric CO_2 concentrations measured at the top of Mauna Loa, a 4,300-meter-high volcano in Hawaii, over a 40-year interval. Mauna Loa was chosen as the measurement site because the air blowing over its summit—clean air from the western Pacific Ocean—is far removed from local sources of pollution. The measurements were begun in 1958 by Charles David Keeling of the Scripps Institute of Oceanography. For this reason, the data are often referred to as the "Keeling curve."

In Figure 1-2 the concentration of atmospheric gas is measured in *parts per million,* or *ppm.* A value of 1 ppm

of a particular gas means that one molecule of that gas is present in every million air molecules. We shall use the abbreviation "ppm" to represent parts per million *by volume* rather than parts per million *by mass.* (In technical literature, *ppmv* is often used for parts per million by volume.) Units of mass and volume are not interchangeable, because a given gas molecule may be heavier or lighter than an average air molecule. Although one part per million may not sound like much, it represents a large number of molecules. A cubic centimeter of air at Earth's surface contains about 2.7×10^{19} molecules, so a 1-ppm concentration of a gas would have 2.7×10^{13} molecules in that same small volume. (If you are not familiar with scientific notation, refer to Appendix I for help.)

As Figure 1-2 shows, the CO_2 concentration in 2001 was about 371 ppm. We say "about" because the atmospheric CO_2 concentration varies slightly from place to place and oscillates seasonally over a range of 5 to 6 ppm. This seasonal oscillation has to do with the "breathing" of Northern Hemisphere forests. Forests take in CO_2 from the atmosphere (and give off O_2) in spring and summer, and they release CO_2 back to the atmosphere during fall and winter. Hawaii is in the Northern Hemisphere (latitude 19° N) and hence is influenced by this cycle. The cycle is reversed in the Southern Hemisphere, but the amount of land area is much smaller, so the magnitude of the CO_2 change is reduced.

Keeling's data show, in addition to this seasonal oscillation, that atmospheric CO_2 levels have increased significantly since 1958. The mean CO_2 concentration that year was about 315 ppm, or 56 ppm lower than today's value. The average rate of increase in CO_2 concentration since then has been 56 ppm/43 yr, or about 1.3 ppm/yr. More-detailed inspection of the curve reveals that the rate of CO_2 increase rose from 0.7 ppm/yr in the early 1960s to 1.7 ppm/yr in the late 1990s. Scientists believe that most of the increase in atmospheric CO_2 has been caused by the combustion of coal, oil, and natural gas but that tropical deforestation is also partly to blame.

The evidence that atmospheric CO_2 is increasing is indisputable. Similar measurements have been conducted at many different stations around the globe. The long-term increase in CO_2 is visible in every set of measurements and is essentially the same as that seen at Mauna Loa. (The range of the seasonal fluctuations, however, varies with the location.) For this reason, both scientists and policymakers agree that the long-term trend in atmospheric CO_2 is real rather than an artifact.

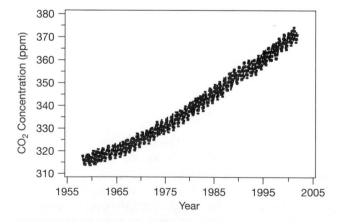

FIGURE 1-2

Measurements of atmospheric CO_2 concentrations at the top of Mauna Loa in Hawaii. These data are known as the "Keeling curve." (Source: C.D. Keeling and T.P. Wharf, Scripps Institute of Oceanography, La Jolla, California. (*http://cdiac.esd.ornl.gov/trends/co2/sio-mlo.htm*)

CO_2 Data from Ice Cores. When did this increase in atmospheric CO_2 begin, and what was the CO_2 level before that time? If we had to rely entirely on measurements made in the modern era, we would not be able to answer these questions. This is where analysis of the record of climate in the past can help. The composition of the atmosphere in the past can be determined by analyzing the composition of air bubbles trapped in polar ice. The bubbles formed as snow at the top of an ice sheet is compacted, and their composition is preserved as they are buried under more snow. The age of the ice can be determined by drilling deep into the ice, removing a section of it, and counting the annual layers of snow accumulation. Figure 1-3 shows results from ice cores—cylindrical sections drilled into the ice—taken at several locations on Antarctica. This graph compares the CO_2 composition of the air bubbles in the ice with a "smoothed" version of the Keeling curve (the dashed curve, from which the seasonal oscillation has been removed). The fact that the ice core measurements match up well with the direct atmospheric measurements in 1958 is convincing evidence that the ice core technique for determining atmospheric CO_2 concentrations yields reliable results.

According to these measurements, the buildup of atmospheric CO_2 began early in the 19th century—well before the dawn of the Industrial Age, which started in earnest around 1850. The rise in CO_2 levels between 1800 and 1850 has been attributed to the deforestation of North America by westward-expanding settlers and is thus known as the *pioneer effect*. The ice core measurements show that the *preindustrial CO_2 concentration* (the value

circa 1800) was about 280 ppm. Evidently, humans have been responsible for almost a 30% increase in atmospheric CO_2 concentration over the past two centuries.

Other Greenhouse Gases. Carbon dioxide is not the only greenhouse gas whose concentration is currently on the rise. Methane (CH_4), nitrous oxide (N_2O), and certain **chlorofluorocarbon** compounds **(CFCs)** have also been increasing as a result of human activities. Also called *freons*, CFCs are synthetic compounds containing chlorine, fluorine, and carbon. Collectively, such gases that are present in the atmosphere in very low concentrations, called **trace gases,** are thought to have contributed almost as much additional greenhouse effect over the past few decades as has CO_2. (Because CO_2 is much less abundant than N_2 or O_2, it is also classified as a trace gas, but it is more than 200 times as plentiful as any of the other gases mentioned here and hence deserves to be in a class by itself.) CFCs have also been implicated in the destruction of stratospheric ozone, as we discuss later in this chapter. For now, we simply note that the evidence for an increase in anthropogenic greenhouse gases is unequivocal: Humans are indeed modifying the composition of Earth's atmosphere. The extent to which we should be concerned about it remains to be determined.

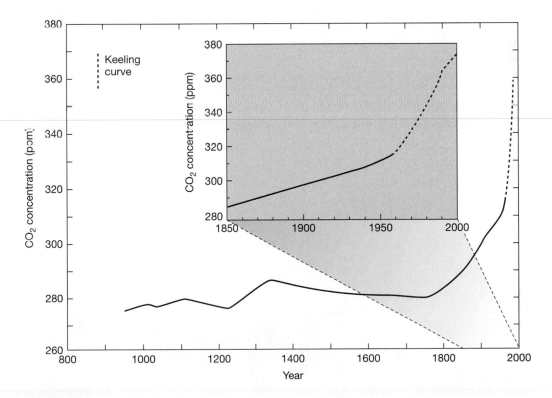

FIGURE 1-3

Atmospheric CO_2 concentrations over the past 1,000 years, as determined from ice cores and from direct atmospheric measurements. (The dashed line is the Keeling curve.) (After *Climate Change,* 1994, Intergovernmental Panel on Climate Change, Cambridge: Cambridge University Press)

Observed Changes in Surface Temperature. The observed rise in greenhouse gases is quite well documented, but what about the effects of this rise? Is there any direct evidence that climate is changing as a result?

The answer to this question is yes and no. Historical data indicate that Earth's surface temperature is on the increase. The data are not as easy to interpret as are the greenhouse gas data discussed earlier, but they are considered to be reliable. At a number of stations around the world, scientists have made accurate atmospheric temperature measurements that date back more than a century. Ocean-crossing ships have also routinely measured sea-surface temperatures during most of this time. Figure 1-4 illustrates the combined data from both types of historical measurements for the entire globe. The mean surface temperature from 1961 to 1990 has been subtracted from the data. The global mean surface temperature has increased from about 0.3°C below this mean value prior to 1900 to about 0.4°C above this mean value today. The overall temperature increase during the 20th century was thus approximately 0.7°C (or 1.3°F). This increase is broadly consistent with the warming expected from a 30% rise in atmospheric CO_2. However, if one compares Figure 1-4 with Figure 1-3, one can see that the surface temperature does *not* increase as uniformly or at the same rate as does atmospheric CO_2. Evidently, the climate is influenced by other factors as well. Problems do exist with these historical temperature data. For example, weather stations located near cities are subject to a well-documented "heat island" effect: As a city grows and as more area becomes covered with dark surfaces such as asphalt, more sunlight is absorbed and the local air temperature can increase by as much as 3°C. This systematic error has been removed from the data shown in Figure 1-4, but it is still a source of uncertainty because it is difficult to remove accurately. (*Systematic* errors exhibit a regular pattern. *Random* errors do not follow any pattern.) Sea-surface temperature measurements are also subject to systematic errors. Prior to the mid-1900s, water temperatures were determined by the "bucket method." A crew member dropped a bucket over the side of the ship, then hauled it back up and measured its temperature with a thermometer. Since then, water temperatures have generally been measured with flow-through devices located on the ship's hull. The two methods do not yield exactly the same results, because the samples may be taken at different water depths and because buckets can warm or cool as they are being examined. Furthermore, the current procedure draws water up through the ship (normally near the engines) and can heat it up. These effects, too, can be corrected for, but not without creating additional uncertainties.

A second problem with the temperature data is that the coverage in time and space is much better in some parts of the world than in others. Populated areas of Europe and North America have been monitored most closely and for the longest time, so the coverage is best in these regions. Most land areas in the Southern Hemisphere have shorter and less-consistent temperature records. And the coverage over some regions of the ocean, particularly remote parts of the Southern Ocean where few ships travel regularly, is sparse indeed. Because sea-surface temperatures can now be monitored from satellites, the oceanic database should improve in the future. However, it may well require several decades of such measurements to establish reliable trends.

Despite such difficulties, climatologists who collect and analyze these surface temperature data are confident that the observed half-degree warming trend over the past century is real. This does not mean, though, that it has been caused by human activities. Evidence shows that the climate was unusually cool between about 1500 and 1850.

FIGURE 1-4

Change in global average surface temperature since 1861. The data are expressed as deviations from the 1961 to 1990 mean value. (Source: Intergovernmental Panel on Climate Change—Third Assessment Report: *http://www.ipcc.ch/pub/wg1TARtechsum.pdf*)

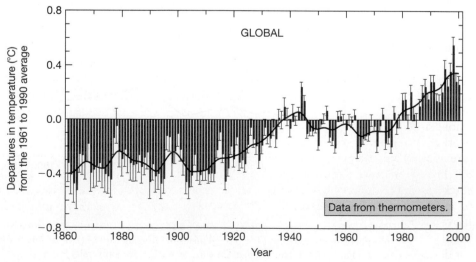
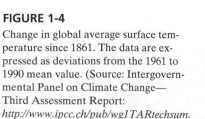

This period has been termed the "Little Ice Age." At least part of the warming since that time may represent a recovery from that naturally cool period rather than warming produced by anthropogenic greenhouse gases. This is another illustration of why it is necessary to understand the past if we want to predict the future.

An additional puzzle in the data shown in Figure 1-4 is that the warming trend seemed to slow, or stop entirely, between about 1940 and 1970. In the Northern Hemisphere, temperatures actually declined by a few tenths of a degree during this period. The decrease over Northern Hemisphere land areas is so pronounced that, by 1970, some climatologists were concerned that Earth might be entering a new glacial period. This worry was heightened by the historical data mentioned earlier that indicated that the present interglacial period might be nearing its end.

One possible explanation for the 1940 to 1970 cooling trend is that it was caused by increased reflection (and thus decreased absorption) of sunlight by *sulfate aerosol particles.* These tiny airborne particles are formed from sulfur dioxide (SO_2) emitted by the burning of coal. Most of the coal burning has taken place in the Northern Hemisphere, so this hypothesis could also explain why that hemisphere cooled more than did the Southern Hemisphere. Recent climate model simulations show that the magnitude of the aerosol effect is sufficient to account for the observed trend. But coal burning also releases CO_2 and hence should contribute to global warming—just the opposite of the observed effect during this 30-year period. This situation is a good example of why it is necessary to understand the whole Earth system in some detail if we are to interpret properly the changes that are occurring.

We cannot assume, however, that even though coal burning may have cooled Earth from 1940 to 1970, it will continue to do so in the future. In the United States, SO_2 is now being removed, or "scrubbed," from smokestack emissions in order to reduce its contribution to acid rain. **Acid rain** is produced when various acids, including sulfuric acid formed from the oxidation of SO_2, dissolve in rainwater. Acid rain can kill fish and damage plants in regions downwind from strong sources of pollution. It has been a problem in parts of the northeastern United States and in eastern Canada because there are many coal-fired power plants along and northward of the Ohio River valley. Other parts of the world, notably Europe, have problems with acid rain as well. Paradoxically, cleaning up smokestack emissions to cut down on acid rain may exacerbate the problem of global warming by reducing sulfate aerosol concentrations in the atmosphere.

Even if we were to quit scrubbing SO_2 out of smokestack gases, the ultimate effect of coal burning would be to warm Earth's atmosphere. Sulfate aerosols are removed from the lower atmosphere by precipitation in a matter of weeks, whereas CO_2 lingers in the atmosphere for decades to centuries. Thus, the CO_2 effect on climate is cumulative, whereas the aerosol effect is not. This example points out the importance of being aware of the time scale on which a global change occurs.

Possible Consequences of Global Warming. Although there is still some debate about whether humans have already altered the global climate, most climatologists agree that we will do so in the future if we continue to consume large amounts of fossil fuel. Should this be a cause for concern? In terms of the change in mean global temperature, we might expect people living in hot places such as India to be worried, whereas those living in Siberia would look forward to the change. But the problem is not quite so simple: A change in temperature might cause other changes as well. A rise in sea level is one frequently mentioned concern. Sea level has already risen by at least 10 cm over the past century. The likely cause is *thermal expansion* of a gradually warming ocean; like most forms of matter, seawater expands when it is heated (unlike pure water which, paradoxically, actually contracts when warmed from 0–4°C). But warmer temperatures could also induce melting of mountain glaciers and ice caps. Increases in sea level on the order of several meters are possible within the next few centuries, and even larger changes are possible in the very long term. Such changes could have serious consequences for people in coastal areas and would be catastrophic for those in small island states. Other climatic changes may also have a broad scale impact on agriculture, including decreases in soil moisture in certain areas and the spread of tropical insect pests. We will return to these possible side effects of global warming later; for now, note simply that the issues are complex and that there are very few simple answers. We also note that this is another reason to study past climate: Earth has been significantly warmer at various times in its past, and we may learn something about what it could be like in the future by examining those past time periods.

Evidence of Ozone Depletion

Global warming is not the only global environmental problem that has caught the attention of the public. Since at least 1985, the potential depletion of stratospheric ozone has also been in the news. (Stratospheric ozone should not be confused with *tropospheric* ozone—ozone near ground level—which is also often in the news because it is a component of *smog.*) The **stratosphere,** where most of Earth's ozone is located, is a layer of the atmosphere that extends from about 10 to 50 km in altitude. Stratospheric ozone is important to living organisms, because it absorbs many of the Sun's harmful ultraviolet rays. Ultraviolet radiation causes skin cancer and other health problems in humans. It adversely affects other organisms as well, notably, microscopic algae that are the base of the food chain in aquatic environments.

The year 1985 was a key one in stratospheric ozone research, because it marked the discovery of the ozone hole above Antarctica. Each year since about 1976, stratospheric ozone levels near the South Pole have fallen by large amounts during October, which is springtime in the Southern Hemisphere. Figure 1-5 shows year-to-year variations of the mean ozone *column depth* above Halley Bay in Antarctica for Octobers between 1957 and 2001. The ozone column depth is the total amount of ozone per unit area above a certain location. The decrease in ozone near the South Pole during October is striking: Ozone levels during October dropped by about half during a short period between 1975 and 1990. Since then, they have remained relatively constant. During the rest of the year, ozone levels in this region have remained close to normal throughout this time period. What has been destroying half the ozone over Antarctica during one particular month?

As soon as the ozone hole was discovered, atmospheric scientists guessed that chlorine compounds were to blame. By 1974, scientists had confirmed that chlorine is capable of destroying stratospheric ozone, and stratospheric chlorine levels have been increasing for the past few decades. Scientists are now fairly certain that the ozone hole is caused by chlorine compounds released from the breakdown of anthropogenic CFCs. The definitive evidence was provided in 1987, when a NASA research plane flew directly into the hole. One of the plane's

instruments measured chlorine monoxide, ClO, which was thought to be a main culprit in ozone destruction; another instrument measured ozone (Figure 1-6). Outside the hole, ozone concentrations were at their normal stratospheric level, and ClO concentrations were very

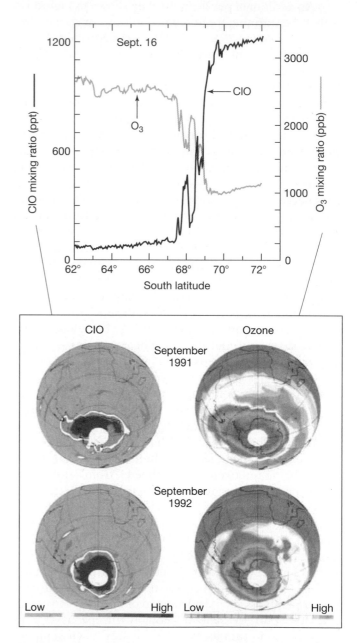

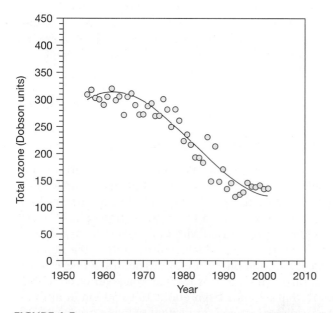

FIGURE 1-5

Mean total ozone over Antarctica during the month of October. The units, called Dobson units, measure the gas per unit area between Earth's surface and the top of the atmosphere (a measurement known as the column depth). One Dobson unit (DU) is equivalent to a 0.001-cm-thick layer of pure ozone at the surface. (Source: *http://www.antarctica.ac.uk/met/jds/ozone/images/zmeanoct.jpg*)

FIGURE 1-6

[See color section] (Top) Simultaneous measurements of ozone (O₃) and chlorine monoxide (ClO) made from a NASA aircraft as it flew into the Antarctic ozone hole in September 1987. The hole was entered at a latitude of about 68° S. The units ppt and ppb stand for "parts per trillion" and "parts per billion," respectively. (Bottom) Contour plots of ClO and O₃ concentrations obtained from spacecraft measurements. These data also show that ozone is low where ClO is high. (From R.W. Christopherson, *Geosystems: An Introduction to Physical Geography, 3/e,* 1997. Reprinted by permission of Prentice Hall, Upper Saddle River, N.J.)

low. Inside the hole, ozone values were more than a factor of two lower, and ClO values were about 15 times higher, than the respective values outside the hole. Faced with such a strong inverse relationship, even scientists who had been skeptical about the connection between stratospheric chlorine and ozone depletion were driven to conclude that the chlorine was directly responsible for destroying the ozone.

The real concern about ozone depletion is not whether it is occurring over Antarctica in October but whether it might occur at hazardous levels over populated regions of the globe. (The few people living down in the far southern portions of Chile and New Zealand are already concerned because they are so close to Antarctica.) So far, nothing as dramatic as the Antarctic ozone hole has been seen elsewhere. However, ozone does seem to be decreasing gradually at mid-latitudes in both hemispheres, perhaps because CFC concentrations in the upper stratosphere are still going up. The good news is that the ground-level concentrations of most CFCs are now decreasing because production of these gases has been reduced or eliminated. Hopefully, the world has acted in time to prevent ozone depletion from becoming a catastrophic problem.

Deforestation and Loss of Biodiversity

Ever since a substantial portion of the human population switched from being hunters and gatherers to being farmers some 10,000 years ago, humans have been altering the land surface. More and more of Earth's land is being "managed" in one way or another—to the extent that it is now fairly difficult to find land areas that are pristine.

Most of these changes have tended to reduce the complexity of the landscape, such as when forested areas (or grasslands) have been cleared and replaced with a single crop species. When the natural vegetation cover is removed, it is not simply the plant species that are lost. With the plants go all the animals (mammals, birds, insects, and so on) and microorganisms that depended on that vegetation in order to live. New species may replace them, but normally the number of species decreases, that is, biodiversity is reduced. When a species is unable to move away or adapt, the change in land use can result in extinction of the species. The genetic information that is shared by—and only by—all the members of that species is thus lost permanently.

Some of the best-known examples of animal species that have gone extinct are the woolly mammoth, the saber-toothed tiger, the dodo bird, and the dinosaurs. Many species that exist today, such as the mountain gorilla and the giant panda, are faced with the threat of extinction. The potential loss of these large mammals represents only the most visible of many similar threats.

The largest, and potentially the most significant, species loss occurring today is taking place in tropical rain-

forests. These warm, moist forests are centered around the Equator. Marked by lush vegetation, they are the most biodiverse habitat on Earth, but they are rapidly disappearing due to deforestation: The trees have been cleared for grazing, farming, timber, and fuel. By 1990, the total area of tropical rainforests had been reduced to less than half the estimated prehistoric cover. The rapidity of deforestation of the Amazon rainforest is illustrated in Figure 1-7. Exactly how fast the tropical forests are

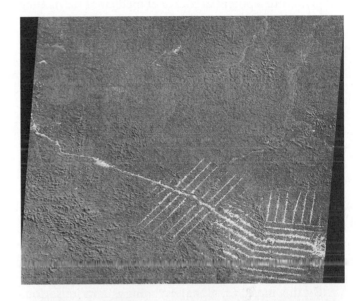

FIGURE 1-7

Satellite photos of Amazonia in 1972 and 1992. (Top: From Earth Satellite Corporation/Science Photo Library, Photo Researchers, Inc. Bottom: From NASA/Science Photo Library, Photo Researchers, Inc.)

disappearing is difficult to determine, but the loss rate is thought to approach 1.8% per year. If deforestation continues at such a rate, by the first quarter of the 21st century almost half the remaining rainforests will be lost, along with 5–10% of all the species on Earth.

Which Changes Should Concern Us the Most?

The concerns about the loss of tropical species are, in some ways, less immediate than the concerns about ozone depletion or global warming. One worry is that the tropical plants are a potential source of medicines for fighting cancer and other diseases. This concern is valid, but it does not have the urgency of the prospect of instantaneous sunburn on exposure to the Sun or of entire states or even entire nations being submerged by a rising sea level.

This does not mean, however, that species loss is not a serious problem. Indeed, in some ways it may be the most serious problem of all. One way of judging the severity of a problem is to estimate how long it would take Earth to recover. If we take this approach, ozone depletion is the least serious problem. The lifetime of chlorofluorocarbons in the atmosphere is on the order of 50 to 150 years, when they are eventually destroyed by solar ultraviolet radiation. This range is long enough to raise serious concerns, but the ozone level should be restored within a few human generations if the preventive measures now in place are continued or strengthened.

By this measure, global warming is a more serious problem because the time scale for recovery could be much longer than 150 years. If we actually do consume an appreciable amount of the fossil fuels that are still available to us, atmospheric CO_2 levels could remain elevated for many thousands of years. Most of the excess CO_2 would be absorbed by the oceans during this time, but even then it would not be completely gone. As we will see in later chapters, it would likely take more than a million years for the excess CO_2 to be removed from the oceans and for atmospheric CO_2 to return to its preindustrial level.

Although this time scale sounds long, it is short in comparison with the time required to restore global biodiversity. Analysis of the fossil record shows that the time scale for recovery of biodiversity after a **mass extinction** (the dying out of many species within a geologically short time interval) is on the order of tens of millions of years. In fact, the system never does recover completely: Although many new species appear and flourish after a mass extinction, they are *different* from the ones that went extinct. That is why humans, instead of dinosaurs, now rule Earth! So, if we do induce a mass extinction of tropical species by deforestation, things will never again be the same.

Global Change on Long Time Scales

We have touched on three major global environmental changes that are occurring in the Earth system today: global warming, ozone depletion, and tropical deforestation. To understand fully the significance of these changes, however, we must understand how the Earth system operated prior to human intervention. Here, we preview three examples of past global change—glacial–interglacial cycles, mass extinction, and changes in solar luminosity—and show how the geologic record provides evidence that allows us to study such changes.

Before we look at these examples of past global change, let us see where they occur on the *geologic time scale* (Fig. 1-8). Geologic time is divided into various intervals at several different levels. *Eons,* at the broadest level, are subdivided into *eras;* in turn, eras are broken down into *periods,* which may be further split into *epochs.* The glacial–interglacial cycles that we discuss, which lasted from about 2.5 million years ago until approximately 10,000 years ago, occurred during the Pliocene and Pleistocene epochs. The mass extinction that we shall talk about occurred at the boundary between the Cretaceous and Tertiary periods, approximately 65 million years ago. A *period,* typically lasting tens of millions of years, is generally a longer unit of geologic time than an *epoch.* Finally, the solar luminosity changes that we discuss have occurred throughout the entire 4.5 billion years of Earth history.

Glacial–Interglacial Cycles: The Vostok Ice-Core Temperature Record

A set of ice cores drilled between the mid-1980s and the early 1990s at Vostok, Antarctica, near the South Pole, has provided a wealth of information about the Pleistocene glaciations. The most important results are shown in Figure 1-9. Figures 1-9a and 1-9c show the estimated range of CO_2 and CH_4 concentrations, respectively; Figure 1-9b shows the deviation in temperature, ΔT, from the present value. (The Greek capital letter delta, Δ, is often used to represent a change in a given quantity.) The value of ΔT represents the change in average air temperature over the Antarctic continent and the surrounding polar oceans. These values are derived from measured values of the deuterium content of the ice. Deuterium is an **isotope** of hydrogen that has both a proton and a neutron in its nucleus. (Normal hydrogen has only a proton.) We talk more about how isotopes are used to estimate temperatures in Chapter 14. The unit of age on the horizontal axis of Figure 1-9 is *kyr B.P.,* or thousands of years before the present.

The section of the Vostok ice core that has been fully analyzed is about 3.35 km deep and it extends back for an extraordinarily long time, almost 420,000 years. (The

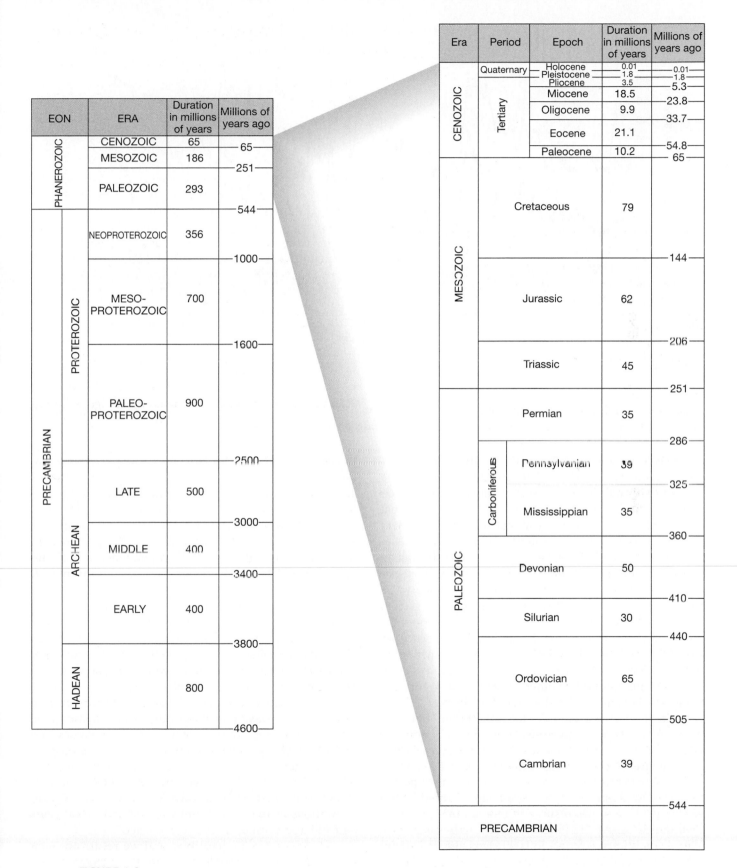

FIGURE 1-8

The geologic time scale. (From W.K. Hamblin and E.H. Christiansen, *Earth's Dynamic Systems, 8/e,* 1998. Reprinted by permission of Prentice Hall, Upper Saddle River, N.J.)

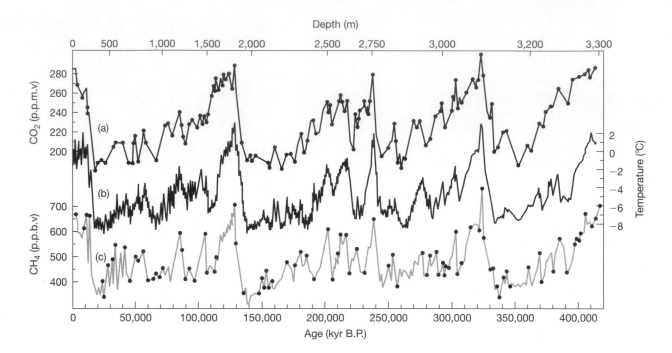

FIGURE 1-9

Measurements of (a) atmospheric CO_2, (b) temperature changes, and (c) atmospheric CH_4 determined from the Vostok ice cores. The temperature changes, DT, were determined from measured values of the deuterium content of the ice. The unit of age on the horizontal axis, kyr B.P., stands for "thousands of years before the present" (where k, for the prefix "kilo-," indicates "thousands"). (After *Climate Change,* 1994, Intergovernmental Panel on Climate Change, Cambridge: Cambridge University Press.)

ice sheet is about 3.7 km thick at this location, but the bottom part has not been drilled because there is a lake at the bottom that we do not wish to contaminate.) The reason that the Vostok record extends so far back in time is that snow accumulates very slowly at that site—the equivalent of only about 2.5 cm of water per year. This value is comparable to the mean annual precipitation over the Sahara Desert. Other parts of the polar ice sheets are approximately as thick but have faster accumulation rates. The short-term CO_2 record shown in Figure 1-3 comes from Siple Station, near the coast of Antarctica; there the snow accumulation rate is equivalent to about 50 cm of water per year. Cores from such locales cover much shorter periods of time than does the Vostok core, even if they are just as deep.

The time interval spanned by the Vostok core extends well beyond the last Ice Age. For the past 2.5 million years, Earth's climate has fluctuated between intensely cold **glacial periods,** in which ice sheets advanced across North America and Europe, and relatively warm **interglacial periods** such as the present, in which the ice sheets retreated. The present interglacial period began—and thus the last Ice Age ended—about 11,000 years ago, as an upward surge in temperature in Figure 1-9b indicates. At 21,000 years ago, Earth was in fullglacial conditions. Around 130,000 years ago, the planet was in the midst of another warm, interglacial period.

Much of the story about the advance and retreat of the glaciers was already known from other sources of data

prior to the drilling of the Vostok ice core. What was new and surprising about the Vostok results was that they showed that atmospheric CO_2 and CH_4 concentrations had varied in concert with surface temperature. The Vostok data show that between 21,000 and 11,000 years ago, atmospheric CO_2 levels rose from about 200 ppm to close to its preindustrial value of 280 ppm, whereas CH_4 increased from about 350 ppb to 650 ppb. The current CH_4 concentration is about 1,700 ppb, or 1.7 ppm. The same abrupt increase in CO_2 and CH_4 concentrations occurred after the previous interglacial period ended, between 140,000 and 130,000 years ago. Indeed, at a finer level, many of the smaller peaks and valleys in the temperature-change curve correspond to specific peaks and valleys in the concentration records of the two gases.

Why would atmospheric CO_2, CH_4, and temperature co-vary in this way? One part of the answer involves the greenhouse effect: As levels of the greenhouse gases CO_2 and CH_4 increased, the magnitude of the greenhouse effect also increased, and the climate became warmer. But what caused atmospheric concentrations of CO_2 and CH_4 to vary in the first place? In particular, why did those concentrations increase so abruptly just after 140,000 years ago and again just after 18,000 years ago?

These are tough questions, and we return to them later. Humans could not have caused these changes. Our ancestors were still making tools out of stone and tending small wood fires—and not burning fossil fuels—when these changes took place. One possible mechanism for driving changes in atmospheric CO_2 levels is a change in the circulation pattern of the deep ocean. As we see in Chapter 5, the deep ocean circulates because cold, salty (and hence, dense) surface water sinks and is replaced by warmer, less dense water from lower latitudes. The deep ocean contains large amounts of dissolved CO_2, some of which is released to the atmosphere when deep water flows upward to the surface. So, the rate at which the deep ocean overturns can affect the concentration of atmospheric CO_2. But the circulation pattern of the deep ocean depends on climate, which is driven by changes in temperature and in evaporation rates at the sea surface. Thus, it would appear that atmospheric CO_2 levels affect climate and that climate, in turn, affects atmospheric CO_2 levels. What we have is a system in which the various components are tightly and intricately coupled. That is why a systems approach is the best way to understand global change.

Mass Extinction: Iridium and the K-T Boundary at Gubbio. Ever since dinosaur bones were first discovered, people have wondered why the dinosaurs disappeared. Dinosaurs flourished for more than 150 million years during an interval called the Mesozoic era, which ended 65 million years ago. At about the same time the dinosaurs disappeared, many other species went extinct as well. Some 60–80% of marine species died, as did numerous species of terrestrial plants and animals. Many possible reasons have been offered for their demise, including changes in climate, changes in vegetation, disease, destruction of the ozone layer by a nearby supernova (an exploding star), volcanic activity, and impact of an extraterrestrial body. No single hypothesis had attracted widespread support, however, until 1980.

That year, Luis and Walter Alvarez, of the University of California at Berkeley, and their colleagues published a paper about a clay layer they had studied in rocks from the mountains near Gubbio, Italy. The clay dated back 65 million years to the K-T boundary. **"K-T boundary"** stands for the transition between two time intervals: the Cretaceous period, abbreviated as "K" (to distinguish it from the Cambrian period, abbreviated as "C"), and the Tertiary period, abbreviated as "T." The Cretaceous period marked the end of the Mesozoic era and was followed by the Tertiary period, part of the Cenozoic era. The dinosaurs and other species disappeared at or just below the boundary between these two periods.

The layer of clay, only a few centimeters thick, was found between thick layers of *carbonate rock* (rock formed from the shells of certain marine organisms). The existence of this clay layer at the K-T boundary (Fig. 1-10) had puzzled geologists for decades. This clay layer had been seen at Gubbio and at numerous other spots around the world, always at the boundary between rocks of the

FIGURE 1-10

The clay layer at the K-T boundary in sediments at Gubbio, Italy. (From Prof. W. Alvarez/Science Photo Library [SPL], Photo Researchers, Inc.)

Cretaceous and Tertiary periods. Walter Alvarez, a geologist, had journeyed to Gubbio in an effort to determine how long it had taken for the clay layer to be deposited.

Luis Alvarez, a physicist (and Walter's father), had a clever idea about how to make that determination. He reasoned that he could calculate the time required to form the clay layer by measuring the abundance of the element *iridium* (Ir). Iridium is a metal in the platinum group of elements, which are very scarce in rocks forming Earth's crust because they are mostly dissolved in its molten iron core. These elements are always raining down on Earth as small particles of debris from asteroids or comets. The rate at which such debris hits Earth is known fairly accurately from measurements of its abundance in cores drilled into the ocean floor. Hence, Luis Alvarez reasoned that he could use the measured iridium abundance in the Gubbio clay layer as a kind of "cosmic clock" to determine the time needed for the clay to have been deposited.

The experiment failed, but it did so for a reason that turned out to be very informative. When the Alvarez team measured the iridium levels at Gubbio, they found the results shown in Figure 1-11. The iridium abundance in the clay layer was up to 10 ppb by mass—more than 100 times higher than what the group expected to find. The amount of iridium in the clay layer was much too large to have been supplied by debris from asteroids or comets. The time required to accumulate that much iridium would have been so long that the signal would have been swamped by the normal deposition of Earth-bound sediments. (Clay accumulates on the ocean floor at a rate of about 1 cm per thousand years as a result of windblown dust that falls on the ocean surface. If the clay layer at the K-T boundary had taken more than a few thousand years to form, it should have contained a large proportion of terrestrial dust and, hence, a relatively small con-

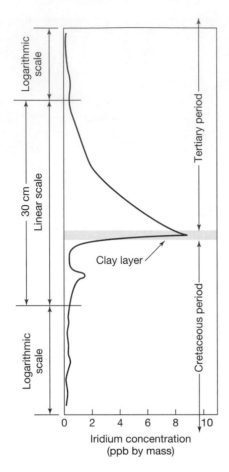

FIGURE 1-11

Iridium concentration versus depth at Gubbio. The middle portion of the depth axis is a linear scale; the upper and lower portions are logarithmic. (After L. Alvarez, *Physics Today*, July 1987.)

centration of iridium.) The Alvarez team reasoned that the iridium must have come instead from the impact of some large, extraterrestrial object, such as an asteroid or a comet. Indeed, by calculating the amount of iridium deposited worldwide, the team estimated the mass of such an incoming body—on the order of 10^{15} kg, which corresponds to a diameter of about 10 km for a rocky asteroid. If the impacting object was a comet, it would have to have been even larger because comets are thought to contain less iridium than do asteroids.

We shall see later on (Chapter 13) that the energy released by an impacting object of this size is enormous—equivalent to about 70 million, 1-megaton hydrogen bombs. Thus, it is plausible that such an event could have triggered extinctions on a mass scale. Since the Alvarezes did their work, additional evidence corroborating a large impact 65 million years ago has been identified, including a deeply buried crater 200 km in diameter underlying the region around Chicxulub, Mexico, on the Yucatan Peninsula. Even this "smoking gun" does not prove that this impact was the cause of the mass extinction. It does

demonstrate convincingly, though, that in the past the Earth system has experienced large shocks from which it has recovered, albeit slowly and in a modified form.

The changes that humans are causing in the Earth system today are less abrupt than those that occurred at the K-T boundary (assuming that the impact theory is correct), but they are still fast compared to most natural changes, and the results could still be catastrophic for certain elements of the biota. We have already noted that large land mammals such as gorillas and pandas are at risk. And with the vast majority of terrestrial species concentrated in the imperiled tropical rainforests, the potential for more widespread mass extinctions is very high. A lesson learned from the K-T boundary crisis, that biodiversity can decrease dramatically over a relatively short time interval, may therefore hold value today.

Changes in Solar Luminosity

All the examples of global change discussed thus far have been based on observational data. Observations, after all, are the cornerstone of science. Not everything of importance is observable, however. For example, we cannot see inside the Sun. Yet scientists believe that the Sun produces its energy through **nuclear fusion,** the joining of two or more light atomic nuclei to form one heavier nucleus. Specifically, four hydrogen nuclei (^{1}H) fuse to form one helium nucleus (^{4}He). This process, which is thought to occur continuously within the Sun, releases large amounts of energy. Even though we cannot observe this phenomenon directly, we are reasonably sure that the fundamental concept is correct.

The fact that the Sun produces energy in this way has important consequences for its long-term evolution. Four hydrogen nuclei take up more space, and therefore exert more pressure, than does one helium nucleus. The pressure in the Sun's core (where nuclear fusion occurs) would therefore be decreasing with time if the fusion of hydrogen into helium were the only process taking place. But what actually happens, models predict, is that the core contracts and heats up slightly as its helium content increases. The temperature rise increases the core's pressure and keeps the core from contracting further, so the Sun remains stable. As the core's temperature increases, so does the rate of nuclear fusion, just as the rates of most chemical reactions increase with increasing temperature. As a result, energy production within the Sun's core rises, and this rise is balanced by an increase in the amount of energy emitted at the surface. The more energy is emitted, the brighter the Sun appears. So, contrary to what we might intuitively expect, the Sun's **luminosity** (brightness) should gradually increase as it depletes its hydrogen fuel.

By how much has solar luminosity changed over the Sun's history? Model calculations performed by a number

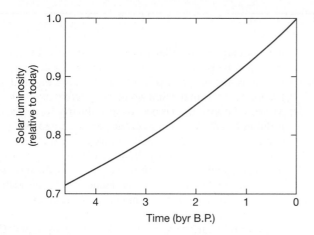

FIGURE 1-12

Estimated change in solar luminosity with time. The unit of age on the horizontal axis, byr B.P., stands for "billions of years before the present." (After D.O. Gough, *Solar Physics* 74, p. 21, 1981.)

of different astronomers have reached essentially the same conclusion. Figure 1-12 shows a typical result, in which the unit of age on the horizontal axis is *byr B.P.*, or billions of years before the present. When the Sun first formed 4.6 billion years ago, it should have been about 30% less luminous than it is today. The Sun's luminosity increased slowly at first and then more rapidly as the buildup of helium in its core continued. At present, the Sun is thought to be brightening by about 1% every hundred million years. By the time the Sun ends its lifetime as a normal star, about 5 billion years from now, it is expected to have brightened by a factor of 2 to 3 as compared to today.

The Effects of Solar Luminosity Changes. How would reduced solar luminosity have affected the early Earth? If all other factors had remained constant, the early Earth should have been colder than it is today. Indeed, calculations (which we will do in Chapter 3) show that the entire ocean should have been ice-covered prior to 2 billion years ago. We know, however, that liquid water has existed on Earth's surface for at least the last 3.8 billion years, because sedimentary rocks (which form from sediments in liquid water) have been forming since that time. And organisms, which require liquid water to survive, have been around for at least 3.5 billion years. The early Earth could not have been a global ice ball, at least not during the time for which a geologic record is available.

This apparent discrepancy is called the *"faint young Sun paradox."* We mention this paradox here because, like the Vostok CO_2 story, it is a problem that can be solved only by considering the Earth system as a whole. The most likely solution is that the level of greenhouse gases in Earth's primitive atmosphere was significantly higher than today. But why should this have been true,

and why would greenhouse gas concentrations have declined as the Sun grew brighter? Does Earth's climate system have some built-in stability mechanism that has kept the mean surface temperature within survivable limits?

The Gaia Hypothesis. James Lovelock, a British biochemist, and Lynn Margulis, an American biologist, have argued that life itself has been responsible for maintaining the stability of Earth's climate. In the process of **photosynthesis**, organisms such as green plants use sunlight, CO_2, and H_2O to produce organic matter and O_2. (*Organic matter* is the carbon-rich material of which organisms are composed.) Through photosynthesis, followed by carbon burial in sediments, Earth's biota may have lowered atmospheric CO_2 levels at just the right rate to counteract the gradual increase in solar luminosity. Alternatively, the biota may have affected the rate at which atmospheric CO_2 is sequestered in carbonate rocks. Carbonate rocks form from reactions of CO_2 with elements (primarily calcium and magnesium) derived from other types of rocks. This process is part of the *carbonate-silicate geochemical cycle,* which we discuss in Chapter 9. In either case, Lovelock and Margulis suggest that Earth has remained habitable precisely because it is in some sense "alive."

This theory of long-term climate stabilization is part of what Lovelock and Margulis called the Gaia hypothesis. In ancient Greek mythology, Gaia (pronounced "guy-ah") was the goddess of mother Earth. In its most basic form, the **Gaia hypothesis** states that Earth is a self-regulating system in which the biota play an integral role. Some proponents of this hypothesis further suggest that the biota manipulate their environment for their own benefit or even, by optimizing the conditions for life, for the benefit of all living things. Such assertions are difficult to justify. Lovelock himself is quick to point out that the biota cannot be expected to cope with all possible disturbances. As an example, we cannot assume that we can safely emit CFCs into the atmosphere because Gaia will somehow protect the stratospheric ozone layer. But it is clear that the Gaia hypothesis is correct at some level: Organisms do play an important role in the overall functioning of the Earth system.

Some form of self-regulation must exist in order for Earth's climate to remain stable over long time scales. Higher greenhouse gas concentrations in the past are the most likely solution to the faint young Sun paradox. But whether the biota are essential to the control mechanism remains controversial. *Abiotic* (nonbiological) *feedbacks* in the carbonate–silicate cycle could have stabilized Earth's climate even if life were not present. Explaining how such a climate control mechanism might work is a recurrent topic in later chapters. Before we attempt to do so, however, we need to look more closely at how the various components of the Earth system function.

Chapter Summary

1. We deal with three main themes: modern global environmental issues, past global change, and the behavior of Earth's systems. To understand present environmental problems, we must know something about Earth's past and something about the way different components of the Earth system interact.
2. Humans are modifying the global environment in several ways.
   a. Global warming may be the most pervasive environmental change that faces us today. The increase in concentrations of greenhouse gases, including carbon dioxide (CO_2), methane (CH_4), nitrous oxide (N_2O), and chlorofluorocarbons (CFCs), in the atmosphere is attributable to human activity. These gases are expected to warm Earth's climate over the next few decades to centuries by enhancing the natural greenhouse effect. They may have already begun to do so: Earth appears to have warmed by about 0.7°C over the past century, on the basis of surface temperature measurements made around the globe. It is still debated, however, whether this temperature rise is a consequence of increased greenhouse gas concentrations or simply a natural fluctuation in the climate system.
   b. The stratospheric ozone layer has already been severely affected by chlorine released from anthropogenic CFCs. The most dramatic impact has been confined to the Antarctic region during October. Strong regulatory steps have already been undertaken to ensure that the ozone layer will be protected in the future. Without such restrictions, the ozone layer's ability to absorb harmful ultraviolet rays from the Sun would be severely diminished.
   c. Massive deforestation is occurring in the tropics today, as it did in North America a century or more ago, when it contributed to the early rise in atmospheric CO_2. Deforestation both increases the buildup of atmospheric CO_2 and significantly decreases biodiversity. The effects of deforestation on biodiversity are permanent and irreversible.
3. Past changes in the Earth system may provide clues to how it will respond to global change in the future.
   a. Variations in surface temperature and atmospheric CO_2 concentrations recorded in ice cores illustrate the coupling between atmospheric CO_2 and climate and show how global warming today fits into the general pattern of glacial–interglacial cycles over the past 2.5 million years.
   b. Studies of the mass extinction at the end of the Cretaceous period 65 million years ago, when the dinosaurs and numerous other species forever vanished from Earth, may shed light on the loss of biodiversity that humans are causing today.
   c. Modeling studies of Earth's response to gradual increases in solar luminosity can help us understand how the climate system remains stable despite large changes in external forcing factors.

Key Terms

acid rain
anthropogenic
atmosphere
biodiversity
biota
carbon dioxide (CO_2)
chlorofluorocarbons
core
crust
deforestation
deuterium
Earth system

fossil fuels
Gaia hypothesis
glacial period
global warming
greenhouse effect
greenhouse gases
hydrosphere
interglacial period
isotopes
K-T boundary
luminosity
mantle

mass extinction
nuclear fusion
ozone (O_3)
ozone hole
ozone layer
photosynthesis
rocks
sediments
solid Earth
stratosphere
system
trace gases

Review Questions

1. a. What is meant by "anthropogenic greenhouse gases"?
   b. Name three such gases that are currently increasing in concentration in Earth's atmosphere.
2. What are the four fundamental components of the Earth system?
3. Explain the difference between global warming and the greenhouse effect.
4. a. By how much has Earth's atmospheric CO_2 concentration increased since the year 1800?
   b. How do we know this?
   c. What are thought to be the primary causes of this increase?
5. Cite two ways in which chlorofluorocarbons can affect the environment.
6. a. How far back in time do direct measurements of Earth's surface temperature extend?
   b. Why is it difficult to determine accurately the long-term temperature trend?
7. How might the burning of coal have had opposing effects on climate during the 20th century?
8. Why is stratospheric ozone important to humans?
9. To what two global environmental problems does tropical deforestation contribute?

10. How are hydrogen isotopes used to infer polar temperature records?
11. How is past surface temperature
    a. determined from the Vostok ice core?
    b. related to atmospheric CO_2 content?
12. Why is iridium a good indicator of impacts by extraterrestrial bodies?

13. a. How has solar luminosity changed during the past 4.6 billion years?
    b. What is the fundamental cause of this change?
14. What is the Gaia hypothesis, and what does it say about the importance of life on this planet?

Critical-Thinking Problems

Write a 1–2 page, typewritten essay on the following questions:

1. Which of the three modern global change problems discussed in this chapter—global warming, ozone depletion, or loss of biodiversity—do you consider to be the most serious? Give reasons for your answer. If you wish, include information drawn from other sources.

2. How do global warming, ozone depletion, and loss of biodiversity compare with other environmental and social problems that the world faces today? You may wish to list the major problems, as you see them, in decreasing order of importance. Justify your answer with an explanation.

Further Reading

General

Lovelock, J. 1995. *Gaia, A New Look at Life on Earth*. Oxford: Oxford University Press.

Schneider, S. H. 1997. *Laboratory Earth: The Planetary Gamble We Can't Afford to Lose*. New York: Basic Books.
Intergovernmental Panel on Climate Change: Third Assessment Report (*http://www.ipcc.ch/*)

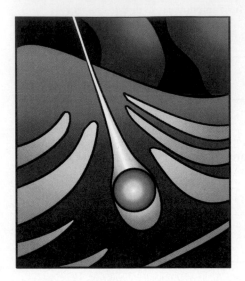

Daisyworld: An Introduction to Systems

Key Questions

- What are systems?
- What are feedback loops?
- What are equilibrium states?
- Does viewing the Earth as a system allow for deeper insight into the interrelationships among the physical and biological worlds?
- Can Earth's climate be self-regulating?

Chapter Overview

In this chapter, we develop the fundamentals of systems theory needed for the study of Earth as a system. First, examples from everyday life are used to introduce the important concepts of systems theory. Then we introduce the simplified climate system of the imaginary planet Daisyworld. Daisyworld is subjected to an increase in solar luminosity even more rapid than that which Earth has experienced over its history. Yet the hypothetical planet is able to counter the tendency for warming by increasing the reflectivity of its surface (by allowing for the spread of white daisies). We will see that this seemingly intelligent response arises without foresight or planning but rather as a natural consequence of interactions within the system. Through such feedbacks, natural systems can remain stable despite disturbances. Although the climate system of Daisy-

world is oversimplified, we suspect that feedbacks like those it exhibits have played an important role in stabilizing Earth's climate over geological time.

The Systems Approach

The systems approach has been used in virtually every area of inquiry, including branches of both the natural and social sciences. Human physiology is a good example of where the systems approach is particularly illuminating. The human body is made up of a number of systems that perform the vital functions of life: a respiratory system that takes in oxygen and eliminates carbon dioxide; a cardiovascular system that circulates the blood, carrying oxygen and carbon dioxide around the body; a digestive system that processes food to fuel all body processes; a nervous system that senses changes in the internal and external environments and controls the activities of the other systems; an endocrine system that regulates ongoing processes such as growth and development; and so on. These systems are interrelated, functioning together to maintain the human body in a healthy state.

The Essentials of Systems

Each **system** is an entity composed of diverse but interrelated parts that function as a complex whole. The individual parts of a system are called **components.** A component can be a reservoir of matter (described by its mass or volume), a reservoir of energy (described by temperature, for example), an attribute of the system (such as body temperature or pressure), or a subsystem (such as the cardiovascular system, one of the interlinked subsystems of the human body). The components of the cardiovascular system itself include blood cells, blood vessels, and the heart.

The **state** of a system is the set of important attributes that characterize the system at a particular time. Body temperature, level of nutrition, and blood pressure are among the attributes that determine the state of the human body. The components of a system interact in such a way that a change in the state of the system is "sensed" throughout the system. In many systems, this linkage allows for the control of important attributes. For example, the endocrine system of the human body is capable of maintaining a nearly constant internal temperature despite large changes in the temperature of the surrounding environment. Suppose that your body temperature starts to rise as the air temperature around you rises. Your hypothalamus, a component of the endocrine system, then directs your sweat glands to increase their production of sweat, which helps cool you. If the ambient temperature then drops, your hypothalamus stops sending the signals to your sweat glands.

Couplings

It is clear from these examples from human physiology that the components of the human body "system" do not exist in isolation. They are linked, allowing for the flow of information from one component to the next. These links are called **couplings.** To understand how couplings allow for system regulation, consider an electric blanket. You set the temperature of the blanket (one system component) by adjusting a temperature controller. You adjust this controller to achieve a body temperature that is comfortable.

A *systems diagram* (Figure 2-1) allows us to keep track of the various couplings within a system. In a systems diagram, couplings are conventionally represented by arrows. There are two types of couplings. In the example of an electric blanket, an increase in blanket temperature causes an increase in body temperature; such a link is called a **positive coupling** (Figure 2-1a). In a positive coupling, a change (increase or decrease) in one component is a stimulus that leads to a change of the same direction in the linked component. When one component increases, a positively coupled component responds by increasing. When the first component decreases, the second component responds by decreasing. A positive coupling is represented by a solid arrow with a normal arrowhead, →.

In contrast, an increase in body temperature above the comfort level would lead you to *decrease* the amount of heat by turning down the controller. This coupling, from body temperature to blanket temperature, is a **negative coupling** (Figure 2-1b). In a negative coupling, a change in one component stimulates a change of the opposite direction in the linked component. When one component increases, a negatively coupled component responds by

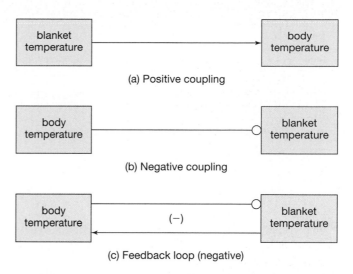

(a) Positive coupling

(b) Negative coupling

(c) Feedback loop (negative)

FIGURE 2-1

Systems diagrams; a negative feedback loop.
(a) An increase (or decrease) in blanket temperature causes an increase (decrease) in body temperature—a positive coupling.
(b) An increase (decrease) in body temperature causes you to decrease (increase) the blanket temperature—a negative coupling.
(c) A negative feedback loop is created by the two couplings.

decreasing. And when the first component decreases, the second component increases. A negative coupling is represented by an arrow with a circular arrowhead, ⎯○.

Feedback Loops

The two couplings we described for the electric blanket create a "round trip," or **feedback loop**, between components (Figure 2-1c). *Feedback* is a self-perpetuating mechanism of change and a response to that change. When you receive feedback from your friends, you are receiving their *re*action to an action of yours. Their reaction now becomes an action, and you *re*act to that action. You may modify your actions by either accentuating or suppressing them, and this modification will affect the nature of the subsequent feedback you receive. Consider Deb, who is Ed's employer. If Deb complains that Ed is dressing inappropriately at work, Ed may respond by dressing more conservatively, or he may instead dress the same or even more inappropriately. Either reaction (which is now an action) will undoubtedly cause a subsequent reaction—praise or criticism—from Deb. In terms of change and response, natural systems with feedback loops behave in a similar manner.

The feedback loop in the electric blanket example is referred to as a **negative feedback loop.** Negative feedback loops tend to diminish the effects of disturbances. An increase in body temperature, however caused, stimulates you to turn down the controller on the blanket. The blanket subsequently radiates less heat, and your body temperature then decreases.

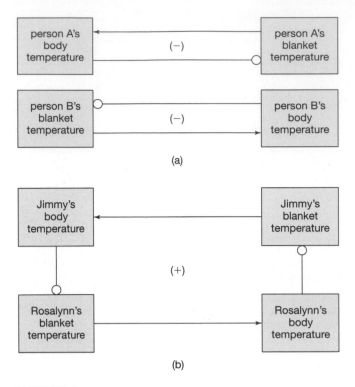

FIGURE 2-2

The consequences of combining feedback loops for a dual-control electric blanket.
(a) Proper usage: Two independent, negative feedback loops.
(b) Improper usage: A single, positive feedback loop formed when the Carters inadvertently exchanged temperature controllers.

In contrast to negative feedback loops, **positive feedback loops** amplify the effects of disturbances. To understand positive feedback loops, consider another electric-blanket example based on a real-life episode in the life of former President Jimmy Carter (Figure 2-2). Jimmy and his wife, Rosalynn, had an electric blanket with dual controls—one for his side of the blanket and another for hers. In his autobiography, President Carter describes the irritation they were suffering:

> During each of the increasingly cold winter nights, we argued about the temperature of our electric blanket. Whenever I said it was too warm, Rosalynn said it was too cold, and vice versa.
>
> One morning I returned from an overnight trip to New York, and she met me at the front door with a warm hug and a smile. "I think our marriage is saved," she said. "I just discovered that our dual blanket controls have been on the wrong sides of the bed, and each of us was changing the temperature on the other's side." [Jimmy Carter, *Living Faith,* New York: Random House, 1996, p. 74.]

If the Carters had been thinking like systems scientists, the reason for their troubles would have immediately been obvious. A systems diagram for the proper use of the blanket is shown in Figure 2-2a. With their own controllers in hand, both persons would adjust their own blanket setting. If either person becomes chilly, he or she turns up the blanket controller and soon returns to a comfortable body temperature.

By inadvertently switching their temperature controllers, the Carters created a single, but more complicated, four-component feedback loop involving both controllers (Figure 2-2b). The new loop is positive, consisting of two positive and two negative couplings. If Jimmy gets a bit warm, he unwittingly turns down Rosalynn's controller. She begins to feel a chill and so turns up what she thinks is her controller (but is actually Jimmy's). Jimmy gets even warmer, and then turns down Rosalynn's controller even further! This runaway response is characteristic of positive feedback loops.

A simple way to identify the "sign" of a feedback loop is to count the number of negative couplings: Negative feedback loops have an odd number of negative couplings; the rules of multiplication apply here. Recall that when two positive numbers are multiplied, the result is a positive number. When two negative numbers are multiplied, the result is also a positive number. But when a positive number and a negative number are multiplied, the result is always a negative number. Thus, if there is an odd number of negative couplings in a feedback loop, the loop is negative. If there are no negative couplings or an even number of them, the loop is positive.

Equilibrium States

The normal electric-blanket feedback loop (Figure 2-1c) acts to maintain body temperature (defining the state of the system) within a comfortable range. If your body temperature is just right, you do nothing to the blanket controller. We refer to this condition as an **equilibrium state** of the system; it will not change unless the system is disturbed. Because this state is created by a negative feedback loop, the equilibrium state is said to be **stable**: Modest disturbances from this state will be followed by system responses that tend to return the system to its equilibrium state. The equilibrium state of comfort for the Carters' feedback loop (Figure 2-2b) was **unstable**: The slightest disturbance from a comfortable state led to system adjustments that carried the system further and further from that state.

To visualize equilibrium states better, we can represent all the possible states of a system as a hilly surface and the present state as a ball that is free to move on that surface (Figure 2-3). The valleys represent stable equilibrium states, and the peaks represent unstable equilibrium states. After a small disturbance, a ball in a stable equilibrium state will roll back down the hill and return to its original state (Figure 2-3a). Valleys, or regions of stability, are defined by the peaks that surround them. A large disturbance—large enough to roll the ball out of its val-

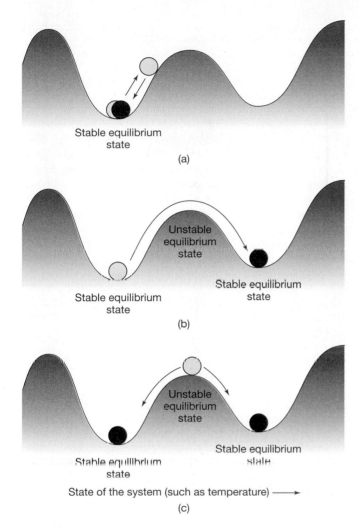

FIGURE 2-3

The equilibrium states of a system, represented as peaks (unstables) and valleys (stables). On disturbance, the system returns to stable equilibrium states but moves away from unstable equilibrium states.

ley and over an adjacent peak—can carry the system to a different equilibrium state (Figure 2-3b). Thus, there are limits to the stability of stable equilibrium states.

In contrast, an unstable equilibrium has no region of stability. A ball disturbed ever so gently from its "resting" point at the top of a peak will roll down the hill and will land in a valley (Figure 2-3c). On its own, the ball will not return to its original state. The slightest disturbance pushes the state of the system toward a new *stable* equilibrium (if one exists). It is unlikely that a given system would remain poised for any length of time at an unstable equilibrium state.

It appears that a system with a single feedback loop has a stable equilibrium state if the feedback loop is negative and an unstable equilibrium state if the feedback loop is positive. This conclusion is usually true, at least for the natural systems we are discussing in this book. (See the box for a discussion of conditions under which positive feedback loops can create stable equilibrium states.) However, in reality, natural systems tend to be combinations of subsystems involving both positive and negative feedback loops. Stability cannot be easily determined by simply inspecting the feedback diagrams of such systems. Rather, they need to be analyzed mathematically.

Perturbations and Forcings

We can learn much by observing how a system responds to disturbances. Our understanding of human physiology, for example, has benefited from the study of patients stricken by illness or accident. The Carters learned about the problem with their electric-blanket system when their body temperatures were disturbed. Similarly, scientists are learning about the Earth system by watching how it responds to disturbances. For example, Earth's climate system is being modified by a variety of natural and anthropogenic factors. One such **perturbation,** or temporary disturbance of a system, is the injection of sulfur dioxide (SO_2) into the atmosphere during volcanic eruptions. Over several weeks, SO_2 reacts to form sulfate aerosol particles (like those formed by the burning of fossil fuels; see Chapter 1) that prevent a small amount of sunlight from reaching Earth's surface. As a result, surface temperatures drop by a bit less than 0.5°C (1°F) globally (Figure 2-4). The climate system recovers from this perturbation several years later as the sulfur is naturally removed from the atmosphere. Because of natural climate variability it is difficult to conclusively ascribe a cool interval following a particular volcanic event to the eruption itself, so Figure 2-4 presents the average climatic response to the five largest eruptions of the last century or so.

A more persistent disturbance of a system is called a **forcing.** In Chapter 1 we mentioned one forcing of Earth's climate: the gradual increase in the amount of sunlight Earth has been receiving over billions of years. How has the climate responded to this forcing? Many scientists argue that the tendency toward surface temperature rise that has accompanied the increased sunlight has been countered by a decrease in atmospheric CO_2 concentrations, reducing the greenhouse effect and thereby cooling the surface.

Understanding the response of the Earth system to forcings such as this is a major focus of this book. Rather than begin with the complex Earth system, however, we consider a much simpler climate system of the hypothetical planet Daisyworld. This planet, whose only life forms are daisies, derives from the creative imaginations of James Lovelock (who, with Lynn Margulis, originated the Gaia hypothesis; see Chapter 1) and his colleague Andrew Watson, an oceanographer. When the Gaia hypothesis was first proposed, a common criticism was that the biota would need to possess the capacity for foresight or planning if the Earth system were to be self-regulating (for example, able to prevent large fluctuations in the

THINKING QUANTITATIVELY
Stability of Positive Feedback Loops

Although most isolated, positive feedback loops are unstable, stable equilibrium states can exist for positive feedback loops if the state of one component depends only on the current state of the other component (i.e., the adjustment of the state of the system is instantaneous rather than incremental). Suppose a child's noisiness increases as the parent gets angrier, and the parent's anger increases as the child's noisiness increases, creating a positive feedback loop. This loop may be stable if the adjustments in anger level and noisiness are instantaneous, modest, and depend only on the current state of one another. For example, the child starts whining when a toy is taken away. The parent will become angry, but the child, having forgotten about the toy and now responding solely to the anger of the parent, may actually become quieter if the anger is moderate. As the child quiets, the parent's anger diminishes, and peace is restored.

However, most natural systems do not behave in this way. The components of natural systems are generally *accumulators* or *reservoirs* of energy or mass, and their response depends not just on the immediate stimulus, but on the accumulation of past stimuli as well. Their response is also time-dependent: Their states do not respond immediately to a stimulus; instead, they do so over some interval of time.

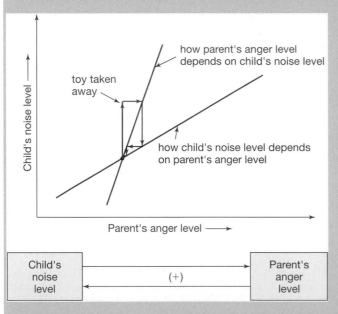

BOX FIGURE 2-1

In this system, the parent and child respond to each other's state instanteously. The system is disturbed from its equilibrium state when the parent takes a toy away from the child. The child's noisiness increases, which causes an increase in the parent's anger level. In response, the child's noisiness actually diminishes, because the parent's response elicits a noise level in the child that is less than its original perturbed level. As the child's noisiness diminishes, the parent's anger level diminishes until equilibrium is restored. The equilibrium state is stable despite being characterized by positive feedback in this special case where the parent and the child respond solely to the instantaneous state of the other. This is not true of most natural systems, and we cannot use this type of diagram to conclusively demonstrate the stability or instability of natural systems.

Equilibrium states characterized by positive feedback in such systems are always unstable. In the example above, if the child's anger had been cumulative, the anger of the parent would simply have made the child noisier, and the situation would have escalated out of control.

As mentioned in the text, a true test of system stability must be performed mathematically. Because most natural systems are time-dependent, their behavior must be described by differential equations. Differential equations are beyond the scope of this book; however, readers who have the required mathematical background are invited to follow the discussion below.

Suppose we have a system of two reservoirs whose states (e.g., amounts of material in the reservoirs) are represented by the variables $A(t)$ and $B(t)$, which are coupled in a feedback loop. Furthermore, suppose that an equilibrium state exists in this system, in which the reservoir sizes are denoted by A_{eq} and B_{eq}. We are interested in how these reservoirs will respond to a disturbance from their equilibrium state. This system can be described by the following two differential equations:

$$dA/dt = a\,(B - B_{eq})$$
$$dB/dt = b\,(A - A_{eq})$$

Here, a and b are constants. The feedback loop is positive if both a and b are positive or if both constants are negative. If a and b have opposite signs, the feedback loop is negative. This follows from our definition of positive and negative couplings. A coupling is positive if component A responds in the same direction as the perturbation to component B; it is negative if the response is in the opposite direction.

The solution to the first of these two coupled differential equations can be shown to be:

$$A(t) - A_{eq} = \left\{ \frac{(A_0 - \beta B_0)}{2} \right\} \exp(\alpha t)$$
$$+ \left\{ \frac{(A_0 - \beta B_0)}{2} \right\} \exp(-\alpha t)$$

Here, A_0 and B_0 are the amounts that A and B are disturbed from their equilibrium values at the initiation of the disturbance, and $\alpha = \sqrt{ab}$ and $\beta = \sqrt{\dfrac{a}{b}}$. The second term on the right-hand side has a negative exponent and thus decays with time, but the first term has a positive exponent and thus will increase without limit as time progresses if α is a real number. Thus, if the product ab is positive, as it must be for a positive feedback loop, the system is clearly unstable. When a and b have opposite signs, though, as they do in a negative feedback loop, then the product ab is negative. The square root of a negative number is imaginary, so α is no longer a real number. In such a case, the system becomes a sinusoidal oscillator. The solution is always bounded, however, thus demonstrating that negative feedback loops are stable. (The solution for $B(t)$ is similar in form to that for $A(t)$.)

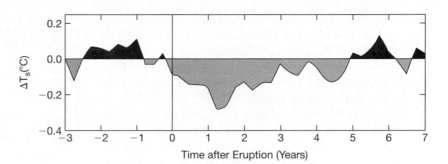

FIGURE 2-4

The average climatic response to the five largest volcanoes of the last 120 years: Krakatau (1883), Santa Maria (1902), Agung (1963), El Chichon (1982), and Pinatubo (1991). Courtesy NASA. (*http://www.giss.nasa.gov/research/intro/hansen-02*)

surface environment). Lovelock and Watson used Daisyworld to demonstrate that natural systems can be self-regulating on a global scale without the need for intelligent intervention. Let us see how.

The Daisyworld Climate System

Imagine that the year is A.D. 2150. We have just determined that there is life on a nearby planet and have sent a manned mission there. On their arrival, the mission scientists observe that the planet is indeed supporting life, but only what appear to be daisies, hence the scientists name the planet Daisyworld. These daisies are unusual, however: They are pure white in color. They appear to

be getting their nutrients and water from the soil; the atmosphere has no clouds and no greenhouse gases. The daisies cover vast regions of the planet's surface; the rest of the surface is mantled in gray soil (Figure 2-5). This means that the amount of sunlight absorbed by the planet depends on the area of darker, bare soil relative to the area of lighter daisy cover. The more sunlight absorbed, the higher the surface temperature. Experiments carried out by the mission scientists show that the growth and spread of daisies across the planet's surface depends only on the temperature around them.

The mission scientists are alarmed because the planet's sun seems to be increasing in luminosity at a much faster rate than is our own Sun. They calculate that the planet will quickly become too hot to support daisy growth.

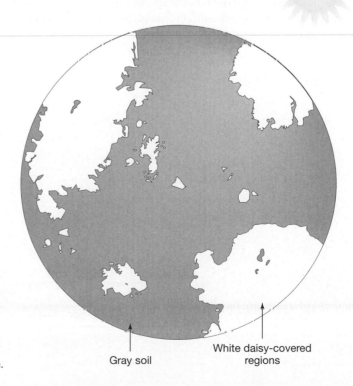

FIGURE 2-5

A view of Daisyworld from outer space.

Gray soil

White daisy-covered regions

However, they make this calculation without considering that the daisies are part of a global climate system in which the reflectivity of the planet is affected by any change in the daisy population. Might a systems approach yield a different prediction for the survival of daisies on Daisyworld in the face of an increasingly luminous sun?

We can represent the Daisyworld climate system on the global scale as a two-component system. One component is the area of daisy coverage, and the other is the average surface temperature of the planet. These two components form a system because they are interdependent: The extent of daisy coverage affects the surface temperature, and the surface temperature affects the growth rate of daisies, which in turn affects the daisy coverage of the planet. Let's explore these interrelationships more fully.

Couplings in the Daisyworld Climate System

Response of Surface Temperature to Changes in Daisy Coverage.
From experience we know that on a sunny day, dark surfaces, such as asphalt roadways, tend to feel warmer than light-colored surfaces, such as concrete sidewalks. Dark surfaces absorb more (that is, they reflect less) of the incoming solar energy than do light surfaces. The reflectivity of a surface is called the surface's **albedo** (Figure 2-6). Albedo is usually expressed as a decimal fraction of the total incoming (*incident*) energy reflected from the surface. Dark soil has a low albedo (0.05–0.15), whereas fresh snow has a high albedo (0.8–0.9). Table 2-1 lists the albedos of some common surfaces.

From the limited amount of information we have about Daisyworld, together with our intuition about albedo, we can graph the relationship between daisy coverage and surface temperature—in other words, the effect that changes in daisy coverage has on temperature. We know that the

TABLE 2-1

Albedos of Some Common Surfaces	
Type of Surface	*Albedo*
Sand	0.20–0.30
Grass	0.20–0.25
Forest	0.05–0.10
Water (overhead Sun)	0.03–0.05
Water (Sun near horizon)	0.50–0.80
Fresh snow	0.80–0.85
Thick cloud	0.70–0.80

surface temperature on Daisyworld is determined by the amount of daisies that cover the surface: The more daisies, the more sunlight reflects off their white petals, the less sunlight absorbed, and finally, the cooler the surface temperature. The graph should have a *negative slope* (that is, "runs downward" from left to right), reflecting the fact that as daisy coverage increases, temperature decreases it (Figure 2-7a). We *cannot* interpret the graph in Figure 2-7a to mean, however, that as surface temperature drops, daisy coverage rises linearly. The way changes in surface temperature affect daisy coverage on Daisyworld is *not* the same as the way changes in daisy coverage affect surface temperature.

This graph can also be expressed as a coupling that links daisy coverage to temperature (Figure 2-7b). The coupling is negative: An increase in daisy coverage causes a decrease in surface temperature, and a decrease in daisy coverage causes an increase in temperature. Note that the "sign" of this coupling—negative—matches the sign of the slope in the graph.

Thus far in our discussion, the albedo of the planet has been a component of the Daisyworld climate system

FIGURE 2-6

A visual comparison of high-albedo and low-albedo surfaces. Light-colored surfaces are more reflective (i.e., have a higher albedo) than dark surfaces, which absorb more sunlight.

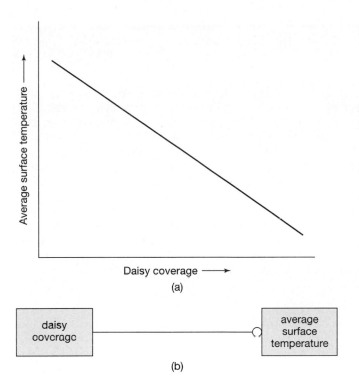

FIGURE 2-7

(a) Graph and (b) systems diagram of the effect of changes in daisy coverage on Daisyworld surface temperature.

that we have treated only implicitly. However, we could treat albedo *explicitly* by adding it as a third component. The coupling that describes the effect of changes in daisy coverage on temperature is now seen as the combination of a positive and a negative coupling that links daisy coverage, albedo, and temperature (Figure 2-8). Decreased daisy coverage leads to a reduction in the average albedo (a positive coupling), and reduced albedo causes an increase in temperature (a negative coupling).

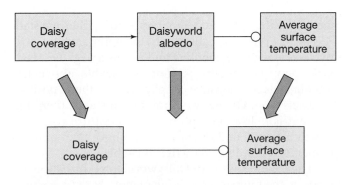

FIGURE 2-8

The same overall coupling as that in Figure 2-7b, but with albedo shown explicitly. A positive and a negative coupling combine to form a negative coupling overall.

When combined, the two couplings in Figure 2-8 form a negative coupling overall, like that shown in Figure 2-7. The rule for combining couplings is the same as the rule for determining the sign of feedback loops. The explicit treatment of albedo, therefore, does not change our conclusion regarding the overall sign of the coupling. For convenience and simplicity, then, we will often treat such couplings implicitly.

Response of Daisy Coverage to Changes in Temperature. In comparison with real daisies, we expect that Daisyworld daisies have an upper and lower temperature limit for survival. They must also have an *optimum,* or most favorable, temperature somewhere in between (let's specify that it is halfway between, for simplicity). A smooth curve drawn through these points is a *parabola* (Figure 2-9). The parabola is a characteristic shape for the temperature dependence of many plants on Earth. It intuitively makes sense that the abundance of an organism would be highest near the organism's optimum temperature and would drop to zero at the upper and lower limits of that organism's temperature range.

The sign of the coupling that reflects the response of daisy coverage to temperature changes depends on temperature, because the relationship is parabolic, as Figure 2-9 shows. If the temperature is below the optimum value for daisy growth, the coupling is positive. If the temperature is above the optimum value, the coupling is negative. This pattern is consistent with the slope of the parabola in Figure 2-9, which has opposite signs on either side of the optimum growth temperature for white daisies.

Equilibrium States in Daisyworld

We can determine the equilibrium states of Daisyworld by combining Figures 2-7 and 2-9. But note that temperature and daisy coverage are on opposite axes, so we cannot simply overlay the plots. Instead, we must invert the axes of Figure 2-7. This inversion does not change the nature of the coupling, it simply interchanges the positions of the two variables so that the axes will match up when we overlay the two graphs.

We can now overlay the two graphs (representing the two couplings); the resulting graph is shown in Figure 2-10a. The curves intersect at two points, labeled P_1 and P_2. These points of intersection are special because they represent the only states of the system that simultaneously fall on the curves, showing both the effect that daisy coverage has on surface temperature *and* the effect that surface temperature has on daisy coverage. For example, at any point other than P_1 or P_2 on the parabola, the effect of temperature on daisy coverage is properly characterized, but the effect of daisy coverage on temperature is not. Conversely, at any point other than P_1 or P_2 on the

USEFUL CONCEPTS
Graphs and Graph Making

Graphs are a powerful way to convey scientific information in an economical way. If a picture is worth a thousand words, a graph can be worth a thousand data points. There are a number of ways in which graphs are used in science, and we present many of them in this textbook. All graphs convey information about the relationship between two (or more) variables. (*Variables* represent any number value.) In a conventional *x–y* graph, data are plotted with the *independent variable* on the *x*-axis and the *dependent variable* on the *y*-axis. The value of the dependent variable depends on the value of the independent variable, but the converse is not true. For example, a graph showing hourly measurements of temperature over the course of several days at a particular place would plot temperature, the dependent variable, as a function of time, the independent variable (Box Figure 2-2a).

Graphs can be used to convey the sense of relationships even when data are not available. We may, for example, convey the notion that temperature varies more or less regularly from day to night with a sketch such as Box Figure 2-2b. We make it clear that no data are being plotted by plotting only a smooth curve; in labeling the axes for such a graph, we do not even show the scale of the graph. In other cases we may wish to put some bounds on the axes, but it would be misleading to place numbers on the axes in a graph for which actual values are not available.

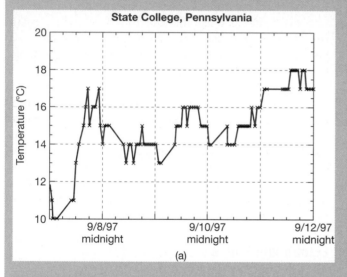

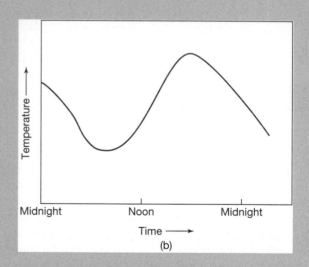

BOX FIGURE 2-2A

Examples of various uses of graphs. (a) Display of data: temperature measurements made at State College, Pennsylvania, during a three-day period. (b) Conveying a concept with no data: sinusoidal daily temperature variation.

straight line, the effect of daisy coverage on temperature is right, but the effect of temperature changes on daisy coverage is not.

Points P_1 and P_2 are the equilibrium states of this system, because they represent the states at which the system is said to be in *equilibrium*. If the system is already in one of these states, it will remain there unless something disturbs it. Note that neither equilibrium state corresponds to the optimum temperature.

But how will these equilibrium states respond to perturbation? We can evaluate the stability of these states by constructing the systems diagrams for Daisyworld (Figure 2-11). Two feedback loops characterize the Daisyworld climate system: one that applies *below* the optimum temperature, to equilibrium state P_1, and another that applies *above* the optimum temperature, to equilibrium state P_2. The diagram applicable below the optimum has a positive and a negative coupling and is thus a negative feedback loop. The feedback loop applicable above the optimum has two negative couplings and is thus a positive feedback loop. Of the two equilibrium states, then, the one that is below the optimum temperature for daisy growth is stable, and the one that is above the optimum temperature for daisy growth is unstable.

Thus, the response of Daisyworld to perturbation depends on the temperature of the planet. At temperatures below the optimum for daisy growth, the system is characterized by negative feedback, which will tend to maintain the temperature and daisy coverage near a stable equilibrium state. (Note that the temperature is below

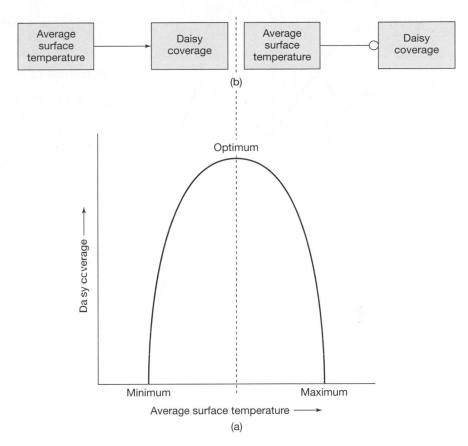

FIGURE 2-9

(a) Graph and (b) systems diagram of the effect of changes in Daisyworld surface temperature on daisy coverage.

the optimum for daisy growth.) If temperatures are perturbed above the optimum, the system will enter a region of positive feedback without a stable equilibrium state. If the perturbation is small, the temperature will return to the cool, stable equilibrium state below the optimum. Larger perturbations will carry the system over the edge of the stable equilibrium state's "valley" (Figure 2-10b), and temperatures will rise above the limits for daisy growth; the daisies will then die.

External Forcing: The Response of Daisyworld to Increasing Solar Luminosity

We have been investigating the behavior of the Daisyworld climate system by perturbing it from its equilibrium states and analyzing its response. Many of the disturbances we will be discussing in later chapters, however, are persistent forcings that can be considered *external* to the system. The response of systems to forcings can be quite different from that to perturbations because the sys-

tem may not be able to return to an original, stable equilibrium state even if negative feedback loops predominate.

The forcing on Daisyworld is the increase in solar luminosity recognized by the mission scientists. How will the climate system respond? Will the temperature rise quickly on Daisyworld, as the scientists predicted, spelling the end for daisies, or will the climate system respond in such a way to forestall the realization of this ultimate catastrophe?

On the basis of our experience about the ability of systems with negative feedback loops to damp perturbations, we might suspect that the system should act to extend the duration of the daisy inhabitation of the planet. Think of how the system would respond to a single, small but permanent increase in solar luminosity. The immediate response would be a warming of the planet's surface. This response, however, would be quickly followed by the spread of daisies, which, by increasing the albedo, would reduce the warming. A new equilibrium state would eventually be achieved at a temperature warmer than the original temperature. Yet it would be cooler than the temperature the planet would have achieved had the daisies not responded to the change in temperature and

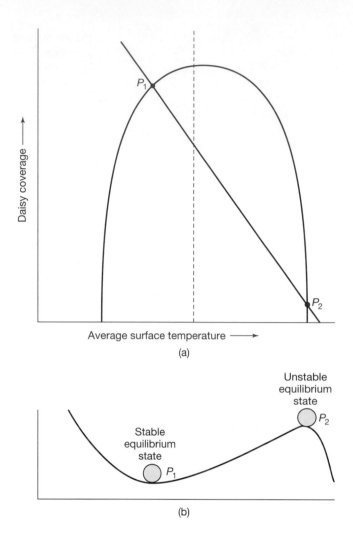

(a)

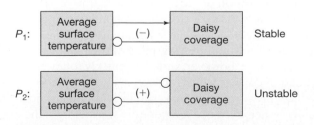

(b)

FIGURE 2-10

(a) The mutual influences of average surface temperature on daisy coverage (the parabola) and daisy coverage on surface temperature (the straight line). The intersection points (P_1 and P_2) are the equilibrium states of the system. (b) The stability of P_1 and instability of P_2.

thereby altered the planet's albedo. Applying this line of reasoning to the problem at hand, we conclude that a persistent trend of increasing solar luminosity should lead to a gradual evolution of the equilibrium temperature of the planet to higher and higher temperatures, but at a rate that is slower than the warming that would otherwise

FIGURE 2-11

Feedback loops appropriate for small perturbations from equilibrium states P_1 and P_2, respectively.

occur in the absence of the feedback between the daisies and their environment.

Response of Daisyworld Couplings to Forcing

To predict the future climate of Daisyworld more accurately, we need to understand how the increasing intensity of Daisyworld's sun will affect the couplings in the system. Because we are assuming that the daisies respond only to temperature changes and thus not to changes in the solar luminosity itself, we would *not* expect modification of the coupling that links surface temperature changes to daisy coverage (the parabola in Figure 2-9). As the sun becomes more intense, the surface temperature will rise, and the percentage of daisy coverage will respond according to the parabolic curve, as before. However, we *would* expect a change in the coupling that relates surface temperature to the extent of daisy coverage (the straight line in Figure 2-10a). For any amount of daisy coverage, the surface temperature will increase as the intensity of the sun increases. This is not to say that the temperature will rise indefinitely. Rather, for any particular value of daisy coverage, the temperature will be higher than expected from Figure 2-7. The graphical result is that the line in Figure 2-7 shifts upward as solar luminosity increases (Figure 2-12).

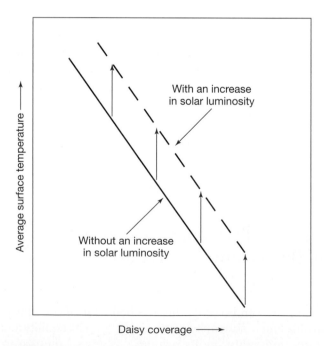

FIGURE 2-12

The effect of an increase in solar luminosity on the dependence of average surface temperature on daisy coverage. If the daisy coverage were fixed at a certain percentage, temperature would simply increase, as shown by the arrows.

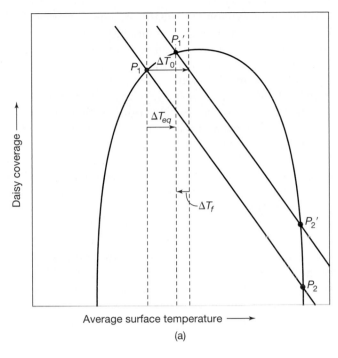

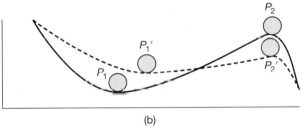

FIGURE 2-13

(a) Response of Daisyworld to an increase in solar luminosity. (b) The stability of P_1 and P_1' and the instability of P_2 and P_2'.

Response of Equilibrium States to Forcing

Let us again consider the response to an incremental increase in solar luminosity. If we combine Figures 2-9 and 2-12, we find, as before, that two equilibrium states exist (labeled P_1' and P_2' in Figure 2-13a). We can guess that, as before, only one of them will be stable, because the diagram has not changed fundamentally. Which one is it? The temperature at point P_1' is below the optimum growth temperature for daisies, so this situation is similar to that for point P_1. The temperature at point P_2' is above the optimum temperature for daisies, so this situation is similar to that for point P_2. Thus, we determine that P_1' is stable and P_2' is unstable.

However, we see that both the temperature and the daisy coverage at the new stable equilibrium are higher. Daisyworld has apparently reacted to increased solar luminosity by increasing the daisy coverage. The accompanying increase in albedo explains why the temperature at the stable equilibrium did not rise as much as it would

have without feedback. Note that the "ridge" that defines the stability limit for P_1' is lower than it was before. This new equilibrium state is apparently less resistant to perturbations; further increases in solar luminosity should eliminate the stable equilibrium state entirely.

We can determine the effectiveness of this feedback mechanism by comparing the equilibrium temperature changes with and without feedback. Without feedback—that is, without any change in the daisy coverage—the temperature change that results from the increase in solar luminosity is large. The temperature change without feedback is represented in Figure 2-13a as ΔT_0. (Recall from Chapter 1 that ΔT means the *change in temperature*.) With feedback, however, the temperature increase is smaller—but not zero. The temperature change of the new equilibrium state (with feedback) is represented as ΔT_{eq}, and the temperature change of the *feedback effect* itself is ΔT_f.

We can express the behavior of the Daisyworld system mathematically:

$$\Delta T_{eq} = \Delta T_0 + \Delta T_f.$$

In other words, the overall temperature change that results from increased solar luminosity is the sum of the temperature change with no feedback and the temperature change due to feedback. In our case, ΔT_{eq} is smaller than ΔT_0; the temperature effect of the feedback ΔT_f is negative. We can see this in Figure 2-13a: The arrow that represents ΔT_f points to the left—the negative direction—instead of to the right. Although we derived this equation for Daisyworld, it is a general relationship that can be applied to any stable equilibrium in a system involving feedback loops: The change in state of a system as it moves from one equilibrium to the next is the sum of the state change that would result without feedback and the effect of the feedback itself.

To quantify the strength of the feedback effect, we can define a value f, called the feedback factor. The **feedback factor** is the ratio of the equilibrium response to forcing (the response with feedback) to the response without feedback. In our example, this ratio is as follows:

$$f = \frac{\text{temperature change with feedback}}{\text{temperature change without feedback}} = \frac{\Delta T_{eq}}{\Delta T_0}$$

Here f is less than 1, because the equilibrium response is smaller with feedback than it would have been without feedback. The value of f is between 0 and 1 whenever the feedback loop is negative but greater than 1 if the feedback loop is positive. As we mentioned previously, systems with positive feedback loops are stable only if they contain negative feedback loops as well. The feedback factor, f, can be defined only for stable systems, such as Daisyworld at point P_1. At point P_2, there is no stable equilibrium, and hence ΔT_{eq} is not defined there.

Climate History of Daisyworld

So far in our presentation of Daisyworld, we have avoided the use of actual values of temperature, daisy coverage, and albedo. However, we can assign reasonable values to the graphs to calculate the climatic response of Daisyworld to increasing amounts of sunlight. To present the details of the calculation would be premature; we shall introduce the physical laws in the next chapter. Suffice it to say that the calculations were based on typical growth curves for real daisies (see Figure 2-9), reasonable albedos for white daisies (0.9) and for gray soil (0.2), and a solar input similar to Earth's that also increases with time.

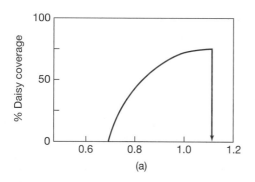

(a)

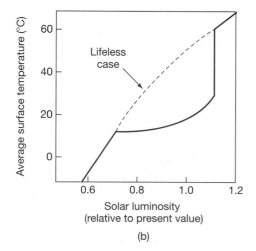

Solar luminosity
(relative to present value)

(b)

FIGURE 2-14

The response of Daisyworld to increasing solar luminosity.
(a) The change in daisy coverage of the planet in response to changes in solar luminosity (relative to the presumed present value).
(b) The change in average surface temperature of Daisyworld in response to increasing solar luminosity (solid line) and the response on a lifeless planet with fixed albedo (dashed line).

Figure 2-14a shows the history of daisy coverage, and Figure 2-14b shows the temperature history of Daisyworld from the time of its formation until the end of daisy inhabitation of the planet. (In these graphs, solar luminosity is plotted on the x-axis instead of time itself; solar luminosity increases more or less linearly with time. Scientists often make such substitutions so that their plots can be generalized; in this case, the plot is correct no matter how fast the change in luminosity occurs.) The solid curve in Figure 2-14b represents the "actual" surface temperature change on the daisy-inhabited planet. The dashed curve in Figure 2-14b shows how that surface temperature would differ if there were no daisies (in other words, no life forms and no feedback). In the early years, temperature increases relatively rapidly. However, once the surface temperature rises above the minimum temperature for daisy survival, the white daisies begin to spread across the planet's surface. Their growth tends to cool the planet by increasing its albedo, and so the rate of warming slows dramatically. The daisy coverage expands rapidly at first, and then more slowly, in response to these increases in temperature, which are much smaller than we would predict for a lifeless (daisy-free) planet or for a planet with daisies but no feedback (and thus constant albedo).

Eventually the daisies run out of space in which to expand, and the planet's temperature begins to increase rapidly. The temperature eventually exceeds the optimum for daisy growth, and further increases in temperature cause reductions in daisy coverage. The feedback loop becomes positive. Once this happens, the system becomes unstable: The surface temperature rises rapidly, and the daisies go extinct. Thereafter, because it is dictated by the lower albedo of the gray soil on the lifeless planet, the temperature overlays the dashed curve in Figure 2-14b.

The Lessons of Daisyworld

By studying the hypothetical planet Daisyworld from a systems perspective, we have learned some interesting things about climate systems in general. First, a planetary climate system is not passive in the face of internal or external influences. There are feedback loops that respond to perturbations and forcings (in this case, solar luminosity). Negative feedback loops in the system counter the external forcings. On Daisyworld, the consequence of this feedback is a longer lifespan for the daisies than one would predict if there were no feedback in the system. We will see in a later chapter that Earth's climate system has negative feedback loops as well that keep its climate relatively stable on both short and long time scales.

Second, the climate regulation system of Daisyworld, and, by analogy, other nonhuman systems that self-regulate, is seemingly intelligent: The response of the daisies is exactly what is needed to counter the solar warming of the planet. Yet no foresight or planning is involved. The daisies simply respond to the increase in temperature, and the planet's temperature responds to the spread of daisies. Such behavior is not restricted to contrived systems like Daisyworld. Indeed, self-regulation is a property common to many natural systems with feedback loops. Lovelock conceived of Daisyworld as a means of demonstrating to his critics that the Gaia hypothesis (which he applied to Earth) did *not* require an intelligent biota. Organisms can be components of self-regulating, natural systems simply because they influence, and are influenced by, the physical environment in which they live.

It is unlikely that the biota would be capable of optimizing their environment for their own good, as seemed to be required by the Gaia hypothesis when it was first proposed. The Daisyworld experiment indicates that it is not necessary for the biota to be capable of optimizing their environment. The Daisyworld system does not *optimize* the temperature for daisies. The stable equilibrium temperature is below the optimum for daisy growth on Daisyworld.

Note that the self-regulation is not perfect. As the sun became more luminous, Daisyworld's climate system responded with a temperature increase, but the increase was much more gradual than would have occurred on a lifeless planet (or one with fixed daisy coverage). Systems like the Daisyworld climate system typically adjust to forcings by a slow but continual modification of their equilibrium states. This response is different from that directed by a thermostat, for instance, which is designed to maintain a constant state (temperature). In a natural self-regulating system, there is no preset state (no optimum value) that the system is programmed to "seek out."

The real Earth is not unlike Daisyworld: Its surface temperature has been maintained within the tolerance limits of living organisms for more than 3 billion years, despite substantial changes in solar luminosity. As on Daisyworld, the reason for the long-term stability of Earth's climate is the existence of strong negative feedback. The feedback loops that operate on Earth are, not surprisingly, more complicated than the one that operates on Daisyworld. Before discussing how Earth's climate system works, we spend the next four chapters learning more about its components and their interactions.

Chapter Summary

1. Components of systems interact in ways that can either enhance or diminish the stability of the system.
   a. The components are linked by couplings, which can be either positive or negative.
   b. When couplings are arranged such that there is a round-trip flow of information, a feedback loop is formed. These feedback loops can be either positive or negative.
   c. Positive feedback loops amplify perturbations or forcings; negative feedback loops diminish them.
2. The presence of feedback loops leads to the establishment of equilibrium states.
   a. Negative feedback loops establish stable equilibrium states that are resistant to a range of perturbations; the system responds to modest perturbations by returning to the stable equilibrium state.
   b. Positive feedback loops establish unstable equilibrium states. A system that is poised in such a state will remain there indefinitely. However, the slightest disturbance carries the system to a new state.
3. The Daisyworld climate system is capable of resisting a warming trend induced by a sun that is becoming brighter with time.

   a. This capacity is the result of a negative feedback loop that involves the feedback between daisy coverage and temperature; it does not require foresight or planning.
   b. The key is the difference in albedo between the white daisies and the gray soil, together with the effect that temperature changes have on daisy growth and coverage.
   c. Daisyworld has two equilibrium states, but only one is stable. This equilibrium state, in general, does not coincide with the optimum temperature for daisy growth.
   d. The temperature response to an increase in solar luminosity can be thought of as a progression from one stable equilibrium state to the next. These equilibrium responses are the sum of the response that would occur without feedback plus the feedback effect itself. In the case of Daisyworld, the temperature change without feedback is larger than that from one equilibrium state to the next: The feedback effect is negative. Thus, the rate at which the planet warms is slower than it would be if there were no feedback between daisy coverage and surface temperature. The interval over which the planet is inhabited by daisies is extended because of the presence of feedback.

Key Terms

albedo
component
coupling
equilibrium state
feedback factor
feedback loop

forcing
negative coupling
negative feedback loop
perturbation
positive coupling
positive feedback loop

stable equilibrium
state
system
unstable equilibrium

Review Questions

1. A perturbation that causes a decrease in component *A* leads to a decrease in component *B*. Is the coupling between these two components positive or negative?
2. What is a feedback loop?
3. Why do negative feedback loops tend to diminish the effect of disturbances?
4. What distinguishes a forcing from a perturbation?
5. Are all equilibrium states stable? Why or why not?
6. What is albedo? How does it influence climate?
7. How are daisies on Daisyworld able to regulate the hypothetical planet's temperature?

Critical-Thinking Problems

1. In the Dysfunctia family, when the children get noisy, the parents get mad. When the parents get mad, the children get noisy. Draw a systems diagram for the Dysfunctia family.
   a. Is the feedback loop negative or positive?
   b. Is the family stable or unstable?
2. Earth's average temperature is determined in part by the amount of CO_2 in the atmosphere, by way of the greenhouse effect. The atmospheric CO_2 content may in turn be affected by the photosynthetic activity of plants, which convert CO_2 into plant tissue. However, the rate of photosynthesis depends on the amount of CO_2 in the atmosphere and on global air temperature. The components of this system—atmospheric CO_2 content, global temperature, and photosynthesis rate—are intimately interconnected. By increasing global photosynthesis rates, plants would tend to lower the atmospheric CO_2 level. In doing so, however, the plants would tend to cool Earth. This cooling, together with the reduced CO_2 level, might tend to reduce the photosynthetic activity of plants.
   a. On the basis of this discussion, draw a systems diagram of the photosynthetic rate–CO_2–temperature system.
   b. How many feedback loops are there?
   c. Are the feedback loops positive or negative?
   d. Describe the response of the system to the following perturbations:
      (i) an increase in atmospheric CO_2;
      (ii) a decrease in temperature.
   e. Extra credit: How might the system respond to a continuous forcing—an increase in solar luminosity through time?
3. Daisyworld has a companion planet that is similar in all ways except that the daisies are black.
   a. What is the effect of an increase in black-daisy coverage on planetary temperature? Express your answer graphically.
   b. Assuming that the effect of temperature on daisy coverage is the same on black-daisy Daisyworld as on white-daisy Daisyworld, draw a *stability diagram*—a diagram analogous to Figure 2-10—for black-daisy Daisyworld. Include two equilibrium states.
   c. Which of the two equilibrium states in part (b) is stable?
   d. Is the stable equilibrium state of part (c) cooler or warmer than that of white-daisy Daisyworld?
   e. How would this system respond to a *decrease* in solar luminosity? Express your answer graphically and in terms of the feedback factor *f*.
   f. Is *f* of part (e) greater than or less than 1?
4. Real daisies have an optimum growth temperature of around 22°C and they die at temperatures 20°C warmer or colder than this value. Let's assume that they cover 100% of Daisyworld at their optimum growth temperature. A parabola that describes this mathematically is:

$$C = 100 - \frac{(T - 22)^2}{4},$$

where C = the percent daisy coverage and T = temperature (see Figure 2-9). Assume that the temperature of (White) Daisyworld is given by $T = 56 - \dfrac{C}{2}$. (This describes a line with a negative slope as shown in Figure 2-7).
   a. Combine these equations (e.g., substitute the second into the first) and solve for the equilibrium solutions, i.e., the values of C and T that satisfy both equations. How many solutions are there? Which solution is stable?
   b. Suppose that the Daisyworld sun brightens so that temperatures increase by 4°C for a daisy coverage of 0 percent, i.e., $T = 60 - \dfrac{C}{2}$. What are the solutions in this case?

Further Reading

General

Lovelock, J. 1991. *Healing Gaia: Practical Medicine for the Planet.* New York: Harmony Books.

Volk, T. 1998. *Gaia's Body.* New York: Copernicus (Springer-Verlag).

Advanced

Saunders, P. T. 1994. "Evolution without Natural Selection: Further Implications of the Daisyworld Parable." *Journal of Theoretical Biology* 166:365–373.

Levins, R. 1974. "The Qualitative Analysis of Partially Specified Systems." *Annals of the New York Academy of Sciences* 231:123–138.

Milsum, J. H. 1968. *Positive Feedback, a General Systems Approach to Positive/Negative Feedback and Mutual Causality.* New York: Pergamon Press.

Global Energy Balance: The Greenhouse Effect

Key Questions

- What are the basic characteristics of electromagnetic radiation?
- What causes the greenhouse effect, and how is the magnitude of this effect determined?
- How do clouds affect the atmospheric radiation budget?
- What are the most fundamental feedbacks in the climate system?

Chapter Overview

Earth is heated by visible radiation from the Sun and cools by radiating infrared energy back to space. Earth's surface temperature depends on the amount of incident sunlight, the planet's reflectivity, and the greenhouse effect of its atmosphere. Certain gases in the atmosphere absorb outgoing infrared radiation and reradiate part of that energy back down to the surface. If this process did not occur, Earth's average surface temperature would be well below the freezing point of water, and life could not survive. Both the greenhouse effect and the amount of absorbed sunlight are strongly influenced by the presence of clouds. Clouds can either warm or cool the surface, depending on their altitude and thickness. Built-in feedback loops involving atmospheric water vapor and the extent of snow and ice cover

are also fundamental aspects of the climate system. All these factors need to be considered in order to predict the response of Earth's climate to future increases in greenhouse gas concentrations.

Introduction

That Earth is suitable for life is largely a consequence of its temperate climate. A fundamental requirement for life as we know it is liquid water, and Earth is the only planet in our solar system that has liquid water at its surface. Venus, our nearest neighbor toward the Sun, has an average surface temperature of 460°C (860°F), hot enough to melt lead. Mars, the closest planet away from the Sun, has an average surface temperature of −55°C (−67°F), the coldest temperatures experienced at the South Pole. Earth's average surface temperature is 15°C (59°F), the same as the mean annual temperature of San Francisco. Earth is not only habitable, it is a relatively pleasant place to live.

Why is Venus too hot, Mars too cold, and Earth just right? This question is sometimes called the "Goldilocks problem" of comparative planetology. Intuition suggests that the answer is that Earth happens to lie at the right distance from the Sun (and hence would receive exactly the right amount of sunlight), whereas Venus and Mars do not (Figure 3-1). A closer look, however, reveals that it is not just the amount of sunlight that a planet receives that determines its surface temperature. A planet's surface is also warmed by the greenhouse effect of its atmosphere. As we saw in Chapter 1, a planet's atmosphere allows sunlight to come in but slows down the rate at which heat is

FIGURE 3-1
Venus, Earth, and Mars, shown roughly to scale. (Left: From NASA/Marten, Photo Researchers, Inc. Center: From NASA/Science Source, Photo Researchers, Inc. Right: From Photo Disc, Inc.)

lost. Without this greenhouse effect, Earth's average surface temperature would be about 33°C (59°F) colder than the observed value. Earth would be an icy, desolate world.

In this chapter, we discuss how the greenhouse effect works. We begin by considering the nature of electromagnetic radiation and why the Sun emits primarily one form of radiation (visible light), whereas Earth emits another (infrared radiation). We show that the incoming solar energy and outgoing infrared energy must be approximately in balance, and we demonstrate how this balance allows us to calculate the magnitude of the atmospheric greenhouse effect. Next, we discuss how both forms of energy are affected by atmospheric gases and by clouds, and we explain why some gases are greenhouse gases but others are not. Finally, we use the systems notation developed in Chapter 2 to introduce real climate feedback mechanisms, and we show why it is necessary to understand these feedbacks in order to estimate the climate changes that are occurring now as well as those that may occur in the future.

Electromagnetic Radiation

What exactly does it mean to say that Earth is heated by radiation from the Sun? From our sense of sight, we know that the Sun emits about 50% of its energy in the form of *visible light*. Let us start by considering what makes up light and other forms of *electromagnetic radiation*.

Properties of Electromagnetic Radiation. A physicist would describe an electromagnetic wave as a propagating disturbance consisting of oscillating (regularly fluctuating) electric and magnetic fields that are perpendicular to each other. For our purposes, we can think of **electromagnetic radiation** as a self-propagating electric and magnetic wave that is similar to a wave that moves on the surface of a pond. A wave of any form of electromagnetic radiation—such as light, ultraviolet, or infrared

radiation—moves at a fixed speed c (the "speed of light"). The numerical value of c for a light wave in a vacuum is 3.00×10^8 m/s. The wave consists of a series of crests and troughs (Figure 3-2). The distance between two adjacent crests is called the **wavelength.** It is typically denoted by the Greek letter λ (lambda). An observer standing at a fixed point in the path of the wave would be passed by a given number of crests in one second. This number is called the **frequency** of the wave. It is represented by the Greek letter ν (nu).

If we neglect complexities like polarization, an electromagnetic wave can be described by these three characteristics: speed, wavelength, and frequency. Not all of these characteristics are independent. The speed of the wave must equal the product of the number of wave crests that pass a given point each second (the frequency) and the distance between the crests (the wavelength). We can express this relationship mathematically as

$$\lambda \nu = c,$$

or equivalently as

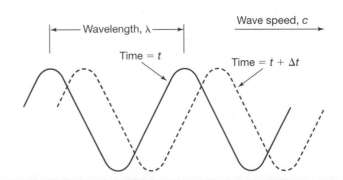

FIGURE 3-2
Simplified representation of an electromagnetic wave, illustrating the concept of wavelength. The solid curve shows the position of the wave at some time t. The dashed curve shows the wave at time $t + \Delta t$.

$$\nu = \frac{c}{\lambda}.$$

The longer the wavelength of an electromagnetic wave, the lower must be its frequency, and vice versa. Conversely, the shorter the wavelength, the higher the frequency.

Photons and Photon Energy

Although we can think of electromagnetic radiation as a wave, at times it behaves more like a stream of particles. A single "particle," or pulse, of electromagnetic radiation is referred to as a **photon.** A photon is the smallest discrete (independent) amount of energy that can be transported by an electromagnetic wave of a given frequency. The energy E of a photon is proportional to its frequency:

$$E = h\nu = \frac{hc}{\lambda},$$

where h is a constant called *Planck's constant*, after the famous German physicist Max Planck. Its numerical value is 6.63×10^{-34} J-s (joule-seconds). Thus, high-frequency (short-wavelength) photons have high energy, and low-frequency (long-wavelength) photons have low energy. This difference in photon energy becomes important when electromagnetic radiation interacts with matter, because high-energy and low-energy photons have very different effects. High-energy photons can break molecules apart and hence cause chemical reactions to occur, whereas low-energy photons merely cause molecules to rotate faster or vibrate more strongly.

The fact that electromagnetic radiation behaves both as a particle and as a wave was one of the great discoveries of physics of the early part of the 20th century. This *wave-particle duality* is not restricted to electromagnetic waves. Rather, it is a general characteristic of matter and energy.

The Electromagnetic Spectrum

The full range of forms of electromagnetic radiation, which differ by their wavelengths (or by their frequencies), makes up the **electromagnetic spectrum** (Figure 3-3). Wavelengths in the visible range are typically measured in nanometers (nm). One *nanometer* is one-billionth of 1 meter. **Visible radiation,** or visible light, consists of a relatively narrow range of wavelengths, from about 400 nm to 700 nm. Within this range, the color of the light depends on its wavelength. Anyone who has observed a rainbow has witnessed this phenomenon. The longest visible wavelengths appear to our eyes as the color red, whereas the shortest wavelengths register as blue to violet. The colors of the rainbow—in other words, the range of component wavelengths of visible light—are referred to as the **visible spectrum.** The term "spectrum" indicates that the light has been separated into its component wavelengths.

About 40% of the Sun's energy is emitted at wavelengths longer than the visible limit of 700 nm in a region referred to as the *infrared* region of the electromagnetic spectrum. Wavelengths of **infrared (IR) radiation** are significantly longer than those of visible light. Hence, it is convenient to keep track of them in units called *micrometers* (μm) rather than in nanometers. One micrometer (formerly called *micron*) equals one-millionth (10^{-6}) of 1 meter. So, 1 μm equals 1000 nm. Infrared wavelengths range from 0.7 μm to 1000 μm. At even longer wavelengths within the electromagnetic spectrum, radiation is transmitted in the form of *microwaves* and *radio waves*. Radio waves can have wavelengths of many meters.

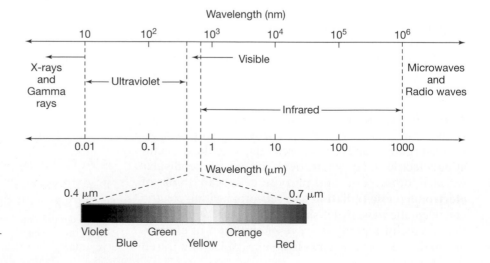

FIGURE 3-3

[See color section] The electromagnetic spectrum.

About 10% of the Sun's energy is emitted at wavelengths shorter than those of visible light as **ultraviolet (UV) radiation.** Wavelengths of the ultraviolet region extend from 400 nm down to about 10 nm. At shorter wavelengths are *X-rays* and *gamma rays*. These high-energy forms of electromagnetic radiation have little effect on our story here, but they do affect the chemistry of the uppermost atmosphere. X-rays, of course, are also used in medicine because they can penetrate skin and muscle tissue and allow us to see the underlying bones.

The regions of the electromagnetic spectrum that are most important to climate are the visible and the infrared. The Sun emits energy in both of these spectral regions. Earth, as we shall see, emits primarily in the infrared. Solar ultraviolet radiation also affects the Earth system significantly by driving atmospheric chemistry. In addition, UV radiation would be lethal to most forms of life were it not almost totally blocked out by oxygen and ozone in Earth's atmosphere.

Flux

We will need one other basic concept from electromagnetic theory in order to proceed: the notion of flux. In general terms, *flux* (*F*) is the amount of energy or material that passes through a given area (perpendicular to that area)

per unit time. In terms of fluid flow, for example, the flux is the volume of fluid that flows perpendicularly into or out of a unit area per unit time. Applied to electromagnetic radiation, **flux** is the amount of energy (or number of photons) in an electromagnetic wave that passes perpendicularly through a unit surface area per unit time.

To demonstrate the concept of flux, let us consider the light given off by an electric lightbulb. A typical, small lightbulb is labeled "60 Watts." A *Watt* (W) is a unit of *power*—formally, the rate at which work is done; informally, the intensity of the bulb in the SI system. One Watt equals one Joule per second. Suppose that a person is standing some distance from such a lightbulb and holds up a sheet of paper directly facing the light (Figure 3-4A). The paper is illuminated by radiant energy from the bulb. The radiation crosses the paper perpendicularly from the lightbulb at a certain flux, or intensity per unit area. That flux is measured in Watts per square meter (W/m^2). The magnitude of the flux depends on how far from the lightbulb the person is standing, but it does *not* depend on how big the paper is because flux is defined as the intensity per *unit* area.

The fact that flux is measured perpendicular to the direction the wave is traveling is important. Suppose that the person is holding the paper at an angle, rather than perpendicularly, to the light (Figure 3-4b). Although the

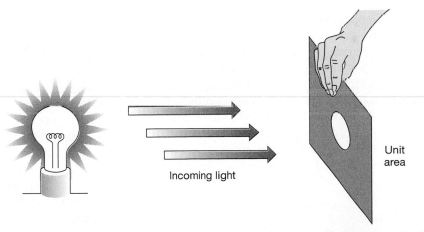

Incoming light

Unit area

Paper is perpendicular to incoming light.

(a)

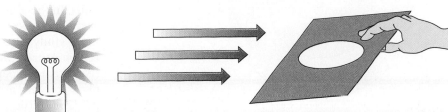

Paper is at an angle to incoming light.

(b)

FIGURE 3-4
Schematic diagram of the concept of flux. The flux of radiation into the paper is reduced when the paper is tilted at an angle to the incoming light.

total area of the paper remains the same, the flux of radiant energy reaching its surface is less because less radiation strikes a given unit area. This simple concept has direct, familiar consequences for Earth's climate. The polar regions are cooler than the tropics because the Sun's rays strike the ground at a higher angle at the poles. Summer temperatures are warmer than winter temperatures because the Sun is higher in the sky during summer. We discuss these fundamental features of climate at greater length in Chapter 4. For now, we simply need to understand the concept of flux.

The Inverse-Square Law

Figure 3-4 demonstrates that the flux of radiant energy from a lightbulb depends on how far away the observer (the person holding the paper) is standing. Likewise, the flux of solar energy decreases as distance from the Sun increases; that is why Venus is illuminated more strongly than Earth. The rate at which this *solar flux* decreases with increasing distance is described by a simple relationship. This relationship, called an **inverse-square law,** is expressed mathematically as

$$S = S_0 \left(\frac{r_0}{r} \right)^2,$$

where S represents the solar flux at some distance r from the source, and S_0 represents the flux at some reference distance r_0 (Figure 3-5).

The inverse-square law has a straightforward physical interpretation: If we double the distance from the source to the observer, the intensity of the radiation decreases by a factor of $(1/2)^2$, or $1/4$. Similarly, if we reduce the distance from the source to the observer by a factor of 3, the radiation intensity increases by a factor of 3^2, or 9.

As an example, consider a hypothetical planet, Planet X, located twice as far from the Sun as is Earth. What would be the solar flux hitting Planet X? Refer to Figure

3-5, and let the Sun be at the center of the two circles. Also, let the inner circle represent Earth's orbit and the outer circle represent the orbit of Planet X. Then r_0 is the average distance from the Earth to the Sun, which is 149,600,000 km, defined as one *astronomical unit* (AU), and S_0 is the solar flux at Earth's orbit, 1370 W/m². (The value of S_0 is determined by satellite measurements.) So according to the inverse-square law, for this example $r = 2$ AU, the solar flux incident at Planet X is

$$S = 1370 \text{ W/m}^2 \left(\frac{1 \text{ AU}}{2 \text{ AU}} \right)^2$$
$$= 342.5 \text{ W/m}^2$$

We would have gotten precisely the same answer if we had expressed the distances in kilometers, but the arithmetic would have been harder.

The inverse-square law is of fundamental importance to the study of planetary climates. It allows us to determine quantitatively why Earth's climate differs from that of Venus and Mars. It also plays a crucial role in our understanding of the causes of the glacial–interglacial cycles of the last 3 million years of Earth's history. As we will see later, small variations in the shape of Earth's orbit, combined with the inverse-square relationship between the distance from the Earth to the Sun and solar flux, cause large changes in the climate of the polar regions and in the size and extent of the polar ice sheets.

Temperature Scales

To understand climate, which is the prevailing weather patterns of a planet or region over time, we must first understand the concept of temperature. *Temperature* is a measure of the internal heat energy of a substance. Heat energy, in turn, is determined by the average rate of motion of individual molecules in that substance. For a solid, these motions consist of regular vibrations, whereas for a gas or liquid they are just random movements of molecules. The faster the molecules in a substance move, the higher its temperature.

Most areas of the world measure temperature (T) by the *Celsius* (formerly, *centigrade*) scale, which is measured in degrees Celsius (°C) and is part of the SI system of units. In the United States, temperature is typically measured in degrees *Fahrenheit* (°F). Scientists, particularly those studying climate, often use the **Kelvin (absolute) temperature scale,** measured in units called *kelvins* (K). (Note that temperatures in the Kelvin scale are given simply as kelvins, not as degrees Kelvin.)

The Celsius temperature scale is defined in terms of the freezing and boiling points of water at sea level (Table 3-1). At sea-level pressure, the freezing point is 0°C, and

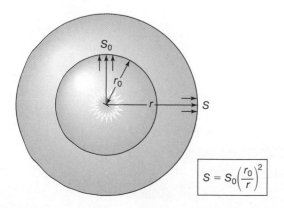

FIGURE 3-5

Diagram illustrating the inverse-square law.

TABLE 3-1

Freezing and Boiling Points of Water by Temperature Scale		
Temperature Scale	*Freezing Point*	*Boiling Point (at sea level)*
Fahrenheit	32°	212°
Celsius	0°	100°
Kelvin (absolute)	273.15	373.15

the boiling point is 100°C. Atmospheric pressure decreases with altitude, as we will see later in the chapter, so it makes a difference where the boiling point is determined. Water boils when its vapor pressure exceeds the overlying atmospheric pressure. Thus, the boiling point decreases with altitude. (This is why it takes longer to hard-boil an egg when you are camping in the mountains. The boiling water is not as hot, so it takes longer to cook the egg.)

The Fahrenheit temperature scale was originally defined on the basis of the temperature of a mixture of snow and table salt (0°F) and the temperature of the human body (about 100°F). Like the Celsius scale, it is defined today in terms of the physical properties of water: The freezing point is 32°F, and the boiling point is 212°F. The following relations allow us to convert temperatures between the Celsius and Fahrenheit scales:

$$T(°C) = \frac{T(°F) - 32}{1.8}$$
$$T(°F) = [T(°C) \times 1.8] + 32$$

Note that converting a temperature *change* from one system of units to the other is easier, because the effect of the different zero points is removed. Thus, a temperature change of 1°C is equal to a change of 1.8°F. Conversely, a change of 1°F is equal to a change of 0.5556 (=1/1.8) °C.

Absolute temperature—that is, temperature on the Kelvin scale—is defined in terms of the heat energy of a substance relative to the energy it would have at a temperature of absolute zero. At *absolute zero,* the molecules of a substance are at rest (or, more precisely, are in their lowest possible energy state). A temperature change of 1 K is equal to a temperature change of 1°C. The zero point of the Kelvin scale is, however, lower than that of the Celsius scale by 273.15°. To convert temperature in degrees Celsius to kelvins, we use the following equation:

$$T(K) = T(°C) + 273.15.$$

Thus, a temperature of absolute zero corresponds to a Celsius reading of −273.15°C.

Blackbody Radiation

In order to fully understand the greenhouse effect, we need one final concept from the world of physics: the concept of blackbody radiation. A *blackbody* is something that emits (or absorbs) electromagnetic radiation with 100% efficiency at all wavelengths. Consider a cast-iron ball (Figure 3-6). At room temperature, the ball looks black because it absorbs most of the light incident on it and gives off little visible radiation of its own. If we heat the ball, however, it begins to glow a dull red. If we heat the ball further, it eventually glows white hot because it radiates at all visible wavelengths. Recall that white light is a mixture of all the colors of the spectrum.

The radiation emitted by a blackbody is called **blackbody radiation.** It has a characteristic wavelength distribution that depends on the body's absolute temperature. This distribution can be described mathematically

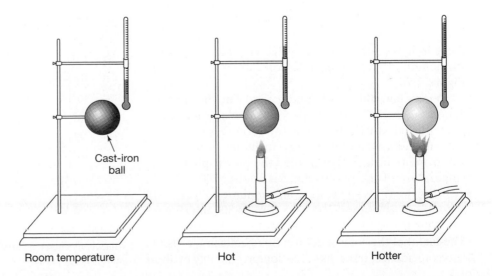

FIGURE 3-6

Change in emitted radiation by a blackbody as it is warmed.

Room temperature   Hot   Hotter

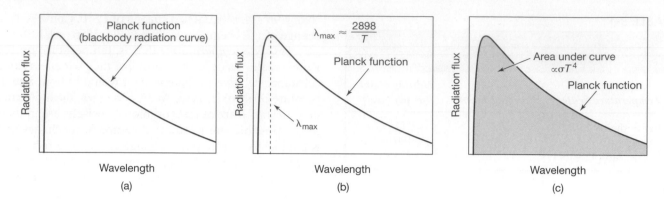

FIGURE 3-7

(a) The Planck function, or blackbody radiation curve; (b) Wien's law; (c) the Stefan–Boltzmann law.

by a relation called the Planck function. The *Planck function* relates the intensity of radiation from a blackbody to its wavelength, or frequency. When shown graphically, this relation is also known as the *blackbody radiation curve* (Figure 3-7a). The Planck function itself is mathematically complicated and is beyond the scope of our discussion here. We can, however, use this relation to derive two simpler rules that are fundamental to an understanding of climate.

Wien's Law

The first rule derived from the Planck function that will assist us in studying climate is called Wien's law. **Wien's law** states that the flux of radiation emitted by a blackbody reaches its peak value at a wavelength λ_{max}, which depends inversely on the body's absolute temperature. According to this rule, hotter bodies emit radiation at shorter wavelengths than do colder bodies. Wien's law may be written as

$$\lambda_{max} \approx \frac{2898}{T},$$

where T is the temperature in Kelvins and λ_{max} is the wavelength of maximum radiation flux in micrometers (Figure 3-7b).

Wien's law allows us to understand why the Sun's radiation peaks in the visible part of the electromagnetic spectrum and why Earth radiates at infrared wavelengths. The Sun emits most of its energy, including visible radiation, from its surface layer, called the **photosphere.** The temperature of the photosphere is about 5780 K. Thus, according to Wien's law, the Sun's radiation flux should maximize at 2898 μm/5780 ≈ 0.5 μm, or 500 nm (Figure 3-8). This is right in the middle of the visible spectrum. (The fact that the solar radiation flux peaks in the visible is no coincidence; our sense of vision presumably evolved

as it did to take advantage of the solar spectrum.) Earth, meanwhile, has a surface temperature of about 288 K, so its radiation peaks at 2898 μm/288 ≈ 10μm—well into the infrared range. In reality, neither Earth nor the Sun is a perfect blackbody, so their emitted radiation flux is not exactly described by the Planck function. Nevertheless, Wien's law is still useful in predicting the wavelength at which most of their radiant energy is emitted.

The Stefan–Boltzmann Law

A second rule derived from the Planck function that will prove useful in climate studies is called the Stefan–Boltzmann law. The **Stefan–Boltzmann law** states that the energy flux emitted by a blackbody is related to the fourth power of the body's absolute temperature:

$$F = \sigma T^4,$$

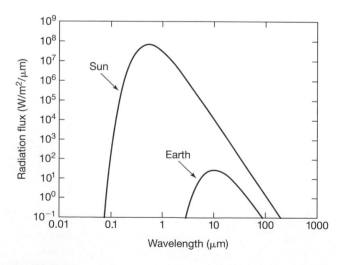

FIGURE 3-8

Blackbody emission curves for the Sun and Earth. The Sun emits more energy per unit area at all wavelengths.

where T is the temperature in kelvins and σ (the lower-case Greek letter sigma) is a constant with a numerical value of 5.67×10^{-8} W/m^2/K^4. The total energy flux per unit area is proportional to the area under the blackbody radiation curve (Figure 3-7c).

As an example of how the Stefan–Boltzmann law can be applied, consider a hypothetical star that has a surface temperature twice that of the Sun. (We shall use stars rather than planets in this example because the radiation emitted from stars is more nearly approximated as blackbody radiation.) Our Sun has a surface temperature of about 5780 K, so the energy flux per unit area is

$$F_{\text{Sun}} = \sigma \, (5780 \text{ K})^4 \approx 6.3 \times 10^7 \text{ W/m}^2.$$

The other star releases energy at a rate of

$$\begin{aligned} F_{\text{star}} &= \sigma (2 \times 5780 \text{ K})^4 \\ &= 2^4 \times \sigma (5780 \text{ K})^4 \\ &= 16 \, F_{\text{Sun}}. \end{aligned}$$

Thus, the amount of energy released per unit area per unit time by the hot star is 2^4, or 16, times greater than that released by the Sun. Evidently, the amount of radiation emitted by a blackbody is a very sensitive function of its temperature.

Planetary Energy Balance

We now have all the tools necessary to analyze Earth's average climate in a quantitative manner. What we need to do next is put them together. The principle that we will apply is that of *energy balance*. To a first approximation, the amount of energy emitted by Earth must equal the amount of energy absorbed. In reality, this cannot be exactly true; if it were, Earth's average surface temperature would never change. We showed in Chapter 1 that the average surface temperature is changing—specifically, it is getting warmer. But it is getting warmer precisely because Earth's energy budget is slightly out of balance: The flux of incoming solar energy exceeds the outgoing IR flux by an almost imperceptible amount (a few hundredths of a percent). The imbalance may be caused by the increase in CO_2 and other greenhouse gases in the atmosphere, or it may be caused by natural fluctuations within the climate system. When the climate system eventually reaches *steady state,* that is, when the surface temperature stops changing, the amount of energy going out will exactly equal the amount of energy coming in.

Physically, Earth's surface temperature depends on three factors: (1) the solar flux available at the distance of Earth's orbit, (2) Earth's reflectivity, and (3) the amount of warming provided by the atmosphere (i.e., the greenhouse effect). The solar flux, S, as mentioned earlier, is the amount of solar energy reaching the top of Earth's atmosphere. Not all this energy is absorbed, however. About 30% of the incident energy is reflected back to space, mostly by clouds. As we saw in Chapter 2, the reflectivity of a planet is called its *albedo*. It is usually expressed as the fraction of the total incident sunlight that is reflected from the planet as a whole. We shall designate albedo by the letter A.

To calculate the magnitude of the third factor, the greenhouse effect, it is convenient to treat Earth as a blackbody even though this is not exactly true. (As we discuss later, the atmosphere radiates and absorbs energy better at some wavelengths than at others because of the presence of gases such as CO_2 and H_2O.) We do this by defining a quantity T_e that represents the **effective radiating temperature** of the planet. This temperature is the temperature that a true blackbody would need to radiate the same amount of energy that Earth radiates. With this definition in place, we can use the Stefan–Boltzmann law to calculate the energy emitted by Earth. By balancing the energy emitted with the energy absorbed, we obtain the following formula (see the Box "A Closer Look: Planetary Energy Balance"):

$$\sigma T_e^4 = \frac{S}{4}(1 - A).$$

This formula expresses the planetary energy balance between outgoing infrared energy and incoming solar energy.

Magnitude of the Greenhouse Effect

What is the significance of the effective radiating temperature? We can think of this quantity as the temperature at the height in the atmosphere from which most of the outgoing infrared radiation derives (see "Critical-Thinking," Problem 4). We can also think of it as the average temperature that Earth's surface would reach if the planet had no atmosphere (assuming that the albedo remained constant). To get a better understanding, let us calculate its value for the present Earth. We can solve the planetary energy balance equation for T_e by dividing both sides of the equation by σ and then taking the fourth root of each side:

$$T_e = \sqrt[4]{\frac{S}{4\sigma}(1 - A)}.$$

If we insert the known values of S (1370 W/m^2), A (30%, or 0.3), and σ (5.67×10^{-8} W/m^2/K^4), we get $T_e \approx 255$ K. Thus, Earth's effective radiating temperature is a relatively chilly $-18°$C, or $0°$F.

A CLOSER LOOK
Planetary Energy Balance

The derivation of the planetary energy balance equation is not difficult, but it does require that we consider the geometry of the Earth–Sun system. The starting point for the derivation is the relation

Energy emitted by Earth = Energy absorbed by Earth.

Let us first calculate the energy emitted by Earth. If we treat Earth as a blackbody with an effective radiating temperature T_e, the Stefan–Boltzmann law tells us that the energy emitted per unit area must be equal to σT_e^4. Earth radiates over its entire surface area, $4\pi R_{Earth}^2$, where R_{Earth} represents Earth's radius (Box Figure 3-1). Thus, the total energy emitted by Earth is

$$\text{Energy emitted} = 4\pi R_{Earth}^2 \times \sigma T_e^4.$$

Now, let us calculate the energy absorbed by Earth. From the Sun, Earth would look like a circle with radius R_{Earth} and area πR_{Earth}^2. Note that it is the area of Earth projected against the Sun's rays that enters here, not half of the surface area of Earth. (Half of Earth's surface area would be $2\pi R_{Earth}^2$, but the Sun's rays do not strike all of this area perpendicularly.) The total energy intercepted must be equal to the product of Earth's projected area and the solar flux (S), or $\pi R_{Earth}^2 S$. The reflected energy is equal to this incident energy times the albedo (A). The difference between these two quantities is the energy absorbed by Earth:

Energy absorbed = Energy intercepted − Energy reflected

$$= \pi R_{Earth}^2 S - \pi R_{Earth}^2 SA$$
$$= \pi R_{Earth}^2 S(1 - A).$$

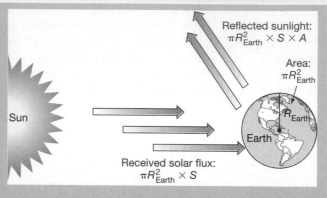

BOX FIGURE 3-1

The amount of sunlight received by and reflected by Earth.

All that remains is for us to equate the outgoing and incoming energy. Using the expressions just calculated, we get

$$4\pi R_{Earth}^2 \times \sigma T_e^4 = \pi R_{Earth}^2 S(1 - A).$$

Cancelling out πR_{Earth}^2 on both sides of this equation and dividing both sides by 4, we obtain the desired equation,

$$\sigma T_e^4 = \frac{S}{4}(1 - A).$$

We saw earlier, however, that the actual mean surface temperature of Earth, T_S, is 288 K, or about 15°C. The difference between the actual surface temperature and the effective radiating temperature is caused by the greenhouse effect of Earth's atmosphere. We can represent this mathematically by letting

$$\Delta T_g = T_s - T_e,$$

where ΔT_g is the magnitude of the greenhouse effect. Thus, $\Delta T_g = 15°C - (-18°C) = 33°C$.

To place this value in context, we can carry out similar calculations for Venus and Mars from known data of the albedos, surface temperatures, and orbital distances of these planets. (See "Critical-Thinking," Problem 2.) The results show that the solution to the Goldilocks problem posed at the beginning of this chapter is more complicated than we might have guessed. Evidently, a planet's greenhouse effect is at least as important in determining that planet's surface temperature as is its distance from the Sun.

We can also apply the planetary energy balance equation to the faint young Sun paradox mentioned in Chapter 1. Recall that solar luminosity, and thus S, is estimated to have been 30% lower early in the solar system's history. It is easy to demonstrate that Earth's average surface temperature would have been below the freezing point of water under such circumstances, if the planetary albedo and the atmospheric greenhouse effect had remained unchanged (see "Critical-Thinking," Problem 5). We have already seen, though, that the early Earth had both liquid water and life on its surface. In later chapters, we discuss ways to resolve this apparent paradox.

THINKING QUANTITATIVELY
How the Greenhouse Effect Works: The One-Layer Atmosphere

Although we calculated that Earth's greenhouse effect provides 33°C of surface warming, our method of obtaining this result (by subtracting the calculated effective radiating temperature from the observed mean surface temperature) provides little insight into the physical mechanism that causes the warming. We can remedy this by doing a simple calculation that demonstrates how the greenhouse effect actually works.

Suppose we treat the atmosphere as a single layer of gas and that this gas absorbs (and re-emits) all of the infrared radiation incident on it (Box Figure 3-2). Let us assume that it absorbs and emits infrared radiation equally well at all wavelengths, so that we can treat it as a blackbody, and that it has an albedo A in the visible spectrum, just like that of the real Earth. What are the temperatures of the gas layer and of the surface beneath it? We will call the layer temperature T_e and the surface temperature T_s, because these quantities are exactly analogous to those discussed in the text.

We can determine the values of T_e and T_s by balancing the energy absorbed and emitted by both the surface and the one-layer atmosphere. Let the amount of sunlight striking the planet be equal to $S/4$ (the globally averaged solar flux). The surface absorbs an amount of sunlight equal to $S/4 \times (1 - A)$, along with a flux of downward infrared radiation from the atmosphere equal to σT_e^4. The atmosphere absorbs an amount of upward infrared radiation from the ground equal to σT_s^4, and it emits infrared radiation in both the upward and downward directions at a rate of σT_e^4. (The real atmosphere also absorbs some of the incoming solar radiation, but we ignore that complication here.) Thus, we can write the overall energy balance in the form of two equations: For the surface,

$$\sigma T_s^4 = \frac{S}{4}(1 - A) + \sigma T_e^4;$$

for the atmosphere,

$$\sigma T_s^4 = 2\sigma T_e^4.$$

(The factor 2 in the second equation arises because the atmosphere radiates in both the upward and downward directions.) If we now substitute the second equation into the left-hand side of the first equation and subtract σT_e^4 from both sides, we obtain

$$\sigma T_e^4 = \frac{S}{4}(1 - A),$$

which is just the familiar energy-balance formula. But dividing the atmospheric energy-balance equation by σ and then taking the fourth root of both sides yields an additional result:

$$T_s = 2^{1/4}\, T_e$$

Thus, the surface temperature is higher than the one-layer-atmosphere temperature by a factor of the fourth root of 2, or about 1.19. For $T_e = 255$ K, as on Earth at present, we get $T_s = 303$ K, and we calculate a greenhouse effect of

$$\Delta T_g = T_s - T_e = 48 \text{ K}.$$

This is higher than the actual greenhouse effect on Earth by about 15 K.

This example is not meant to be realistic. The real atmosphere is not perfectly absorbing at all infrared wavelengths, so some of the outgoing IR radiation from the surface leaks through to space. This effect tends to make ΔT_g smaller. Conversely, a more-accurate calculation would subdivide the atmosphere into a number of different layers. Including more layers tends to make ΔT_g bigger and is the reason why a thick atmosphere, like that of Venus, can produce a really huge amount of surface warming. The calculation does, however, illustrate the basic nature of the greenhouse effect: By absorbing part of the infrared radiation radiated upward from the surface and re-emitting it in both the upward and downward directions, the atmosphere allows the surface to be warmer than it would be if the atmosphere were not present.

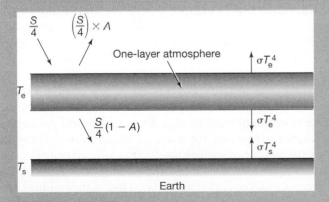

BOX FIGURE 3-2

The greenhouse effect of a one-layer atmosphere.

Atmospheric Composition and Structure

Atmospheric Composition

To understand the greenhouse effect in more detail, along with other aspects of climate and Earth's radiation budget, we must learn a few fundamental facts about the composition and structure of Earth's atmosphere. Table 3-2 lists the main constituents of Earth's present atmosphere and their relative abundances.

As Table 3-2 indicates, the three most abundant constituents of our atmosphere are nitrogen, oxygen, and argon. Nitrogen is a relatively *inert* (chemically unreactive) gas, but when split into its constituent atoms, it plays an important role in biological cycles. Oxygen, which is highly reactive, is the essential gas that all animals must breathe; it is required by many other life forms as well. Argon is almost completely inert; it is the product of the radioactive decay of potassium, K, in Earth's interior. These three constituents—nitrogen, oxygen, and argon—are not greenhouse gases. In other words, they do not contribute to Earth's greenhouse effect.

Although they appear at the bottom of Table 3-2, water vapor and carbon dioxide are two of the most important atmospheric constituents. Besides being directly used by organisms, they are also strong greenhouse gases. We will soon see what makes a particular gas a greenhouse gas.

In addition to the major constituents listed in Table 3-2, Earth's atmosphere also contains a number of minor (or "trace") constituents that affect climate. The most important of these are methane, nitrous oxide, ozone, and freons. Their concentrations are generally much lower than those of the major constituents. Despite their low concentrations, these trace gases are important greenhouse gases. Table 3-3 lists the major greenhouse gases. (Note that water vapor and carbon dioxide are repeated here.) It is convenient to keep track of these gases in units

TABLE 3-3

Important Atmospheric Greenhouse Gases	
Name and Chemical Symbol	Concentration (ppm by volume)
Water vapor, H_2O	0.1 (South Pole)–40,000 (tropics)
Carbon dioxide, CO_2	370
Methane, CH_4	1.7
Nitrous oxide, N_2O	0.3
Ozone, O_3	0.01 (at the surface)
Freon-11, CCl_3F	0.00026
Freon-12, CCl_2F_2	0.00054

of parts per million (ppm), which we defined in Chapter 1. Take a moment to convince yourself that the 0.00001–4% value of water vapor and the 0.037% value of CO_2 given in Table 3-2 are equivalent to the 0.1–40,000 ppm and 370 ppm values of water vapor and of CO_2 given in Table 3-3.

Table 3-3 is by no means a complete list of greenhouse gases. Several other gases affect climate to some extent or are otherwise important in atmospheric chemistry. The gases listed in Table 3-3, however, are the ones that are most important to the modern problem of global warming, and hence they are the ones on which we focus.

Atmospheric Structure

How Atmospheric Pressure Varies with Altitude. Other characteristics of Earth's atmosphere that influence climate and the radiation budget are its pressure and temperature structure. *Pressure* may be defined as the force per unit area exerted by a gas or liquid on some surface with which it is in contact. The pressure exerted by the atmosphere at sea level is defined as one *atmosphere* (atm). A pressure of 1 atm is equivalent to about 15 lb/in^2 in the English system and to 1.013 *bar,* or 1013 *millibars* (mbar), in the metric system. (The pressure unit in the SI system is the *Pascal* (Pa) but this unit is cumbersome in atmospheric work: $1\ Pa = 1 \times 10^{-5}\ bar \approx 9.9 \times 10^{-6}\ atm$.) An instrument used to measure atmospheric pressure is called a *barometer,* the name of which derives from the metric unit of measure "bar."

At higher levels in the atmosphere, the pressure decreases markedly (Figure 3-9a). This change in pressure is what makes your ears pop in an airplane. (The cabin is pressurized, or the popping would be much worse.) The decrease in altitude follows the **barometric law,** which states that atmospheric pressure decreases by about a factor of 10 for every 16-km increase in altitude. Thus, the pressure is about 0.1 bar at 16 km above the surface, 0.01 bar at 32 km, and so on. In more precise terms, the barometric law says that pressure decreases *exponentially* with altitude. Note from Figure 3-9a that the exponential de-

TABLE 3-2

Major Constituents of Earth's Atmosphere Today	
Name and Chemical Symbol	Concentration (% by volume)
Nitrogen, N_2	78
Oxygen, O_2	21
Argon, Ar	0.9
Water vapor, H_2O	0.00001 (South Pole)–4 (tropics)
Carbon dioxide, CO_2	0.037*

*In 2002

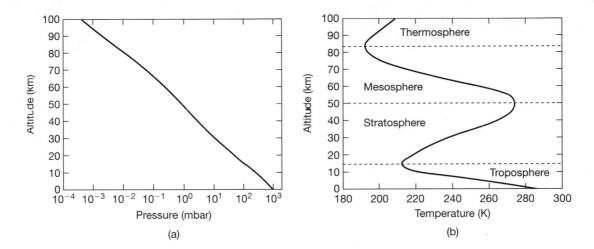

FIGURE 3-9

(a) How pressure varies with altitude in Earth's atmosphere. (b) How temperature varies with altitude in Earth's atmosphere. The different regions of the atmosphere, determined by temperature regimes, are labeled.

crease in pressure appears almost like a straight line when pressure is plotted on a logarithmic scale. The slight deviation from linearity is caused by the variation in temperature with altitude (discussed next). Pressure decreases faster with height in regions where the air is colder.

How Atmospheric Temperature Varies with Altitude.
The vertical temperature structure of the atmosphere is more complicated than the vertical pressure structure (Figure 3-9b). This temperature profile is the basis for distinguishing four regions within Earth's atmosphere: the troposphere, the stratosphere, the mesosphere, and the thermosphere. Temperature decreases rapidly with altitude in the lowermost layer of the atmosphere, the **troposphere,** which extends from the surface up to 10–15 km (higher in the tropics, lower near the poles). Immediately above the troposphere is the **stratosphere,** which is located from about 10–15 km to 50 km above the surface and in which temperature increases with altitude. Above the stratosphere, temperature decreases with altitude in the **mesosphere** (from about 50–90 km) and then increases once again in the uppermost layer, the **thermosphere** (above about 90 km). These temperature-based "spheres" overlap with atmospheric layers based on other characteristics. For example, the *ionosphere* (a layer that reflects radio waves) includes parts of both the thermosphere and the mesosphere. The very outermost fringe of the atmosphere, where the gas is so tenuous that collisions between molecules become infrequent, is often termed the *exosphere.*

THE TROPOSPHERE The atmospheric layers that are most important to climate studies are the two lowermost ones: the troposphere and the stratosphere. The tropo-

sphere is where most of the phenomena that we call weather—such as clouds, rain, snow, and storm activity—occur. It differs from the other atmospheric layers in that it is well mixed by convection. **Convection** is a process in which heat energy is transported by the motions of a *fluid* (a liquid or a gas). Such motions are generated when a fluid in a gravitational field, like that of Earth, is heated from below. A familiar example is the convective motion that occurs when a pot of water is heated on the burner of a stove (Figure 3-10). The warm water at the bottom of the pot is less dense than the cool-

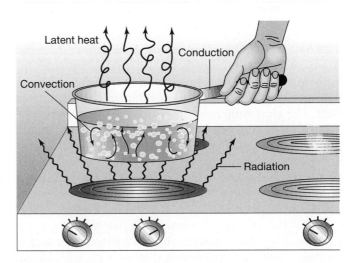

FIGURE 3-10

A pot of water on a stove, illustrating convection. The fluid circulates because it is heated from below. (From R.W. Christopherson, *Geosystems: An Introduction to Physical Geography, 3/e,* 1997. Reprinted by permission of Prentice Hall, Upper Saddle River, N.J.)

er water at the top. As a result of this imbalance, the fluid overturns (it circulates, or convects) and will continue to do so as long as the pot is being heated. If the water were heated uniformly or from above, convection would not occur.

We note for completeness that a third mode of heat transfer (in addition to radiation and convection) is **conduction**. Conduction is the transfer of heat energy by direct contact between molecules. The coils of the electric burner shown in Figure 3-10 heat the bottom of the pot by conduction. Conduction plays little role in atmospheric (or oceanic) heat transfer, however, so we will make no further mention of it.

The troposphere is convective because roughly half the incoming sunlight is absorbed by the ground and by the ocean surface. The energy from this light is eventually reradiated to space as IR radiation, but it cannot make its way directly from the surface in this form because IR radiation is absorbed by atmospheric greenhouse gases and by clouds. So, the energy is instead transported by fluid motions until it reaches an altitude where the atmosphere is more transparent to IR radiation. Only then can the heat energy radiate away from Earth.

As we shall see in Chapter 4, the upward convection of warm, moist air plays a major role in the global energy balance. Convection of heat in a moist atmosphere is more complicated than that in a dry atmosphere, because water can condense or evaporate. When water is evaporated from the ocean surface or from rivers and lakes, energy is taken up by the resulting vapor. This energy is referred to as the latent heat of vaporization. When the water vapor condenses to form clouds, the same amount of latent heat is released to the atmosphere. In more general terms, **latent heat** is the heat energy released or absorbed during the transition from one phase—gaseous, liquid, or solid—to another.

THE STRATOSPHERE. The stratosphere differs from the troposphere in several respects. The pressure is substantially lower in the stratosphere, in accordance with the barometric law. The two layers differ in composition as well. The stratosphere contains most of Earth's ozone. Stratospheric air is also very dry, containing less than 5 ppm of water vapor on average. Thus, condensation of water vapor does not occur, and so clouds and precipitation are absent. (An exception occurs in the polar regions during winter, where tenuous *polar stratospheric clouds* can form. These clouds play a key role in the development of the Antarctic ozone hole, as we will see in Chapter 17.) Stratospheric air is not convective and is therefore less well mixed than tropospheric air. Indeed, the name "stratosphere" derives from the word "stratified," which means layered.

THE VERTICAL TEMPERATURE PROFILE. Why does the vertical temperature profile in Figure 3-9b exhibit

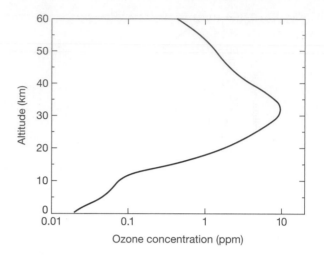

FIGURE 3-11

An approximate profile of the vertical variation of ozone concentration in Earth's atmosphere.

all those curves? The reason has to do primarily with where the atmosphere is heated—that is, where solar energy is absorbed. The high temperatures near the ground are caused by the absorption of sunlight at Earth's surface, which then heats the atmosphere above it. The high temperatures near 50 km are caused by the absorption of solar UV radiation by ozone. The ozone concentration actually peaks some 20 km lower, in the middle stratosphere (Figure 3-11), but the heating rate is highest in the upper stratosphere because more UV radiation is available at those altitudes. The vertical heating distribution also explains why the stratosphere is not convective: The maximum heating occurs at the top of the layer, so there is no tendency for the air to rise. Above 50 km, both the ozone concentration and the heating rate decline, so the temperature decreases with altitude in the mesosphere. Finally, the temperature rise above 90 km in the thermosphere is caused by the absorption of short-wavelength UV radiation by molecular oxygen, O_2.

Physical Causes of the Greenhouse Effect

We determined earlier that Earth's greenhouse effect warms the surface by some 33°C compared with the temperature we would expect if there were no atmosphere. This warming has been attributed to the presence of greenhouse gases, especially H_2O and CO_2. Why do some gases contribute to the greenhouse effect whereas others, such as O_2 and N_2, do not?

Molecular Motions and the Greenhouse Gases H_2O and CO_2

The defining property of a greenhouse gas is its ability to absorb or emit infrared radiation. Gas molecules can absorb or emit radiation in the IR range in two different ways. One way is by changing the rate at which the molecules rotate. The theory of *quantum mechanics* describes the behavior of matter on a microscopic scale—that is, the size of molecules and smaller. According to this theory, molecules can rotate only at certain discrete frequencies, just as most house fans can operate only at certain speeds. The rotation frequency is the number of revolutions that a molecule completes per second. Consider one photon of an electromagnetic wave that is incident on an individual molecule (Figure 3-12). If the incident wave has just the right frequency (corresponding to the difference between two allowed rotation frequencies), the molecule can absorb the photon. In the process, the molecule's rotation rate increases. Conversely, the rotation rate slows down when the molecule emits a photon.

The frequency (or wavelength) of the radiation that can be absorbed or emitted depends on the molecule's structure. The H_2O molecule is constructed in such a manner that it absorbs IR radiation of wavelengths of about 12 μm and longer. This interaction gives rise to a very strong absorption feature in Earth's atmosphere called the **H_2O rotation band.** It can clearly be seen in Figure 3-13, which shows the percentage of radiation at different wavelengths that is absorbed during vertical passage through the atmosphere. Virtually 100% of infrared radiation longer than 12 μm is absorbed, although some of this absorption is caused by CO_2 (see below). The H_2O rotation band extends all the way into the microwave region of the electromagnetic spectrum (above a wavelength of 1000 μm), which is why a microwave oven is able to heat up anything that contains water.

A second way in which molecules can absorb or emit IR radiation is by changing the amplitude with which they vibrate. Molecules not only rotate, they also vibrate—their constituent atoms move toward and away from each other. Again consider an electromagnetic wave that is incident on a molecule. If the frequency at which the molecule vibrates matches the frequency of the wave, the molecule can absorb a photon and begin to vibrate more vigorously. (Similarly, a vibrating tuning fork will induce vibrations in a second tuning fork if the pitches of the two instruments are the same. The pitch is proportional to the frequency of the sound wave.)

The *triatomic* (three-atom) CO_2 molecule can vibrate in three ways. We need to concern ourselves only with the *bending mode* of vibration (Figure 3-14). This vibration has a frequency that allows the molecule to absorb IR radiation at a wavelength of about 15 μm. It gives rise to a strong absorption feature in Earth's atmosphere called the **15-μm CO_2 band.** The 15-μm CO_2 band overlaps the H_2O rotation band and, hence, is hard to distinguish in Figure 3-13. It is, however, easily seen by satellites that look down at Earth's atmosphere from above. Because it occurs fairly near the peak of Earth's outgoing radiation, this absorption band is particularly important to climate. Earth's surface emits strongly in this wavelength region, but very little of this radiation is able to escape directly to space because it is absorbed by CO_2 molecules in the atmosphere. This is why CO_2 is such an important contributor to the greenhouse effect.

Other Greenhouse Gases

Water vapor and CO_2 are the most important greenhouse gases in Earth's atmosphere, but several other trace gases—notably CH_4, N_2O, O_3, and freons—also contribute to greenhouse warming (Table 3-3). These gases have more of an effect on outgoing radiation than their small concentrations would suggest because they absorb at different wavelengths than do H_2O and CO_2. Freons, for example, have absorption bands within the 8- to 12-μm *window region*, where both H_2O and CO_2 are poor absorbers (see Figure 3-13). Thus, one molecule of Freon-11 contributes much more to the greenhouse effect than does one CO_2 molecule. Ozone also has an absorption band in this region centered at 9.6 μm. Thus, O_3 is a good greenhouse gas as well.

Now, recall that we asked the question, Why are O_2 and N_2 poor absorbers of IR radiation and, thus, do not contribute significantly to the greenhouse effect? We are now ready to answer that question. *Diatomic* (two-atom) molecules can rotate and vibrate just like the more complicated molecules, H_2O and CO_2, discussed earlier (Figure 3-15). The O_2 and N_2 molecules, however, are perfectly symmetric: Both of their constituent atoms are identical. Hence, there is no separation of positive and negative electric charges within the molecule. As noted earlier, an electromagnetic wave actually consists of oscillating electric and magnetic fields. To a first approximation, these fields cannot interact with a totally symmetric molecule; the electromagnetic wave passes by

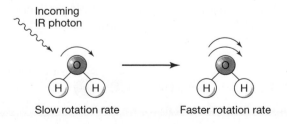

FIGURE 3-12

The rotation rate of an individual H_2O molecule increases when the molecule absorbs a photon of infrared radiation.

Incoming IR photon

Slow rotation rate

Faster rotation rate

FIGURE 3-13

Percentage of radiation absorbed during vertical passage through the atmosphere. Absorption of 100% means that no radiation penetrates the atmosphere. The nearly complete absorption of radiation longer than 13 μm is caused by absorption by CO_2 and H_2O. Both of these gases also absorb solar radiation in the near infrared (wavelengths between about 0.7 μm and 5 μm). The absorption feature at 9.6 μm is caused by ozone. (From data originally from R. M. Goody and Y. L. Yung, *Atmospheric Radiation, 2nd ed.*, New York: Oxford University Press, 1989, Figure 1.1.)

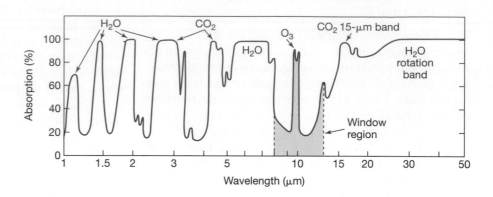

such a molecule without being absorbed. (Note that CO_2 is a symmetric molecule, because the three atoms are, on average, arranged in a line. However, the symmetry is broken when the molecule bends, allowing 15-μm radiation to be absorbed or emitted.)

Effect of Clouds on the Atmospheric Radiation Budget

Gases are not the only constituents of the atmosphere that affect its radiation balance; that balance is also influenced by the presence of clouds and aerosols. We will postpone the discussion of aerosols until Chapter 10, because their climatic effects are usually rather small and short term. (Sulfate aerosols, for example, cooled Earth by about 0.5°C (1°F) for a year or two after the Mt. Pinatubo eruption, as described in Chapter 2.) The effects of clouds, though, are large and cannot be ignored (Figure 3-16). Unfortunately, these effects cannot always be calculated reliably either, and this leads to significant problems in climate prediction.

Types of Clouds

The effect of clouds on Earth's radiation budget is difficult to calculate quantitatively, partly because there are many different types of clouds (Figure 3-17). *Cumulus clouds* are the familiar puffy, white clouds that look like balls of cotton. They are composed of droplets of liquid

water and are formed in convective updrafts. *Cumulonimbus clouds* are big, tall cumulus clouds that give rise to thunderstorms. *Stratus clouds* are grey, low-level water clouds that are more or less continuous. They cover much of the eastern United States during winter. *Cirrus clouds* are high, wispy clouds composed of ice crystals rather than liquid water, because the temperature of the upper troposphere is well below the freezing point.

Opposing Climatic Effects of Clouds

Have you ever noticed that cloudy days are relatively cool, yet cloudy nights are relatively warm? That is because clouds affect planetary energy balance in two opposing ways. Clouds cool Earth during the daytime by reflecting incident sunlight back to space. We noted earlier that at present the planetary albedo is about 0.3. A large fraction of this is caused by clouds. In fact, without clouds, Earth's albedo would probably be closer to 0.1. According to the planetary energy balance equation, reducing the albedo from 0.3 to 0.1 would raise the effective radiating temperature of Earth by about 17°C (30°F). The increase in surface temperature on a cloud-free Earth would be smaller than this, however, because clouds also absorb and re-emit outgoing infrared radiation and, thus, contribute significantly to the greenhouse effect. This effect dominates at night and helps keep cloudy nights warm.

To complicate matters further, the effect of any particular cloud depends on its height and thickness. Low, thick clouds, such as stratus clouds, generally cool the surface because their primary influence is to reflect incoming solar radiation. High, thin clouds, such as cirrus clouds,

FIGURE 3-14

The bending mode of vibration of the CO_2 molecule.

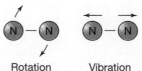

FIGURE 3-15

Rotation and vibration for a diatomic molecule, such as N_2 or O_2.

FIGURE 3-16

Photo of Earth from space, showing clouds. (From NASA Headquarters.)

tend to warm the surface because they contribute more to the greenhouse effect than to the planetary albedo. The reason for the difference is twofold: First, the elongated ice crystals of which cirrus clouds are composed allow much of the incident solar radiation to pass through but absorb most of the outgoing IR radiation. In contrast, stratus clouds reflect much of the incoming visible radiation, in addition to absorbing radiation at IR wavelengths. Second, cirrus clouds occur higher in the troposphere than do stratus clouds and are therefore colder (Figure 3-18). According to the Stefan–Boltzmann law, cirrus clouds therefore radiate less IR energy to space. Because they absorb the upward-directed IR radiation from the warm surface and reradiate it at a lower temperature, cirrus clouds make a large contribution to the atmospheric

greenhouse effect. Lower-lying stratus clouds do this as well, but their radiating temperature is higher and so their contribution to the greenhouse effect is not as large.

Earth's Global Energy Budget

The various factors that we have just discussed can be combined to calculate a global energy budget for Earth, as in Figure 3-19. The incident solar flux in this diagram is normalized to 100 arbitrary "units" of radiation. These 100 units of incoming energy are balanced by 30 units of reflected solar energy (25 reflected by the atmosphere and 5 reflected by the surface) and 70 units of outgoing infrared radiation. About half the incident solar radiation makes it down to the surface; the other half is either reflected or absorbed by the atmosphere. Within the atmosphere, energy is transported by a combination of radiation, convection, and the latent heat associated with the evaporation and condensation of water vapor. This latter process is a very important one: Roughly half of the solar energy absorbed by the surface (24 out of 45 units) goes directly into evaporating water.

The greenhouse effect is shown here as an additional 88 units of downward-directed infrared radiation. Thus, the total energy flux absorbed by the surface is 133 units (= 45 units of solar radiation + 88 units of IR radiation). This value is almost twice the net amount of energy absorbed by the Earth (70 units). The reason is that infrared radiation is absorbed and re-emitted multiple times within the atmosphere, so the internal fluxes can actually be higher than the net input of energy. *At the top of the atmosphere, however, the net downward solar radiation flux (incoming minus reflected) must equal the outgoing infrared flux. This statement is the principle of planetary energy balance.*

(a)

(b)

FIGURE 3-17

Photos of (a) stratus (From Claudia Parks, The Stock Market) and (b) cirrus clouds. (From G.R. Roberts/Nelson Riwaka, Photo Researchers, Inc.)

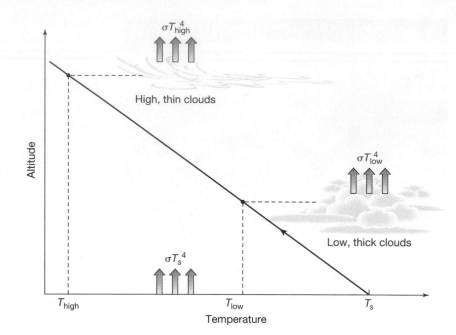

FIGURE 3-18

The different effects of high and low clouds on the atmospheric radiation budget. High, thin clouds are more transparent to incoming sunlight and radiate at a lower temperature than do low, thick clouds. The expressions σT_{high}^4, σT_{low}^4, and σT_s^4 represent the radiation flux at the temperature of high, thin clouds, at the temperature of low, thick clouds, and at the surface temperature, respectively.

Introduction to Climate Modeling

How can we utilize our knowledge of Earth's current energy budget to predict what Earth's surface temperature might have been in the past or how it might vary in the future? Because the climate system is complex, we need some sort of computer model to keep track of all the intricacies. Instead of estimating the magnitude of the greenhouse effect by subtraction, as we did earlier in this chapter, we need to be able to calculate it directly from the measured or predicted concentrations of greenhouse gases.

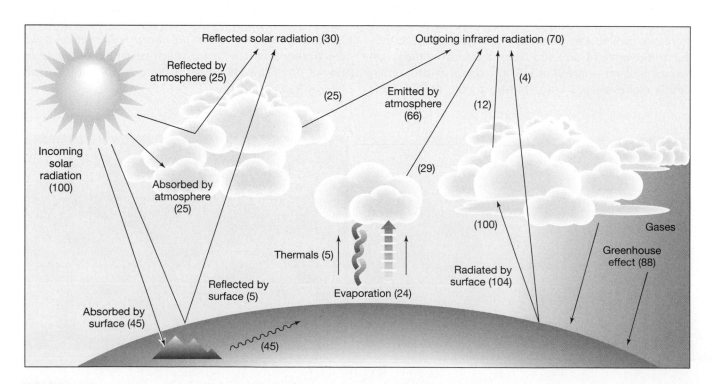

FIGURE 3-19

Earth's globally averaged atmospheric energy budget. All fluxes are normalized relative to 100 arbitrary units of incident radiation. (From S. Schneider, *Climate Modeling,* Scientific American, 256:5, 72–80, 1987.)

Such a calculation must take into account the rotational and vibrational absorption bands of all the different greenhouse gases. Doing so quantitatively requires that we combine the predictions of quantum mechanics with laboratory measurements of the strengths of different absorption bands. Fortunately for climate modelers, a great deal of effort has gone into obtaining the required parameters. As a result, a voluminous database of information on the absorption characteristics of various molecules of atmospheric interest is now available.

Armed with these data, the climate modeler must next decide how to incorporate them into a computer model of the atmosphere. The most complete type of model is called an atmospheric **general circulation model (GCM),** also sometimes referred to as a *global climate model.* These elaborate computer models include a three-dimensional representation of the atmosphere (or oceans) that simulates winds (or currents), moisture transport, and energy balance. Atmospheric GCMs are replete with clouds, winds, snow, rain, and most of the other phenomena that we call weather. Thus, they are capable of predicting how climate varies on a regional basis. But GCMs have a number of drawbacks, the most serious of which is that they require large amounts of runtime on the world's fastest computers to simulate even a few years of global climate. We describe GCMs in some detail in Chapter 6 because they play a central role in climate policymaking today. We begin here, however, with models that are somewhat less complicated and, hence, easier to understand.

One-Dimensional Climate Models—RCMs

For many purposes it is sufficient to construct simpler climate models that require less effort to program and less computer time to run. We have already seen one such model—the one-layer atmosphere model described in the box earlier in this chapter. But that model did not produce a good, quantitative estimate of the greenhouse effect, nor did it account for the contributions of different greenhouse gases. The simplest model that is capable of doing both of these things reliably is called a **radiative-convective model (RCM).** In an RCM, the climate system is approximated by averaging the incoming solar and outgoing IR radiation over Earth's entire surface. The vertical structure of the atmosphere (Figure 3-9) is taken into account (unlike in the one-layer model), but horizontal variations are ignored. Thus, such models are sometimes called one-dimensional climate models, in contrast with the three-dimensional GCMs. The vertical dimension (altitude) is then divided into a number of layers. The RCM calculates the temperature of each layer by taking into account the amount of energy received or emitted in the form of radiation, along with the effects of convection and latent heat release in the lowermost layers.

Radiative Effect of Doubling Atmospheric CO_2. Although RCMs are quite simple compared with the real climate system, they allow us to estimate the magnitude of the greenhouse effect as a function of the concentrations of various greenhouse gases in Earth's atmosphere. These models correctly predict that the greenhouse-induced temperature difference ΔT_g for the present atmosphere is 33°C, in agreement with the estimate derived earlier by the subtraction $\Delta T_g = T_s + T_e$. (This is *not* a trivial result. We would have to work through a lot of the relevant physics to come up with this answer.) More importantly, RCMs allow us to predict the average surface temperature increase that should result from an increase in the concentration of greenhouse gases. A commonly cited benchmark is the temperature change that would result from a doubling of the atmospheric CO_2 concentration from 300 ppm (its value near the turn of the 20th century) to 600 ppm. RCM calculations show that, all other factors being equal, such a change in CO_2 would produce an increase of about 1.2°C (2.2°F) in the global average surface temperature. In the terminology developed in Chapter 2, this value is the temperature change ΔT_0 that would result in the absence of any feedbacks in the climate system.

In reality, we would expect other factors in the climate system to change as atmospheric CO_2 increases, and so a temperature change of +1.2°C is not the best estimate we could make of the effect of CO_2 doubling. To obtain a better estimate, we must consider what those climate feedbacks might be and how strongly they affect our answer.

Climate Feedbacks

Climate feedbacks are extremely important because they can either amplify or moderate the radiative effect of changes in greenhouse gas concentrations. That is why we devoted most of Chapter 2 to explaining how they work. There, we dealt with an imaginary feedback system involving the percentage of daisy cover on the hypothetical planet Daisyworld. Here, we discuss several feedback processes that affect climate on Earth.

The Water Vapor Feedback

One of the most important feedbacks in the climate system involves the concentration of atmospheric water vapor. As noted earlier, water vapor is an excellent absorber of IR radiation and, hence, a good greenhouse gas. Unlike CO_2, however, water vapor is typically close to its *condensation point*—the temperature at which a vapor condenses to form a liquid. If Earth's surface temperature were to decrease for some reason, water vapor would condense out in the form of rain or snow, leaving less water vapor behind in the atmosphere. This reduction in atmospheric water vapor would cause a corresponding decrease in the greenhouse effect, which, in turn, would lower the surface temperature still further. Conversely, an increase in surface temperature would cause an increase in the rate at which water vapor evaporates from the oceans. This would increase the concentration of water vapor in the atmosphere, thereby increasing the greenhouse effect and further warming Earth's surface.

The net result of this interaction between water vapor abundance and Earth's surface temperature is a positive feedback loop that tends to amplify small temperature perturbations (Figure 3-20). This feedback loop can be incorporated in RCMs by assuming a fixed *relative humidity* profile in the troposphere. **Relative humidity** is the concentration of water vapor in an air parcel divided by the concentration that would be present if the air parcel were *saturated* with water vapor (i.e., on the verge of condensation). When such a calculation is performed, the RCM predicts that the equilibrium change in surface temperature for CO_2 doubling, ΔT_{eq}, is about twice that which would have occurred otherwise. Recall from Chapter 2 that we can write

$$\Delta T_{eq} = \Delta T_0 + \Delta T_f$$

where ΔT_0 is the temperature change with no feedbacks and ΔT_f is the change caused by the feedback. For the problem of CO_2 doubling, $\Delta T_0 = 1.2°C$ (2.2°F), $\Delta T_{eq} \approx 2.4°C$ (4.4°F), so the temperature change caused by the water vapor feedback is approximately 1.2°C. Furthermore, the feedback factor f is given by

$$f = \frac{\Delta T_{eq}}{\Delta T_0} = \frac{2.4°C}{1.2°C} = 2$$

A feedback factor of 2 indicates that this is a strong, positive feedback on the climate system.

Snow and Ice Albedo Feedback

A second feedback loop that is expected to have some impact on modern global warming, but is especially important for glacial–interglacial variations, involves albedo

changes caused by snow and ice. As Earth's climate cools, the extent of wintertime snow and ice cover increases in temperate regions. On longer time scales, the permanent ice cap in the northern polar regions expands toward the equator, resulting in the periods of glaciation known as the Ice Ages. Snow and ice have a much higher albedo than does land or water (refer to Table 2-1). Therefore, increases in snow and ice cover should cause further decreases in surface temperature. The result is a positive feedback loop that tends to amplify induced changes in Earth's surface temperature (Figure 3-21). As snow and ice cover are restricted to middle and high latitudes, modeling this feedback loop quantitatively requires the use of two-dimensional or three-dimensional computer models.

The IR Flux/Temperature Feedback

Both of the feedbacks discussed so far are positive. But systems that contain only positive feedback loops are unstable. Does this mean that Earth's climate is unstable? No. Earth's climate system contains a very strong negative feedback that is so basic that it is often overlooked. The feedback loop that stabilizes Earth's climate on short time scales is the relationship between surface temperature and the flux of outgoing IR radiation (Figure 3-22). (We have already hinted in Chapter 1 that there is another feedback loop that stabilizes Earth's climate on long time scales, but that is not what we are talking about here.) If Earth's surface temperature were to increase for some reason, the outgoing IR flux from the top of the atmosphere would also increase. But if the outgoing IR flux were to increase, the surface temperature would tend to decrease, because more energy would be lost from the Earth system. This feedback loop might appear to be trivial, but it is not; there are situations in which it can fail. In particular, the positive correlation between surface temperature and the outgoing IR flux can break down if the atmosphere contains a very large amount of water vapor. This, we think, is what happened to our sister planet, Venus, and it led to what is sometimes called a *runaway greenhouse*. But we will save that story for Chapter 19.

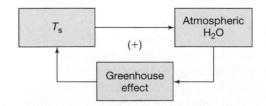

FIGURE 3-20
Systems diagram showing the positive feedback loop that includes atmospheric water vapor.

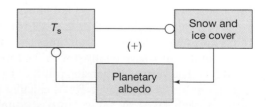

FIGURE 3-21
Systems diagram showing the positive feedback loop that includes snow and ice cover.

FIGURE 3-22

Systems diagram illustrating the negative feedback loop between surface temperature and the outgoing flux of infrared radiation. This feedback is the fundamental reason that Earth's climate is stable.

The Uncertain Feedback Caused by Clouds

Another important feedback process in the climate system is that provided by changes in clouds. Unfortunately, this feedback process is not as easy to quantify as the ones just discussed. You already know that clouds can either warm the surface, or cool it, depending on their height. This alone should provide a hint that estimating their feedback effect might be difficult. In addition to this problem, clouds are inherently three-dimensional: they form at some locations and not at others because of the way the winds blow. Hence, we will postpone our discussion of cloud feedback until Chapter 6. Keep in mind, though, that cloud feedback is one of the greatest uncertainties in the study of global warming.

In summary, we have now examined Earth's climate system in enough detail to understand how the atmospheric greenhouse effect warms the planet and how the planet's average surface temperature may respond to a human-induced increase in greenhouse gases. But Earth's climate cannot be described by just its average surface temperature. The term "climate" includes many other related factors, such as latitudinal and seasonal temperature gradients, winds, and precipitation. To study these phenomena, we need to broaden our spatial perspective and consider the Earth system from a three-dimensional perspective. The next two chapters describe how the transport of heat from one location to another by the atmosphere and oceans determines these other important features of Earth's global climate.

Chapter Summary

1. Earth is warmed by the absorption of visible radiation from the Sun and is cooled by the emission of infrared radiation to space.
   a. Much of the infrared radiation emitted by Earth's surface is absorbed and re-emitted by atmospheric gases.
   b. The result is a greenhouse effect that warms the surface by about 33°C. Without this natural greenhouse effect, Earth would be too cold to support life.
2. Only certain atmospheric gases, most importantly H_2O and CO_2, contribute to the greenhouse effect. These gases absorb infrared radiation by changing the rate at which individual molecules rotate or vibrate. Other trace gases, such as freons, can contribute substantially to the greenhouse effect by absorbing radiation at different wavelengths than do H_2O and CO_2.
3. Clouds affect the atmospheric radiation budget both by reflecting incident sunlight and by contributing to the greenhouse effect. Low, thick clouds tend to cool the surface; high, thin clouds tend to warm it.

4. Earth's climate system contains several well-understood feedbacks that play important roles in regulating climate change.
   a. The climate system is stabilized by a strong negative feedback loop between surface temperature and the outgoing infrared flux.
   b. The system is destabilized by a positive feedback loop involving atmospheric water vapor. Because it acts on short time scales, this feedback is likely to play an important role in contemporary global warming. Climate models predict a surface temperature response to CO_2 doubling that is twice that of models in which this feedback is neglected.
   c. The system is also destabilized by a positive feedback loop involving the extent of snow and ice cover due to the effect of albedo.
   d. Clouds may also contribute to climate feedback, but their effect is not well understood.

Key Terms

barometric law
blackbody radiation
conduction
convection
effective radiating temperature
electromagnetic radiation
electromagnetic spectrum
15-μm CO_2 band
flux
frequency
general circulation model

H_2O rotation band
infrared radiation
inverse-square law
Kelvin temperature scale
latent heat
mesosphere
photon
photosphere
radiation
radiative-convective model
relative humidity

Stefan–Boltzmann law
stratosphere
thermosphere
troposphere
ultraviolet radiation
visible radiation
visible spectrum
wavelength
Wien's law

Review Questions

1. How are the wavelength and frequency of an electromagnetic wave related?
2. What is a photon?
3. What physical law describes the manner in which the intensity of sunlight changes as the observer moves away from the Sun?
4. Name two physical laws that apply to blackbody radiation. What do these laws tell us about the nature of the emitted radiation?
5. What is the major contributor to Earth's albedo?
6. What are the three most abundant gases in Earth's atmosphere?

7. List the four layers of Earth's atmosphere. How are they defined?
8. Name three mechanisms by which heat energy can be transferred. Which two are important in Earth's global energy budget?
9. Identify two physical processes by which gases can absorb infrared radiation. Give examples of each process.
10. Why are O_2 and N_2 not greenhouse gases?
11. Describe the different ways in which climate is affected by high and low clouds.
12. Identify two positive feedback loops in Earth's climate system. Why is Earth's climate stable despite these destabilizing, positive feedbacks?

Critical-Thinking Problems

1. a. Given that a 300-K blackbody radiates its peak energy at a wavelength of about 10 μm, at what wavelength would a 600-K blackbody radiate its peak energy?
   b. If the two bodies in part (a) were the same size, what would be the ratio of the heat emitted by the hotter object to the heat emitted by the colder one?
2. a. Venus and Mars orbit the Sun at average distances of 0.72 AU and 1.52 AU, respectively. What is the solar flux at each planet?
   b. Venus has a planetary albedo of 0.8, and Mars has an albedo of 0.22. Using the answer to part (a), determine the effective radiating temperatures of these planets.
   c. How do the effective radiating temperatures determined in part (b) compare with the value for Earth, and why is this result surprising?
   d. The mean surface temperatures of Venus and Mars are 730 K and 218 K, respectively. Using the answer to part (b), determine the magnitude of the greenhouse effect on each planet.
   e. How do the results of (d) compare with the magnitude of the greenhouse effect on Earth?
3. a. The Sun radiates at an effective temperature of 5780 K and has a radius of about 696,000 km. Remembering that 1 AU = 149,600,000 km, derive the approximate value of the solar flux at Earth's orbit.
   b. Compare your answer with the value given in the text.
4. The tropospheric lapse rate (the rate at which temperature decreases with altitude) is approximately 6°C (11°F) per kilometer. Given that the mean surface temperature of Earth is 288 K and the effective radiating temperature is 255 K, from what altitude does most of the emitted radiation derive?
5. Solar luminosity is estimated to have been 30% lower than today at the time when the solar system formed, 4.6 billion years ago.

   a. If Earth's albedo was the same as it is now ($A = 0.3$), what would have been its effective radiating temperature at that time?
   b. If the magnitude of the greenhouse effect had also remained unchanged ($\Delta T_g = 33$ K), what would Earth's average surface temperature have been? How does this compare with today's value?
6. For atmospheric CO_2 concentrations not too different from the present value, the radiative forcing of CO_2 can be expressed by the formula

$$\Delta F = -6.3 \ln\left(\frac{C}{C_0}\right),$$

where $C_0 = 300$ ppm is the CO_2 concentration near the turn of the 20th century, C is the CO_2 concentration at some other time, and ΔF is the change (in watts per square meter) in the outgoing infrared flux caused by the change in CO_2 concentration. The function $\ln(x)$ denotes the natural logarithm of a given number x. Any scientific calculator has this function key.

   a. By how much would the outgoing infrared flux decrease if the atmospheric CO_2 concentration were increased from 300 ppm to 600 ppm (i.e., if $C = 600$ ppm)?
   b. By how much would surface temperature have to increase in order to bring the radiation budget back into balance in part (a), assuming that the planetary albedo and the amount of water vapor in the atmosphere do not change? (*Hint:* Use the planetary energy balance equation to calculate how much T_e would have to change to balance the radiation budget. Remember that the left-hand side of this equation represents the outgoing infrared flux. The quantity T_s will change by the same amount as T_e if the amount of water vapor is held constant.)

The Atmospheric Circulation System

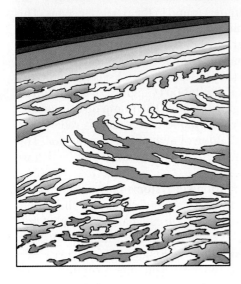

Key Questions

- Why does air move?
- Are the movements of the winds random across the surface of the Earth, or do they follow regular patterns?
- What implications do these circulatory systems have for global climate?
- What other factors govern the geographic and seasonal distributions of temperature and rainfall?

Chapter Overview

Earth's climate is a central theme of ours. We focus on the role climate plays in the Earth system and explain how Earth's climate works, how climate has changed through time, and how it may change in the future. An important element of Earth's climate is the atmospheric circulation. In Chapter 3 we described the global energy budget and showed that if we average the radiation fluxes around the globe and over a few years there is a balance. Earth emits as much energy as it receives, aside from the issue of anthropogenic increases in the greenhouse effect. If we look at regions smaller than the globe and over time periods of less than a year, however, the situation is very different. There is a significant imbalance in the distribution of energy at various latitudes:

The tropics receive a surplus of radiative energy, whereas the poles run a deficit. This imbalance causes an equator-to-pole temperature gradient that results in density and pressure differences in the atmosphere. The density and pressure differences cause air to move in a global scale pattern of wind belts, which are modified by Earth's rotation and by the distribution of land and water. The net effect is to restore the latitudinal energy balance by moving surplus energy away from the tropics to cancel out the deficit at the poles. In the process, energy is used to evaporate water from the land and ocean surfaces, water vapor is carried by wind, and energy is released when the vapor condenses to form clouds. Thus, there are close interactions between the transport of energy and of water by means of circulating air. In other words, Earth's atmospheric circulation has a direct impact on the global distributions of temperature and precipitation.

The Global Circulatory SubSystems

Anyone who has felt wind blow, watched clouds move, and seen rain fall is aware that large parts of the Earth system are in constant motion. Even the continents and oceans, despite their apparent permanence, are continuously moving. The island of Iceland in the North Atlantic, for example, is spreading, and its two sides are moving away from each other fast enough to be measured by today's instruments. Although these movements may sometimes appear random, they form part of a well-ordered circulation of energy and matter throughout the Earth system.

Like the circulatory system of humans (part of the cardiovascular system), Earth's circulatory subsystems work to maintain the planet in a thermal and chemical balance. The human circulatory system transports dissolved gases, nutrients, and hormones throughout the body; carries away waste products; helps regulate the acidity of body fluids; and is a vital part of the body's thermoregulatory system, carrying warm blood from one area to another. Although the human circulatory system is not an exact analogy to Earth's circulatory subsystems, these systems do have much in common. Essential gases and nutrients are transported throughout the Earth system, and waste products are removed from their area of production. All of Earth's circulatory subsystems act in some way to help regulate the global temperature: The winds and ocean currents redistribute the energy received from the Sun, and the motions of the solid Earth redistribute carbon and help regulate the CO_2 level of the atmosphere. The circulations within the solid Earth are discussed in Chapter 7; here and in Chapter 5 we examine those circulations that occur within the fluid part of the Earth system: the atmosphere and oceans.

The purpose of this chapter is to describe the major characteristics of the atmospheric circulation, to explain why they occur, and to illustrate the way in which they affect the transport of energy and materials around the globe. In Chapter 3 we described the energy input and output from the Earth system as a whole; now we take that system apart and examine some of its internal workings—specifically those related to climate. In doing so, we have two primary objectives. The first is to explain why weather and climate vary across the globe. The second is to emphasize that because of the internal workings of Earth's climate system, the response to global-scale processes and changes may not be uniform around the globe. Organized movements of the atmosphere occur over many different time and space scales. These movements range from centimeter-scale swirls, or *eddies,* to global-scale motions of the wind belts. All of these are important in one way or another, but we limit our discussion to processes that are global in extent and that have the greatest influence on the transport of energy and mass through the Earth system. One of Earth's most important constituents is water. Cycling continuously among the atmosphere, the oceans, and the land surface, water carries with it energy, dissolved nutrients, and other matter—all vital for maintaining an environment suitable for life.

In the same way that the functioning of the cardiovascular system in the human body ultimately depends on the ability of the heart to keep pumping, the functioning of Earth's circulatory subsystems rely on several different pumps. Each of these pumps drives a different circulatory mechanism, and each works at a different speed. Over shorter time scales (years to decades), the most important pump is found in the tropical oceans. This pump is responsible for the movements of the air and the surface ocean over most of the globe. The energy source that drives this pump is radiation from the Sun. Over longer time scales (about 1,000 years), a second pump drives the deep-ocean circulation (Chapter 5). The ultimate energy source is again the Sun. The pump operating over the longest time scales (millions of years) is radioactive decay and the production of heat in Earth's interior. This pump causes the movements of the continents, which we discuss in Chapter 7.

All these circulation subsystems play a vital role in the operation of the Earth system. We know that in humans the cardiovascular system can maintain stability only if the blood keeps moving. Similarly, the Earth system can maintain stability only as long as its circulatory subsystems continue to function. The long-term pump (that is, the processes of internal heat production and plate tectonics) ceased to function on Mars. As we will see in a later chapter, the planet's inability to support an environment suitable for life may be at least partly a result of the failure of this circulation mechanism.

The Atmospheric Circulation

Recall from Chapter 3 that the troposphere is the lowermost layer of the atmosphere. Most of the processes we are interested in take place in the troposphere, so we limit our discussion here to that layer. Although the circulation of the stratosphere plays a role in the depletion of stratospheric ozone, we will save that discussion for Chapter 17.

The Movement of Air

Air moves over Earth's surface because there are horizontal differences in pressure. Air also moves vertically either because it is forced to rise mechanically (e.g., when it encounters a mountain range) or because there are changes in *buoyancy.* **Buoyancy** is the tendency of an object to float in a fluid. Buoyancy is controlled by differences in *density* between the object and the fluid, where density is given by the mass of a substance within a unit volume. (The greater the mass within a given volume, the greater the density.) Ultimately, all of these horizontal and vertical movements (except those due to mechanical forces) can be attributed to differences in temperatures across the globe. To explain how these movements occur, we need to understand how pressure and density are related to temperature.

Vertical Movement. It is easiest to picture these relationships by thinking of vertical and horizontal movements separately. Imagine the situation with a hot-air balloon. Remember from Chapter 3 that heating causes molecules to move faster. In this case, the faster the air molecules move, the more they collide with each other and with the interior of the balloon. These collisions exert a force (i.e., air pressure) on the interior surface. If the balloon was a fixed container (one that could not expand), there would be an increase in the air pressure within the container. Thus we see a connection between the temperature and pressure of a gas: As the temperature increases, the pressure increases. But the balloon *is* expandable. As the pressure starts to increase, the air pushes outward on the interior of the balloon, causing it to expand. So in a balloon, it is the volume rather than the pressure that increases (see the Box "A Closer Look: The Relationships between Temperature, Pressure, and Volume—The Ideal Gas Law").

If we begin with the balloon partially inflated, it will contain a certain number of air molecules. As these are heated, the number of molecules does not change, but they move faster, increasing the pressure on the interior of the balloon; this causes the balloon to expand. We now have the same number of air molecules as before (the mass [m] hasn't changed), but they occupy a greater volume [V]. This means that the density of the air [ρ] must decrease ($\rho = m / V$). Because the air in the balloon is less dense than the air surrounding it, the balloon becomes *positively buoyant,* and it rises. The balloon will continue to rise until the density of the air outside the balloon matches that inside (*neutral buoyancy*). If the air in the balloon is more dense than the surrounding air, the balloon would have *negative buoyancy,* and it would sink. Exactly the same processes occur in the atmosphere when we heat the surface below a parcel (column) of air. The surface heats the parcel of air at the bottom of the column; the air parcel expands, its density decreases, and the parcel rises through the air column. Cooling a parcel of air higher in the column causes it to become more dense than the surrounding air, and the parcel sinks.

Horizontal Movement. How do horizontal movements occur? We saw that warmer air has a lower density than cooler air. If we consider two adjacent columns of air, one warmer than the other, the cooler column would have a greater density than the warmer column. This difference in density would cause the air to move horizontally from the region of higher-density cool air to the region of lower-density warmer air—the air moves down the den-

A CLOSER LOOK
The Relationships Between Temperature, Pressure, and Volumes—The Ideal Gas Law

We can see from the text discussion that we can define specific relationships among the temperature (T), pressure (P), and volume (V) of a gas. (In this case, volume refers to the space occupied by a fixed mass of gas molecules; density is the inverse of volume for a fixed mass, so we need not consider it separately here.) First, we know that if the temperature is held constant, then an increase in gas pressure results in a decrease in volume, and a decrease in pressure results in an increase in volume. In other words, pressure is inversely proportional to volume: The product of pressure and volume is a constant. Mathematically, we can write

$$P_{initial}\, V_{initial} = P_{final}\, V_{final}, \text{ (Boyle's law)}$$

where $P_{initial} V_{initial}$ is the product of the initial pressure and volume, and $P_{final} V_{final}$ is the product of the final (new) pressure and volume. This relationship, known as **Boyle's law,** was discovered in 1662 by the British chemist Robert Boyle.

If, instead, gas pressure is held constant, then we know that an increase in temperature results in an increase in volume and that a decrease in temperature results in a decrease in volume. In other words, volume is directly proportional to temperature: The quotient of volume and temperature is a constant. Mathematically, we can write

$$\frac{V_{initial}}{T_{initial}} = \frac{V_{final}}{T_{final}}, \text{ (Charles's law)}$$

where $V_{initial}/T_{initial}$ is the quotient of the initial volume and temperature, V_{final}/T_{final} is the quotient of the final volume and temperature, and temperature is in Kelvins. This relationship, known as **Charles's law,** was discovered in 1787 by the French physicist Jacques Charles.

All gases are found to behave the same way over a wide range of conditions, and Boyle's Law and Charles's law can be combined to give the ideal gas equation:

$$PV = mRT$$

where m is the mass and R is a constant called the gas constant for 1 kg of gas (the value of R depends on the particular gas concerned).

It is important to note that these relationships apply to idealized gases, that is, gases in which there are no attractive forces between molecules. In reality, gases may not respond to environmental changes exactly as described here. Nevertheless, these relationships do represent a very close approximation of how gases behave.

sity gradient. The atmospheric pressure is a force [**F**] determined by the mass [*m*] of the air column and the acceleration [**a**] due to gravity (remember Newton's Second Law of Motion: **F** = *m***a**). Averaged globally and through time, the atmosphere exerts a pressure of 1013 mb on Earth's surface. Thus 1013 mb is considered to be one atmosphere of pressure (1 atm). The pressure decreases as you rise up in the atmosphere (because there is less air above you) until, at the top of the atmosphere, the pressure reduces to zero. The actual pressure recorded at any point on the surface or in the atmosphere, however, can be highly variable under different conditions of elevation and temperature. So, for adjacent columns of air with similar volumes, the colder high-density air has a higher atmospheric pressure. Hence, as you see on weather charts, air flows from high pressure regions to regions of low pressure.

From this discussion we can establish two important points that will help explain why air moves:

1. Air tends to move from an area of higher pressure to an area of lower pressure until the two pressures are equalized. In other words, air (wind) will move horizontally in the lower troposphere from higher to lower pressure. Pressure differences among air masses are typically related to the distribution of surface temperatures.

2. If an air mass is heated until its density is lower than that of its surroundings, the lower-density air will rise. This phenomenon is a form of convection. (We discussed this phenomenon in a more general sense in Chapter 3 when we demonstrated the convection

of a fluid that is heated from below.) Conversely, if an air mass is cooled until its density is higher than that of the underlying air, it will sink. This phenomenon is referred to as **subsidence**.

The Driving Force: The Global Energy Distribution

We learned in Chapter 3 that the average global temperature is determined by the balance between the solar energy absorbed by Earth and the infrared radiation emitted to space. However, neither the radiation received from the Sun (our primary energy source) nor the infrared emission from Earth is distributed uniformly across Earth's surface. The incoming solar energy varies with latitude and with season, whereas the outgoing terrestrial radiation depends on the temperature of the surface and atmosphere at each location.

The distribution of the incoming solar radiation changes with latitude as a result of the change in surface area presented to the Sun's rays as Earth's surface curves (Figure 4-1). The energy from the Sun radiates outward in all directions; however, by the time the Sun's rays reach Earth, they are essentially parallel to each other. This means that the flux of solar energy passing perpendicularly through the plane *A-B* in Figure 4-1 will be the same at any point. For example, the three "beams" in the diagram are equal in solar flux when they pass through the plane. Because of the curvature of Earth, however, when these beams reach the top of Earth's atmosphere, the same amount of light is spread over a much larger area at the poles than at the equator. Consequently, each square meter of surface receives proportionate-

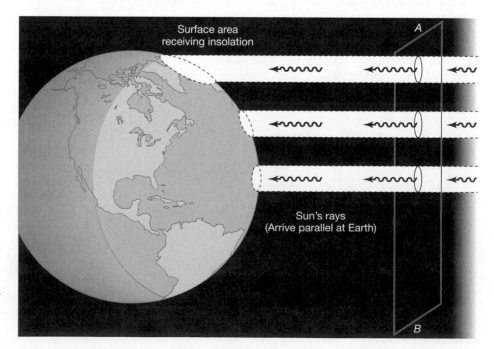

FIGURE 4-1

Variation of incoming solar energy with latitude. The radiation reaching Earth is spread over larger and larger areas as we move from the equator to the poles. Each square meter of the surface receives proportionately less energy as we move to higher latitudes. (From R.W. Christopherson. *Geosystems: An Introduction to Physical Geography*, 3/e, 1997. Reprinted by permission of Prentice Hall, Upper Saddle River, N.J.)

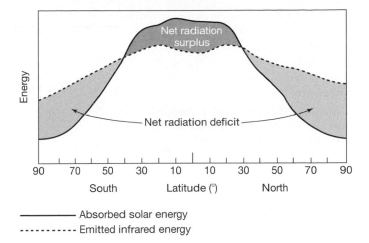

FIGURE 4-2

The distribution of absorbed solar and emitted infrared radiation with latitude. There is a surplus of energy in the tropics, where incoming radiation is greater than outgoing, and a deficit at high latitudes, where more radiation is emitted than is received.

ly less energy at the higher latitudes, and the incoming solar flux thus decreases from the equator toward the poles. (Recall the lightbulb and sheet of paper experiment in Chapter 3.)

The solar radiation absorbed at the surface follows the same general pattern, although the actual amount absorbed varies with cloud cover and atmospheric absorption. This equator-to-pole gradient in the energy absorbed at the surface exerts a primary control on Earth's climate. Figure 4-2 shows this gradient (solid curve) as a function of latitude (i.e., the amount averaged around each latitude band). As we might expect from the previous discussion, the maximum absorbed solar energy is found in the tropics, and the available solar energy decreases rapidly as we move toward the poles. This gradient in absorbed solar energy is the single most important control on temperature. More energy is generally available at the equator than at the poles, so we can assume that temperatures should be highest in the tropics and lowest at high latitudes. Figure 4-2 also shows the latitudinal distribution of infrared radiation emitted from Earth to space (dashed curve). The higher emissions in the tropics are a result of the high surface temperatures there and the correspondingly high temperature in the middle troposphere, from whence the outgoing radiation is emitted. Again, you can refer back to the discussion of the IR flux–temperature feedback described in Chapter 3.

The difference between the incoming solar radiation and the outgoing terrestrial radiation is referred to as *net radiation*. Referring again to Figure 4-2, note that the energy absorbed exceeds the energy emitted in the tropics (net radiation is positive); near the poles, the reverse is true (net radiation is negative). This distribution of available energy is a permanent feature of Earth's climate system. The gradient seems to imply that the tropics should get warmer while the poles get progressively colder. Clearly this does not happen; other processes must be operating to ensure an energy balance at each latitude. In reality, the latitudinal energy gradient produces atmospheric temperature and density differences that force the atmosphere to circulate, carrying warmer air toward the poles and colder air toward the equator. These circulations move energy from regions where there is a surplus to regions where there is a deficit. Most of what we experience as weather and climate is this response of the atmosphere to the unequal latitudinal distributions of energy.

The General Circulation of the Atmosphere

From our description of the energy distribution and our discussion of how air movements occur, we can build a picture of what we would expect the global-scale circulation of the atmosphere to look like. This circulation involves several characteristic features that we will discuss in turn. Taken together, these circulation features represent a negative feedback loop as the atmosphere responds to the temperature gradient by transferring energy latitudinally to reduce the gradient and restore an energy balance. The continuous addition of energy from the sun of course means that the energy distribution is never balanced.

Convergence. We begin with the heating in the tropics. The large solar input to the tropics heats the surface (primarily ocean), which in turn heats the overlying air. As we saw earlier, when heated from below, air will rise by convection. The tropical air near the surface rises, creating a low-pressure region there. But we saw that air tends to move horizontally from an area of higher pressure to an area of lower pressure. Thus, the rising air is replaced by surface air moving equatorward into the region of low pressure from regions of higher pressure (Figure 4-3). The merging of air masses that are moving inward toward a low-pressure region is called **convergence**. The converging air masses that meet at the tropics and rise make up the **intertropical convergence zone (ITCZ)**.

The surface heating produces evaporation in addition to convection. As the convecting air rises, it cools, and the evaporated water (water vapor) in the convecting column condenses to form clouds. As a consequence, the

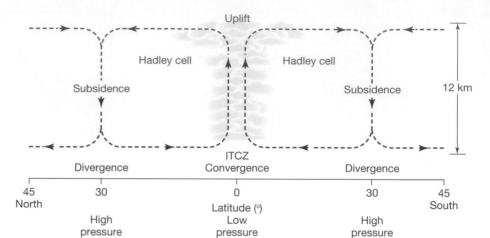

FIGURE 4-3

Convergence, divergence, and the Hadley circulation in the tropics. There is a Hadley cell on either side of the intertropical convergence zone (ITCZ), located over the equator. Rising air in the ITCZ is replaced by inflowing air (convergence) at the surface. Outflowing air (divergence) in the upper troposphere sinks at about 30° N and 30° S, completing the circulations in the two cells.

ITCZ is characterized by extensive areas of cloud cover and heavy precipitation. We talk more about evaporation, condensation, and rainfall later in the chapter.

Divergence. The top of the troposphere, located at about 12–15 km in the tropics, forms a barrier to further uplift. Remember that temperatures generally increase in the stratosphere and that the higher temperatures produce a stable structure that limits convection from below. The air that rises in the ITCZ, upon reaching this barrier, is forced to diverge poleward. **Divergence**, in this case, refers to the movement of air outward from a region in the atmosphere. This poleward-moving air subsides at about 30° N and 30° S latitude, replacing the air that is moving equatorward at the surface (Figure 4-3). The air warms as it sinks, which prevents condensation from occurring and clouds from forming. As a result, these regions are characterized by clear skies and low rainfall amounts. If you check an atlas, you will find that such areas coincide with some of the world's largest deserts (e.g., the Sahara and Arabian deserts and the Great Australian Desert). The subsiding air also leads to an area of high pressure and divergence at the surface.

Hadley Circulation. This pattern of air movement, with convergence occurring in the tropics and divergence and subsidence some 30° away in one large convection cell, is called **Hadley circulation**. This circulation pattern was named for George Hadley, the British meteorologist who first explained the phenomenon. The convection cells on either side of the equator, referred to as *Hadley cells*, represent the dominant north–south mode of circulation between 30° N and 30° S latitude. Note, however, that the Hadley cells—and the ITCZ—are not continuous around the globe. The circulation takes place in individual cells of rising and subsiding air, and the pattern is further broken up by land–ocean contrasts. The ITCZ is most obvious in the Atlantic and Pacific Oceans and is readily observed in satellite images such as that shown in Figure 4-4. The

large-scale circulation in Southeast Asia and the Indian Ocean is dominated by the monsoon, which is described later in this chapter.

The convection cells in the ITCZ result directly from surface heating in the tropical oceans. In fact, although solar heating provides the fuel for tropospheric circulation, the actual pump that drives the circulation is the release of latent heat during convection. (We discussed latent-heat release in Chapter 3.) The energy of solar radiation, used to evaporate water from the ocean surface, is converted to latent heat, and the latent heat is released to the atmosphere in huge towers of convective cloud clusters within the ITCZ (Figure 4-5). It is this release of latent heat that pumps the air around each Hadley cell.

Midlatitude and High-Latitude Circulation. Thus far we have discussed the atmospheric circulation only between the equator and 30° N or S. What about atmospheric circulation from there to the poles? The very low temperatures at the poles, particularly in winter, result in increased air density near the surface and, thus, in higher pressures than occur in the tropics. The higher density and pressure lead to divergence and a general movement of cold air outward at the surface, that is, toward the equator. The divergence is accompanied by subsidence from above. The equatorward-moving cold air meets the warm air moving poleward from the subtropics, producing a zone of steep temperature gradients called the **polar front zone** at approximately 60° N and S latitude. The two air masses do not mix easily: The warm air is less dense than the cold air, which therefore sinks below the warm air when the two air masses meet (Figure 4-6). The polar front zone, therefore, slopes poleward with increasing altitude in the atmosphere. Note that, because of dynamic processes that come into play when air moves over a curved surface, this frontal zone forms a wavelike structure around the hemisphere. The actual latitude at which the front is located, therefore, varies from place to place.

FIGURE 4-4

Satellite image of the eastern Pacific and Central America. These images are obtained from geostationary satellites, which orbit over the equator at an altitude of about 35,000 km and at an orbital speed that keeps pace with Earth's rotation; thus the satellite appears to remain stationary over the same spot on the equator. This image was obtained by the National Oceanographic and Atmospheric Administration (NOAA) during Northern Hemisphere summer. A line of convective clouds marks the ITCZ just north of the equator. The clear areas to the north and south of the ITCZ mark the descending arms of the Hadley cells. (GOES infrared image for 9 August, 1996, 21:00 UCT from the U.S. National Oceanic and Atmospheric Administration website at *http://lwf.ncdc.noaa.gov/servlets/GoesBrowser/* Courtesy of NDAA/ National Climatic Data Center.)

When we put Figures 4-3 and 4-6 together, we see an alternating pattern of northward- and southward-moving air at the surface (Figure 4-7). Such north–south movement is called *meridional* circulation. If we look at Figure 4-7 from above, we might expect to see a general pattern of surface winds such as those depicted in Figure 4-8. We would expect surface winds to blow out of the high-pressure zones at the poles and at about 30° N and S, and to blow toward the low-pressure zones at the equator and at about 60° N and S. The actual pattern, however, is more complicated because winds tend to blow in east–west directions as well. Indeed, the east–west motions are considerably greater than the north–south motions. We know that differences in solar heating cause the equator-to-pole movement we have been discussing. What causes the east–west movements?

The Coriolis Effect

East–west movements of surface winds are the result of the Coriolis effect. The **Coriolis effect** (named for Gaspard Gustav de Coriolis, the French mathematician who in 1835 proposed that the concept applies to surface winds) is the apparent tendency for a fluid (air or water) moving across Earth's surface to be deflected from its straight-line path. (Some texts refer to a *Coriolis force* in relation to this effect. This force, however, is only an apparent force due to the observer's frame of reference, not a real force due to an identifiable source, such as the gravitational pull of a planet.) Viewed from Earth, a north–south moving object appears to be deflected to the east or west. Viewed from space, the same object is in fact seen to move in a straight line. The apparent curve that we see is the result of our frame of reference—we normally view the object's movement from *within* the system.

The Coriolis effect applies to any object moving on a rotating body. To visualize this, let us first consider Earth rotating on its axis. The two vertical lines in Figure 4-9a represent the distance moved in a given time interval, and the arrows represent the rotation speed of Earth's sur-

FIGURE 4-5

Convective towers in the ITCZ. Solar heating evaporates large amounts of water from the tropical oceans. The air cools and condenses as it rises, releasing the energy used for evaporation as latent heat. The release of latent heat in these convective towers is the pump that drives the Hadley circulation. (From NASA Science Source, Photo Researchers, Inc.)

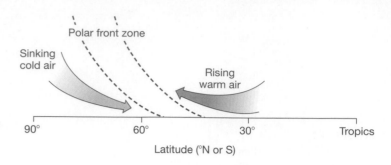

FIGURE 4-6

Mixing of air in the mid-latitudes. The lower-density warm air from the tropics rises above the colder air moving equatorward from high latitudes. These contrasting air masses do not mix very easily. This zone is characterized by large temperature contrasts over very short distances.

face at different latitudes over that interval. The speed of rotation is greatest at the equator (approximately 464 m/sec), and it decreases as we move north—or south—until it becomes zero at the poles.

Now imagine an object, such as an air mass, that is apparently stationary at a point on Earth's surface. Although this air mass is not moving relative to the surface, it is traveling eastward at Earth's rotation rate for that location. (For example, an object that is stationary at any one of the points marked by the left-hand edges of the rotation arrows in Figure 4-9a would, to an observer in space, appear to move along the arrow's path.) Because the rotation rate changes with latitude, an air mass moving northward from the equator—from point *A* to point *B* in Figure 4-9b—will appear to curve off to the right of its straight-line path, arriving at *X* rather than at *B'*. Why does this happen?

In the time it would take the air mass to travel from *A* to *B*, the Earth rotates from *A* to *A'* (and *B* moves to *B'*). Remember that the air mass at *A* is moving not only northward but also eastward at Earth's speed of rotation (represented by the distance *A-A'*). As long as it is

between the equator and point *B*, the air mass is moving from west to east faster than Earth is rotating at *B*. Thus, the air mass will "gain" on the ground below it and will arrive at point *X* instead of at *B'*. Although it is difficult to visualize, the air does in fact move in a straight line; if we were watching from space rather than from Earth, that is what we would see.

We have seen how an air mass (or any object) moving northward in the Northern Hemisphere is deflected to its right. Following the same reasoning, we can see that an air mass moving southward in the Northern Hemisphere also curves to its right (relative to the direction of initial movement), because now the air mass is moving eastward more slowly than Earth's surface immediately underneath. The easiest way to keep track of this is to think of the deflection direction relative to the direction of initial motion of the object—it is always to the right of the direction of initial motion in the Northern Hemisphere and to the left in the Southern Hemisphere.

What happens when the initial direction of movement is due east or due west? As it happens, the Coriolis effect still comes into play, but for a different reason. When an

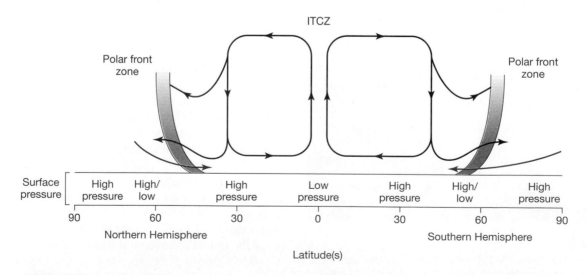

FIGURE 4-7

The north–south (meridional) circulation of the troposphere. The tropical circulation is dominated by the Hadley circulation, whereas mid-latitude circulation and weather are controlled by the location of the polar front zone and the mixing of cold polar air with warm air from the tropics.

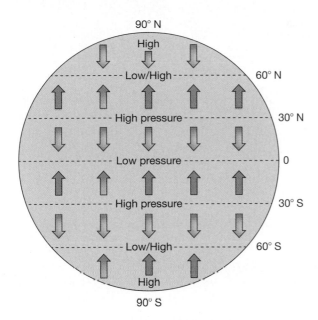

FIGURE 4-8

A possible model of the surface winds obtained by plotting, on a globe, the pattern of surface winds that would be deduced from Figure 4-10. Surface winds blow out of the high-pressure zones at the poles and at 30° N and 30° S and blow toward the low-pressure zones at the equator and in the mid latitudes.

object is set in motion along a circular path, *centrifugal force*—another apparent force—tends to push the object away from the center of rotation. (This is the same phenomenon that forces your car off the road if you try to turn a corner too fast.) If an air mass in the Northern Hemisphere is moving eastward faster than Earth is rotating at that latitude, that air mass will experience an apparent centrifugal force that pushes it directly away from Earth's spin axis.

We can break this apparent force down into two components: one component that is acting perpendicular to the surface, and one that is horizontal (parallel) to the surface (Figure 4-10). For an eastward-moving wind in

the Northern Hemisphere, the horizontal component is to the south; the wind would curve to the south, or to the right. For a westward-moving wind, the horizontal component is to the north; the wind still curves to the right. In the Southern Hemisphere, the effects would be opposite: Both eastward-moving and westward-moving winds would curve to the left.

Although these descriptions of the deflection effect for north–south and east–west moving objects appear to be very different, mathematically the deflections are identical, and both are referred to simply as the Coriolis effect. The Coriolis effect increases as the speed of the object increases. And whereas the speed at which Earth's surface is moving due to its rotation is zero at the poles and a maximum at the equator, the Coriolis effect is zero at the equator and increases with latitude. The only place on Earth's surface where the Coriolis effect does not come into play is at the equator. An air mass moving eastward or westward around the equator is not deflected from its original path. Such an air mass is not changing latitude, so there is no Coriolis effect due to the difference in rotation rate with latitude. Nor is there a horizontal component to the centrifugal force, so again there is no Coriolis effect.

Distribution of Surface Winds

We can now modify the simplistic pattern of northward moving and southward-moving surface winds in Figure 4-8 to obtain the more realistic pattern of surface winds shown in Figure 4-11. There the winds are deflected to the right and left of the paths of initial motion in the Northern and Southern Hemispheres, respectively. This deflection of the winds due to the Coriolis effect gives rise to easterly winds at high latitudes. Meteorologists refer to winds in terms of the direction from which they blow. In other words, an "easterly" wind is a wind that blows from east to west. The mid-latitudes are characterized by westerly flow, and the tropics by easterly winds called the *northeast* and *southeast trade winds*. Winds at the equator

FIGURE 4-9

The Coriolis effect. (a) As Earth rotates, the speed of the surface is greatest at the equator and is zero at the poles. (b) At the equator, Earth has rotated from A to A'. Points A and B have moved to A' and B. If the effects of Earth's rotation are ignored, air moving from A to B would actually curve to the right and arrive at point X.

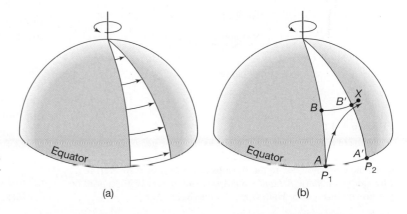

(a)   (b)

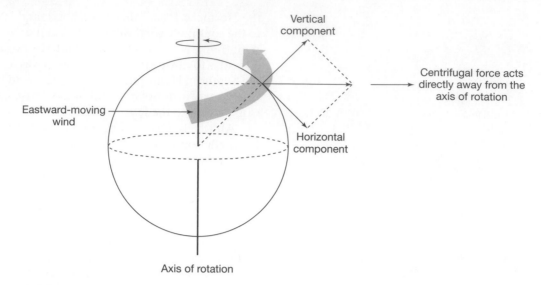

FIGURE 4-10

The Coriolis effect produced by the centrifugal force acting on eastward- or westward-moving winds.

tend to be highly variable in direction. This region where winds are light and frequently change direction is referred to as the *doldrums*.

Considerations of pressure differences, buoyancy, and the Coriolis effect have led us to a good first approximation of the general circulation of the troposphere.

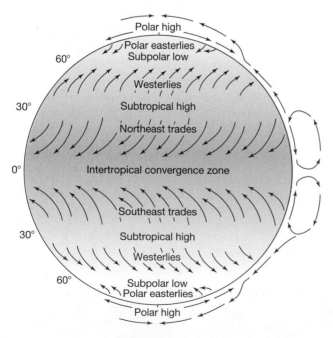

FIGURE 4-11

The pattern of surface winds. This shows the same general pattern of winds as Figure 4-8, but the wind directions have been changed to include the deflection due to the Coriolis effect. (From T. McKnight. *Physical Geography: A Landscape Appreciation*, 6/e, 1999. Reprint by permission of Prentice Hall, Upper Saddle River, N.J.)

This pattern, however, is still a little too simplistic. In reality the indicated winds, for example, do not blow continuously, and they are not continuous around the globe. As we noted earlier, uplift in the ITCZ takes place in clusters of convective cells rather than in two giant cells, one on each side of the equator. The rising air moves poleward and, under the influence of the Coriolis effect, turns to the right in the Northern Hemisphere (and to the left in the Southern Hemisphere). It thus becomes a westerly flow in the upper troposphere. Some of this air subsides near 30° N or S latitude to form the subtropical high-pressure belt (Figure 4-11), but the subsidence too is concentrated in localized areas. The locations of these high-pressure systems vary with season, although they are always found in these approximate locations. The trade winds blow from the equatorward side of these semipermanent high-pressure systems. Similarly, poleward-moving air from the subtropical high-pressure zone curves due to the Coriolis effect, producing a generally westerly flow in the mid-latitudes. The actual flow pattern, however, is highly variable from day to day.

The pressure and wind patterns in the mid-latitudes depend on the location of the subtropical highs as well as on the distribution and movement of temporary areas of high or low pressure that form in association with the steep temperature gradients in the polar front zone. Small areas of low pressure (on the order of 1000 km wide) form in this zone in part due to the surface-temperature gradient, but also because of dynamic processes occurring higher in the troposphere. As air blows into these regions, it curves to the right (in the Northern Hemisphere), producing a localized circular flow pattern referred to as *cyclonic flow*. Air flowing out of a high-pressure region (referred to as an *anticyclone*) will also curve to the right in the Northern Hemisphere, creating an *anticyclonic,* or clockwise, flow. (The direction of air flow around cyclones and anticyclones is reversed in the Southern Hemisphere.)

Low-pressure systems that form outside the tropics are referred to as *extratropical cyclones.* The circular flow mixes warm air from the equatorward side of the system with cooler air from the high latitudes. As we noted earlier, the warm and cool air masses do not mix easily; hence these systems are characterized by extensive uplift as the warmer, less dense air rises above the cooler and denser air mass. As we will see later in the chapter, this rising air results in the formation of rain or snow. These circulation features move along the polar front, bringing low pressure as they move over a region, which is then replaced by higher pressure as they move past. These transient high-pressure and low-pressure systems are characteristic features of mid-latitude climates and account for much of the day-to-day variability in weather in these regions. At high latitudes, the polar easterlies are most clearly developed in winter (when the coldest surface temperatures occur and when this area of low temperatures is most extensive). However, low-pressure systems from the mid-latitudes often migrate into the polar regions in summer, breaking down the surface high-pressure systems and disrupting the easterly flow.

Upper Level Flow. Referring back to Figures 4-7 and 4-8, we see that the pressure distribution at the surface results in alternating regions of poleward and equatorward moving air. At higher levels, however, Figure 4-7 suggests that all of the air is moving from the ITCZ toward the poles. Why the difference?

Think back to the description of the latitudinal energy balance and the discussion of pressure, temperature, and volume relationships. On the large scale, we have a planet where the troposphere has warm air in the tropics and relatively cooler air at the poles. As the warmer air expands and the cooler air contracts, then the depth of the troposphere changes with latitude (which is also suggested by Figure 4-7). Consequently, we have the situation shown in Figure 4-12a. As the troposphere is thicker in the tropics than at the poles, then the change in pressure with height must be slower in the tropics, as shown in Figure 4-12b. If we join these tropical and polar pressure surfaces, we get the situation shown in Figure 4-12c. Notice we have given a slight slope to the pressure surfaces to reflect the decreasing temperatures toward higher latitudes.

Now, away from the surface where local temperature and pressure changes can be large, if we take any line of constant altitude in the diagram (e.g., *A-A'* or *B-B'*) we see that at any point along these lines the pressure is always higher on the equatorward side than the poleward side. Air will flow down the pressure gradient from high to low pressure and, on average, the flow at higher levels in the troposphere is from the tropics to the pole. You

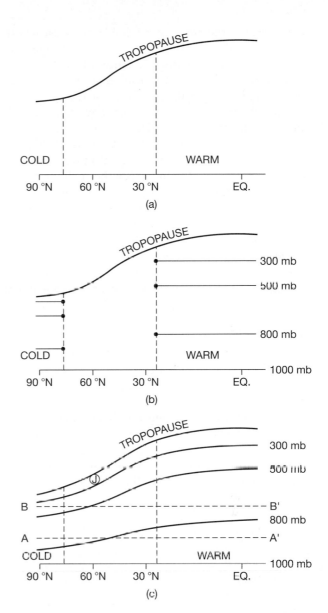

FIGURE 4-12

Pressure change with height in the troposphere. The higher temperatures in the tropics causes the air to expand, raising the height of the tropopause compared to the poles (a). As the atmospheric pressure is the same in both regions, the decrease in pressure with height must be slower in the tropics compared to the poles (b). If the tropical and polar pressure surfaces are joined, we see the steepest pressure gradients in the mid-latitudes. At any height in the atmosphere, the pressure is higher in the tropics than it is at the poles (c). As wind speeds are greatest where the pressure gradient is steepest, the highest wind speeds (the jet streams) will be found high in the troposphere in the mid-latitudes—the *J* in Figure 4-12c.

can imagine that the wind speed will be greatest where the pressure gradient is the steepest, which happens in the upper troposphere in the mid-latitudes. Belts of high windspeeds that we see in this location are referred to as *jet streams* (indicated by the *J* in Figure 4-12c).

What if we take a horizontal view? We know that if the air is moving poleward, and thus changing latitude, it must come under the influence of the Coriolis effect. Consequently, the air will curve to the right in the Northern Hemisphere and to the left in the Southern Hemisphere. In other words, there will be a westerly component to the flow in both cases. The force that is pushing the air down the pressure gradient is referred to as the pressure gradient force. This force is balanced by the Coriolis effect—such that the air actually flows at right angles to the gradient and the resulting movement is referred to as the *geostrophic wind* (Figure 4-13). Other forces (centripetal and centrifugal forces) come into play if the air is following a curved trajectory around centers of higher or lower pressure, where the greater the curvature, the more the flow departs from being geostrophic. Friction also plays a role, with its greatest effect being close to the surface. Friction between the moving air and the surface reduces the wind speed and counters some of the Coriolis effect. In this case, rather than flowing at right angles to the gradient (like the geostrophic wind), the air flows down the gradient at an angle less than 90°. This is why surface winds spiral into low-pressure centers and out of high-pressure centers rather than circling them (which would

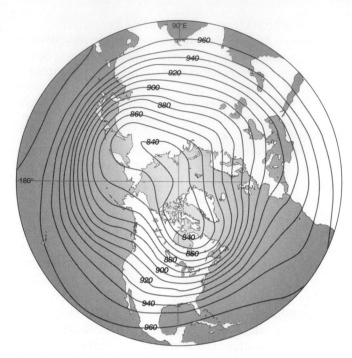

FIGURE 4-14

Northern Hemisphere mean January 300 mb geopotential heights. The heights are in decameters (1 decameter = 10 m). Essentially, the map shows the height of the 300 mb surface. The surface slopes down from the tropics to the Arctic and follows a wave-like structure around the hemisphere. (Map provided by the NOAA-CIRES Climate Diagnostics Center, Boulder, Colorado, from their website at *http://www.cdc.noaa.gov/*)

be the case if the flow was geostrophic). This spiraling effect is clearly seen in Figure 4-20.

Moving back to the upper troposphere, where friction is not an important issue, we see that the flow is close to being geostrophic. It flows normal to the gradient, but it follows a wavelike path around the globe (Figure 4-14). For dynamical reasons that are more complex than we want to get into here, when a fluid moves over a rotating surface, it follows a wavelike trajectory—in this case curving toward the equator and back toward the poles in several large waves that extend around each hemisphere. These waves were first described mathematically by Carl G. Rossby in 1938, and are now referred to as *Rossby waves*. The number of waves, their location, and how well developed they are varies from day to day. For those of us living in the mid-latitudes, these waves steer the low- and high-pressure systems that produce our day-to-day weather. This is why television weather forecasts often show you the locations of the jet streams—as they control the path that low-pressure systems will follow.

Seasonal Variability. The simple pattern of global winds shown in Figure 4-11 is also modified by seasonal variations. As well as changing with latitude, the distribution

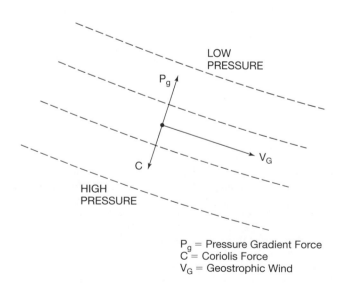

P_g = Pressure Gradient Force
C = Coriolis Force
V_G = Geostrophic Wind

FIGURE 4-13

The geostrophic wind results from the balance of the pressure gradient force and the force due to the Coriolis effect. The dashed lines are isobars—lines of equal pressure. The pressure gradient force (P_g) acts perpendicular to the isobars in the direction of the low pressure. This is balanced by the force due to the Coriolis effect (C), resulting in the geostrophic wind (V_G) blowing parallel to the isobars and to the right of the pressure gradient force (in the Northern Hemisphere).

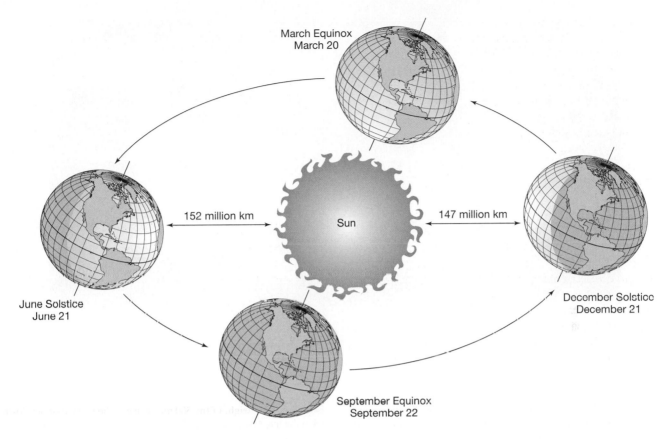

FIGURE 4-15
The seasons. The seasons are controlled by Earth's obliquity and Earth's orbit around the Sun. The hemisphere that is "tilted" toward the Sun experiences summer while it is winter for the hemisphere that is "tilted" away from the Sun. The equinoxes (when the Sun is directly overhead at the equator) mark the transition seasons: fall and spring. (From T. McKnight. *Physical Geography: A Landscape Appreciation*, 6/e, 1999. Reprinted by permission of Prentice Hall, Upper Saddle River, N.J.)

of solar energy varies with the seasons. Figure 4-15 shows the seasonal pattern of Earth's orbit around the Sun. The time when Earth is closest to the Sun is referred to as *perihelion;* Earth is farthest from the Sun at *aphelion.* This difference in distance from the Sun affects the seasonal distribution of temperature. More important to seasonality is the *tilt,* or **obliquity,** of the Earth. Obliquity refers to the angle of the Earth's spin axis relative to a line drawn perpendicular to the plane of the planet's orbit around the Sun. Each planet has a different angle of tilt. Earth's axis is tilted 23.5° from the perpendicular. On human time scales, the obliquity remains constant as the Earth revolves around the Sun (Figure 4-15). On somewhat longer time scales, the obliquity varies by about ±1° (see Chapter 14).

For six months of each year, the Northern Hemisphere faces the Sun and the Southern Hemisphere faces away; for the other six months, it is the Southern Hemisphere that faces the Sun while the Northern Hemisphere faces away. The hemisphere that faces the Sun receives much more solar energy than does the other hemisphere. It is this factor that determines the seasons. Consider the

June 21 solstice (the day with the longest period of sunlight in the Northern Hemisphere) in Figure 4-15. Imagine Earth spinning around its axis. The North Pole will remain in sunlight the whole time, while the South Pole will remain in darkness. The opposite is true at the solstice on December 21 (the day of shortest sunlight duration in the Northern Hemisphere). The result is six months of sunshine and six months of darkness at the poles.

The greatest heating occurs where the Sun is directly overhead. Due to Earth's obliquity, the latitude at which this occurs varies continuously throughout the year: from 23.5° S (called the Tropic of Capricorn) on December 21 to 23.5° N (called the Tropic of Cancer) on June 21. There are also two days on which the Sun is directly overhead at the equator—the *vernal equinox,* March 20, and the *autumnal equinox,* September 22. On these dates, daytime and nighttime are equal in length. (These dates vary by a day or so because the calendar is adjusted every four years to reconcile it with Earth's revolution around the Sun.) Notice, though, that within a narrow band near the equator the Sun is always close to being overhead. Hence there is a large input of solar radiation to the tropics at all

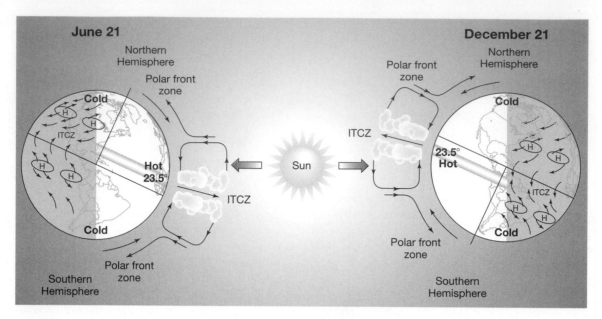

FIGURE 4-16

Seasonal migration of the atmospheric circulation patterns. The ITCZ is found in the summer hemisphere, where the circulation is weaker and the patterns are shifted toward the pole. The subtropical high-pressure cells that mark the descending arms of the Hadley circulations are denoted by H.

times. The difference between daytime and nighttime temperatures in this region, in fact, is usually much greater than the seasonal difference there.

The seasonal variability in incoming energy shifts the atmospheric circulation patterns northward and southward as the seasons change (Figure 4-16). The hemisphere experiencing summer has less of a temperature gradient between the tropics and the pole than does the opposite hemisphere. The fact that the Sun shines continuously for six months at each pole compensates for the fact that the poles do not receive as much solar energy per unit area as do the tropics. This reduced temperature gradient weakens the strength of the atmospheric circulation. At the same time, because the Sun is directly overhead somewhere away from the equator, the maximum solar energy is directed somewhere poleward. Consequently, the steepest temperature gradients are shifted toward the poles, and the circulation patterns are also shifted poleward. In the winter hemisphere (with six months of darkness at the pole), the equator-to-pole temperature gradient is much stronger and the steepest gradients are shifted equatorward. As a result, the atmospheric circulation is more intense and the circulation patterns are shifted toward the equator.

The ITCZ also moves northward and southward as a result of these seasonal shifts in insolation. The ITCZ will reach its maximum northward location late in the Northern Hemisphere summer. (There is a time lag in all of these shifts between the time that the solar heating occurs and the resulting shift in the circulation pattern). It will then migrate southward, crossing the equator in the fall and reaching its most southern location late in winter (Southern Hemisphere summer). The upper tropospheric circulation is similarly affected, with more intense wind speeds in winter and the jet streams shifting north and south with the seasons. Note that when the ITCZ is located over the equator, the poleward-moving air in the upper troposphere produces westerly winds in both hemispheres. As the ITCZ shifts northward, however, the southward-moving air will turn to the right, producing easterly winds in the equatorial region, before curving back to the west when it reaches the Southern Hemisphere. The same thing will happen in reverse when the ITCZ is south of the equator. So, while most of the upper tropospheric flow is westerly, there are frequently narrow bands of easterly winds in the equatorial regions.

Global Distributions of Temperature and Rainfall

In the first part of this chapter, we described some of the main features of the global-scale atmospheric circulation. For the remainder of the chapter, we look at the effect this circulation has on other parts of the Earth system, specifically, the global temperature and rainfall distributions.

As we have learned, the ultimate cause of the atmospheric circulation is the distribution of available energy. More interesting for our purposes is the fact that the interaction between temperature and circulation is not a one-way process. As we indicated earlier, the circulation itself is an important component of Earth's thermoregulatory system, transporting energy (heat) from areas where there is a surplus to areas where there is a deficit.

The transport of water is also strongly affected by the atmospheric circulation. The distribution of water about the globe is important in that organisms require a sufficient supply of water to maintain life. That distribution is also important for the transport of dissolved materials. As we will soon see, evaporation and precipitation are strongly influenced by temperature and, therefore, by the distribution of energy. Furthermore, the transport of water in its various forms (liquid, water vapor, and ice) also modifies the temperature distribution by affecting the radiation budget (Chapter 3) and thus feeds back to affect the circulation. Hence, we see that temperature, precipitation, and the atmospheric circulation are all closely linked and that interactions and feedbacks exist among all three of these components of Earth's climate.

Land–Ocean Contrasts

Beyond the latitudinal distribution of energy, global temperature patterns are also strongly influenced by the distribution of land and ocean. Recall from Chapter 2 (Table 2-1) that the albedo of the ocean surface is considerably lower than the albedo of most land surfaces. In consequence, oceans absorb more of the available solar energy than do land surfaces at the same latitude.

Land and ocean surfaces also behave very differently in what they do with that energy. An ocean surface rapidly transfers heat downward by turbulent mixing and to the atmosphere above by convection. Part of the contrast between land and ocean is due to differences in their thermal properties. The land surface rapidly loses heat to the atmosphere by convection, but it transfers heat downward relatively slowly by conduction. How easily this transfer occurs depends on the physical and chemical properties of the material; the rate at which this occurs is described by its *thermal conductivity*. More formally, thermal conductivity is the rate at which heat energy passes through a column of material that has a temperature gradient along the column of 1 K, or 1° C, per meter. Water has a high thermal conductivity, whereas land surfaces have low thermal conductivities. Furthermore, we can consider the *heat capacity* of the two types of surfaces. Heat capacity at constant volume is the energy required to raise the temperature of a unit mass of a substance by 1 K or 1° C without changing its volume. In other words, heat capacity is a measure of how much energy must be added to an object to change its temperature. The heat capacity of water is about three to four times that of dry soil. Thus, the input of a given amount of energy will raise land temperatures much more than it will raise sea-surface temperatures.

More important is that the ocean surface transfers heat rapidly downward by turbulent mixing. As we will see in Chapter 5, the surface layers of the ocean are well mixed; when the ocean surface is heated, the heat is mixed downward within the surface layers. The amount of material that must warm (or cool) is very large, so the temperature change is very slow. A further factor is differential absorption of the two surfaces. Whereas all the solar radiation falling on the land surface is reflected or

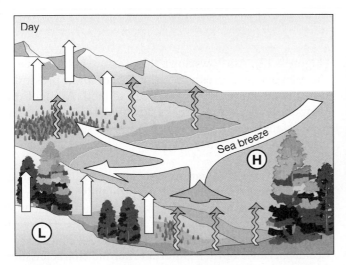
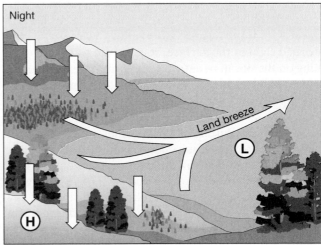

FIGURE 4-17

The sea breeze. The heating of the land during the day causes localized convection with low pressure at the surface. This convection establishes a pressure gradient from the ocean toward the land that results in onshore wind. At night, the rapid cooling of the land (relative to the ocean) causes this circulation to break down and may even reverse the flow, causing the wind to blow from the land toward the water. (From T. McKnight. *Physical Geography: A Landscape Appreciation*, 6/e, 1999. Reprinted by permission of Prentice Hall, Upper Saddle River, N.J.)

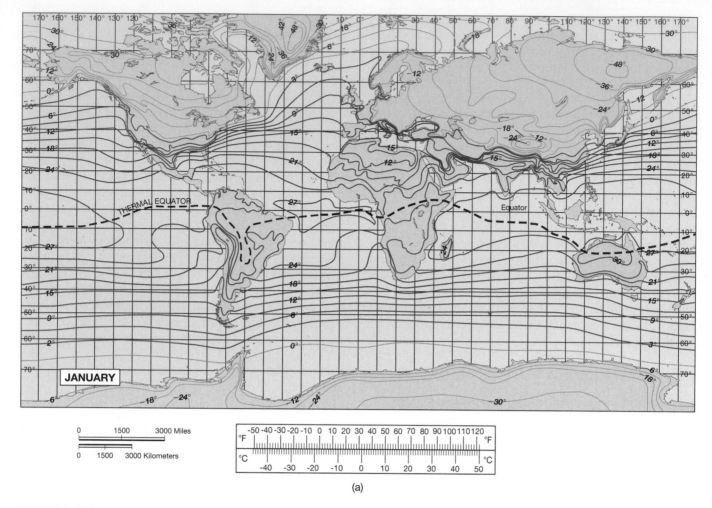

(a)

FIGURE 4-18A

Global temperature distributions in degrees celsius for (a) January, (b) July, and (c) the annual range (difference between summer and winter) (From R.W. Christopherson. *Geosystems: An Introduction to Physical Geography*, 3/e, 1997. Reprinted by permission of Prentice Hall, Upper Saddle River, N.J.)

absorbed right at the surface, some of the solar radiation falling on the ocean penetrates and is absorbed below the surface. Hence, energy is transferred downward even more rapidly in water than on land.

The Sea Breeze. Putting all this together, we see that with equal amounts of incoming energy, land surfaces will heat up much more rapidly than do ocean surfaces but will also cool down much more rapidly once the input of energy is reduced. Land surfaces heat up quickly during the day and cool quickly at night, whereas ocean surfaces warm slowly in the day, and the temperature drops very little at night. The sea breeze that occurs near coastlines is a direct result of this diurnal variability. The heating of the land surface during the day warms the overlying air and gives rise to small areas of low pressure and uplift; cooler temperatures over the ocean result in relatively higher pressures and subsidence of the cooled air above

(Figure 4-17). Air flows down the pressure gradient (from the area of higher pressure to the area of lower pressure), creating on-shore flow from ocean to land. At night this temperature structure breaks down, and the atmospheric circulation weakens. If the land cools sufficiently, the circulation pattern may reverse.

Continentality. We can see the same effect on a larger scale in terms of seasonal climate variability. As we noted earlier, the seasonal variation over mid-latitudes and high latitudes is much greater than in the tropics. This variability is also much greater over land surfaces than over the oceans because of their different thermal characteristics. This property is referred to as *continentality*. The more continental the climate, the more it is characterized by seasonal temperature extremes. Land surfaces are much warmer than ocean surfaces in summer and much colder in winter. The effect of continentality on global tempera-

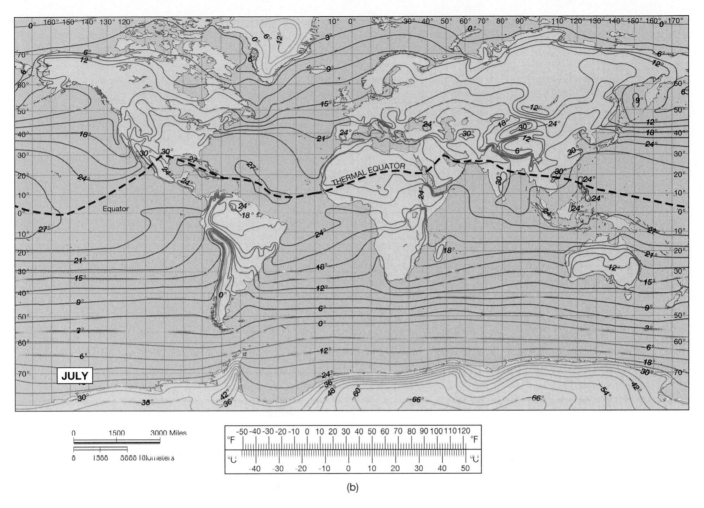

(b)

FIGURE 4-18B *Continued*

tures is visible in Figure 4-18c. The greatest seasonal vari-ability is found in the interior of large continental masses, and the lowest variability is over the tropical oceans. The oceans provide a moderating effect in coastal regions that reduces the temperature extremes. The temperature dif-ference between ocean and land surfaces also affects the mean sea-level pressure distribution (again feeding back to affect the circulation). Figure 4-19 shows the average atmospheric pressure that would be found over land if the land surfaces were at sea level. In Northern Hemisphere winter (Fig. 4-19a), the very cold surface temperatures of interior North America and Asia cool the lower layers of the atmosphere, producing high surface pressures over those land masses. The North Atlantic and North Pacific are characterized by low-pressure zones produced by the low-pressure systems forming along the steep tempera-ture gradient in the polar front zone. In summer this gra-dient decreases (Figure 4-19b). The low-pressure zones are less well developed and are displaced poleward, the subtropical highs expand, and the continental regions of high pressure are replaced by regions of low pressure.

The Southern Hemisphere (Figures 4-19a and 19b), with its huge expanse of ocean and very little land area, presents a less complicated picture. The mid-latitude air flow is much more *zonal*—that is, the air circulation more closely follows lines of latitude—than it is in the Northern Hemisphere. The seasonal temperature change causes the air flow patterns to shift north and south slightly and caus-es pressure gradients to change, but the distribution of land and oceans produces much less variability of winds around the Southern Hemisphere than around the Northern Hemisphere.

We see, therefore, that the broad pattern of global temperatures is determined by the latitudinal distribution of net radiation, so higher temperatures occur in the trop-ics and lower temperatures at the poles. This distribution varies with season such that the seasonal range of tem-peratures is slight in the tropics and increases poleward. Beyond this, we see from the preceding discussion that the seasonal variability is strongly modified by land–ocean contrasts, the interior of continents having a much greater seasonal range than do coastal locations.

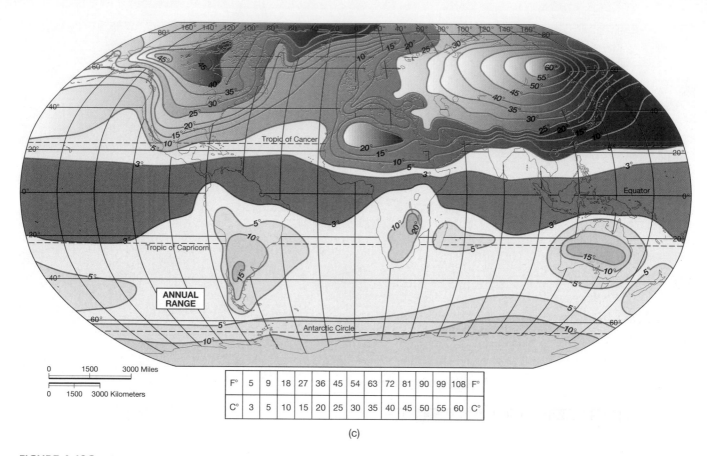

F°	5	9	18	27	36	45	54	63	72	81	90	99	108	F°
C°	3	5	10	15	20	25	30	35	40	45	50	55	60	C°

(c)

FIGURE 4-18C   *Continued*

As we noted earlier, the circulation systems described here are very generalized; they represent averaged conditions with much of the day-to-day variability removed. On any particular day, the circulation and the resulting wind field will be much more complex. As an example, Figure 4-20 shows the surface wind field over the Pacific Ocean measured by a satellite-borne radar system on a September day. The white arrows show the wind direction. The ITCZ is clearly seen where the NE and SE trade winds converge just north of the equator. Note that the southern hemisphere trade winds are blowing from the southeast–as they move equatorward toward the ITCZ they curve to the left. As they cross the equator (because the ITCZ is located north of the equator at this point) they move northward and, as they move away from the equator and come under the influence of the Coriolis effect again, they recurve toward the right. This is more apparent in the Eastern Pacific. The subtropical high-pressure cells are also well depicted—you can clearly see the air spiraling out from the center of the cells.

Monsoons.   The most extreme consequence of the seasonal variability due to differential heating of land and ocean surfaces is the monsoon regime of Southeast Asia. The **monsoon** is a seasonal reversal in the surface winds. In summer the large Asian land mass, with its high elevations in the Tibetan Plateau of central Asia, causes high surface temperatures, low atmospheric pressures, and intense convection of air above the surface. The rising air is replaced by air moving in from the high-pressure region over the Indian Ocean to the south (Figure 4-21a). The moist air drawn in from the Indian Ocean cools as it rises above the mountains of southwest India and over the Himalayas. In both instances the rising air produces clouds and heavy rainfall (the monsoon rains). In winter the pattern reverses: high elevations and persistent snow cover enhance the continentality, producing even lower temperatures. This results in high atmospheric pressure and subsidence of air over the continent and a southward flow of air (Figure 4-21b). A similar feature, but on a much smaller scale, is found in the southwestern United States, where a "monsoonal" flow from July through mid-September brings moist air in from the Gulf of California and the Eastern Pacific.

Global Precipitation Patterns

In addition to transporting energy, the circulation of the troposphere also involves the movement of material

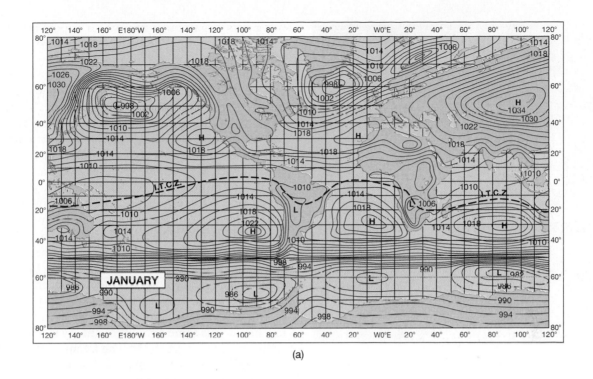

(a)

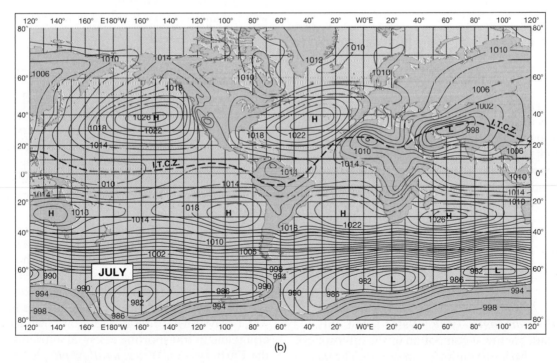

(b)

FIGURE 4-19

Average sea-level pressure patterns: (a) January and (b) July. Units are in millibars (mb), where are atmosphere (atm) equals 1.013 bar (1013 mb). Low-pressure areas are designated by L and high-pressure areas by H. (From R.W. Christopherson. *Geosystems: An Introduction to Physical Geography*, 3/e, 1997. Reprinted by permission of Prentice Hall, Upper Saddle River, N.J.)

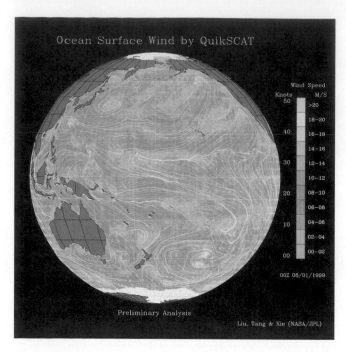

FIGURE 4-20

[See color section] The surface windfield over the Pacific Ocean. The data were derived from a satellite-borne radar system and the white arrows show the direction of air movement at 00Z on August 1, 1999. (Image provided by NASA's Jet Propulsion laboratory from their website at *http:// winds.jpl.nasa.gov/images/olga_pacific_high.jpg/*)

across Earth's surface. We can imagine how effective the atmosphere is at transporting material from the fact that pollution from mid-latitude industrial sources has been found on both polar ice caps.

The most important substance transported in the atmosphere is water, in the form of water vapor and clouds. Both water vapor and clouds are important for several reasons: They play a dominant role in the global energy balance, they are a significant factor in determining the distribution of fresh water around the globe, and they are highly variable in time and space (making them difficult to predict).

The Global Hydrologic Cycle. Water is not only the most important chemical compound in the atmospheric circulation but also in the entire Earth system. The human body is 60% water by weight, and all organisms require some water in order to live. From space we see that Earth is a planet dominated by water (Figure 4-22). Seventy percent of Earth's surface is covered by oceans. The poles are encased in extensive sheets of ice that either float on the surface of the ocean (*sea ice*) or form glaciers several kilometers thick over land. Clouds, which are made of condensed water vapor, swirl across the surface, continuously changing in size, shape, and location but always covering about 50% of the globe at any time. The water

that exists in gaseous form (water vapor) also varies in amount across the globe—from near zero over the ice caps to about 7% in the tropics.

Water is unique in the Earth system for a variety of reasons, not the least of which is that it is the only naturally occurring substance that can exist in all three phases (solid, liquid, and gas) at the temperatures found on Earth's surface. Because water changes so readily from one phase to another, it cycles easily among all the system components. In so doing, water plays a vital role in many Earth system processes. Changing from a solid to a liquid or from a liquid to a gas requires a large addition of energy. That energy is stored in the water molecule in the form of latent heat. The **latent heat of vaporization**, which we described in Chapter 3 as the energy needed to convert liquid water to water vapor, is 2260 kJ/kg at 100°C. (This much energy must be added to each kilogram of boiling water to convert it to water vapor.) When the process is reversed and the water changes from a gas back to a liquid, this same amount of energy is released to the environment (Figure 4-23). The **latent heat of fusion**, or the energy needed to convert ice to liquid water, equals 335 kJ/kg at 0°C. When the process is reversed and the water changes from a liquid back to a solid, this amount of energy is again released to the environment. To raise the temperature of liquid water from 0°C to 100°C requires 419 kJ/kg. To convert ice to water vapor thus takes 3,014 kJ/kg (= 2260 + 419 + 335).

These values apply to water at sea level, and they will vary slightly as atmospheric pressure changes. If these conversion processes occur in different locations, then there is a net transfer of energy from one place to another. Therefore, the distribution and movement of water in its various phases has important consequences for the transfer of energy and the global pattern of surface temperatures.

We saw in Chapter 3 that water vapor and carbon dioxide are the most important of the present-day greenhouse gases. Without them, much of Earth would be too cold to support life. We saw also that clouds have a major impact both on Earth's albedo and on the emission of terrestrial radiation to space. And we will see in Chapter 7, water plays a vital role in breaking down rocks (weathering) and in transporting essential nutrients throughout the Earth system. *Water, in all its phases, is the primary medium by which energy and matter are circulated among the Earth system components.*

Water in the Earth system is concentrated in several major reservoirs:

1. The oceans, where the water exists in the form of seawater;

2. The land surface, in the form of ice sheets, glaciers, snow, lakes, and rivers, and the land subsurface, in the form of groundwater; and

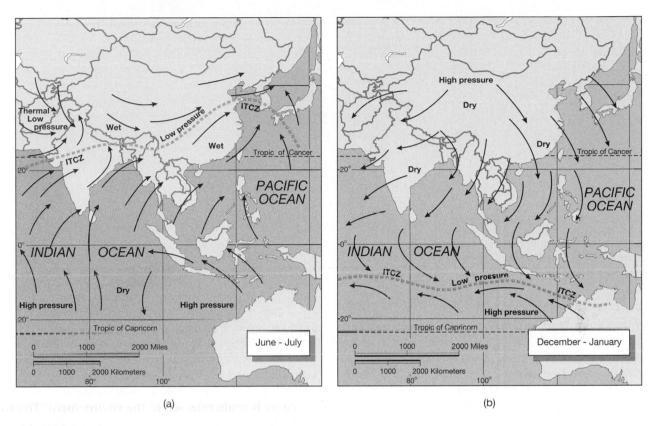

(a)   (b)

FIGURE 4-21

The monsoon flow over southeast Asia. (a) Summer heating of the Tibetan Plateau produces intense convection and low surface pressures, drawing in moist air from the Indian Ocean to the south. (b) The reverse occurs in winter, when low temperatures and extensive snow cover on the plateau produce high surface pressures, subsidence, and outflowing air. (From R.W. Christopherson. *Geosystems: An Introduction to Physical Geography*, 3/e, 1997. Reprinted by permission of Prentice Hall, Upper Saddle River, N.J.)

FIGURE 4-22

[See color section] Earth, viewed from space at about 37,000 km (23,000 mi), is dominated by water. (From NASA Headquarters.)

3. The atmosphere, in the form of water vapor and clouds.

These reservoirs and the pattern of water storage and movement throughout the system comprise the global **hydrologic cycle** (Figure 4-24a). Most of Earth's water—about 97%—is stored in the first reservoir, the oceans (Figure 4-24b). Almost 3% is on or in the second reservoir, land. Of this amount, three-quarters is trapped in the polar ice sheets (in Greenland and Antarctica). Should the Greenland ice sheet melt, it would raise the global sea level by about 6 m; should all of Antarctica melt, the sea level would rise by about 60 m. A small amount of water exists in mountain glaciers as well. Most of the remainder occurs as **groundwater**, or water that penetrates through soil and rock and collects below the surface. Water stored in rivers, lakes, and the soil accounts for less than 1% of all the water found on land. Almost two-thirds of this amount is stored in lakes and reservoirs, about one-third occurs in the soil, and a tiny fraction occurs in rivers. The third reservoir, the atmosphere, contains less than 0.001% of all the water on Earth. Figure 4-24b also gives the annual exchange of water among the three major reservoirs.

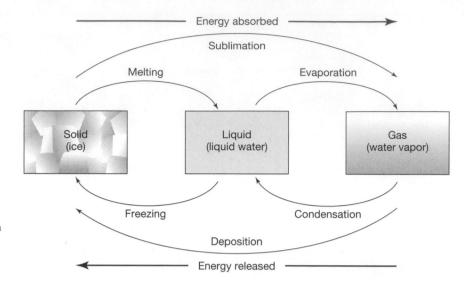

FIGURE 4-23

Schematic diagram of the different phases of
water. Energy is absorbed as water changes from
a solid to a gas (moving left to right in the diagram) and is released as water changes from a
gas to a solid (from right to left).

Precipitation and Saturation Vapor Pressure. The transfer of water between the land/ocean surface and the
atmosphere takes place through evaporation and *precipitation.* Precipitation occurs when atmospheric water
vapor condenses to form small droplets of liquid water.
When the water droplets reach sufficient size, they fall
because of gravity. If they do not evaporate before they
reach Earth's surface, we experience them as rain. If atmospheric temperatures are below freezing, the droplets
fall instead as snow or sleet.

One way of expressing the amount of water vapor
present in the atmosphere is to measure the contribution
that water vapor makes to the atmospheric pressure. We
saw in Chapter 3 that air is composed of numerous different gases. Each gas exerts its own pressure. What is
measured as the atmospheric pressure is the sum of all
the **partial pressures** of the individual gases—that is, the
pressure each gas would exert if it were the only gas present. The pressure exerted by water vapor is referred to
as the *vapor pressure.*

Imagine a body of water. Water molecules at the surface that have a little more energy than do their neighbors
can overcome the attractive forces that hold the molecules
together and thereby escape as water vapor molecules into
the air above. This is the process of **evaporation.** Some of
these water vapor molecules that subsequently come in
contact with the water surface would lose energy, be
"caught" by the liquid water molecules, and become liquid water again. This is the process of **condensation.** Once
the rate of condensation equals the evaporation rate—that
is, as many molecules leave the gas as are added to it—
the gas is at equilibrium. At this point, the vapor pressure
of water is referred to as the *saturation vapor pressure.* In
this scenario, the saturation vapor pressure depends only
on the rate at which molecules are transferred from liquid to gas and back again. This rate depends on the ener-

gy of the molecules, which means it depends on temperature. (Recall that the higher the energy, the higher the
temperature.) Therefore, as temperature increases, the
saturation vapor pressure increases.

Figure 4-25 is a graph of saturation vapor pressure
versus temperature for water. In general, we can think of
clouds as forming when the air is at the saturation vapor
pressure for water. Further evaporation adds water vapor
molecules to the air, where they condense to form water
droplets in clouds. When these droplets become large
enough to overcome the upward motion of the air, they
fall as precipitation. Assume that an air mass is at the temperature and vapor pressure indicated by point *P* in
Figure 4-25. Point *P* is not on the curve; the air is not at
the saturation vapor pressure for that temperature, hence
clouds would not form and precipitation could not occur.
We can bring that air mass to the *saturation point*—that
is, to equilibrium at a point on the temperature versus
saturation vapor pressure curve—in two ways. First, we
could add more water vapor through increased evaporation from the surface and thereby increase the saturation
vapor pressure. That action would move the air mass from
point *P* up along the vertical dashed line toward the curve.
Second, we could reduce the air temperature, which
would move the air mass from point *P* to the left along the
horizontal dashed line toward the curve.

Because the saturation vapor pressure varies with
temperature, knowing just the vapor pressure on a given
day does not give a good indication of when clouds will
form. Consequently, we normally think in terms of *relative
humidity*—the ratio of the actual vapor pressure to the
saturation vapor pressure at that temperature. (We are
using vapor pressure as a measure of the amount of water
present; hence this definition is equivalent to that presented in Chapter 3.) The relative humidity is usually expressed as a percentage; a relative humidity of 100%

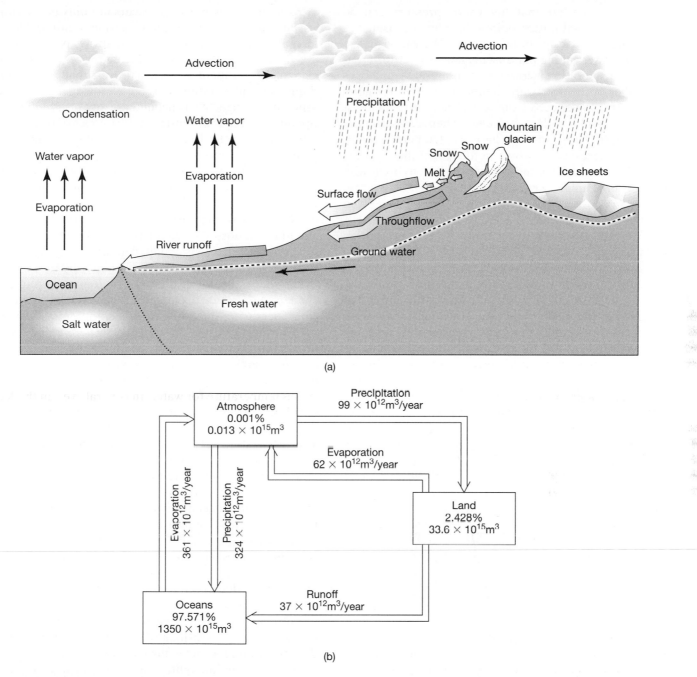

(a)

(b)

FIGURE 4-24

The global hydrologic cycle. (a) Schematic diagram of how water, in its various phases, is stored and moved throughout the Earth system. (b) The sizes of the major reservoirs of water and the rate at which water is transferred between them. (From T. McKnight. *Physical Geography: A Landscape Appreciation*, 6/e, 1999. Reprinted by permission of Prentice Hall, Upper Saddle River, N.J.)

represents air at the saturation vapor pressure. In general, water vapor will condense to form water droplets and clouds when the air is fully saturated. But in fact, very clean air may have greater than 100% relative humidity (i.e., it can be *supersaturated*) without condensation taking place. Condensation is facilitated by impurities in the air—microscopic particles (solid or liquid) that are small

enough to remain in suspension in the air. Such particles are known as *cloud condensation nuclei* (CCNs) when they are used in cloud formation. These nuclei can come from many sources, both natural and anthropogenic. It is likely that many of the clouds that form from CCNs over land surfaces derive from human-produced sources, such as sulfates.

We said earlier that the vapor pressure can be brought to the saturation point either by increasing the vapor pressure or by cooling the air. We can visualize both processes taking place as air moves over different surfaces: evaporation increasing as unsaturated air moves over lakes or the ocean, and temperatures decreasing as the air moves over cooler surfaces. The largest and most rapid changes take place, however, when air rises in the troposphere. As we saw in Chapter 3, temperature generally decreases with altitude in the troposphere, where most clouds exist. Thus most rainfall situations occur with some form of **uplift**, or rising of air masses. Uplift is a general term denoting any process by which air at a given level in the atmosphere is lifted to a higher altitude.

We can get some appreciation for how precipitation is distributed around the globe, therefore, by recognizing that most precipitation takes place as air cools when it is forced to rise. We have already mentioned two of the processes that result in uplift: first, large-scale uplift that occurs with the mixing of air masses of different densities; second, uplift due to convection. Consequently, there is heavy precipitation along the polar front zone in the mid-latitudes and in the vicinity of the ITCZ (Figure

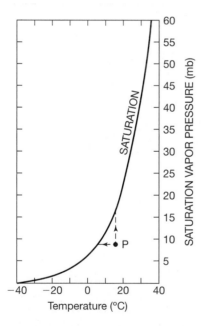

FIGURE 4-25

Saturation vapor pressure versus temperature for water. The curve shows the temperatures and vapor pressures at which the air becomes saturated.

4-26). Convection, however, occurs not only in the tropics, but wherever there is intense surface heating. Therefore, although convection does not always produce rain, it is the dominant rainfall-producing process over warm land masses in summer. A third process that forces air to rise is the confrontation between a moving air mass and a mountain range. Such encounters cause *orographic* precipitation on the windward (upwind) slopes of mountains. (Orography is the branch of geography that involves mountains and mountain systems.) For example, orographic precipitation commonly occurs on the western slopes of the Sierra Nevada mountain range and on the southern slopes of the Himalayas.

Under these circumstances, precipitation is enhanced due to the atmospheric circulation. Under other circumstances, precipitation is inhibited in certain areas. We call such areas **deserts**, and we can examine why they are located where they are (Figure 4-27). Remembering that condensation results from uplift (which cools the air) or from increasing the amount of available moisture, we would expect to find deserts in areas where uplift is suppressed or where there is an inadequate moisture supply.

In general, precipitation is low in the interior of large land masses, simply due to the distance from moisture supplies. Deserts are located in the vicinity of the descending arms of the Hadley cells, as we noted earlier, and on the leeward (downwind) slopes of mountains. They occur also, perhaps unexpectedly, on the west coasts of large continents in areas that lie equatorward of the mid-latitude low-pressure systems. For reasons that we will discuss in Chapter 5, these regions are characterized by cold offshore ocean currents. The cold ocean currents reduce evaporation and cool the air that moves over them, a combination that inhibits convection and precipitation over the adjacent coastline. The deserts that form in this way are called *littoral* (alongshore) deserts. In fact, one of the driest deserts in the world is the Namib Desert along the coast of southwest Africa. The desert of Baja, California (actually in Mexico), is another example. Although you generally think of deserts as being hot, some deserts are also located in the cold polar regions. The low temperatures inhibit uplift, and, where precipitation does occur, quantities are small because of the low saturation vapor pressures. The central part of Antarctica is, in fact, a desert. There is enough ice on Antarctica to raise global sea levels by 60 m, but the average annual snowfall on the plateau is the equivalent of less than 51 mm (2 in.) of liquid water.

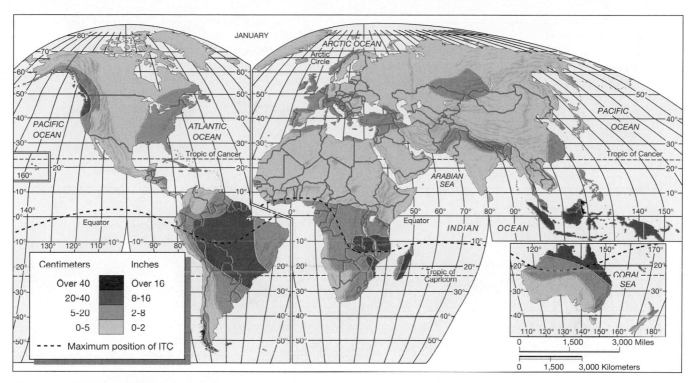

MODIFIED GOODE'S HOMOLOSINE EQUAL-AREA PROJECTION

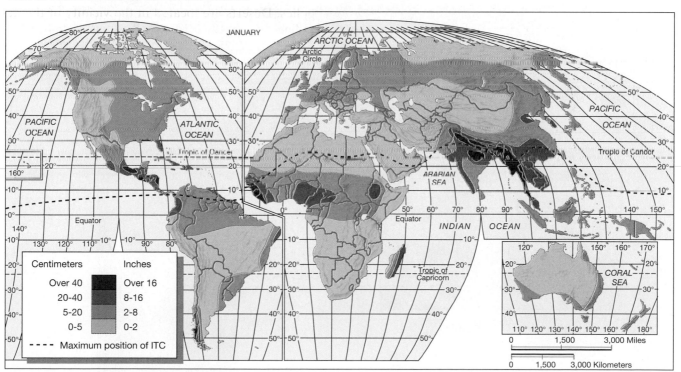

FIGURE 4-26

Global distributions of precipitation over land in (a) January and (b) July. (From T. McKnight. *Physical Geography: A Landscape Appreciation*, 6/e, 1999. Reprinted by permission of Prentice Hall, Upper Saddle River, N.J.)

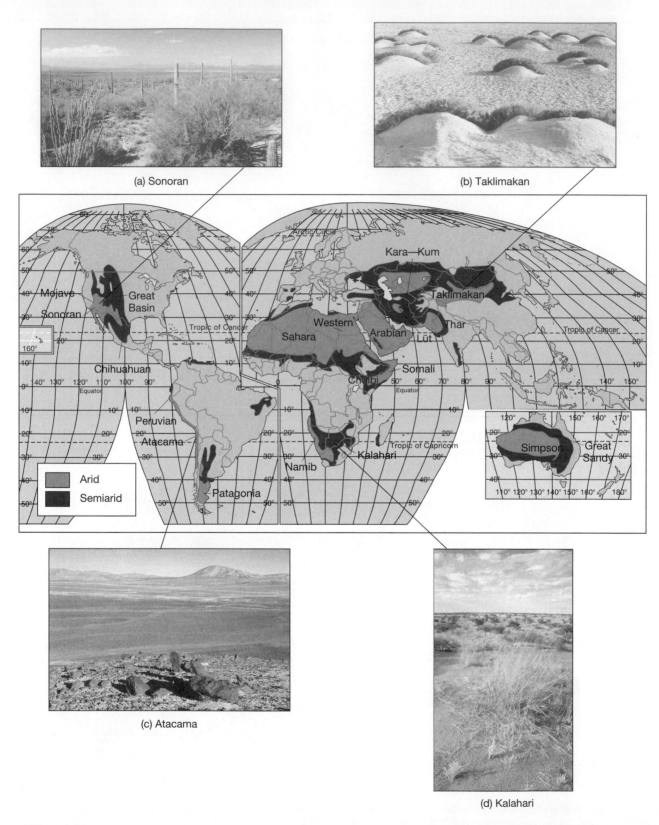

(a) Sonoran

(b) Taklimakan

(c) Atacama

(d) Kalahari

FIGURE 4-27

The distribution of the world's major deserts. (From R.W. Christopherson. *Geosystems: An Introduction to Physical Geography*, 3/e, 1997. Reprinted by permission of Prentice Hall, Upper Saddle River, N.J.)

Chapter Summary

1. The driving force for the atmospheric circulation is the global distribution of energy.
    a. The angle at which the Sun's rays strike the Earth changes from the equator toward the poles. The result is that incoming solar radiation decreases with latitude. More solar radiation is received in the tropics than at the poles, resulting in an equator-to-pole temperature gradient.
    b. This temperature gradient drives the atmospheric circulation because of the inverse relationship between the temperature and the density of a gas: Higher temperatures produce lower densities. Differences in the distribution of global temperatures cause differences in air density and, therefore, pressure.
    c. Air tends to move from areas of high pressure to areas of low pressure. These large-scale movements of air produce the global windbelts.
    d. These windbelts are significantly modified by the Coriolis effect, which is caused by the Earth's rotation.
2. The net effect of these atmospheric movements is to redistribute available thermal energy. There is a negative feedback between the energy gradient and the circulation.

Energy is moved from low latitudes, where Earth is hot, toward the poles, where it is cold.
3. The atmospheric circulation exerts a major control on global temperature patterns. The movement of the air carries water vapor from one region to another. Because evaporation and condensation are largely a function of temperature, the redistribution of water around the globe is also strongly tied to temperature distributions and to the atmospheric circulation.
    a. Precipitation is enhanced wherever the circulation promotes uplift and is inhibited in areas dominated by subsidence.
    b. Precipitation amounts are also affected by continentality and the distance from moisture sources.
    c. The distribution of land and ocean affects the distribution and variability of surface temperatures. Variability increases as the distance from the ocean increases.
    d. The circulation, temperature, and precipitation distributions are modified by seasonal variations in incoming solar energy caused by Earth's obliquity and Earth's orbit around the Sun.

Key Terms

Boyle's law	evaporation	monsoon
buoyancy	general gas law	obliquity
Charles's law	groundwater	partial pressure
condensation	Hadley circulation	polar front zone
convergence	hydrologic cycle	subsidence
Coriolis effect	intertropical convergence zone	uplift
deserts	latent heat of fusion	
divergence	latent heat of vaporization	

Review Questions

1. What are the functions of the global circulatory system?
2. Explain why the distribution of solar energy varies with latitude.
3. a. Draw a graph showing the variation of incoming solar energy and outgoing infrared radiation with latitude.
    b. Indicate the regions of energy surplus and energy deficit.
    c. Explain why this distribution is important for the atmospheric circulation.
4. Explain why heating an air mass causes it to rise.
5. Use a diagram to describe Hadley cells. Why does the Hadley circulation change seasonally?
6. What is the Coriolis effect? How does the Coriolis effect help determine the global pattern of winds?
7. Explain why Earth experiences different seasons throughout the year. Which parts of Earth experience the greatest seasonal variability, and which parts experience the least? Explain why.
8. Contrast the different roles of turbulent heat transfers and conduction in modifying the thermal response of a land surface and an ocean surface.
9. Use map sketches to explain the processes that drive the Southeast Asian monsoon.
10. What is latent heat? Explain why latent heat is important for the redistribution of energy.
11. a. What is meant by saturation vapor pressure?
    b. Draw a graph that plots saturation vapor pressure as a function of vapor pressure and temperature.
    c. Explain why the information shown in the plot is useful for understanding the relationships between atmospheric circulation and precipitation.
12. Describe three processes that produce uplift in the atmosphere and are important in causing precipitation.

Critical-Thinking Problems

1. Sketch a map of India. Locate the major mountain ranges. Show which areas you think would have high rainfall and which areas you think would have low rainfall, and explain why.
2. In this chapter, we discussed several types of deserts, including polar deserts. The center of the Antarctic ice sheet, for example, receives very little precipitation each year and is regarded as a desert, although it does not match the customary idea of what a desert is. From this chapter we saw that:
   - Precipitation generally decreases as temperature decreases (because saturation vapor pressure is much lower in cold air than in warm air).
   - Much mid-latitude and high-latitude precipitation occurs in extratropical storm systems that move along the polar front zone.
   - The polar front zone is located in the latitudes where the temperature gradient is greatest. (This will be equatorward of the ice margin, where cold air draining off the ice cap moves equatorward to meet the warm air that is blowing poleward from the subtropical highs.)
   a. Put this information together in a systems diagram that has two feedback loops: one that links ice extent, albedo, temperature, and snowfall; and one that links ice extent, temperature, the location of the polar front zone, and snowfall.
   b. Are these feedback loops positive or negative?
   c. What implications do the feedback loops in part (a) have for the long-term growth of an ice sheet?
3. Indicate on two world maps the areas where you would expect to find relatively high rainfall and where you would expect to find relatively low rainfall, or even deserts, in (a) July and in (b) January. (c) Explain these distributions.

Further Reading

General

Robinson, P. J., and A. Henderson-Sellers. 1999. *Contemporary Climatology* (2nd edition), Harlow: Longman.

Advanced

Hartmann, D. L. 1994. *Global Physical Climatology*. San Diego: Academic Press.

The Circulation of the Oceans

Key Questions

- Why do ocean currents form?
- How can the circulations of both the surface ocean and the deep ocean basins be driven by solar radiation and be closely linked, yet operate at very different time scales?
- What role does ocean circulation play in the global climate system?

Chapter Overview

We continue our discussion of Earth's circulatory subsystems by describing the processes that drive the circulation of the world's oceans. The movement and circulation of the oceans is tied very closely to the circulation of the atmosphere: Both are ultimately driven by the distribution of available solar energy, and their motions are linked by friction at the sea surface. In Chapter 4, we described an imbalance in the latitudinal distribution of energy that produces an equator-to-pole temperature gradient at the surface—the driving force for the pattern of Earth's surface wind. These wind patterns are responsible for the circulation of the ocean surface and the formation of the world's major ocean currents. As with the atmosphere, once the ocean starts to move, it comes under the influence of the Coriolis effect, which plays a significant role in the resulting circu-

lation patterns. The oceans are vertically stratified, with more dense water at the bottoms of the major ocean basins and less-dense water near the surface. The density is controlled by the temperature and by the salt content (*salinity*) of the water. The deep ocean water is separated from the surface layer of the ocean by a transition zone with sharply defined density, temperature, and salinity gradients. This deep-ocean water moves as a response to small changes in density that occur over wide areas, and the movement is largely independent of the surface-ocean circulation. Together, however, both types of ocean circulation contribute to the redistribution of available energy in the Earth system, albeit over very different time scales. And both play a major role in the distribution of nutrient supplies in the oceans.

Winds and Surface Currents

Chapter 4 showed that the circulation in the troposphere is caused by atmospheric pressure gradients that result from vertical or horizontal temperature differences. We saw that from a global perspective, these temperature variations are caused by latitudinal differences in solar heating. But ocean surfaces are also heated by incoming solar radiation. Do the oceans, therefore, circulate for the same reason as the atmosphere? The answer is no, because the solar heating of the ocean takes place at the *upper* surface of the fluid, whereas the solar heating of the atmosphere occurs largely at the *lower* surface of the fluid near Earth's surface. Solar heating results in warmer water at the surface of most of the world's oceans. But the Sun's rays warm only the top few hundred meters of the ocean; 90% of the radiation that penetrates the surface is absorbed in the first 100 m. The warmer water is less dense than the cooler water below, which is not affected by the surface heating. This situation is inherently stable, so there is very little vertical movement. It is similar to the situation in the stratosphere. Recall from Chapter 3 that the atmosphere at this level is stable because the maximum solar heating occurs high in the stratosphere, the site of peak absorption of ultraviolet radiation by ozone. Where temperature increases with height, there is no density imbalance, and convection cannot take place. The fluid—water or air—remains well stratified. The true situation in the ocean is actually more complicated than this, as we will see, because the density of seawater is also affected by its salt content. It remains true, however, that the ocean overturns very slowly.

At the same time, temperature changes in the ocean occur slowly. Remember from Chapter 4 that the oceans have a high heat capacity—it takes a considerable amount of heat to produce just small changes in temperature. Slight differences in incoming solar radiation from place to place thus have little impact on the surface temperature of the ocean, so lateral temperature and density differences are slight over large areas. Unlike the troposphere, therefore, the surface ocean does not circulate as a *direct* response to the surface heating. Instead, surface temperature plays a more indirect role: The surface temperature influences the atmospheric circulation, and the resulting pattern of global winds determines the circulation of the upper ocean.

The movement of the wind over the ocean causes friction at the surface. As a result of friction, the wind drags the ocean surface with it as it blows, thus setting up a pattern of surface-ocean *wind-drift* currents. The force of the wind acting on the surface is referred to as *wind stress*. The water movement is usually confined to the top 50 to 100 m of the ocean, although well-developed currents such as the Gulf Stream in the North Atlantic and the Kuroshio Current in the North Pacific may extend as much as 1–2 km below the surface. The Coriolis effect influences ocean currents just as it does winds, so the water is deflected to the right of the path of the wind in the Northern Hemisphere (and to the left of the wind's path in the Southern Hemisphere). Observations show that this deflection tends to be approximately 20–25° from the wind direction. Thus, as a first approximation of what the ocean circulation should look like, we can take the surface winds of Figure 4-11 and add a large ocean bounded by continents on the east and west. In doing so we obtain the surface-ocean circulation pattern shown in Figure 5-1. The trade winds produce westward-flowing currents in the tropics. When these currents reach the western continental boundary, they are deflected northward and southward. They then come under the influence of the westerlies, which cause the currents to flow eastward in the mid-latitudes. When these currents reach the eastern land mass, some water is deflected to the pole and some toward the equator. The waters that flow toward the poles are replaced by equatorward flow along the western land mass. The waters that flow toward the equator come back under the influence of the trade winds and are blown westward again. The currents complete a large, circular

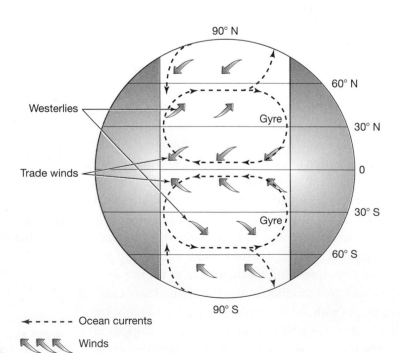

FIGURE 5-1

A simplified view of the surface-ocean circulation.

circulation pattern (called a **gyre**) in the subtropical oceans. The circulation of these gyres is clockwise in the Northern Hemisphere and counterclockwise in the Southern Hemisphere.

Compare the simplified model in Figure 5-1 with the observed distribution of ocean currents in Figure 5-2. The same general circulation features are apparent. Figure 5-2 shows counterclockwise gyres in the Southern Hemisphere and clockwise gyres in the Northern Hemisphere. The figure identifies the world's major ocean currents and designates them as "warm" or "cool," labels that we will explain later in this chapter. The pattern in the real world is more complicated because the distribution of land and water is not as simple as it is in Figure 5-1. In the Southern Hemisphere, the westerlies result in an eastward-flowing current—the *West-Wind Drift*—that extends around the globe because, in the real world, there is very little land in the middle and higher latitudes to deflect the water back toward the west.

Convergence

There are further differences between Figure 5-1 and the actual circulation of the ocean surface that are not apparent from the diagram. To begin with, although our predicted gyres are present, the explanation of their occurrence is a little more complicated than we previously suggested. In order to explain why they form, first we need to describe a few more processes that take place at the ocean surface. If we consider the circulation pattern shown in Figure 5-1, we might expect that water would pile up as it reached the coasts. In the Northern Hemisphere, therefore, we would expect to find water piling up in the northeast and southwest portions of the gyre. This does not happen; rather, water piles up (or converges) in the *middle* of the gyre. This *convergence* results from the combined effects of the wind-driven surface-ocean currents, Earth's rotation, and, ultimately, friction.

The Norwegian explorer Fidtjof Nansen made a key observation that led to a better understanding of convergence during an expedition across the Arctic Ocean in the 1890s. His ship, the *Fram,* was frozen into the ice at the beginning of winter and drifted with the ice for over a year. It had long been thought that the surface-ocean currents were produced by the winds. Among many observations made during the expedition, however, Nansen noted that the ice (and the ship) did not drift *with* the wind but at 20–40° to the right of the surface wind path.

Walfrid Ekman, a Swedish physicist, first made the connection between wind-driven currents and Earth's rotation and derived a mathematical explanation of Nansen's observations. Due to friction between wind and the water surface, some of the kinetic energy of the air is transferred to the top layer of the water. As that layer moves, it drags along the water just below it, which in turn drags along the water just below that, and so on. The water appears to move as many thin, coupled layers, and kinetic energy is transferred down the water column.

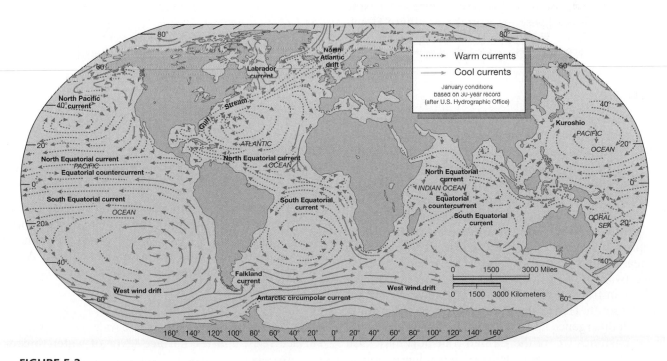

FIGURE 5-2

The major surface-ocean currents. (From R.W. Christopherson, *Geosystems: An Introduction to Physical Geography,* 3/e, 1997. Reprinted by permission of Prentice Hall, Upper Saddle River, N.J.)

However, as the energy is transferred downward, friction causes some of the energy to be dissipated in the form of heat, so each level moves more slowly than the level above. At some depth below the surface, the effects of the wind-induced movement disappear. However, as each layer moves, it is again subject to the Coriolis effect. Once a layer starts to move, the water is deflected to the right of the path of the layer above (or the wind path, for the surface layer) in the Northern Hemisphere and to the left in the Southern Hemisphere. The deeper below the surface, the farther each layer is deflected to the right or left of the surface layer, producing a spiraling effect known as the **Ekman spiral** (Figure 5-3a).

Ekman's theory predicts, under strong and persistent winds in the open ocean, (1) that the surface current will flow at 45° to the surface-wind path, (2) that the flow will be reversed at approximately 100 m below the surface (that is, the current at 100 m will flow in a direction opposite to the surface current), and (3) that it will also be considerably reduced in speed. In practice, there are few observations of a well-developed Ekman spiral, but observations do show that the surface flow is to the right of the surface-wind path (although usually at an angle less than 45°). The observations also bear out a further prediction from the theory—that when the movements of all the individual layers of water in the spiral are added, the net direction of transport within the water column is at a right angle (90°) to the wind direction. This net movement of water is referred to as **Ekman transport.** In a clockwise gyre in the Northern Hemisphere, the effect of Ekman transport is to push water into the center of the gyre (Figure 5-3b). Note that the counterclockwise gyres in the Southern Hemisphere will produce exactly the same result, because the Coriolis effect deflects the water to the left.

Divergence

Just as there are parts of the ocean where convergence occurs, there are also parts of the ocean where divergence occurs. In the equatorial Atlantic of the Northern Hemisphere, for example, the Northeast trades (which, remember, blow from the northeast toward the southwest) result in a westward-flowing surface current, the North Equatorial Current. The net Ekman transport is 90° to the right of the wind, which means that the bulk of water transport is directed almost due north. Conversely, the southeast trades (in the Southern Hemisphere) produce the westward-flowing South Equatorial Current, and the net Ekman transport is to the left of the wind flow, toward the south. Hence, divergence occurs near the equator. Both divergence and convergence can also be found along coastlines where the Ekman transport may push water toward or away from the coast, depending on the direction of wind movement and the surface current. Important areas of divergence occur off the southwest coast

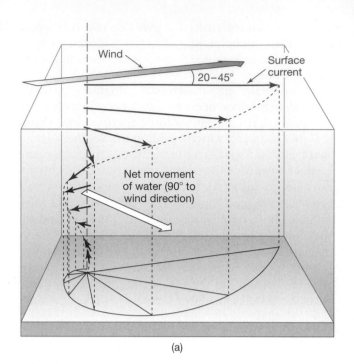

(a)

FIGURE 5-3A

The Ekman spiral.

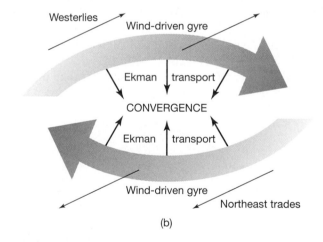

(b)

FIGURE 5-3B

Convergence in the center of a subtropical gyre due to Ekman transport.

of North America and the west coast of North Africa due to the easterly winds and southward-moving currents in these regions. Similar areas of divergence are found off the west coasts of South America and southern Africa, where northward-moving currents have the same effect.

Upwelling and Downwelling

In areas of convergence, the surface water piles up, the sea surface rises, and the surface layer of water thickens (Figure 5-4a). In areas of divergence, the surface water moves away, the sea surface drops, and the surface layer thins. Where convergence occurs, the accumulation of

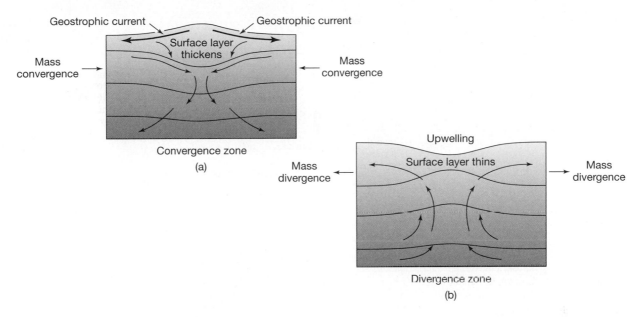

FIGURE 5-4

Schematic representation of zones of convergence and divergence. (a) Surface water accumulates in convergence zones, increasing the surface elevation (very exaggerated in the diagram) and thickening the surface layer. (b) The opposite happens in divergence zones—there is a decrease in surface elevation, and the surface layer thins.

water causes it to sink in a process known as **downwelling.** Conversely, where divergence occurs at the surface, water must rise from below to replace it (Figure 5-4b). Water at depth is cooler than water at the surface. The rising of cooler water to the surface to replace warm, divergent surface water is referred to as **upwelling.** As we will see later in this chapter, these deeper waters also tend to be rich in nutrients. Upwelling, therefore, brings these nutrient-rich waters to the surface.

Geostrophic Flow

Having explained Ekman transport and convergence, we now come to the real reason why the circulation in the subtropical oceans takes the form of very distinct oceanic gyres. Areas of convergence and divergence produce slight variations in sea-surface elevation across the ocean basins, so the sea surface actually slopes from one point to another. This difference in elevation is very slight—on the order of a few meters over 10^2 to 10^5 km (that is, slopes of 1 in 10^5 to 1 in 10^8). Yet these slight elevation gradients are sufficient to cause a downslope force on the water due to gravity. If we consider the subtropical ocean in the Northern Hemisphere, for example, we have already seen that the northeast trade winds produce a westward-flowing ocean current near the equator, whereas the prevailing westerly winds in the mid-latitudes result in an eastward-flowing current. The circulation is completed by the deflection of water along the coastlines at the ocean margins. Ekman transport in the surface layers causes convergence and the pile-up of water in the middle of the ocean (Figures 5-3b and 5-5a).

The sea surface is only about 50 cm higher in the center of the gyre than at the edges, but gravity acting on this pile of water results in a force (referred to as the *pressure-gradient force*) that pushes outward, down the gradient, from the center. As the water flows, however, it is deflected by the Coriolis effect until that effect balances the pressure-gradient force acting down the slope. The result of the two forces acting in opposition is to cause a flow of water off to the side—to the right in the Northern Hemisphere and to the left in the Southern Hemisphere (Figure 5-5b). Thus we end up with a circular flow of water around the gyre that is approximately parallel to the ocean slope (Figure 5-5c). Note the similarity to the description of the geostrophic wind in Chapter 4. In this case, the resulting current is called a **geostrophic current,** which flows around the gyre clockwise in the Northern Hemisphere (and counterclockwise in the Southern Hemisphere), in the same direction as the original wind-driven flow. In practice, the flow is a little less than 90° to the slope, so in fact the water tends to spiral inward as it moves around the gyre and the convergence in the gyre results in downwelling.

Boundary Currents

Ocean gyres are a prominent feature of the surface circulation. Figure 5-2, however, gives the impression that the flow around these gyres is not symmetric. Indeed, the flow in the western part of the gyre is confined to a narrow path with a fast-flowing current (a *western boundary current*), which in the east is more diffuse, spread over a

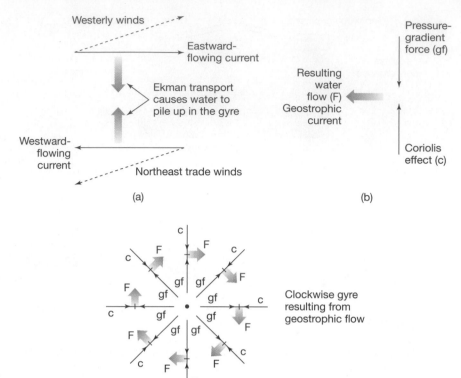

FIGURE 5-5

(a) The subtropical gyres are formed by geostrophic currents that occur when Ekman transport from the wind-driven currents causes water to pile up in the center of the gyre. (b) There is a force due to gravity, acting down the gradient of the surface slope, that is opposed by the Coriolis effect. The net effect is a flow of water at approximately 90° to the slope. (c) The result is a geostrophic current that flows approximately perpendicular to the slope of the sea surface around the gyre.

much larger area and with much-reduced current speeds (an *eastern boundary current*). Eastern boundary currents also tend to be divergent; the Ekman transport is away from the continent, thinning the surface layer along the coastline. The thinner surface layer and the divergent flow promote upwelling in these regions.

The most-studied western boundary current is the Gulf Stream in the western North Atlantic. The Gulf Stream begins as a narrow (50–75 kilometers wide), fast-flowing stream of warm water (20° C or higher) in the Florida Current, flowing northward between Bermuda and Cuba. The current can reach depths of more than a kilometer, with surface speeds between 3 and 10 km/hr. This current follows the coast northeastward to Cape Hatteras, North Carolina, where it continues across the North Atlantic as the Gulf Stream (see Figure 5-2). Moving northeastward across the Atlantic, the Gulf Stream decreases in speed and the flow broadens into the North Atlantic Drift, which eventually flows into the Arctic Basin north of Norway. On reaching Europe, the North Atlantic Drift splits north and south; the southward-flowing component becomes an eastern boundary current, the Canary Current. The Canary Current moves much more slowly than the Gulf Stream, is shallower (reaching depths of only about 500 m), and is much broader—up to 1000 km across. As the water flows back toward the tropics, it comes under the influence of the northeast trade winds

that push it to the west in the Northern Equatorial Current to complete the gyre. This elliptical flow of water essentially isolates the area in the center of the gyre, the Sargasso Sea. The Sargasso Sea is named for the extensive cover of seaweed often found in this area. Low current speeds and light, variable winds made this region difficult to traverse back in the days of sailing ships. The ancient mariners were also afraid of being entangled by the huge mats of seaweed that covered the surface.

Our discussion of wind-driven currents illustrates how wind stress, the Coriolis effect, and the pressure-gradient force serve to produce convergence, geostrophic flow, and gyres in the subtropical oceans. However, our discussion still does not account for the asymmetric nature of the gyres and the very different modes of flow in the eastern and western boundary currents. This pattern is caused by dynamic forces that operate when fluids tend to move in a rotary motion. How this happens is explained in the Box "A Closer Look: Vorticity."

Ocean Circulation and Sea Surface Temperatures

The large-scale surface-wind pattern produces gyres in the surface layer of the mid-latitude oceans. In the Northern and Southern Hemispheres, the Coriolis effect and Ekman transport cause a net movement of water into the center of the gyres. The higher surface elevation in the

A CLOSER LOOK
Vorticity

The text explains why large-scale gyres form in the subtropical oceans. But to explain the asymmetric pattern of the gyres and the differences between eastern and western boundary currents, we need to introduce an additional concept—that of vorticity. **Vorticity** describes the tendency of a fluid to rotate. A tendency to rotate in a counterclockwise direction is referred to as positive vorticity, whereas a tendency for clockwise rotation is negative vorticity. We refer to vorticity as being the *tendency* to rotate (rather than the actual rotary motion of the fluid) because different forces could impose both a positive and a negative vorticity on the same mass of water at the same time, and the actual amount and direction of the rotation would then depend on the net effect of the different forces.

So what produces vorticity? Imagine yourself standing exactly on the North Pole. You would be spinning around a vertical axis at the rate of one rotation every 24 hours (actually, 23 hours and 56 minutes). In other words, you would be experiencing a counterclockwise rotation about your vertical axis. At the equator you would experience no angular rotation, because you would be standing at exactly 90° to Earth's axis of rotation (Box Figure 5-1). Anywhere between the equator and the pole you would experience some fraction of the pole's angular rotation. This angular rotation about a vertical axis

at Earth's surface, brought about because of Earth's rotation, produces vorticity that is referred to as **planetary vorticity**. Mathematically, the planetary vorticity is identical to the Coriolis effect. Like the Coriolis effect, planetary vorticity acts in the opposite direction in the Southern Hemisphere.

A tendency for rotary motion can be created by a number of factors other than planetary rotation. Surface waters being driven by cyclonic or anticyclonic circulations in the atmosphere (i.e., low-pressure or high-pressure systems) will produce positive or negative vorticity. Similarly, *current shear,* in which the speed of the current changes across the current, will also produce vorticity. Representing the speed of the current by the length of the arrows in Box Figure 5-2a, for example, would produce negative (clockwise) vorticity; the faster-moving water tends to curl in toward the slower part of the current. The current shown in Box Figure 5-2b, conversely, would produce positive (counterclockwise) vorticity. Current shear can be quite dramatic where friction with a coastal boundary slows the edge of currents that flow parallel to the coastline. The vorticity produced by the pattern of surface winds and by current shear is referred to as **relative vorticity**. The **absolute vorticity** experienced by a body of water is then simply the sum of the planetary and relative vorticities.

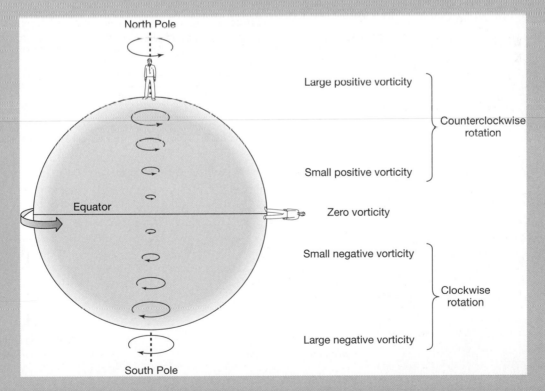

BOX FIGURE 5-1
Planetary vorticity, increasing from zero at the equator to a maximum at the poles. (After Open University, *Ocean Circulation*, New York: Pegamon Press, 1989.)

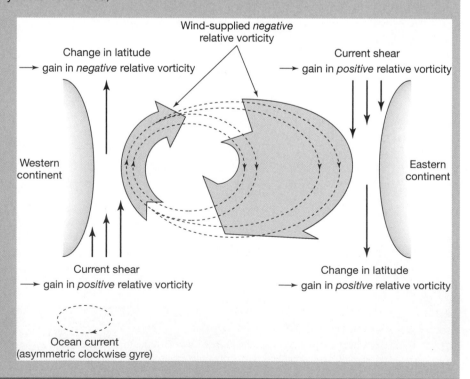

BOX FIGURE 5-2

Schematic diagram of current shear producing negative (clockwise) and positive (counterclockwise) relative vorticity. The lengths of the arrows represent relative speeds in the current.

How does this help us explain the difference between western and eastern boundary currents? The answer lies in the fact that, like mass and energy, absolute vorticity is a conserved property. In Box Figure 5-3, we can see the various factors that contribute to vorticity around a gyre. To begin with, the anticyclonic surface-wind pattern produces *negative relative vorticity* all around the gyre. At the eastern boundary current, this wind-supplied negative relative vorticity is balanced:

- The negative relative vorticity is balanced in part by a small increase in positive relative vorticity due to friction (current shear) along the coast. This factor is not large; because of the width of the current, only a small portion interacts with the coastline.
- Positive relative vorticity also arises from the southward flow of the water. The water is moving into an area of lower positive planetary vorticity; for vorticity to be conserved,

there has to be a decrease in negative relative vorticity (or, looked at another way, a gain in positive relative vorticity).

The situation is very different on the western boundary:

- Current shear again produces positive relative vorticity.
- The change in latitude, however, leads to a gain in *negative relative vorticity* (the water is moving into regions of large positive planetary vorticity and so must acquire negative relative vorticity). This gain in negative relative vorticity reinforces rather than offsets the negative relative vorticity imparted by the surface-wind pattern.

The only way balance can be achieved in the western boundary current is by increasing the friction along the boundary, which is achieved with a narrower, deeper, and faster-flowing current. Hence the asymmetric nature of the gyre.

BOX FIGURE 5-3

Contributions to the changes in relative vorticity around an asymmetric subtropical gyre. (After Open University, *Ocean Circulation*, New York: Pergamon Press, 1989.)

center of the gyres causes a geostrophic current to flow around the gyres in the same direction as the wind-driven flow, thus reinforcing the surface circulation. The shape of the gyre and the nature of the eastern and western boundary currents are then determined by the need to balance the forces that produce a tendency for water to rotate differently in different parts of the gyre.

The resulting circulation pattern has a significant impact on the redistribution of energy around the globe and on regional temperature (see Figure 5-2). The equatorial currents are warmed by the large input of solar radiation at low latitudes. When these currents are deflected poleward, they carry warmer water to middle and high latitudes. However, these currents lose heat as they travel poleward. When they are deflected northward and southward by the eastern land mass, the water moving poleward is warmer than the polar ocean, whereas the water moving equatorward is colder than the tropical ocean. At the same time, the surface water that originates in the polar oceans and moves equatorward is also colder than the mid-latitude oceans. Ocean currents thereby aid in the latitudinal redistribution of energy: They move warmer water toward the poles and cooler water toward the equator.

Consider the warm North Atlantic Drift as an example. This northwest-flowing current brings fairly warm waters to northern Europe (see Figure 5-2). Due to the predominantly westerly flow of air in the mid-latitude troposphere, most of northern Europe is warmed by the waters from the west and ultimately from the south. Contrast the seasonal variability and much milder conditions in southern Scandinavia with the more extreme conditions on the Labrador coast at the same latitude (see Figure 4-18). Southern Scandinavia benefits from the warmth of the North Atlantic Drift. The air passing over Labrador, however, comes from the cold interior parts of Canada, and a cold offshore current (the Labrador Current) brings sea ice down to this area in winter. The result is lower temperatures and a much greater seasonal temperature range in that area. Notice that we now also have the explanation for the cold offshore currents, such as the Benguela and Humboldt Currents, that are responsible for the Namib and Atacama deserts on the west coasts of South Africa and South America, respectively.

The Circulation of the Deep Ocean

Salinity

In contrast to the surface-ocean circulation, which is driven by atmospheric winds, the deep-ocean circulation is driven by differences in water density. These differences are caused by variations in temperature and in **salinity,** or the salt content of a water mass. Salinity is measured in terms of the proportion of dissolved salt to pure water. The salinity of the ocean is a measure of the quantity of different elements—sodium and chlorine (Na, Cl) being the most common—dissolved in a given mass of seawater (*g* salt/*kg* seawater). Oceanographers, until recently, expressed salinity in parts per thousand (‰), or *per mil.* (Now they express it without units.) A salinity of 1%, or 1 out of a possible 100 parts salt, is equivalent to 10‰, or 10 out of a possible 1000 parts salt. The average salinity of the world's oceans is approximately 35‰, but there is some variability from ocean to ocean.

The primary constituents of sea salt are the ions chloride (Cl^-), sodium (Na^+), sulfate (SO_4^{2-}), magnesium (Mg^{2+}), calcium (Ca^{2+}), and potassium (K^+) (Table 5-1). Except for calcium, which shows some variability from place to place, these elements are found in nearly constant proportions around the globe. Many of the minor constituents—but not all—show a similar uniformity. Most of the variability occurs in those constituents that are utilized by marine organisms.

The salts contained in seawater are largely the result of the breakdown of crustal rocks, or *weathering.* Weathering occurs when rocks are altered by physical or chemical processes. When water flows over or through rocks, it removes soluble materials (ions). Rivers eventually carry the soluble ions to the ocean. In fact, it is estimated that rivers deliver between 2.5×10^{12} and 4×10^{12} kg (about 4 billion tons) of dissolved salts to the oceans each year. The oceans are much saltier than the river water because, when ocean water evaporates, the salt is left behind (increasing the salt concentration) in the

TABLE 5-1

Salt Content of the Earth's Oceans		
Salt Ion	**Grams per Kilogram (g/kg) of Ion in Seawater**	**Ion by Weight (%)**
Chloride (Cl^-)	18.980	55.04
Sodium (Na^+)	10.556	30.61
Sulfate (SO_4^{2-})	2.649	7.68
Magnesium (Mg^{2+})	1.272	3.69
Calcium (Ca^{2+})	0.400	1.16
Potassium (K^+)	0.380	1.10
Bicarbonate (HCO_3^-)	0.140	0.41
Bromide (Br^-)	0.065	0.19
Boric acid (H_3BO_3)	0.026	0.07
Strontium (Sr^{2+})	0.013	0.04
Flouride (F^-)	0.001	0.00
Total	34.482	99.99

Source. PINET, P. R., *Oceanography*, St. Paul, MN: West Publishing Co., 1992.

A CLOSER LOOK
The Salt Content of the Oceans and the Age of Earth

Following ideas first expressed by British astronomer Sir Edmund Halley in 1715, Irish scientist John Joly attempted to calculate the age of Earth on the basis of estimates of the salt content of the ocean and the rate of delivery of salts to the ocean. Two hundred years after Halley, Joly calculated Earth to be 80–89 million years old. However, we now know that Earth is approximately 4.6 billion years old. So where did Joly go wrong?

Joly assumed that the ocean had simply been accumulating all the salts delivered to it by rivers at a constant rate since Earth first formed. Joly neglected the various processes that remove salts from seawater (see the accompanying text). Repeating Joly's calculations but using current estimates of ocean volumes and salinities, we obtain the following:

- The total amount of salt in the oceans is approximately 5×10^{19} kg.
- The rate at which rivers deliver salt is 4×10^{12} kg/yr.
- Therefore, the "age" of Earth is $5 \times 10^{19}/4 \times 10^{12} = 13 \times 10^{6}$ yr.

Thirteen million years is somewhat less than Joly calculated with his knowledge of the world's river discharge, chemical composition, ocean volume, and salt content. The "age" that we have calculated is, in fact, the average length of time salt remains in the ocean. As we will see in Chapter 7, the length of time a substance remains in a given reservoir is called the *residence time*.

ocean. Some of that evaporated fresh water falls on the land and eventually runs back to the ocean. It weathers rocks along the way and thus delivers a new supply of salts that accumulate over time to increase the saltiness of the ocean.

If such great volumes of salts reach the oceans each year, are the oceans still getting saltier with time? The answer is no, because many processes also remove salts from seawater. These processes include the following:

1. Evaporation of seawater from shallow seas. The remaining salts are concentrated and precipitate from solution as **evaporite deposits**, such as halite (table salt, NaCl) and gypsum ($CaSO_4 \cdot 2H_2O$).

2. Biological processes. For example, some marine microorganisms remove the elements calcium or silicon from seawater to form their shells, some of which are eventually deposited in ocean sediments.

3. Chemical reactions between seawater and newly formed volcanic rocks on the sea floor.

4. The formation of sea spray. As small droplets of seawater become airborne, salts, especially sodium and chlorine, are removed when the spray is deposited on land. These salts are eventually returned to the oceans via rivers.

Overall, salts are removed from seawater at a rate that essentially equals the rate of input, when averaged over geologic time scales (millions of years). In other words, the present salt content of the oceans does not represent the result of continuous accumulation but simply a balance between the rates of input and output of salts (see the Box "A Closer Look: The Salt Content of the Oceans and the Age of Earth").

Variations in salinity are caused by regional differences in evaporation, precipitation, sea-ice formation and melt, and river runoff. Surface salinities increase where evaporation exceeds precipitation. We see this effect in such areas as the Mediterranean Sea, the Red Sea, and the Arabian Gulf. In contrast, the Gulf of Bothnia in the Baltic Sea, which also has little exchange of water with the open ocean but experiences much greater precipitation, has relatively low salinities. A similar effect is seen in the Chesapeake Bay on the Atlantic coast of the United States.

Thermohaline Circulation

Because deep-ocean circulation depends on temperature and salinity, this circulation is referred to as **thermohaline circulation** (*thermo* is Greek for "heat," and *haline* comes from the Greek *hals,* for "salt"). In discussing atmospheric circulation, we showed that large horizontal (in particular, latitudinal) changes in temperature and pressure lead to steep pressure gradients and a relatively rapid air circulation. In the deep oceans, horizontal changes in density are small, whereas vertical changes can be larger. But the densest water is at the bottom, so the structure is very stable. Consequently, the movement of water through the deep ocean is relatively slow. Although density-driven movements are much slower than the surface currents, they are no less important in shaping Earth's climate on time scales of hundreds to thousands of years.

The Vertical Structure of the Oceans. The vertical structure of the deep oceans is determined by water density: The highest densities tend to occur in the deepest layers, while the lowest densities are typically found near the surface. Water density, in turn, is controlled by temperature and salinity: Usually, density increases as salinity increases

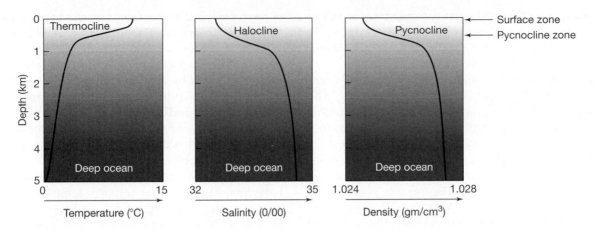

FIGURE 5-6

Generalized profiles of temperature, salinity, and density in the mid-latitude ocean basins. These diagrams show the ocean to be divided into three layers: the surface zone, where there is little change in temperature, pressure, and density with depth; the pycnocline zone, where density increases rapidly (the pycnocline) and where there is an increase in salinity (the halocline) and a rapid decrease in temperature (the thermocline); and the deep ocean, where salinity generally increases slowly with depth, temperatures gradually decrease with depth, and there is little change in salinity.

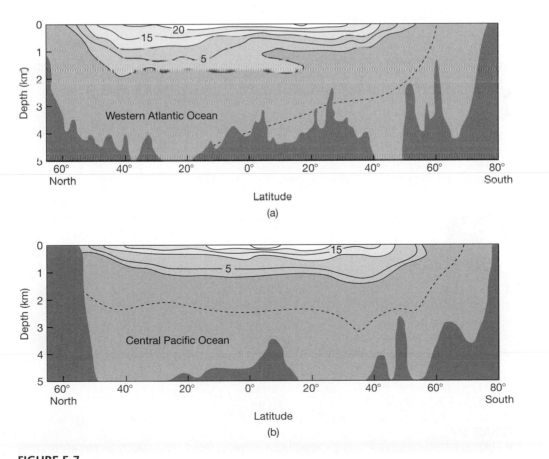

FIGURE 5-7

Vertical distribution of temperature in the (a) Atlantic and (b) Pacific Oceans. The thermocline separates deep water from the surface layer in the tropics, but deep water extends to the surface at high latitudes. Temperature values are given in degrees Celsius.

or as temperature decreases; density decreases as salinity decreases or as temperature increases. However, water is an unusual fluid in that the density of fresh water *increases* as temperature increases from 0°C to 4°C. After that point, water acts more like other fluids in that its density *decreases* as temperatures continue to increase. The temperature at which maximum density occurs varies as salinity changes. Maximum density occurs at 4°C for fresh water (0‰), 2°C for a salinity of 10‰, and at the freezing point for a salinity of 24.6‰. As salinities continue to increase, the temperature of maximum density continues to decrease (it actually stays at the freezing point, which also decreases as salinity increases).

The lower-density zone, which occurs in the top 60–100 m of the ocean, is called the *surface zone,* a layer that interacts with the overlying atmosphere. This interaction takes place through evaporation, precipitation, exchanges of kinetic energy (the effect of winds and friction), radiative exchanges (the absorption of solar radiation and the emission of long-wavelength radiation), and the exchange of heat. This zone is well mixed by wind action, and so the surface zone is often referred to as the **mixed layer**.

The transition zone between the surface zone and the deep ocean is on the order of a kilometer in thickness and is characterized by a rapid increase in density with increasing water depth. The very sharp increase in density is called the **pycnocline**; the transition zone is referred to as the *pycnocline zone* (Figure 5-6). In some regions this density gradient is dominated by salinity changes, and salinity rises rapidly with increasing depth. In this case, the salinity gradient is specifically referred to as the **halocline**. In most other regions, temperature changes dominate the density gradient, and temperature drops rapidly with increasing depth. There the transition is called the **thermocline**. In either case, the steep density gradient makes this layer very stable. This stability limits vertical movements and insulates the deep ocean from seasonal changes in temperature and salinity. The deep ocean below the pycnocline (typically 1–5 km depth) contains about 80% of the volume of all oceans. Deep-ocean water is further stratified, with the highest densities at the

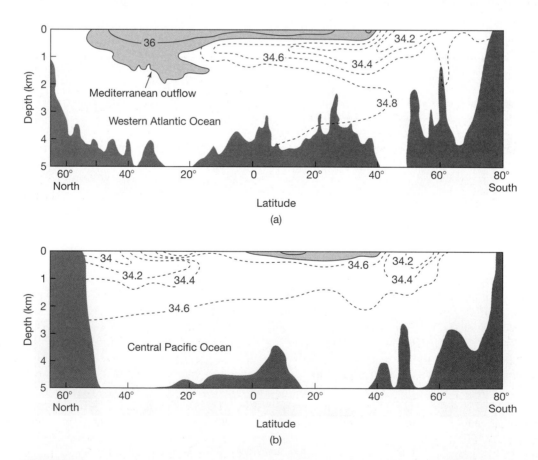

FIGURE 5-8

Vertical distribution of salinity in the (a) Atlantic and (b) Pacific Oceans. The salinity profiles are more complicated than the temperature distributions in Figure 5-7. In the deep ocean, salinity tends to increase in the deeper waters. There is a salinity maximum at the surface in the tropics, however, due to evaporation. When water evaporates, the salts are left behind. Where evaporation exceeds precipitation, there is a net loss of water from the surface layer, and the remaining water has a higher salt concentration.

sea floor. The water column within the deep ocean, therefore, is also stable, and little vertical movement takes place. The movement that does occur is subhorizontal, along sloping layers of equal density (*isopycnals*).

Figures 5-7 and 5-8 show temperature and salinity profiles, respectively, from the Atlantic and Pacific ocean basins. We see that the simple vertical structure outlined above applies to most of the world's oceans, with the exception of those in high latitudes. The high-latitude seas are characterized by low temperatures and relatively low salinities at the surface, similar to the waters of the deep oceans. We will see next that there is a connection between these high-latitude surface waters and the deep oceans and that the formation of very dense surface water near the poles is, in fact, the primary driving force for the deep-ocean circulation.

Bottom-Water Formation. Deep-ocean circulation begins with the production of dense (cold and/or salty) water at high latitudes. This very dense water can be produced by several processes. For example, cooling and increased salinity may result from a large difference between evaporation and precipitation or from the formation of sea ice. Forming along the sea-ice margin in just a few regions in the polar oceans, **bottom water** constitutes the densest water produced in the oceans. Near the poles, the surface waters are cooled below the normal freezing point (−1.9°C in some areas) by contact with the cold overlying atmosphere. (The freezing point is lower than that of pure water because of the presence of salt.) When that water freezes, it forms a layer of sea ice several meters thick that floats on the surface of the polar oceans. When the ocean surface freezes, most of the sea salt is excluded, because the salt does not fit into the crystal structure of the ice. As a result, the water just beneath the sea ice becomes saltier, and an underlayer of very cold, highly saline water forms. The combination of low temperatures and high salinity results in very dense water that sinks and flows down the slope of the basin and spreads toward the equator as the bottom layer of water in the deep-ocean basins.

One of the major sites of bottom-water formation is the Weddell Sea off Antarctica (Figure 5-9). The water formed there, called **Antarctic Bottom Water (AABW)** circles Antarctica and flows northward as the deepest layer in all three major ocean basins (Atlantic, Pacific, and Indian). Although few direct measurements of deep-ocean circulations have been made, reconstructions using observations of the distributions of temperature and salinity have identified Antarctic Bottom Water as far north as 45° N in the North Atlantic and 50° N (at the Aleutian Islands) in the North Pacific. Typical speeds for deep-ocean currents are only 0.03 to 0.06 km/hour, yet Antarctic Bottom Water has traveled more than 10,000 km from its formation site in the Weddell Sea, a trip that has taken some 250 years.

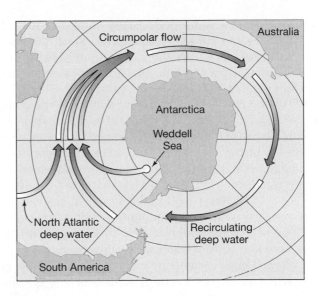

FIGURE 5-9

The Weddell Sea. Part of the Weddell Sea is occupied by an ice shelf (a mass of ice several hundred meters thick that flows from the West Antarctic ice cap). The remainder of the sea is covered by sea ice in winter. The ice forms near the coast and is pushed northward by persistent winds blowing off the ice cap. As the ice is pushed away from the coast, open water is exposed that freezes rapidly in the very cold temperatures. This ice, in turn, is pushed northward, allowing even more ice to form in a continuous process throughout the winter. This region is thus an ice-making factory. The result is the formation of very cold, highly saline water at the surface, which sinks to produce Antarctic Bottom Water. (After W. S. Broecker, and T.-S. Peng, *Tracers in the Sea*, New York: Eldigio Press, 1982, Figure 7-17.)

Similar masses of cold, dense water form in the Arctic Ocean—off the coast of Greenland—and flow south at depth into the western North Atlantic. These water masses are referred to as **North Atlantic Deep Water (NADW)**. The processes of NADW formation are not entirely clear. Warm waters flowing northward in the southwest North Atlantic are evaporated and some portion of the water vapor is carried westward across Central America, where it falls as precipitation in the Pacific. This has the effect of slightly diluting (making less saline) the waters of the eastern Pacific, while producing higher salinities in the western Atlantic (the salt is left behind when the water evaporates). The Gulf Stream carries this more saline water northward where it cools, and the low temperatures and high salinities cause it to sink. At the same time, sea ice formation in the Greenland and Norwegian Seas—like in the Weddell Sea—further increases the salinity and density of the surface waters in these regions. The North Atlantic Deep Water provides approximately half the input of deep water to the world's oceans, and the remainder comes from the Weddell Sea. The NADW that forms to the west of Greenland in the Labrador Sea sinks directly into the western Atlantic; the NADW that forms in the Norwegian Basin subsides and is dammed behind the *sills* (undersea ridges) that connect Greenland to Iceland and Iceland to the

British Isles. This water periodically flows over the sills and cascades into the deep basins of the North Atlantic.

North Atlantic Deep Water flows southward through the Atlantic Ocean and joins the Antarctic Circumpolar Current, which flows around Antarctica. There the NADW and the AABW combine and circle the continent. They then proceed to branch off into the Indian and Pacific Oceans (Figure 5-10). Some of the water completes the circle, reentering the Atlantic or continuing around for another circuit. The time scale over which this occurs is indicated in Figure 5-11, which shows the age of the water at various places in this flow. The map actually shows the change in the amount of radioactive carbon (^{14}C) present in the water masses, which represents the time since that body of water sank below the mixing layer and was no longer exchanging carbon dioxide with the atmosphere (see the two Boxes "Useful Concepts: Isotopes and their Uses" and "A Closer Look: Carbon-14—A Radioactive Clock"). The decreasing proportions of ^{14}C indicate the pattern of flow. You can see the youngest water in the North Atlantic, getting progressively older as it flows south to the Southern Ocean. There is a further addition of young water off the Antarctic coast, and again the water gets progressively older as it flows around the Southern Ocean and up into the Indian Ocean or the northeastern Pacific Ocean. Some mixing of the deep water with surrounding water masses occurs, as does some

biological addition of ^{14}C. Consequently, the ^{14}C age does not necessarily reflect the true age of the water masses. However, from these data it is possible to compute flow rates into the various basins, from which it is possible to determine the replacement (residence) time for deep water and the upwelling rates (how long it takes to get back to the surface). Taking residence times and upwelling rates into account, a parcel of Antarctic Bottom Water will re-emerge at the surface in the Indian Ocean (on average) after 335 years, or in the Pacific Ocean after 595 years. The average residence time for the entire deep ocean is approximately 500 years.

The Thermohaline Conveyor Belt

We have now described how surface waters in the polar regions sink and spread at depth throughout the world's oceans. Ultimately these waters must return to the surface to complete the circulation (Figure 5-12). Water is only slightly compressible (we cannot cram more and more of it into the same space), so if water is sinking at high latitudes, it must be rising somewhere else. There must also be some flow of surface water into the high latitudes to replace the water that is subsiding and moving equatorward. This return flow is even more difficult to monitor than the flow of bottom water. It takes place very slowly through the pycnocline over the whole ocean

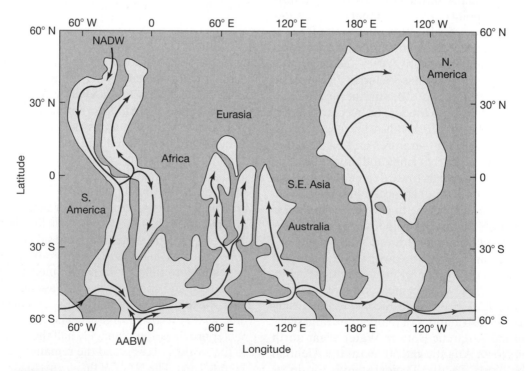

FIGURE 5-10

Flow pattern of the North Atlantic Deep Water and the Antarctic Bottom Water. This diagram represents the flow at a depth of 4000 m; the strange-looking continent/ocean configuration is what we would obtain if the oceans were drained to this depth. (After W. S. Broecker, and T.-S. Peng, *Tracers in the Sea,* New York: Eldigio Press, 1982, Figure 1-12.)

Near Bottom $\Delta^{14}C‰$ Values

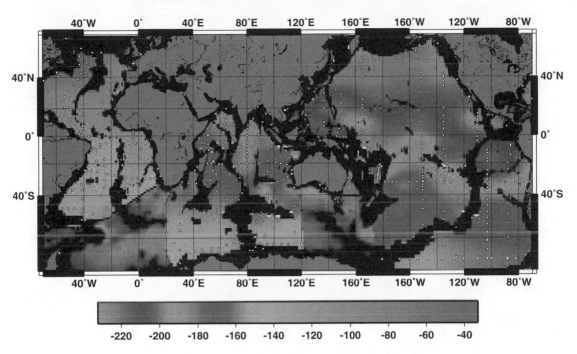

FIGURE 5-11

[See color section] ^{14}C difference values for the near bottom waters of the world's oceans. The values represent the change in the amount of radioactive carbon (^{14}C) present in the water body compared to present-day surface waters (see "A Closer Look, Carbon-14—A Radioactive Clock"). The smallest values represent waters where the ratio of radioactive to stable carbon are most similar to the present-day ocean values (i.e., the youngest water bodies). Regions with the largest difference values show the oldest water masses. The ^{14}C acts as a tracer that shows the path of water movement in the deep oceans. (Diagram courtesy of Robert M. Key, Princeton University.)

USEFUL CONCEPTS
Isotopes and Their Uses

Much of what we know about the present, and especially the past, Earth system comes from the use of isotopes. **Isotopes** are atoms of a given element that have different numbers of neutrons in their nuclei. Isotopes have the same **atomic number**—that is, the same number of protons in the nucleus—but a different **mass number**, which is the total number of protons plus neutrons in the nucleus. Carbon is a good example because it has several isotopes that are used for all sorts of different purposes in studying the Earth and its biota. All carbon atoms have 6 protons. ^{12}C also has 6 neutrons, while ^{13}C has 7 neutrons and ^{14}C has 8 neutrons. The superscript preceding the element's symbol denotes the mass number.

Some isotopes of a given element are **stable isotopes**, which means that they do not spontaneously change into other isotopes or into atoms of another element. **Unstable isotopes** spontaneously change into other isotopes or elements by a process called **radioactive decay**. ^{12}C and ^{13}C are both stable, while ^{14}C is unstable and radioactive. Some 98.89% of all carbon is ^{12}C. ^{13}C constitutes most of the remaining carbon (~1.11%), while ^{14}C occurs only 0.0000000001% of the time. In other words, there are 10^{12} atoms of ^{12}C for every one atom of ^{14}C.

The stable and unstable carbon isotopes are both useful but for entirely different reasons. ^{13}C is used for studying the behavior of the carbon cycle over long time scales. Various microorganisms, especially photosynthetic ones, take up the different isotopes of carbon at different rates. Generally, the heavier isotope is taken up more slowly than the lighter one. We'll see in Chapter 11 that this leaves a useful record for understanding rates of organic carbon burial and variations in atmospheric oxygen in the geologic past. ^{14}C is also taken up more slowly by microorganisms, but that effect is dwarfed by the fact that it is also radioactive. Thus, the main use of ^{14}C is for **radiometric age dating**. The accompanying Box, titled "A Closer Look: Carbon-14—A Radioactive Clock," describes how this technique is used to study deep-ocean circulation.

A CLOSER LOOK
Carbon-14—A Radioactive Clock

The radioactive isotope (or *radioisotope*) of carbon (^{14}C) is produced in the upper atmosphere through bombardment by cosmic rays from distant sources in the galaxy. This bombardment breaks apart atoms producing neutrons, which may then collide with other atoms. Nitrogen atoms (^{14}N) have 7 protons and 7 neutrons. When a nitrogen atom is struck with one of these "cosmic ray" neutrons, the neutron replaces one of the protons in the nucleus. The atom now has 6 protons and 8 neutrons—in other words, the ^{14}N becomes ^{14}C. The ^{14}C is unstable and so it decays back to nitrogen. Radioactive decay is exponential— half occurs in the first 5,730 years, half of the remainder in the next 5,730 years, and so on. This time period, in which half of the initial quantity of radioactive isotope decays, is referred to as the isotope's **half-life** (see Box Figure 5-4).

The ^{14}C is rapidly oxidized to $^{14}CO_2$ and is distributed through the atmosphere. Production of ^{14}C occurs at a relatively constant rate, so the proportion of ^{14}C to stable carbon in the atmosphere remains constant. Living organisms take up the unstable carbon and, although the ^{14}C immediately begins to decay, it is replenished by more ^{14}C from the atmosphere, maintaining an equilibrium that matches the proportions in the atmosphere. Once the organism dies, however, metabolic activity ceases so the ^{14}C continues to decay radioactively, but can't be replenished. Knowing the rate at which this decay occurs, it is possible, by looking at the ratios of ^{14}C to ^{12}C, to determine how long ago the organism lived. If you have a piece of wood, for example, that has half the amount of ^{14}C (in relation to ^{12}C) than we find in living trees, then we know that the piece of wood came from a tree that died roughly 5,700 years ago. This process, called **radiocarbon dating**, has been used to date materials back to 50,000 years ago and is used extensively in archeology and for reconstructing past climates (more on this in Chapter 15).

^{14}C is also useful as a tracer of ocean circulation. Because the atmosphere exchanges CO_2 with the ocean surface, the surface waters of the ocean have nearly the same ratio of ^{14}C to ^{12}C as does the atmosphere. When surface waters sink, however, the ^{14}C that is present begins to decay, and it cannot be replenished. Consequently, the ratio of ^{14}C to ^{12}C in the deep waters gives a measure of how long it has been since the water was near the ocean surface: low $^{14}C/^{12}C$ ratios indicate "older" deep water. By measuring $^{14}C/^{12}C$ ratios in different localities, we can trace the time it takes for water to flow around the globe. The youngest water is found in the Weddell Sea near Antarctica and in the Norwegian/Barents Sea between Norway and Greenland. These are places where bottom water is being formed. The oldest deep water is found in the Northeast Pacific. By combining these and other data, the path and rate of the thermohaline circulation can be determined (Figure 5-11).

BOX FIGURE 5-4

The graph of radioactive decay is exponential. In other words, half of the radioactive parent is left after one half-life. After a second half-life, a quarter of the parent is left, and so on. (From J.P. Davidson, W.E. Reed, and P.M. Davis, *Exploring Earth: An Introduction to Physical Geology*, 1997. Reprinted by permission of Prentice Hall, Upper Saddle River, N.J.)

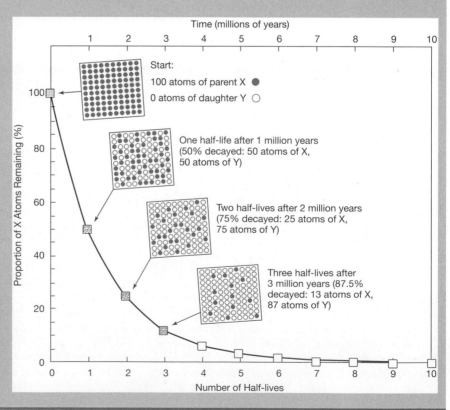

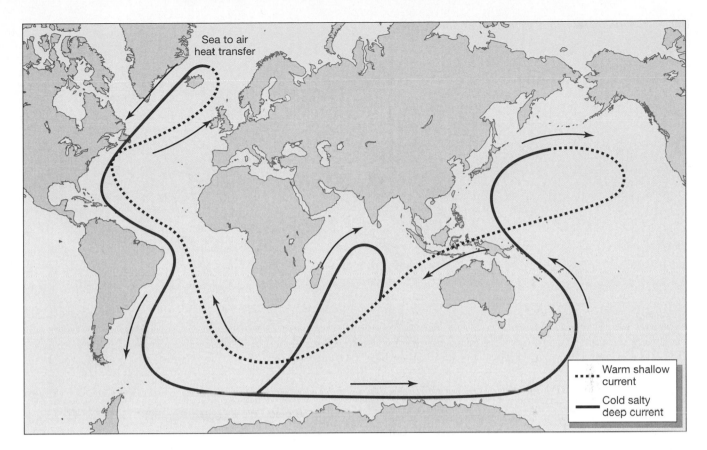

FIGURE 5-12

An idealized map of the deep-water flow (solid lines) and the returning surface circulation (dashed lines). This circulation has been described as a global conveyor belt. The deep water flows out of the North Atlantic, mixing with warmer water to the south. It is recooled by mixing with the cold surface water that subsides around Antarctica. Joining with the Antarctic Bottom Water, it flows around Antarctica in the Antarctic Circumpolar Current. Branches then flow back into the Atlantic as well as the Pacific and Indian Oceans, where upwelling brings the cold waters to the surface. The water eventually returns via the surface currents to the North Atlantic to complete the circulation. (From W.K. Hamblin and E.H. Christiansen, *Earth's Dynamic Systems*, 8/e, 1998. Reprinted by permission of Prentice Hall, Upper Saddle River, N.J.)

FIGURE 5-13

[See color section] Nimbus-7 Coastal Zone Scanner Data obtained from the SeaWiFS project website (*http://seawifs.gsfc.nasa.gov/ SEAWIFS.html*). The Coastal Zone Scanner carried on the Nimbus-7 satellite was one of the first instruments to record ocean color. The satellite detected the pigments from chlorophyll in phytoplankton and so measures the phytoplankton concentrations in the near-surface waters. The light shading shows the regions with the highest productivity.

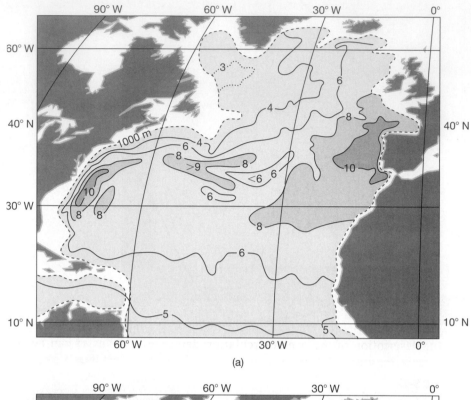

(a)

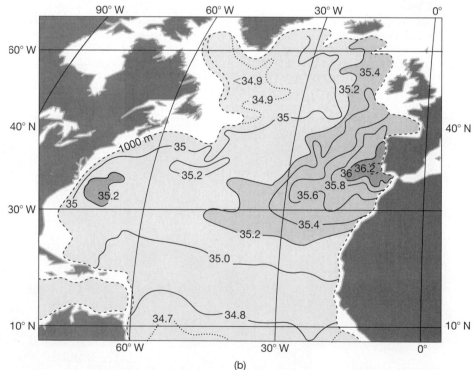

(b)

FIGURE 5-14

The distribution of (a) temperature (in degrees Celsius) and (b) salinity (per mil) at 1000 m depth in the North Atlantic, showing the spread of Mediterranean Sea water. (After Open University, *Ocean Circulation*, New York: Pergamon Press, 1989.)

and more rapidly through upwelling in currents along the eastern boundary of ocean basins and other regions of upwelling and deep mixing. Once the former deep water has reached the surface, the surface circulation that we discussed earlier returns the water to the polar regions. According to Wallace Broecker, a geochemist at Columbia University, this complete cycling of ocean water that

is driven by thermohaline circulation can be likened to a giant conveyor belt.

The thermohaline conveyor belt is a significant feature of the Earth system in several respects: It plays a dominant role in the recycling of ocean nutrients, and it has a major impact on Earth's climate. Much of the life that exists in the oceans can be found in the near-surface layers, utilizing sunlight for photosynthesis—phytoplankton, for example—or living off the animals that feed on phytoplankton. These plants and animals use the nutrients in ocean water, so the surface layers become relatively depleted in nutrients. When these organisms die, they sink through the water column, decompose, and release the nutrients back into the water. The deeper ocean, therefore, is relatively rich in nutrients. The thermohaline circulation transports these nutrient-rich waters around the globe, returning the nutrients to the surface in areas of upwelling, primarily along the continental margins. Consequently, the concentrations of marine life are greatest in these upwelling regions. This is illustrated in Figure 5-13, a satellite image showing productivity in the oceans. The satellite measures the ocean color and the information is converted to concentrations of chlorophyll pigment in phytoplankton, so the image is showing where high concentrations of phytoplankton are found near the ocean surface. Note the high productivity in the North Atlantic and in upwelling coastal zones. In contrast, note the low concentrations in the middle of the primary ocean gyres.

Our description of the ocean circulation depicts a complex system of surface-wind-driven currents overlying a deep ocean with a relatively simple circulation driven by bottom-water formation and surface divergence. In reality, the deep oceans are much more complex. Clearly distinguishable water masses can be identified at different depths and in different geographic locations where variations in temperature and salinity impart different characteristics to the water bodies. For example, high evaporation rates and low rainfall (together with little river runoff) produce relatively warm, highly saline water in the Mediterranean Sea. This water flows out of the Mediterranean at depth through the Straits of Gibraltar and is clearly recognizable as a plume of warm, saline water spreading out into the mid-Atlantic at a depth of about 1000 m (Figure 5-14). We need not concern ourselves with these added complexities, but it is worth noting that our knowledge of the deep-ocean circulation is limited and that much scientific investigation remains to be done in order for us to fully understand what is going on. This understanding is particularly important because the oceans play such a significant role in global climate.

Ocean Circulation and Climate

As we discussed earlier, the ocean circulation has a strong influence on global temperatures. The transport of warm surface water toward the poles, to replace the bottom water that forms near the sea-ice margin, is one mechanism by which excess solar energy is transferred poleward. Figure 5-15 shows the Northern Hemisphere poleward heat transport in the atmosphere and ocean. The total heat transport data are derived by calculating the heat transfer necessary to balance the radiation budget at each

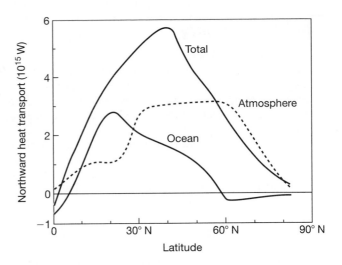

FIGURE 5-15

Poleward heat transport in the Northern Hemisphere. (After Open University, *Ocean Circulation*, New York: Pergamon Press, 1989.)

latitude (see Figure 4-2) and estimating how much of this transfer can be accomplished by the atmosphere. The ocean heat transport is then obtained by subtracting the atmospheric transport from the total. The estimates indicate that the ocean (1) provides almost as much poleward heat transport as does the atmosphere and (2) transports more heat than does the atmosphere at low latitudes, whereas the atmospheric transport dominates at middle to high latitudes.

At the same time, the oceans represent a vast reservoir of heat, absorbing heat from the atmosphere in some areas and releasing it in others. Because water heats up and cools down relatively slowly, pools of water that are cooler than normal or warmer than normal will cool or warm the atmosphere on time periods of months to seasons or years—the time needed for the pools of water to heat up or cool down. On much longer time periods, however, the average effect of the oceans on the atmosphere is determined by the overall temperature of the oceans. Most of the water in the oceans lies in the deep oceans, and its temperature is largely determined by the process of bottom-water formation and by the transport of bottom water around the ocean basins. If the process of bottom-water formation changes, the ocean temperatures will change—and so will climate. Estimates of the rate of bottom-water formation under the present climate, together with measurements of ocean volume, indicate that it would take about 1,000 years to recycle all of the deep water in the oceans. Hence, we can anticipate that the thermohaline circulation could moderate climate over time periods of about 1,000 years. However, we also have geologic evidence indicating that brief interruptions or changes in the thermohaline circulation can also have rapid and large impacts on regional climates, as we will see in Chapter 15.

Chapter Summary

1. As with the atmosphere, the driving force for the oceanic circulation is the global distribution of energy. Unlike the atmosphere, however, the oceanic circulation is driven indirectly by temperature differences: The surface-ocean circulation is, in fact, driven by the circulation of the atmosphere.
   a. Due to friction, wind blowing over the ocean surface drags the surface waters along, producing ocean currents.
   b. The pattern of surface-ocean currents is modified by the Coriolis effect, a consequence of Earth's rotation, and by the distribution of land and oceans.
2. The thermohaline circulation of the deep oceans results from temperature and salinity variations, which control the density of ocean waters.
   a. Cold, saline water is formed in the North Atlantic and in the Weddell Sea off Antarctica.

   b. The combination of low temperatures and high salinities produces very dense water that sinks to the ocean floor and flows as bottom water throughout the world's oceans.
3. Bottom water eventually rises to the surface in zones of upwelling and returns in surface currents back to the high-latitude source regions, completing a vast oceanic conveyor belt.
   a. In combination with the atmospheric circulation, the net effect of these oceanic circulations is to redistribute thermal energy from low latitudes, where Earth is hot, toward the poles, where it is cold.
   b. The thermohaline conveyor belt and the associated zones of upwelling and downwelling play a significant role in climate and in the distribution of nutrients in the oceans.

Key Terms

absolute vorticity
Antarctic Bottom Water
bottom water
downwelling
Ekman spiral
Ekman transport
evaporite deposit

geostrophic current
gyre
half-life
halocline
mixed layer
North Atlantic Deep Water
planetary vorticity

pycnocline
relative vorticity
salinity
thermocline
thermohaline circulation
upwelling
vorticity

Review Questions

1. What effects does the surface-wind pattern have on the circulation of the oceans?
2. Why do ocean currents not move in exactly the same direction as the wind?
3. What is the Ekman spiral? Explain why Ekman transport occurs.
4. What is upwelling? Where does upwelling occur?
5. What is meant by a geostrophic current?
6. Explain the different characteristics of western and eastern boundary currents.

7. Where does the salt in the oceans originate? Are the oceans getting saltier and saltier with time? If not, then why not?
8. Define thermohaline circulation. What are the processes that drive the circulation of the deep oceans?
9. Explain the differences among the pycnocline, the halocline, and the thermocline.
10. What is bottom water? Where and how does bottom water form?
11. What is meant by the term *thermohaline conveyor belt?*
12. Explain what effects the ocean has on modifying the global temperature distribution.

Critical-Thinking Problems

1. Explain what is meant by Ekman transport and what role it plays in producing oceanic gyres in the surface waters of the subtropical oceans.
2. Use a rough map sketch to help explain the role that the oceans play in determining the climates of southern South America and southern Africa, poleward of 20° S.

3. Water is a very unusual substance in that it reaches maximum density between the freezing point and 4°C, depending on salinity. As you cool the water surface to these temperatures it becomes more dense and sinks (rather than immediately freezing). This means that you have to cool the whole water body (the lake or the surface, mixed, layer of

the ocean) to this temperature before you can cool the surface layer enough to freeze, which is why some lakes can remain unfrozen even when the air temperature drops well below freezing. When you have cooled the surface layer to the freezing point, water is again unusual in that its solid form (ice) is actually less dense than the liquid, so ice floats.

Consider how different the world would be if water behaved like most other substances and continued to increase in density down to the freezing point, and if ice was denser than liquid water. Speculate on what this might have meant for life on the planet.

Further Reading

General

Open University. 1989. *Ocean Circulation.* Oxford: Pergamon Press.
Perry, A. H., and J. M. Walker. 1977. *The Ocean-Atmosphere System.* New York: Longman.

Advanced

Wunsch, C. 2002. What is the thermohaline circulation? *Science,* 298, 1179–1181.
Rahmstorf, S. 2003. Thermohaline circulation—The current climate. *Nature,* 421, 699.

CHAPTER 6

Modeling the Atmosphere–Ocean System

Key Questions

- What are Global Climate Models and why do we need them?
- What types of variables do Global Climate Models take into account?
- What can Global Climate Models tell us about Earth's climate?
- What affects the accuracy of Global Climate Models?
- Can we rely on what Global Climate Models tell us about the greenhouse effect?

Chapter Overview

In this chapter we continue the discussion of numerical models that we began in Chapter 3. Three-dimensional Global Climate Models (GCMs) are the primary tool used for projecting potential future climate changes that may occur from increasing concentrations of atmospheric greenhouse gases. We begin by presenting a brief history of the development of GCMs. We then describe, in very general terms, how these models are constructed and what information these models include. In particular we focus on the vast array of variables these models must take into account to reflect the complexities of the earth system. Once we know what variables they account for, we can begin to understand how the models produce reasonable simulations of the present-

day climate. We then illustrate the way these models have been used for climate change experiments. Finally, we discuss some of the limitations of the models and assess what they can and cannot tell us about our possible future.

Why Numerical Models?

We have talked in earlier chapters about systems and the way we can study them. We have described Earth's energy budget, the latitudinal distribution of net radiation, and the resulting circulations of air and water over Earth's surface. We have seen how some very basic physics allows us to make deductions of how the climate system should operate, and we have seen how these deductions are borne out by climate observations. From this it is possible to account for the very broad-scale distributions of temperature and precipitation across the planet and explain how they vary geographically and seasonally. While there is still undoubtedly much more to learn, we have a reasonably good knowledge of the physical and chemical processes at work in our atmosphere and oceans. None of this knowledge, however, allows us to predict how the system will change in the future.

To be able to predict the future, we need a model of the system. In our simple systems model (Figure 6-1) we need to be able to change the input, or some part of the process, and see how this plays out to affect the climate or the model output. We can construct a variety of different types of models. The systems diagrams you have seen,

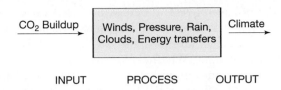

FIGURE 6-1

Simple system model for CO_2 build-up and climate.

and will see, throughout the book are *conceptual models*. A conceptual model is a qualitative description of a system that provides some understanding of how the system works but does not allow one to describe its detailed behavior. An example of such a model is the Daisyworld system described in Chapter 2. The feedback diagrams shown there allow us to find the stable and unstable equilibrium points in the system, but they do not allow us to calculate the actual planetary surface temperature or how the system would evolve in time as solar luminosity changes. This kind of detail can be added to the model, the results of which are shown in Figure 2-14. Doing so, however, requires the use of mathematics.

Consider also the ice–albedo feedback described in Chapter 3. This is an important feedback in the Earth system so you will see this diagram again in later chapters. It is simple to see that if we increase temperatures we reduce the ice cover, and we reduce the albedo, which further increases temperatures—a positive feedback. But by how much? Unless we can quantify those linkages in some way, we can't really tell what will happen. Will just a slight change in temperature cause a change in ice cover? Is there some threshold that has to be overcome first? If we change the temperature from $-40°C$ to $-38°C$ over the central Arctic, presumably this will not have the same effect as changing the temperature from $0°C$ to $+2°C$ at the ice margin. We can quantify these linkages statistically.

That is, we can derive an empirical (observable) relationship between the system components—an *empirical* or a *statistical model*. A statistical model involves mathematics but in a limited way that is based on observations rather than fundamental physical principles. Based on observations we could discover, for example, that the sea ice edge advances or retreats 100 km for every $1°C$ temperature change (remember from Chapter 4 that sea ice is the thin layer of ice that forms over the polar oceans).

This helps, but it is still not enough. Other feedbacks come into play (Figure 6-2). If the ice retreats 100 km it will expose more ocean surface. The ocean is usually warmer than the air, so there will be more heat transferred from the ocean to the atmosphere. We could consider that there will be more evaporation from the ocean, so atmospheric water vapor (a greenhouse gas) will increase. Maybe this will result in more clouds, which both cool the system because of their albedo and warm the system because they absorb longwave radiation (see Chapter 3). What will be the net effect of all these changes? It would be impossible to develop a simple conceptual model or a statistical model that can answer the question. As we broaden the scope to look at the whole climate system we realize that there are hundreds of different variables interacting in many interconnected systems that may operate differently over different parts of the globe. To build a statistical model you would need observations of every possible combination and change, and if such an incredible database existed, we still couldn't answer the question of what the future may hold. The past climates for which data could exist may not include the climate that may develop in the future.

The only realistic possibility for predicting the climate future is to construct a numerical model based on the physics of the system—a model that takes into account all of the processes and feedback loops that we know about. If we can build such a model, and if it produces a realistic simulation of the present climate, then

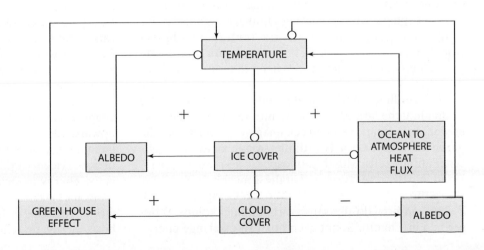

FIGURE 6-2

Feedback diagram for ice-ocean-atmosphere interactions at the ice margin.

we can change the model slightly (e.g., by increasing the greenhouse gas composition of the atmosphere) and expect it to provide a reasonable approximation of how the climate system will respond to the slight change we introduced. In the next section, we introduce models that do so, discuss how they are constructed, and show how we can use them to learn more about the future.

General Circulation Models

What Variables Are Accounted For in the Model?

Chapter 3 described a one-dimensional, radiative-convective climate model (RCM), which took into consideration the vertical energy exchanges in the atmosphere and at the surface. Such models average the incoming and outgoing energy fluxes over the entire globe. They are useful for computing the globally averaged greenhouse effect, but they have nothing to say about temperature variations between different parts of the surface. These models can be made somewhat more sophisticated and realistic by averaging conditions around a set of different latitude bands (e.g., 0–10° N, 10–20° N, 20–30° N, etc.). We could do that for all latitude bands and have a model that not only includes vertical energy exchanges, but also energy transfers between adjacent latitude bands, so that we now have a two-dimensional model (height and latitude). Such models are often described as **energy-balance climate models** (EBMs). They are useful because they allow for additional feedbacks that cannot be calculated in the one-dimensional models (the RCMs). One use of these models is to examine cloud feedback processes by allowing clouds to vary as a function of latitude. They can also be used to explore the ice–albedo feedback referred to above—again by having the albedo change as ice cover extends further equatorward or retreats poleward. You will see the results of some energy balance climate models in the discussion of "Snowball Earth" in Chapter 12. These EBM models cannot be used to make detailed predictions about modern climate change, however, because the atmospheric winds and waves that are responsible for latitudinal energy transfer are inherently three-dimensional, not two-dimensional. If we want to make accurate predictions of future climate, we need to move up to three dimensions.

Three-dimensional climate models take the next step of dividing the latitude bands up into longitudinal blocks as well. In these models you can imagine the surface of the globe divided into a two-dimensional array of longitude/latitude cells, with the atmosphere above each cell divided into discrete layers so that the atmosphere is divided into a three-dimensional grid of boxes. Using these models we can now do the same sorts of calculations as described in Chapter 3, but each box can exchange energy with the boxes above and below, and on all four sides (east–west and north–south). We can include not only the effects of different latitudes, but we can also take geography into account. Whether, for example, we are over land or water, mountains or plains, deserts or tropical forests, has an effect on the local climate. Putting all of these together allows for a much more realistic simulation of the global climate and how it is distributed.

Early versions of these three-dimensional models were mainly atmospheric models: Their purpose was to simulate the primary fields of motion in the atmosphere and they were referred to as **General Circulation Models** (GCMs). Similar developments then took place in efforts to model the circulation of the oceans. To differentiate between the two types of models we labeled them *Atmospheric General Circulation Models* (AGCMs) and *Ocean General Circulation Models* (OGCMs). When they were later combined into a single model it became an *Atmosphere–Ocean General Circulation Model* (AOGCM). Then, as if to confuse everybody, as these models became even more sophisticated and attempted to capture the whole climate system (we talk further about this later in the chapter), they were again referred to as **Global Climate Models** (GCMs)! In this book we use the names AOGCM and GCM interchangeably. If a model includes only the atmosphere or only the ocean, we will refer to it as an AGCM or an OGCM.

Figure 6-3 gives an impression of the information that is included in the model and the types of calculations made by the model. You can imagine that solar radiation enters a grid box at the top of the atmosphere. The model calculates how much is reflected back to space by air molecules, how much is absorbed in the cell, and how much is transmitted down to the next level. The same calculations are made as the radiation is transmitted down through each level until it reaches the surface. There, radiation is either reflected back upward, or absorbed. The absorbed radiation heats the surface, which then emits longwave radiation upward as a function of its temperature. This is an application of the Stefan-Boltzmann law again (Chapter 3). Some of this radiation will be absorbed by clouds, water vapor, and other greenhouse gases, which raises the temperature of the atmospheric grid boxes. Each box will, in turn, emit longwave radiation both upward and back down toward the surface (i.e., the greenhouse effect).

The radiation absorbed at the surface also results in evaporation and the transfer of *sensible* and latent heat upward into the atmosphere. Recall from Chapter 3 that latent heat is the energy associated with phase transitions of water (ice to liquid and liquid to vapor). **Sensible heat** is the energy associated with the thermal motion of air molecules (which is to say with air temperature). As the model develops differences in temperature, vertically or horizontally, this causes changes in air density that result

Modeling the Climate System

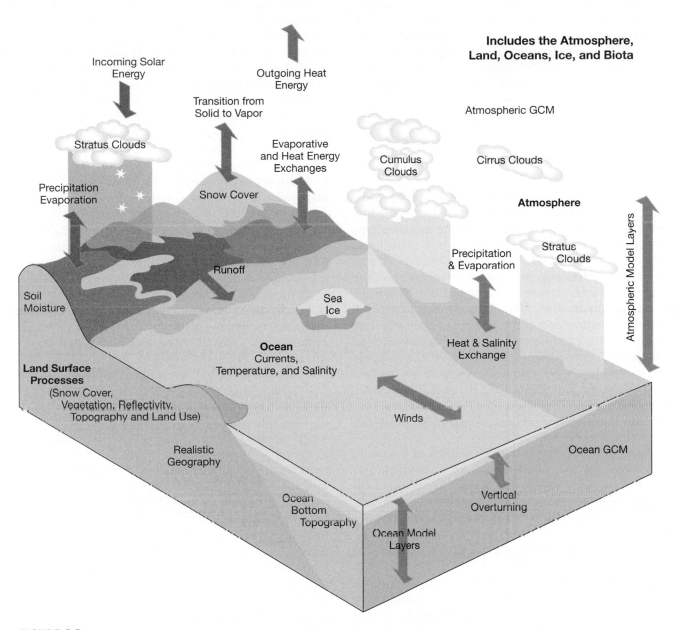

FIGURE 6-3

Some of the processes included in Global Climate Models. (From the National Assessment Synthesis Team, *Climate Change Impacts on the United States: The Potential Consequences of Climate Variability and Change,* U.S. Global Change Research Program.) (*www.usgcrp.gov*)

in the movement of air between adjacent boxes on each side (horizontal winds) or vertically (uplift or convection, and subsidence). As the air moves between these boxes it carries with it energy, mass (including water vapor and aerosols), and momentum. Depending on the temperature and humidity of the box, the water may stay as a gas or it may condense to form clouds. The clouds produce rain that falls through the atmosphere. It may evaporate as it falls or reach the surface as precipitation, which will occur as rain or snow depending on the temperature. The changing characteristics of the box (in terms of water vapor, aerosols, temperature, clouds) all change the radiative properties of the box, which further changes the transmission, absorption, and reflection of the solar and terrestrial radiation fluxes and the transfer of sensible and latent heat.

All of this sounds complicated enough, but present-day Global Climate Models are much more sophisticated than this. The variables seem endless. The surface cells can be land or water. If they are land cells they have an elevation, so the model takes into consideration the surface topography, as well as soil and vegetation cover. When rain falls on the surface, some evaporates, some infiltrates the soil, and some runs off over the surface. How much depends on the soil characteristics, the surface temperature, how much water is already present, and the surface slope. The soil is divided into layers, allowing for water and heat to be transferred up and down in the soil and for water to flow laterally between adjacent boxes below the surface. Each model grid cell has a characteristic surface cover that includes a predominant vegetation type.

The vegetation type determines how deeply roots penetrate the soil (which influences water infiltration rates and evapotranspiration), and each vegetation type has a characteristic stem and leaf structure (which also affects evapotranspiration). In addition, the type of vegetation helps determine the surface albedo and also the surface roughness. How "rough" the surface is affects the friction between the low-level wind and the surface, which influences air movement (turbulence) and affects the rate at which heat is exchanged between the surface and the atmosphere. If precipitation falls as snow, this not only changes the albedo, but may also smooth out the surface as the vegetation becomes covered by snow (thus changing heat fluxes). Each of these processes and interactions is described by an equation or series of equations that can be solved by the model.

Early models were less sophisticated in regard to oceanic data and used observed sea surface temperatures for the ocean cells, or allowed the ocean surface to exchange water and energy with the overlying atmosphere. In this case, many models included a 50–65m mixed layer that allowed for energy exchange at the surface and some seasonal heat storage. Some of these models allowed heat to be transported across adjacent cells in a pattern that matched the present-day distribution of ocean currents. Today, Global Climate Models include ocean circulation models that are as detailed as the atmospheric component. The ocean is divided into layers. Energy, mass, and momentum are exchanged at the ocean surface. The wind drives ocean currents and mixes the surface layers, and the model tracks temperature and salinity changes that determine water density and drive the deep ocean (thermohaline) circulation. There is convergence and divergence resulting in downwelling and upwelling water that connects the deep ocean and the surface layers. Sea ice forms where the surface ocean temperature is at or below the freezing point. Again, within each cell, the ice is divided into layers, heat is transferred through the ice, and the ice thickens and thins through the seasons as the water and air temperatures change.

This still does not give a full description of all the processes and calculations that are included in a Global Climate Model. Hopefully, however, this description does illustrate the point that these are very sophisticated models that take into account all of the processes described so far in Chapters 3, 4, and 5, plus others that go beyond the level of detail we have discussed in the text. Let's think about the number of calculations involved in running such a model. A model with a 2.5° × 2.5° longitude/latitude grid size will have over 20,000 cells distributed around the globe. If we divide the atmosphere into 20 layers, that gives 400,000 grid boxes in which all of the atmospheric calculations have to be made (including all of the transfers with the adjacent cells). Six thousand of the surface cells are over land and include at least a five-layer soil model (another 30,000 sets of calculations). Remember that there are many different calculations going on within and between each cell: calculations of radiation fluxes, heat transfers, momentum transfers winds, convection, subsidence, cloud formation, rainfall, and so on.

To account for time passing, these calculations need to be updated every 10 to 30 model minutes (the model time step) depending on the spatial resolution (grid size) of the model. The higher the resolution (the smaller the grid squares), the more frequently the calculations need to be updated. This is for reasons of numerical stability—the results become completely unphysical if the time step is too large compared to the grid spacing. There are millions of calculations that have to be made for every day of the model climate, and given current computer capabilities, today's models may need to be run for several hundred model years to simulate the past 100 years of climate and projecting out 100 years into the future. The latest *parallel* computers, running a relatively high resolution 2.8° × 2.8° latitude/longitude model, may simulate three or four model years for each 24 hours of computing time. Parallel computers run a task on multiple processors—the problem is broken down into pieces and each processor works on a separate piece. The world's more powerful and expensive parallel computers with 32 processors might take 50 days of continuous processing to do a single model run. As you might expect, there are only a relatively small number of these models in the world. They are extremely expensive to develop, and costly to run. The Yokohama Institute for Earth Sciences took delivery of the world's fastest supercomputer in March 2002. With 5,120 processors, the computer is expected to achieve peak performances of 40TFLOPS (40 trillion floating-point operations per second). This computer is being dedicated to developing an "Earth Simulator" and the system is currently testing new atmospheric and oceanic circulation models.

There is one more wrinkle we must discuss here. In the past, these models were constructed essentially the way we described above—by dividing the globe into boxes and calculating all of the transfers across the adjacent cell boundaries. All of the model equations were expressed as finite difference equations; that is, the transfers of energy, mass, and momentum between boxes were calculated as some function of the difference in various quantities (e.g., temperature) between the boxes. These models are, therefore, referred to as **finite difference models**. Because the atmosphere is unbounded at the sides (i.e., it is continuous around the globe) it is possible to express all of these equations very differently, as wave functions. In a model that uses wave functions, the number of waves resolved determines the spatial resolution of the model—the equivalent to the grid size in the finite difference models. These models are referred to as **spectral models**. Most current GCMs are spectral models; their output, however, is still usually presented as a gridded product.

Using Models for Climate Experiments

For the rest of this chapter we focus on learning more about conducting climate change experiments with these models, but it is worth noting that GCMs have many other uses besides projecting greenhouse warming. These types of models also allow us to test our understanding of how the climate system works. We can perform analyses of climate sensitivity by changing some of the parameters and seeing what effect it has on the outcome. For example, we can have a fixed cloud cover or let the cloud cover vary to see how important clouds are (we'll talk more about this further on). We can change the vegetation cover, what happens to global climate if we replace the Amazon rainforest with grassland? You'll find out in Chapter 18. In Chapters 4 and 5 you saw how mountain chains and the distribution of land and ocean affects climate, and in Chapter 7 you will see how the distribution of continents has changed with time. We can enter previous land–ocean distributions into the GCM and determine what effect this had on past climates (*paleoclimate modeling*), or we can model the effects of future and past volcanic eruptions on climate (Chapter 15). The science and research applications are numerous, and we use versions of these models everyday for weather forecasting.

The first attempts at numerical climate modeling did, in fact, begin as experiments in weather forecasting. Lewis Fry Richardson, a British ambulance driver in France during World War I, spent his off-duty time solving the basic equations of atmospheric motion using a mechanical calculator. His objective was to develop a numerical weather forecast for parts of Europe. He published the results in 1922 but, as you might expect, there was little further progress in numerical modeling until the development of electronic computers in the 1940s. Weather forecasting was indeed one of the first challenges worked on with the new computer at the Princeton Institute for Advanced Studies in the late 1940s. Today, almost all weather forecasts are made with numerical models of one sort or another.

Efforts to transform weather prediction models into climate models began in the late 1950s, and the development of ocean circulation models began in the 1960s. Since then, Global Climate Models have steadily grown in size and complexity in response to the ever-increasing computing power that has become available. Figure 6-4, taken from the Third Assessment Report (TAR) of the International Program on Climate Change (IPCC), published in 2001, shows the history of climate model development. It is interesting to note that different parts of the model were developed independently and later incorporated into the GCMs. We see that these models now include atmosphere, ocean, land, and sea ice components, as well as various aerosol and carbon cycle models. Looking ahead, the next developments will likely be the incorporation of dynamic vegetation and atmospheric chemistry models (i.e., as climate changes and those changes result in land use and vegetation changes, these, in turn, will modify atmospheric chemistry and surface characteristics, which will further change the climate).

Originally, researchers typically ran GCMs by inputting present conditions such as the solar constant, the rotation rate of the Earth, the distribution of continents and oceans, topography, vegetation cover, and gas composition of the atmosphere (these fixed conditions are referred to as the model **boundary conditions**) and then putting the model into motion until it reached an equilibrium state. This involved running it for about 15 years (model time) until the model developed a stable climate (in which the mean climate didn't change with time) and then running it for 15 to 30 more years to obtain some measure of the climate variability. Once the models reached the steady state point, scientists would change one of the boundary conditions (e.g., the concentration of greenhouse gases in the atmosphere) and run the model again. The difference in the two model runs indicated the climate change that would be expected for the boundary condition, in this case the change in greenhouse gases. Most experiments doubled or quadrupled the greenhouse gas concentrations. Because the models are run out to equilibrium conditions each time before boundary conditions are changed, these are referred to as **equilibrium climate experiments**.

Equilibrium Climate Change Experiments Early atmospheric GCM experiments during the 1970s held the cloud cover constant (the observed global cloud distribution was set as a boundary condition of the model). A climate change experiment was then conducted to see what the effect of four times the (then) present level of atmos-

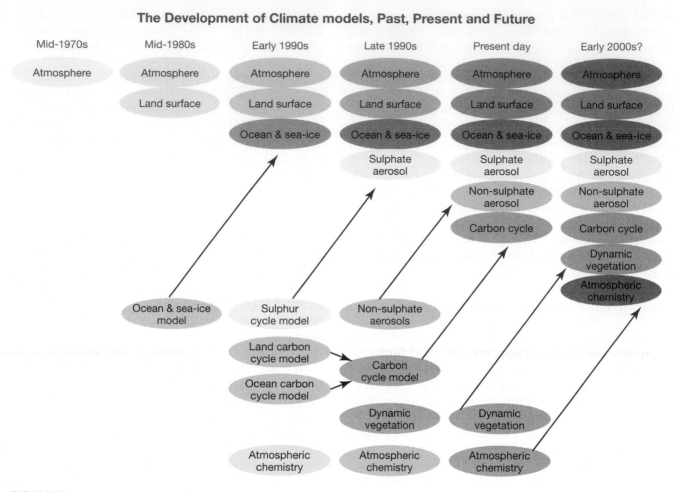

FIGURE 6-4

History of climate model development. (From the Intergovernmental Panel on Climate Change, Climate Change 2001, *The Scientific Basis: Technical Summary of the Working Group I Report.*) (*http://www.grida.no/climate/ipcc_tar/wg1/010.htm*)

pheric CO_2 would be on the climate. The result was a 4°C increase in global temperature. A later version of the same model with variable cloud cover produced a 4°C mean global temperature increase when the amount of atmospheric CO_2 was only doubled. The difference between the two experiments illustrates the important role that clouds play in the climate system and further illustrates the need for numerical climate models. One might expect that surface warming due to increased levels of atmospheric CO_2 would increase evaporation and increase cloud cover. You learned in Chapter 3 that clouds act to both warm and cool the system, and in the present climate these two effects almost cancel each other out—with the cooling (albedo) effect being slightly greater than the warming effect. One might think, therefore, that increasing cloud cover might have little impact, or might enhance slightly the negative feedback that dominates at present. In fact, the results were exactly opposite to this—the clouds had a strong positive feedback that enhanced the

CO_2 warming. The increased temperatures produced deeper (higher) moist convection, which produced an increase in the amount of high-level cirrus clouds. You will also remember from Chapter 3 that the greenhouse effect for high clouds is larger than the albedo feedback, and so high clouds cause surface warming.

A number of model experiments were carried out in the 1980s with model resolution varying from 5° × 7.5° to 8° × 10° latitude/longitude grids, variable cloud cover, a mixed-layer ocean, and a prescribed oceanic latitudinal heat transport. These models were all equilibrium models that produced mean global temperature increases of approximately 2.5°C to 5°C when atmospheric CO_2 was doubled. Toward the end of this time period there were also several higher resolution model runs that produced temperature changes of about 3.5°C. If we compare these to the one-dimensional model results presented in Chapter 3 (which were all for equilibrium conditions) we see that the direct radiative forcing (with no climate feedbacks) resulted in a 1.2°C surface temperature increase. When the water vapor feedback was added to the one-dimensional model, it doubled the response to about 2.5°C. The three-dimensional models that include geography and additional feedback processes increased the response still further. The indication is that as we make the models more realistic they appear to be dominated by feedback processes that amplify the initial change in climate. The more complex and realistic the model, the greater the resulting changes in climate.

Transient Climate Experiments. More recent projections have used **transient model experiments**. In this case, rather than running the model to equilibrium conditions before changing the boundary conditions, we start the model at the present or at some point in the past and then run it forward to assess how the climate changes with time. To do this we need to input data that will indicate how the greenhouse forcing will change with time. Various scenarios have been developed assuming a wide range of future socioeconomic states and the resulting changes in greenhouse forcing. These are discussed further in Chapter 16.

These transient experiments require a more sophisticated treatment of the ocean than the simple mixed layer models discussed earlier. We have seen that the atmosphere responds quickly to any change, while the oceans respond much more slowly. Time-scales of change are on the order of weeks in the atmosphere, seasons for the land surface and upper ocean, and hundreds of years for the deep ocean. Ocean models thus take considerably longer to reach equilibrium than atmospheric models. To save computational time, the atmosphere and ocean components are usually run separately before they are coupled together. Most techniques for coupling the two models, however, result in some degree of climate "drift" in the model away from the observed climate. Modelers are faced with the choice of letting the drift occur (and therefore not reproducing the observed climate), or adding in some corrective adjustment (referred to as a *flux adjustment*). Recently this problem may have been solved in some coupled model runs that did not exhibit this sort of climate drift.

A common approach with early transient model experiments was to increase atmospheric CO_2 by 1% per year. This results in a doubling of atmospheric CO_2 in about 70 years. The model would be run until it reached present-day equilibrium and then atmospheric CO_2 would be increased at a compounded rate of 1% per year. A comparison of 19 of these runs produced a warming of between 1.1°C and 3.1°C with a mean of 1.8°C (most were in the 1.5°C to 2.5°C range) by the time atmospheric CO_2 doubled. For many of these models, only about 60–65% of the equilibrium change had taken place by the time the atmospheric CO_2 doubled. This reflects the large thermal inertia introduced into the systems by the oceans. However, the equilibrium climate response remains unchanged. In other words, once the atmospheric CO_2 has doubled in the transient experiment, if the model calculation is extended forward in time with that level of atmospheric CO_2, it develops an equilibrium climate similar to the equilibrium models reported earlier. This is illustrated in Figure 6-5, which shows the results of a coarse resolution Atmosphere–Ocean General Circulation Model (AOGCM) running with a 1% per year increase in atmospheric CO_2. The CO_2 increase is stabilized at the doubling point (lower curves), and when it is quadrupled (upper curves). The colored lines show the results with a simplified model that allows no energy exchange with the deep ocean (so the difference shows the effects of ignoring these ocean exchanges). The transient surface temperature change is about 2°C. If a doubled CO_2 level is maintained, there would still be a further increase of approximately 1.5°C in global mean temperature as time passes in the model run.

Transient GCM experiments have also been run with a range of other forcing scenarios. These use time-dependent forcing (forcing that changes over time) in which the simulations start at some point in the past (usually the middle of the 19th century) and are run with estimates of the actual forcing during the 20th century. These models can be validated against the observed climate record, and are then extended into the future using a projection of future atmospheric radiative forcing. Some of these models have also been run with projections of future sulfate aerosol concentrations (see Chapter 16), which in several ways act to cool the surface by increasing the albedo. In these experiments, CO_2 doubling occurs by about 2060. Comparing 2021–2050 with 1961–1990, models that just include the CO_2 forcing (and no future sulfate aerosol concentrations) show an increase of 1.6°C with a range from 1.0°C to 2.1°C. (This actually is only a little lower than the numbers presented in Figure 6.5 when you take into account they are from a time interval well before doubling occurs.) When sulfate aerosols are included, the mean temperature increase is 1.3°C with a range from 0.8°C to 1.7°C.

Can We Trust What the Models Tell Us?

GCMs compute the three-dimensional exchanges of energy, mass, and momentum within the atmosphere and oceans and at the interface between the atmosphere and the land or ocean surface. The rules that govern these exchanges can be expressed as equations derived from the basic physical laws that describe these processes. Deal-

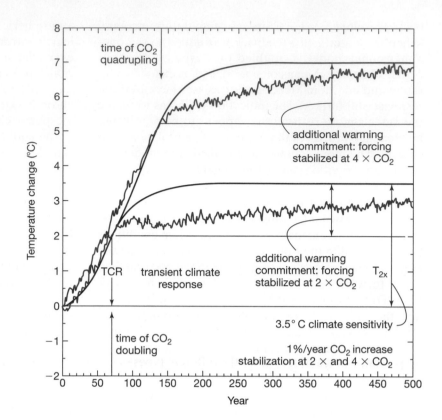

FIGURE 6-5

Global mean temperature change for 1%/yr CO_2 increase with subsequent stabilization at 2 × CO_2 and 4 × CO_2. (From the Intergovernmental Panel on Climate Change, Climate Change 2001, *The Scientific Basis: Projections of Future Climate Change.*) (*http://www.grida .no/climate/ipcc_tar/wg1/338.htm*)

ing as we are with basic principles of physics that are reasonably well understood, it would seem that there should be no inherent reason why these models should not produce an accurate representation of the climate system. As we will see below, they do in fact produce a reasonable simulation of the present climate. However, there are problems, most of which are due to the resolution at which the calculations are made.

Subgrid-Scale Processes

Many of the physical processes that take place in the climate system occur on spatial scales that are too small for the grid-scale of the model. A good example of this is cloud cover. A typical grid cell in today's models may be 250 km across, so the cell may represent an area of over 62,000 km^2. In effect, the model has to treat this as a uniform region with a single temperature, rainfall rate, cloud cover, and so on. However, you only have to look upward to see that clouds do not form as rectangular boxes, 250 km on a side—which is what the model produces. The physical processes that give rise to localized convection in the atmosphere, cause clouds to form and dissipate, and cause ice crystals to grow and raindrops to fall, all occur on spatial scales that are much smaller then the model grid cell. While we can develop mathematical models of cloud formation, we cannot include them in GCMs

because the computing power simply isn't available to do the calculations at the *resolution* needed.

Resolution in this context refers to the effective grid size of the model (spatial resolution) and the model time step (temporal resolution). Refer back to the earlier discussion of the number of calculations made by GCMs. If we take the 2.5° grid cell we used as an example before, that gave us 400,000 grid boxes for a 20-layer model. If we doubled the resolution to 1.25°, that means four times the number of grid boxes for which these computations have to be made, plus four times the number of land cells so four times the number of soil calculations, and so on. In actual fact, the number of extra calculations will be even greater because, for any given model, increasing the grid's resolution usually means you also have to increase the number of vertical layers and shorten the time step, so more calculations for each day of model climate. A model run that took 50 days at the previous resolution may now take 250 days at the new resolution—and you still do not have a model that can resolve the cloud cover. As another example, think of the landuse/vegetation cover. At present, most models can only be assigned one type of vegetation in a grid cell, so we use the predominant vegetation type. It may, for example, be deciduous forest. In reality there will be a mix of landuse or vegetation cover in any 62,000-km^2 region. If nothing else, there are likely to be variations in forest density, plus there may be roads, lakes, maybe some towns, and so on—all of which have very different effects on the energy and water exchange at the surface. Even if the vegetation cover was completely uniform, the actual energy and moisture exchanges occur at the level of individual trees, roots, and leaves, which certainly can't be resolved by the model.

Where there are processes for which we can't explicitly write and solve a set of physical equations, that is, all of these processes that take place at a finer time or space scale than the model can resolve (**subgrid-scale processes**), we have to **parameterize** the relationships involved. Parameterizations are simply empirical or statistical functions that relate two variables based on observed relationships between those variables (like the statistical models referred to earlier). Clouds, for example, are generated in the model as some function of humidity and temperature, based on what we observe in today's climate. Different models may use different parameterization schemes, and how models parameterize these subgrid-scale processes accounts for much of the variability in the output from different models. This is why we present a range of results for the different types of climate experiments described earlier. The various ways models parameterize cloud cover accounts for a large part of the differences in model results, and is probably the most important factor leading to uncertainties and errors in the atmospheric component of the GCMs.

Dependence on Initial Conditions

Starting the climate model from almost identical, but still slightly different, conditions can also result in differences in the model output. This is illustrated in Figure 6-6 from the IPCC Third Assessment Report. In the figure there are three realizations of temperature distributions made with the same model and same greenhouse gas and aerosol forcing. The maps show the temperature difference between the period 1975–1995 and the first decade of the 21st century. Each of the simulations began with slightly different initial conditions a century earlier. You can see that the model results are essentially the same when looking at broad global patterns, but the sensitivity to initial conditions does result in some differences in regional detail. This example demonstrates that any single

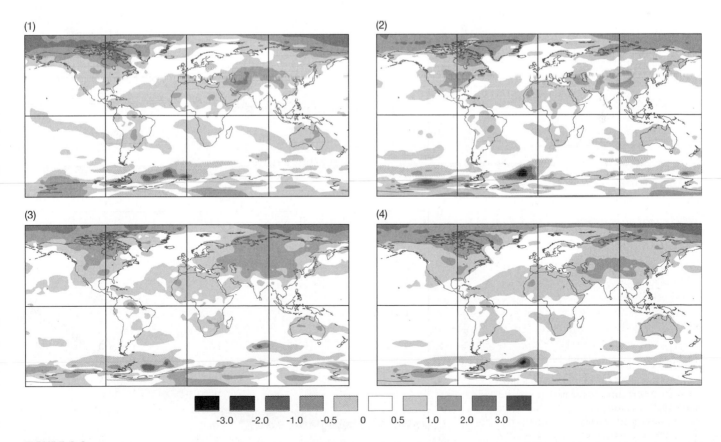

FIGURE 6-6

Ensemble model runs showing three realizations of a transient experiment with identical atmospheric forcing, but with slightly different initial conditions. The temperatures show the differences from the first decade of the 21st century to the period 1975–1995. The fourth map is the ensemble mean. (From the Intergovernmental Panel on Climate Change, Climate Change 2001, *The Scientific Basis: Projections of Future Climate Change.*) (*http://www.grida.no/climate/ipcc_tar/wg1/338.htm*)

run of the climate model simply represents only one of a range of possible outcomes. This problem is addressed in present climate experiments by running models several times from slightly different starting points and presenting the model runs as an ensemble. The ensemble runs can then be averaged to give a single ensemble mean. In Figure 6-6 the fourth map represents the ensemble mean.

Are the Model Results Usable?

There are other problems with the models and their projections of present and future climates that we do not need to go into here. We simply want to demonstrate that the models are not perfect and to show that one of the main reasons why there is uncertainty in the results is due to the fact that many important climate processes operate on time and space scales that the models cannot resolve. This does not mean that the model results are invalid. These models do a very good job of projecting ahead the mean global climate. There are differences between models primarily because of the different ways they treat sub-grid-scale processes. These differences account for the spread (the differences) in the model results that we examined earlier in the chapter. However, there is also extensive agreement between the models. Positive feedback processes dominate in all of the global warming projections; all show similar circulation patterns and all agree, in a broad sense, on the global pattern of temperature changes. The changes are greatest at high latitudes and least in the tropics and are greater over the continents compared to the oceans at the same latitude.

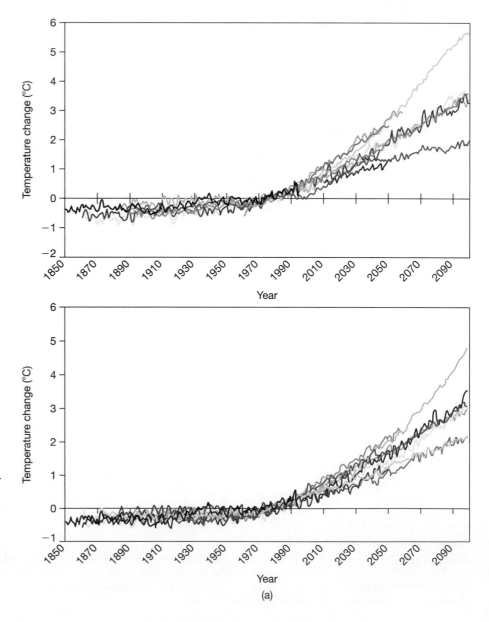

FIGURE 6-7

Time evolution of globally averaged temperature change relative to the period 1961–1990. The top graph shows the results of greenhouse gas forcing, the bottom graph shows the results of greenhouse gas forcing plus aerosol forcing. (From the Intergovernmental Panel on Climate Change, Climate Change 2001, *The Scientific Basis: Projections of Future Climate Change.*) (*http://www.grida.no/climate/ipcc_tar/wg1/338.htm*)

(a)

Figure 6-7 from the IPCC Third Assessment Report shows the time evolution (change over time) of the temperature change produced for 10 different model climates—all using the same atmospheric (greenhouse and aerosol) forcing. The temperatures presented represent the difference between the annual globally averaged value of each year and the global mean for the period 1961–1990. Each wavy line shows the results of a different model run. The solid black line shows the observed temperature change. The top graph shows the results with just the greenhouse forcing, while the bottom graph includes both greenhouse gas and aerosol forcings. Figure 6-8 shows the same results, but for precipitation. Notice that the model spread (the differences between individual models results) increases into the future as the forcing becomes larger, but all of the models are very close to each other and to the observations for the 150 years of the observed record. This indicates that the models (even if their results differ slightly from each other) do a reasonable job of simulating the present-day climate system. The present generation of GCMs, which are coupled ocean–atmosphere climate models, are effective at simulating the climate system at subcontinental scales (i.e., for large regions like Western Europe) and over time periods of seasons to decades. However, there are still large uncertainties when it comes to using these models to assess regional climate change over briefer periods of time.

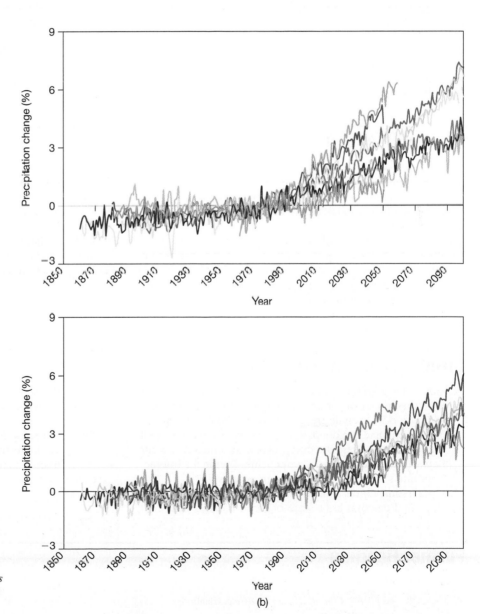

FIGURE 6-8

Time evolution of globally averaged precipitation change relative to the period 1961–1990. The top graph shows the results of greenhouse gas forcing, the bottom graph shows the results of greenhouse gas forcing plus aerosol forcing. (From the Intergovernmental Panel on Climate Change, *Climate Change 2001, The Scientific Basis: Projections of Future Climate Change.*) (*http://www.grida.no/climate/ipcc_tar/wg1/338.htm*)

(b)

Chapter Summary

1. The climate system is inherently complex. Multiple climate variables interact over a broad range of time and space scales through interconnected positive and negative feedback processes. The complexity is such that mathematical models of the physical processes are the only feasible approach to projecting climate change.

2. Researchers employ a range of models from simple one-dimensional globally averaged radiative-convective models (described in Chapter 3), to more complex three-dimensional general circulation models (GCMs). GCMs include linked models of the atmosphere, ocean, soils, vegetation, and sea ice cover. These models are used for a variety of applications—to study how the climate system works, to forecast weather, to recreate past climates, and to assess the poten-

tial for future climate change. Modern GCMs run in a transient mode and reflect the time history of atmospheric forcing and projections of future greenhouse gas emissions.

3. The biggest weakness of these models is their relatively coarse spatial resolution. The inability to resolve subgrid-scale processes, many of which are important to the climate system, produces some uncertainty in the model results. The way in which different models parameterize these subgrid-scale processes is one of the primary reasons for the spread in climate change projections produced by different GCMs.

4. Despite these weaknesses, the models do an effective job of simulating the observed global climate record. They produce climate change projections that are valid at the subcontinental level and for time periods of seasons to decades.

Key Terms

boundary conditions
energy-balance climate models
equilibrium climate experiments
finite difference models

general circulation models
global climate models
parameterize
resolution

spectral models
subgrid-scale processes
transient model experiments

Review Questions

1. What is meant by a conceptual model? Give an example.
2. What is meant by a statistical model?
3. What is the difference between a one-dimensional radiative-convective model and a two-dimensional energy balance model?
4. What is a general circulation model?
5. What are some of the science and research applications for GCMs?
6. What is meant by the term *model boundary conditions*? Give examples of the boundary conditions that might be used in a climate model.

7. What is meant by the term *equilibrium climate experiment*?
8. What is the difference between an equilibrium experiment and a transient experiment?
9. Why are subgrid-scale processes a problem for global climate models?
10. Explain what is meant by parameterization.
11. What is the purpose of running an ensemble model?
12. Why are ensemble model runs necessary?

Critical Thinking Problems

1. Construct a systems diagram of a global climate model. Show each subcomponent (atmosphere, oceans, etc.) as a separate box. Include the ocean as two boxes: the mixed layer ocean and the deep ocean. Use arrows to connect all of the boxes and describe the types of interactions that occur between the different subcomponents. Divide the interactions into short-term and long-term processes.

2. Obtain some topographic maps for your local region. If possible, find one that has a scale of about 1:50,000. Also find

one that shows about 250 km on each side of the same region. Divide the large area map into 250-km squares and visually estimate the average elevation in each square. Then divide the squares into quarters and repeat the exercise. Compare the results and then compare them to the topography in the higher resolution map. Discuss what the implications of this exercise are for climate models.

Further Reading

Advanced

Trenberth, K. (ed.). 1992. *Climate System Modeling.* Cambridge: Cambridge University Press.

Circulation of the Solid Earth: Plate Tectonics

Key Questions

- How do the physical and chemical characteristics of Earth change with depth toward its center?
- What is plate tectonics?
- What provides the energy that drives plate tectonics?
- How can we relate the surface features of Earth to plate tectonics?
- What is the rock cycle?
- How have the geographical positions of the continents changed through time as a result of plate tectonic activity?

Chapter Overview

Our understanding of solid Earth processes has taken a great leap forward since the mid-1960s with the development of the theory of *plate tectonics*. Yet there is still much to learn about the composition and dynamics of the solid Earth. Earthquakes and volcanoes demonstrate that the interior of Earth is not a static place. Rather, like the oceans and atmosphere clinging to its surface, the solid Earth is in motion. The energy that drives the circulation of the solid Earth derives not from the Sun but from Earth's interior. Convective currents in the interior are coupled to the rigid rocks that form the continents and sea floor, putting the continents in motion. Where continents collide, huge mountain belts form; where oceanic blocks collide with each other or with continents, deep-sea trenches and volcanoes form. These plate tectonic forces join with the surface processes of rock weathering and erosion to generate landscapes and recycle elements from solid Earth reservoirs into the soils, hydrosphere, and atmosphere, making those elements available to the biota once again. Thus, plate tectonic activity is critical to the maintenance of a biologically active planet.

We will study the circulation of the solid Earth in the way we studied atmospheric and oceanic circulation. First, we explore the anatomy of the planet, from its exterior to its greatest depths. The tools used to reveal Earth's internal structure give us clues about the temperature and compositional variations in the interior, but very little direct information exists. Then we discuss how the heat flux from the interior produces the motions within the solid Earth and how these motions form and modify Earth's major surface features: mid-ocean ridges and mountains, deep-sea trenches, and transform faults. We then trace the history of these motions over the past 3 billion years, during which time the continents have drifted apart and joined together again and again in a global tectonic cycle.

Introduction

The German meteorologist Alfred Wegener is largely credited with establishing the fundamental underpinnings of the theory that we now call **plate tectonics.** According to this theory, Earth's surface is divided into rigid plates of continent and ocean floor that move relative to each other through time. (*Tectonics* is the study of Earth's crust and the processes that deform it.) Wegener was fascinated by the near-perfect fit between the coastlines of Africa and South America and by the correspondence among the geological features, fossils, and evidence of glaciers on these two separate continents. Could all the continents once have been assembled into a *supercontinent?* Wegener believed so, as had others before him. He, however, was the first to put together all the diverse evidence in support of that concept. He named the proposed land mass **Pangea,** meaning "all Earth." He proposed that Pangea began to break apart just after the beginning of the Mesozoic era, about 200 million years ago, and that the continents then slowly drifted into their current positions. This theory is called **continental drift.** Wegener's maps, produced in 1924, are remarkably similar to the best global paleogeographic reconstructions available today (Figure 7-1).

Although it was welcomed by paleontologists because of its consistency with the terrestrial fossil record, Wegener's theory of continental drift was not well received by the geophysicists of his day. The British scientist Sir Harold Jeffreys presented calculations in 1925 demonstrating that the continents could not possibly plow through the rigid sea floor, as the theory seemed to require. Other scientists were unconvinced because Wegener could not propose a physical mechanism for driving the motion of the continents. Indeed, many of Wegener's own calculations and proposed mechanisms were found to be in error and untenable.

The acceptance of Wegener's theory of continental drift awaited a better understanding of the structure and operation of the solid Earth. That understanding has come about largely as the result of geophysical exploration, which has revealed the complex anatomy of Earth's interior and has led to the theory of *seafloor spreading.*

Anatomy of Earth

Seismic Probing of Earth's Interior

For centuries geologists have been probing Earth's surface, cataloguing the variety of rocks exposed and studying the processes that have led to their formation. The study of material that has risen to the surface has given

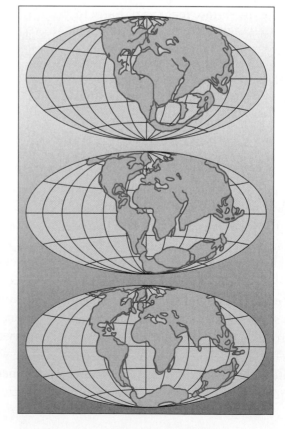

FIGURE 7-1

Wegener's reconstructions of the positions of the continents in the geological past. (From W.K. Hamblin and E.H. Christiansen, *Earth's Dynamic Systems*, 8/e, 1998. Reprinted by permission of Prentice Hall, Upper Saddle River, N.J.)

Late Carboniferous
(about 300 million
years ago)

Eocene
(about 50 million
years ago)

Pleistocene Glacial
(about 1 million
years ago)

some insight into the chemical and *mineralogical* composition of the shallow interior, but virtually everything we know about Earth's deep interior has been derived by indirect methods. Paramount among these methods is the science of *seismology,* the study of earthquakes and related phenomena.

Earthquakes. An **earthquake** is the sudden release of stored energy as a result of rapid movement between two blocks of rock. This energy radiates away from the earthquake in the form of vibrations. The site of energy release, known as the *focus,* can be anywhere from very near Earth's surface to as deep as 700 km below. This area, Earth's uppermost shell, is rigid; when it deforms, it does so *elastically.* This means that the material recovers its shape after the force that is tending to deform it is removed, unless it is deformed to the point of fracture and the original shape cannot be recovered. If there is differential movement on either side of the break, the fracture is called a *fault.* The *epicenter* of an earthquake is the position on Earth's surface directly above the focus (Figure 7-2).

Seismic Waves. Just as a person jumping into a swimming pool produces waves and ripples in the water, earthquakes create vibrations called **seismic waves** that ripple through Earth's interior, away from the earthquake's focus, as a result of that deformation. Two types of seismic waves are generated: body waves and surface waves. Both types spread outward from the focus. As we might expect, **body waves** travel through Earth's interior, whereas **surface waves** travel only across the surface. Surface

waves transmit earthquake energy along Earth's surface, where movement is unconstrained vertically. The motion is much like that of a water wave, easily seen by watching a cork bobbing up and down: Particles are displaced upward, backward, downward, and then forward in a circular motion. There is no net movement of the particles, but energy is transmitted away from the center. Body waves are categorized as either P waves or S waves on the basis of their mode of propagation through Earth. **P waves,** or primary waves, result from the compression of material in Earth's interior. The material is alternately compressed and, as the wave travels away, stretched. Thus, a P wave travels as a series of compressions and expansions in the overall direction of wave movement, similar to the way sound travels or to the response of a spring or a Slinky® (Figure 7-3a). **S waves,** which are also called secondary or shear waves, are transmitted as displacements perpendicular to the overall direction of wave travel. An analogy is the movement of a spring swung from side to side (Figure 7-3b). Although both solids and fluids can

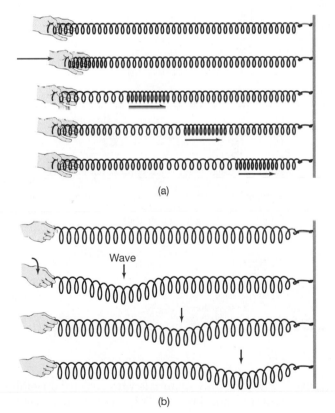

(a)

(b)

FIGURE 7-3

The movement of seismic body waves. (a) P waves alternately compress and expand materials, like the transfer of a compressive force imposed on a spring. (b) S waves move material from side to side, perpendicular to the overall direction of wave motion, just as a spring responds to repeated side-to-side shaking. (From J.P. Davidson, W.E. Reed, and P.M. Davis, *Exploring Earth: An Introduction to Physical Geology,* 1997. Reprinted by permission of Prentice Hall, Upper Saddle River, N.J.)

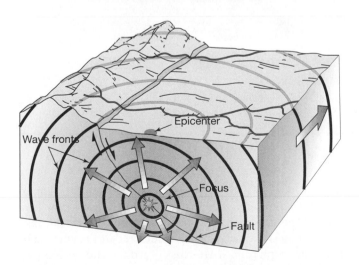

FIGURE 7-2

An earthquake's focus is located at depth. The point on the surface that directly overlies the focus is the epicenter. (From J.P. Davidson, W.E. Reed, and P.M. Davis, *Exploring Earth: An Introduction to Physical Geology,* 1997. Reprinted by permission of Prentice Hall, Upper Saddle River, N.J.)

A CLOSER LOOK
The Principle of the Seismograph

In the simplest sense, seismographs consist of a recording drum, which is anchored to bedrock and situated below a suspended weight with a pen attached. The weight has sufficient inertia to remain vertical as the ground (and the recording drum) vibrate beneath it (Box Figure 7-1). During an earthquake, the recording drum vibrates back and forth, allowing the seismologist to determine the amount of displacement that occurs as the seismic wave passes through the ground below.

The *amplitude,* or size, of seismic waves is related to the amount of energy released during an earthquake. Although seismologists now have more sophisticated ways of assessing earthquake amplitude, the public has grown accustomed to the use of the *Richter scale.* On the Richter scale, the lower limit of detection is arbitrarily assigned a value of 0. For every tenfold increase in seismic-wave amplitude (and approximately 30-fold increase in energy), the Richter scale increases by one unit. Thus, an earthquake with a magnitude

Support

Wire

Heavy mass

Rotating drum

Horizontal earth motion

BOX FIGURE 7-1

A seismograph. (From S. Judson and S.M. Richardson, *Earth: An Introduction to Geologic Change,* 1995. Reprinted by permission of Prentice Hall, Upper Saddle River, N.J.)

(a)

BOX FIGURE 7-2a

Aerial view of collapsed section of the Cypress viaduct on Interstate 880, Oakland, California, after the 1989 Loma Prieta earthquake.

transmit P waves, only materials with structural rigidity (solids) can transmit S waves. This fact has proved to be important in the characterization of Earth's interior, allowing us to identify particular regions as fluids rather than solids, as we will see later.

Eventually the path of all body waves intersects Earth's surface. There they can be detected and recorded by a *seismograph,* which is a sensitive instrument that detects slight vertical and horizontal displacements of Earth's surface (see the Box "A Closer Look: The Principle of the Seismograph"). The rate at which seismic

body waves travel through Earth depends on the properties of the material in Earth's interior. If we know how much time it takes waves to travel from the earthquake source to a site where they are detected at the surface by a seismograph, and if we can determine the path a particular seismic wave has taken, then we can calculate an average wave speed along that path. For a single earthquake event, a seismograph near the wave source records waves that traveled very shallowly through Earth, whereas a seismograph far from the source of the earthquake receives seismic waves that may have traveled through

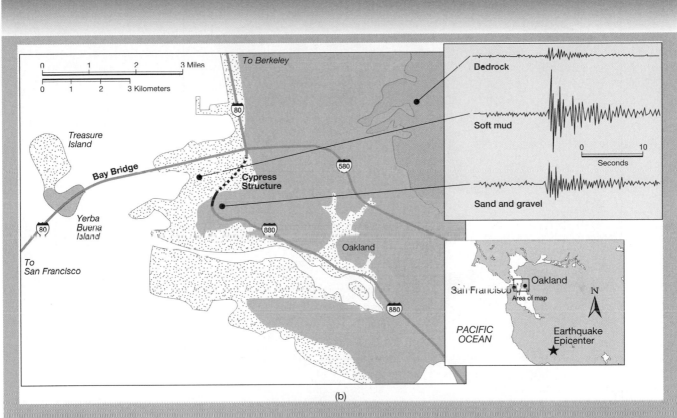

(b)

BOX FIGURE 7-2b

Seismograms of the Loma Prieta earthquake recorded by seismographs situated on bedrock, sand and gravel, and soft mud. Notice how the amplitude and duration of ground shaking is greater in soft mud than in bedrock. (After S. Hough, et al., 1990, *Nature,* 344, pp. 853–855; see also *http://geopubs.wr.usgs.gov/fact-sheet/fs176-95).*

of 5 on the Richter scale releases 30 times as much energy as one with a magnitude of 4.

Seismographs from around the world recorded the October 17, 1989, Loma Prieta earthquake with an epicenter 16 km north-east of Santa Cruz, California (Box Figure 7-2b). The earthquake registered 7.1 on the Richter scale and created the first major rupture along the San Andreas Fault since the famous 1906 San Francisco earthquake. Sixty-eight peo-

ple died as a result of the Loma Prieta event, and nearly four thousand were injured. Thousands of homes and businesses were damaged or destroyed, with an estimated dollar loss on the order of $7 billion. Structures built on mud and sand suffered larger amplitude and longer duration shaking than those on bedrock. One of the lasting images of the earthquake is the collapsed sections of the Cypress viaduct on Interstate 880 (Box Figure 7-2a).

Earth's center. Thus, by comparing several seismographic records from various places around the world for a particular event, we can construct a fairly detailed three-dimensional view of the paths along which seismic waves travel through Earth. This process is called *seismic tomography* (Figure 7-4).

Generalized Structure of Earth. The gross picture revealed by seismic imaging is of a layered Earth comprising a *crust,* a *mantle* (consisting of an upper mantle and a lower mantle), an *outer core,* and an *inner core* (Figure 7-5), defined on the basis of contrasts in seismic wave velocities. The shallowest of these transitions is the crust/ mantle boundary, first discovered by the Croatian

seismologist Andrija Mohorovičić who was investigating shock waves traveling through Earth from Zagreb (the former Yugoslavia) in the early 20th century. The boundary is now known as the *Mohorovičić discontinuity,* or the **Moho,** in his honor. This boundary is defined by a sharp increase in seismic wave velocities; P-wave velocities in the crust of around 5–6 km/sec increase to uppermost mantle velocities of about 8 km/sec. Beneath the continental crust, the depth to the Moho ranges from as much as 75 km in young mountain belts to 20 km in areas undergoing extension and crustal thinning. Beneath the oceanic crust, the Moho is at a nearly constant depth of around 7 km below the ocean floor.

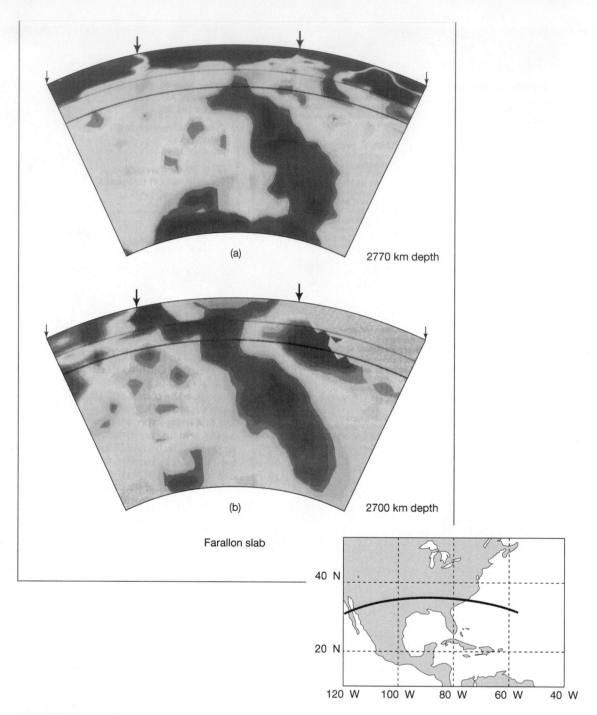

FIGURE 7-4

[See color section] Tomographic image of the mantle's (a) P-wave and (b) S-wave velocity variations underneath North America, along the transect line shown in the insert. Blue colors (see the full-color plate) indicate regions of fast seismic velocities, while reds indicate slow seismic velocities. The blue region cutting across the center of the diagram is the downgoing Farallon slab, which has been subducting under North America for 100 million years. (From W.K. Hamblin and E.H. Christiansen, *Earth's Dynamic Systems*, 8/e, 1998. Reprinted by permission of Prentice Hall, Upper Saddle River, N.J.)

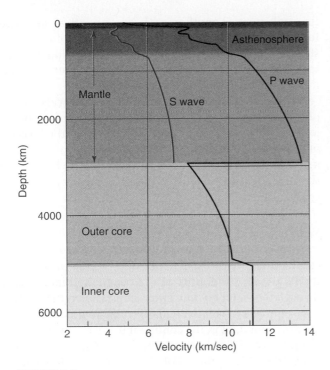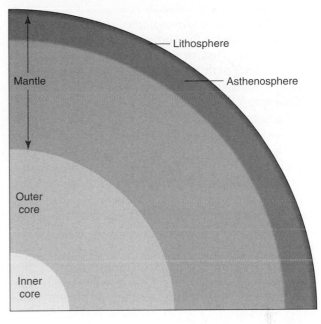

FIGURE 7-5

Internal structure of Earth, showing the distribution of seismic wave speeds. The speed changes with depth, defining the boundaries between the crust, mantle, outer core, and inner core. (From W.K. Hamblin and E.H. Christiansen, *Earth's Dynamic Systems*, 9/e, 2001. Reprinted by permission of Prentice Hall, Upper Saddle River, N.J.)

Below the Moho, the velocities of both P waves and S waves generally increase with depth through the mantle, although a *low-velocity zone* (or LVZ) exists at a depth of between 80 and 300 km (see Figure 7-5). Velocities then increase again through the transition zone between the upper and lower mantles, but the increase is not smooth: It occurs in a stepwise fashion, which indicates some sort of incremental change in mantle properties. Seismologists think that this change is related to a transformation of the minerals present to more compact, denser forms. Seismic velocities then increase more gradually with depth through the lower mantle.

The boundary between the lower mantle and the outer core, at a depth of 2900 km, is distinguished by a significant drop in P-wave velocities and by the disappearance of S waves. Because S waves travel only through solids, geophysicists infer that the outer core is a metallic fluid. At a depth of 5150 km, P-wave velocities increase markedly. This increase in velocity measurably deflects P waves from their anticipated path, confirming that the inner core is solid. Tremendous pressure at Earth's center is thought to be responsible for converting fluids to solids there.

This description of Earth's interior structure, although essentially correct, is greatly simplified. Seismic tomography reveals a more heterogeneous distribution of seismic wave velocities (see Figure 7-4), perhaps related to the exchange of materials between Earth's interior and its surface. Clearly Earth's interior is a dynamic place. Plate tectonic activity has altered the chemical and thermal structure of the mantle and may have operated in response to influences conveyed through the mantle from the core. Let us explore these heterogeneities in more detail, utilizing information not only from seismology but also from *petrology,* the study of the origin and evolution of the chemical and mineralogical compositions of rocks.

The Crust

Earth's uppermost layer, the crust, is not homogeneous; rather, it varies in both thickness and composition. The most pronounced differences are between the continental and oceanic crust. *Continental crust* underlies the continents, whereas *oceanic crust* underlies the ocean basins. As delineated by variations in Moho depth, continental crust is thicker than oceanic crust. It is also less dense and on average older. The two types of crust differ in chemical and mineralogical compositions as well. To understand these differences, we must first introduce a genetic classification of crustal rocks.

Igneous, Sedimentary, and Metamorphic Rocks. All rocks are composed of **minerals,** defined as naturally occurring inorganic solids of definite crystal structure and chemical composition. Geologists recognize three major types of rocks: *igneous, sedimentary,* and *metamorphic.* **Igneous rocks** form by the cooling and solidification of **magma,** which is molten, or liquid, rock. If the magma solidifies beneath Earth's surface, the rocks are called *intrusive* igneous rocks. **Granite** is a well-known intrusive rock. If the magma is carried to Earth's surface at a volcano and erupts, it is called *lava* and cools rapidly, forming *extrusive* igneous rocks. **Basalt** is an abundant type of extrusive igneous rock. Both intrusive and extrusive igneous rocks vary in composition, especially in the amount of the mineral quartz (SiO_2) they contain. *Felsic* igneous rocks (e.g., granite or *rhyolite,* its extrusive analogue) are quartz-rich, light-colored, and less dense than *mafic* igneous rocks (such as basalt or *gabbro,* its intrusive analogue).

Rocks (of any type) that are exposed at Earth's surface tend to decompose, or *weather,* into finer materials called **sediments**—layers of unconsolidated mineral matter that is transported by water, wind, or gravity. As new sediments are deposited on top of existing sediments, the underlying sediments become compacted, expelling water from the pores between sediment grains. The remaining pores may become filled with mineral cements precipitated from subsurface fluids. Compaction and cementation are processes that contribute to **lithification**, the conversion of loose sediments into thick, cohesive, layered deposits known as **sedimentary rocks**. Sedimentary rocks formed from sand-sized grains (>63 μm in diameter) are called *sandstone,* whereas rocks composed of finer grains are called *mudstones.* Finely layered mudstones are commonly called *shales.* Sedimentary rocks may instead form chemically or biochemically. For example, some marine organisms form a shell or skeleton by precipitating calcium carbonate ($CaCO_3$) minerals. When the organisms die, the shells and skeletons accumulate on the sea floor and ultimately lithify into sedimentary rocks known as *limestones.*

Rocks (of any type) that are exposed to high temperatures, high pressures, chemically active fluids, or any combination of these agents are transformed in mineralogical and chemical compositions. As long as no melting is involved, the altered material is said to have been *metamorphosed* and is called a **metamorphic rock.** (When melting occurs, the resulting rock is igneous.) *Marble* is metamorphosed limestone, and *slate* is metamorphosed shale.

Major Rock-Forming Minerals. Both continental and oceanic crust are composed primarily of rocks made of **silicate minerals**—that is, minerals rich in silicon and oxygen. *Feldspars* are the most abundant minerals in the continental crust. They are silicate minerals with aluminum, sodium, calcium, and potassium in their structures. Quartz also is an abundant silicate mineral in the upper continental crust. Together with certain other minerals, feldspars and quartz form granite and *granodiorite* (a slightly less quartz-rich rock), the hard, erosion-resistant rocks that create the high peaks of many mountain ranges, including the Sierra Nevadas of California.

Magnesium and iron-rich silicate minerals such as *olivine* and *pyroxene* characterize the basalts of the oceanic crust. The Hawaiian Islands are composed of basaltic materials. The high relative abundance of these dense minerals in the oceanic crust accounts for the higher density of mafic oceanic crust in relation to felsic continental crust.

Sedimentary Cover. Sediments and sedimentary rocks overlie the basalts and granites/diorites of the oceanic and continental crust, respectively. The source of these materials is easy to identify in the ocean: Sediments settle through the water column and accumulate on the sea floor as a sequence of relatively flat layers. Sedimentary rocks are also abundant in the continental crust, however. Some sedimentary rocks accumulated in basins on the continents themselves, but most were originally deposited as sediments on the sea floor and may have been deeply buried. Subsequent tectonic activity transported these sediments onto the continents. There they became exposed in mountain belts where the once-flat layering has typically become highly deformed through uplift. The oldest parts of the continents were once sedimentary rocks that have become significantly deformed and altered through many cycles of tectonic activity and metamorphism.

The Mantle

Beneath the crust is the mantle, which extends from the Moho to the top of the fluid outer core. The exact structure and composition of the mantle is a hotly debated topic among geologists, largely because the mantle is very difficult to observe. The samples of deeper mantle material that are available at the surface were brought up during rare geological events, such as the formation of kimberlites. *Kimberlites* are long, pipe-shaped igneous bodies that were emplaced after having passed rapidly from the upper mantle to the near-surface. They are remarkable in that they contain diamonds that formed under the high-pressure conditions of the mantle.

Most of what we know about the mantle is inferred from seismology. The velocity structure so determined indicates that the mantle is relatively uniform in composition and formed of silicate minerals. However, as depth increases so too do pressures and temperatures, causing changes in the structural and mineralogical composition of these silicates. These changes affect seismic wave velocities.

The recognition of the seismic low-velocity zone from depths of 80–300 km in the upper mantle proves to be an important link in plate tectonic theory. Most geologists accept that the low seismic wave velocities are the result of the presence of some molten rock at this depth. There need not be much; the data can be explained if only 1% or less of the rock is molten. Yet the small amount of melt present is critical, because it allows the crust and upper mantle to move relative to the underlying mantle—a basic tenet of plate tectonics.

A transition zone of rapidly increasing seismic wave velocities from depths of 370–650 km separates the upper and lower mantles. Geologists disagree on the reason for the transition zone. Many conclude from seismic evidence and theory that the transition zone is the consequence of mineralogical changes, whereas others conclude that differences in elemental composition are the cause. We return to this controversy later in our discussion of mantle convection.

The Core

The abundance of most elements in the crust and mantle can be explained by using meteorite compositions as a basis for comparison. Meteorites are thought to be fragments of larger *planetesimals* (bodies tens to hundreds of kilometers in diameter), many of which now reside in the asteroid belt of the solar system. Their parent bodies formed at the same time that the Sun did from the solar nebula. (This process is discussed further in Chapter 10.) A particular class of meteorites, the *carbonaceous chondrites*, is made of material that is thought to be essentially unaltered from the original nebular composition. Thus, carbonaceous chondrites are thought to be representative of the average abundances of elements in the solar system, including silicon, sodium, magnesium, calcium, and oxygen—all the basic rock-forming elements. Earth contains much less water and other *volatile* (easily vaporized) *compounds* as a consequence of its high-temperature formation, but we expect that its composition should otherwise be chondritic. By comparison with chondritic meteorites, the mantle and crust are significantly depleted in iron. This deficiency is made up for in the core, which is believed to be dominated by iron, along with small amounts (about 6%) of nickel, and approximately 8–10% of some unknown light element, which could be oxygen, sulfur, hydrogen, or silicon. (The light element has to be there; otherwise, the core would be even denser than it is observed to be.) The iron–nickel core is much denser than the overlying mantle, which, together with changes in seismic wave velocity, explains why seismic waves reflect off the core–mantle boundary.

Although the core is far removed from Earth's surface, it affects surface conditions because it is the source of Earth's magnetic field. Like a simple bar magnet, Earth

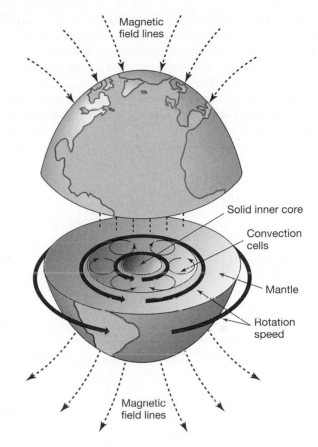

FIGURE 7-6

Earth's magnetic field is like that of a bar magnet, except that this field is generated electromagnetically by convection in the outer core. Dashed arrows indicate lines of force of the magnetic field. (From W.K. Hamblin and E.H. Christiansen, *Earth's Dynamic Systems*, 8/e, 1998. Reprinted by permission of Prentice Hall, Upper Saddle River, N.J.)

has a magnetic field with north and south poles. Unlike a bar magnet, though, a magnetic dynamo (Figure 7-6) generates Earth's magnetic field. A **magnetic dynamo** is a mechanism that transforms energy from fluid motions (convection) into electrical currents that create a magnetic field. In a dynamo, the convecting fluid (liquid iron, in Earth's case) must be a conductor of electricity. The outer core convects because a temperature gradient is established across it by heat loss from the solid, inner core.

Why does the liquid outer core convect? As we discussed in Chapters 3 and 4, thermal convection occurs when a fluid is heated from below. Our example was the troposphere, which is heated by the warm Earth's surface. The ultimate source of this energy is sunlight. No sunlight makes it down to Earth's core, of course, so we must look for a different energy source there. One possibility is radioactive decay (which we cover later in this chapter), but the elements responsible for this heat source are preferentially concentrated in the crust and mantle rather than in the core. More likely, the energy required to drive convection in the outer core is derived from the gradual growth of the inner core. As Earth's interior cools, liquid iron slowly freezes out to form particles of solid iron. This "freezing" process releases heat, just as the freezing of water to form ice does. The newly formed particles of solid iron also heat the outer core frictionally as they settle down to join the inner core. The heat released by both of these processes is thought to be what powers outer core convection and, thus, the magnetic dynamo.

The Theory of Plate Tectonics

Sea-Floor Spreading

As we discussed in the introduction to this chapter, the theory of continental drift lay more or less dormant for several decades after the publication of Wegener's ideas, largely because of the lack of acceptance by geophysicists. It is ironic that the resurgence of interest in Wegener's theory was the result of information obtained in the 1960s by geophysicists investigating the topographic and magnetic features of the sea floor. (*Topography* refers to the configuration of a surface, in particular the position and elevation of its features.) During and just after World War II, an intensive period of mapping took place that revealed intriguing details of the sea floor (Figure 7-7). This work gave the world the first evidence of chains of subsea volcanic mountains running down the centers of the ocean basins; we now call these features **mid-ocean ridges.** A *rift*, or narrow valley, runs down the center of such ridges. In the early 1960s, scientists proposed that these linear volcanic chains represent new sea floor that is extruded along the mid-ocean ridges. Once it forms, the new sea

FIGURE 7-7

Ocean sea-floor topography inferred from high-resolution satellite measurements of anomalies in the gravitational field. Note transform faults extending across the Atlantic, especially near the equator and across the eastern Pacific. (From the National Oceanic and Atmospheric Administration [NOAA].)

floor spreads to the sides of the ridges, generating the central rift, and is replaced at the *ridge axis*—the middle of the rift—by even younger new sea floor. This process was named **sea-floor spreading.** Having just crystallized from the magma, newly formed sea floor at sea-floor spreading centers is hot and expanded. As it spreads to either side of the plate boundary at the ridge axis, the material cools and contracts, and the sea floor subsides. This process occurs symmetrically across the axis of spreading and so creates symmetrical undersea mountain belts some 1000–4000 km wide that rise 2–3 km from the sea floor.

The real key to the origin of these features came from a better understanding of the magnetic characteristics of the sea floor. The *magnetic* **polarity** is the geographic orientation of the North and South Poles. From studies of volcanic rocks extruded on land, scientists knew that magnetic polarity has flipped numerous times in Earth's history. The reasons are not well understood, but have to do with the complex behavior of the convecting liquid outer core. As the lava that formed these volcanic rocks cooled beyond a critical temperature of about 570°C (called the *Curie point*), the rocks became magnetized in the direction of Earth's magnetic field at the time of cooling. More-recent flows record switches in the magnetic field, with the North magnetic pole roughly coincident with the South geographic pole, and vice versa (Figure 7-8). Radiometric dating (see below) provided the ages of volcanic flows.

Because the basaltic rocks of the sea floor were known to be of volcanic origin, in the late 1950s oceanographic expeditions were designed to map the magnetic character of the sea floor. As a result, a startling observation was made: The sea floor has a striped magnetic pattern, with the stripes running essentially parallel to the mid-ocean ridges. The stripes reflect alternating bands of polarity. Today's magnetic polarity is considered to be

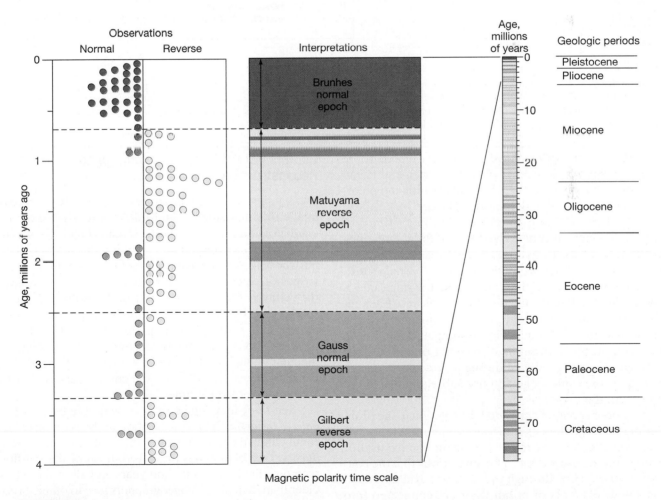

FIGURE 7-8

Magnetic reversals as recorded in volcanic rocks preserved on land for the last 75 million years. The last 4 million years of reversals are highlighted at the left. The pattern of change over a few million years is distinctive, and can be used as a signature for establishing the age of a sequence of rocks elsewhere in the world. (From W.K. Hamblin and E.H. Christiansen, *Earth's Dynamic Systems*, 9/e, 2001. Reprinted by permission of Prentice Hall, Upper Saddle River, N.J.)

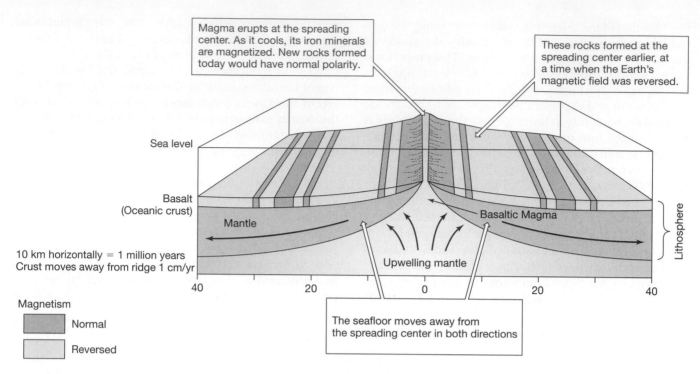

Sea level

Basalt
(Oceanic crust)

Mantle

Basaltic Magma

Lithosphere

10 km horizontally = 1 million years
Crust moves away from ridge 1 cm/yr

Upwelling mantle

40   20   0   20   40

Magma erupts at the spreading center. As it cools, its iron minerals are magnetized. New rocks formed today would have normal polarity.

These rocks formed at the spreading center earlier, at a time when the Earth's magnetic field was reversed.

The seafloor moves away from the spreading center in both directions

Magnetism

Normal

Reversed

FIGURE 7-9

Magnetic stripes develop as new crust is added to the ocean floor at mid-ocean ridges and cools, becoming magnetized according to the magnetic field that exists at the time. As this material moves away from the axis, new sea floor is created, and its magnetization may be reversed if Earth's magnetization has reversed polarity in the intervening time. (From S. Judson and S.M. Richardson, *Earth: An Introduction to Geologic Change*, 1995. Reprinted by permission of Prentice Hall, Upper Saddle River, N.J.)

normal, and the opposite polarity is considered to be *reversed* (Figure 7-8). Stripes on one side of a mid-ocean ridge were matched to others of similar width and polarity on the opposite side of the ridge (Figure 7-9).

Geophysicists soon concluded that this pattern of magnetic stripes must be caused by sea-floor spreading. At the time of its formation, the new sea floor locks in the contemporaneous magnetic direction. It is then transported in opposite directions away from the ridge axis as new, molten rock is extruded from the volcano. Each reversal of the magnetic field produces a magnetic stripe on the sea floor. New material forming at a mid-ocean ridge is thereby differentiated from the older sea-floor material that was produced during a previous magnetic interval at the same ridge and has subsequently drifted away from the ridge axis.

Sea-floor spreading provided the solution to the problem plaguing geologists interested in continental drift since the days of Wegener: How could the continents drift through the rigid sea floor? The answer was that the continents do *not* plow through the sea floor. Rather, continents and segments of ocean floor are connected into plates that continuously move away from one another at mid-ocean ridges. Geologists have used a variety of types of evidence to reconstruct this drift of the continents throughout Earth's history, largely confirming the notion of Pangea proposed by Wegener decades earlier.

Continental Drift and Paleogeographic Reconstructions

The symmetrical magnetic stripes on the sea floor increase in age to either side of the Mid-Atlantic Ridge, recording the opening of the Atlantic Ocean (Figure 7-10). In a sense, we can reverse time by rolling the sea floor back into the Mid-Atlantic Ridge, bringing together once again stripes of equal age at the ridge axis. In this technique, the Atlantic Ocean closes first in the Southern Hemisphere and then in the Northern Hemisphere, as South America slips into place alongside Africa. The close fit between the two continents, especially at the edges of their continental shelves, was part of what convinced Wegener in the early 1900s that the continents have drifted apart over geological time.

Sea-floor magnetic stripes provide the best tool for *paleogeography*, the reconstruction of the positions of the continents in the past. However, the tectonic process of *subduction* (see below) has destroyed much of the sea-floor record of the past 200 million years and all of the record older than that. So, paleogeographers turn to other sorts of evidence to determine ancient continental positions.

Sedimentary rocks prove very useful in this regard. Glacial deposits generally form at high latitudes (poleward of about 45°), so we assume that glacial sediments indicate high *paleolatitudes*—that is, the latitudes at which

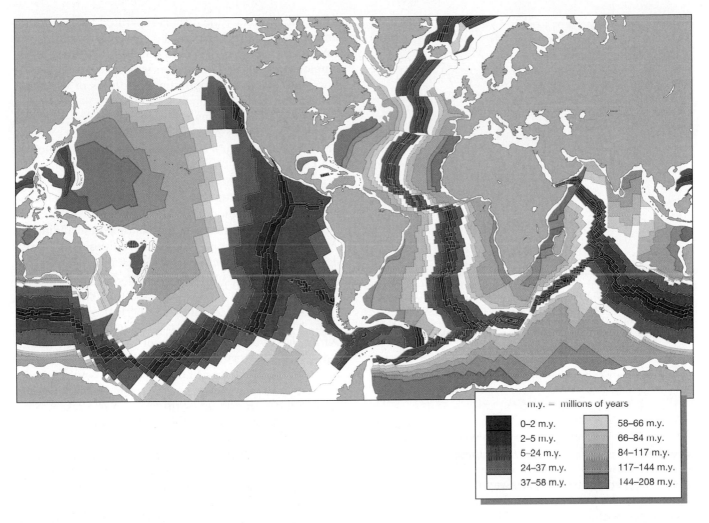

m.y. = millions of years

0–2 m.y.	58–66 m.y.
2–5 m.y.	66–84 m.y.
5–24 m.y.	84–117 m.y.
24–37 m.y.	117–144 m.y.
37–58 m.y.	144–208 m.y.

FIGURE 7-10

[See color section] The age of the ocean floor is shown as bands of different color on the basis of the magnetic striping developed during sea-floor spreading. The youngest ocean floor is near the mid-ocean ridge, while the oldest is furthest away. (From R.W. Christopherson, *Geosystems: An Introduction to Physical Geography*, 3/e, 1997. Reprinted by permission of Prentice Hall, Upper Saddle River, N.J.)

the rocks formed. (We discuss an important exception to this rule in Chapter 12—the so-called "Snowball Earth" episodes of the Paleoproterozoic and Neoproterozoic, for which it appears that ice sheets extended into tropical latitudes.) Similarly, because coral reefs are located in the tropics today, we assume that reef limestones indicate tropical paleolatitudes. Salt deposits indicate subtropical paleolatitudes because they form preferentially in arid regions underlying the descending branches of the tropical Hadley cells. In addition, similar fossils on two continents indicate that the continents were in close proximity, or joined, at the time the organisms lived, allowing their migration. One must be cautious, however, in applying these paleolatitude indicators, because they are based on the assumption that the present is the key to the past. There may be times, however, when this assumption does not hold. Thus, geologists are motivated to look for more reliable paleolatitude indicators. The least ambiguous pa-

leolatitude indicator comes from magnetism in rocks. By measuring the angle of the preserved magnetic field with respect to the sedimentary layering of the rocks (which is presumed to have originally been horizontal), geologists are able to estimate the angle of the magnetic field with respect to the horizontal. This angle gives the latitude at which the sedimentary deposit formed (Figure 7-11). Rocks in which the magnetic field is nearly parallel to the bedding plane must have formed near the equator; rocks in which the magnetic field is perpendicular to the bedding plane must have formed near the poles. These inferences, of course, are based on the assumption that Earth's magnetic field has always had two poles (North and South) and that the magnetic poles have always been approximately aligned with the geographic poles. One might rightfully point out that this is also a case of using the present to interpret the past. However, in this case, the assumption seems to be well-founded.

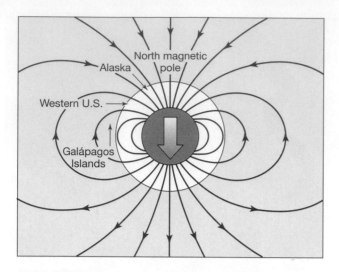

FIGURE 7-11

Earth's magnetic field, indicating that the angle of the field lines with respect to Earth's surface varies from horizontal at the equator to vertical at the poles. Paleogeographers use this feature to determine paleolatitudes of rocks, on the basis of the magnetic field orientation they acquired at the time of their formation. (From W.K. Hamblin and E.H. Christiansen, *Earth's Dynamic Systems*, 8/e, 1998. Reprinted by permission of Prentice Hall, Upper Saddle River, N.J.)

During the past several million years when the continents could not have drifted very far, the magnetic poles have remained close to the geographic poles. Magnetic dynamo theory also suggests that, far away from its source (the outer core), the magnetic field should always be more or less aligned with the planet's spin axis.

Unfortunately, most of the tools paleogeographers use constrain only the latitude, not the longitude, of the continents. Longitudinal positioning is much more difficult. Similarity of fossil assemblages on adjacent continents is taken to indicate proximity, whereas large differences between the types of fossils found on two con-

tinents imply that wide oceans separated the continents and prevented the dispersal of organisms. However, features other than oceans, mountains and deserts for example, can also prevent dispersal. Thus, there are large uncertainties in the longitudinal positions of the continents prior to 200 million years ago.

The sequence of maps shown in Figure 7-12 shows how the continents are thought to have drifted over the past 500 million years, based on all available evidence. In the Cambrian and Ordovician Periods (540–440 million years ago), the continents were widely dispersed along the equator. Over the next 300 million years, the continents drifted together and collided. The collisions created large, Himalayan-style mountain ranges, including the Appalachians of the eastern United States. By 300 million years ago, the continents were assembled into the gigantic supercontinent Pangea, centered on the equator. Pangea eventually began to disassemble about 200 million years ago. The Atlantic Ocean formed as a rifting apart of continents first between North America and Africa, next between South America and Africa, and finally between North America and Europe. By 120 million years ago, Africa, Antarctica, India, and Australia had begun their separate paths. North America rotated and drifted from the low latitudes into its present position. Most impressive was the long journey of India as it detached from Antarctica and ultimately collided with Asia some 50 million years ago. We will see later on (Chapter 12) that this ongoing collision between India and Asia may be responsible, at least in part, for our present, relatively cool global climate.

The paleomagnetic and geological evidence of continental positions before 600 million years ago is sparse. Available paleomagnetic data seem to indicate that between about 900 and 600 million years ago the continents were assembled into another Pangea-like supercontinent. Prior to this time the data are so scarce that paleogeographic reconstructions are not currently feasible.

FIGURE 7-12

Paleogeographic reconstructions from the Ordovician to the latest Cretaceous. (From R.W. Christopherson, *Geosystems: An Introduction to Physical Geography*, 3/e, 1997. Reprinted by permission of Prentice Hall, Upper Saddle River, N.J.)

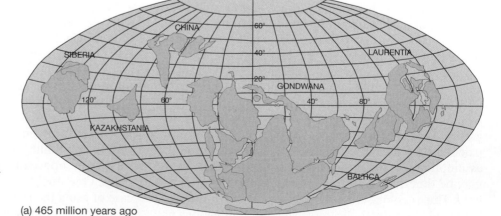

(a) 465 million years ago

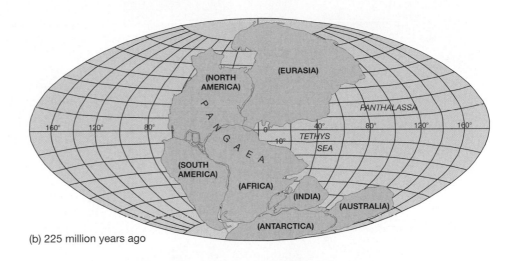

(b) 225 million years ago

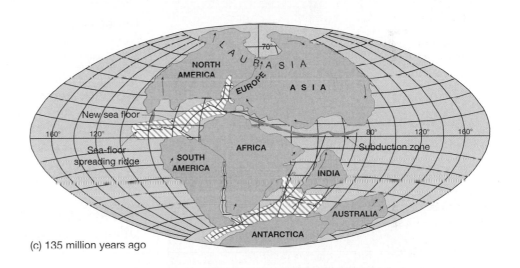

(c) 135 million years ago

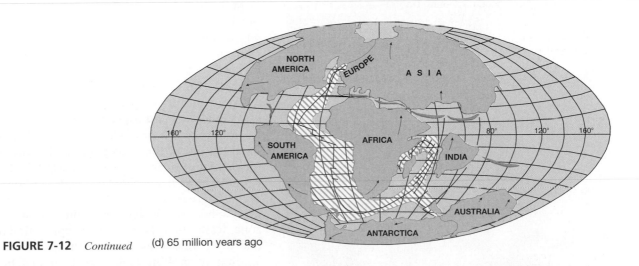

FIGURE 7-12   *Continued*   (d) 65 million years ago

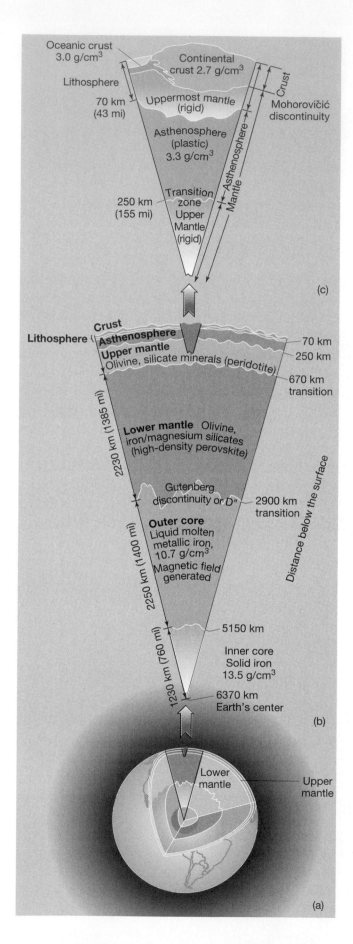

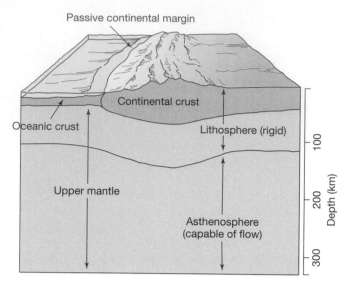

FIGURE 7-13

(a) Internal structure of Earth, comparing the traditional classification by seismic wave velocities with the plate tectonic classification by material strength. (From R.W. Christopherson, *Geosystems: An Introduction to Physical Geography*, 3/e, 1997. Reprinted by permission of Prentice Hall, Upper Saddle River, N.J.) (b) A cross-section of the upper mantle and crust showing the relative positions of the lithosphere (crust plus uppermost mantle) and asthenosphere. (From J.P. Davidson, W.E. Reed, and P.M. Davis, *Exploring Earth: An Introduction to Physical Geology*, 1997. Reprinted by permission of Prentice Hall, Upper Saddle River, N.J.)

New Structural Categories: Lithosphere and Asthenosphere

As Wegener learned long ago, the conventional separation of the solid Earth into the core, the mantle, and the crust on the basis of seismic wave velocities is inadequate in view of plate tectonic theory. To explain the drift of continents, the mobile plates need to be distinguished from the lubricating layer below. To do so, the mantle and crust are best recategorized according to material strength (Figure 7-13). The plates extend through the crust and into the uppermost mantle; we call this outermost sphere the **lithosphere.** The upper, crustal part of the lithosphere is *brittle*, that is, it fractures in response to stress. Below the lithosphere is the **asthenosphere,** a region of the upper mantle that acts more like a fluid than a solid. The asthenosphere is *ductile*—it flows plastically, or deforms easily, in response to stress. The top of the asthenosphere is coincident with the mantle's low-velocity zone. Recall that the preferred explanation for low seismic wave velocities there is the presence of a small amount of molten rock. The asthenosphere extends through the mantle's transition zone to a depth of around 700 km. Below this depth, the lower mantle is thought to be much less ductile because of the effects of very high pressure.

Plates and Plate Boundaries

According to the theory of plate tectonics, the lithosphere is divided into about 20 rigid plates (Figure 7-14). The crustal portions of some plates are entirely oceanic, whereas other plates include both oceanic and continental crusts. *Oceanic lithosphere* describes a plate that is topped by oceanic crust; *continental lithosphere* refers to a portion of a plate topped by continental crust.

We now know that tectonic activity (such as earthquakes or volcanism) is concentrated at plate boundaries; there is little activity within a plate. This activity is the result of plate motion: The plates move relative to each other at average speeds of a few centimeters per year. As a result of friction between the plates, there are alternating periods of stasis (during which stresses build) and periods of movement (when they are released) both at the plate boundary and near the surface. (Seismic and satellite measurements indicate that at greater depths or farther from the plate boundary, the motions are more continuous.) After a period of stasis, pent-up energy is released suddenly as the plates jump past each other, causing earthquakes. As predicted, the distribution of earthquakes at Earth's surface follows plate boundaries quite closely (compare Figures 7-14 and 7-15).

There are three types of plate boundaries (or margins): divergent, convergent, and transform (Figure 7-16). At *divergent margins,* lithospheric plates are moving away from each other. At *convergent margins,* plates are moving toward each other. At *transform margins,* plates are slipping past each other. Each boundary type is represented differently at Earth's surface. In other words, each type of plate margin is reflected in distinctive surface features: mid-ocean ridges, deep-sea trenches, and transform faults, respectively.

Divergent Margins

Divergent margins are regions where stresses are pulling apart the lithosphere. The mid-ocean ridges, described above, represent most of the divergent plate boundaries on Earth. Divergent boundaries that occur on land represent sites of continental fragmentation, or *rifting,* where the continental crust stretches. Tensional forces pull the continent apart. In the process, faulting occurs and flat-bottomed valleys called *rift valleys* form. During the breakup of Pangea some 200 million years ago, these types of plate boundaries were much more common than they are today. Today the rift valleys of East Africa, along with the Gulf of Aden and Red Sea to the north, are our best example of a divergent continental boundary in the making (Figure 7-17). If one studies this area in more de-

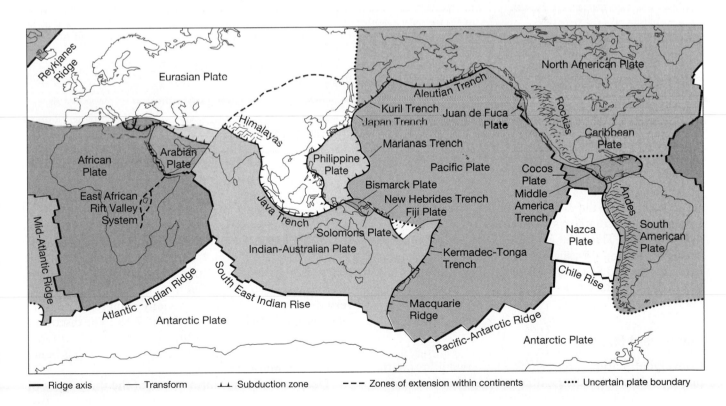

FIGURE 7-14

The lithosphere is divided into rigid plates. (From S. Judson and S.M. Richardson, *Earth: An Introduction to Geologic Change*, 1995. Reprinted by permission of Prentice Hall, Upper Saddle River, N.J.)

A CLOSER LOOK
Deep-Sea Life at Mid-Ocean Ridge Vents

The axial portion of a mid-ocean ridge is marked by volcanic activity, earthquakes with a shallow focus, and the venting of hot fluids rich in dissolved metals and hydrogen sulfide. Seawater drawn into the ridge along its flanks flows through cracks in the oceanic crust and is expelled through vents in the axis. Along the way, the seawater is heated, and chemical reactions with basalt alter its composition: Magnesium is removed and sulfate is reduced to sulfide, and calcium and trace metals are added. The circulation of seawater through the mid-ocean ridge also alters the chemical composition of the oceans. While exiting through the vents, iron sulfide minerals precipitate from the altered seawater and give the plumes a black coloration. For this reason, the vents through which the fluids exit are called *black smokers* (Box Figure 7-3a). They are also known as *hydrothermal vents*, because they release heated seawater.

The fluids released by black smokers sustain a unique community of organisms. These organisms synthesize organic matter with the help of bacteria that carry out **chemosynthesis** rather than photosynthesis. In other words, these bacteria utilize energy from inorganic matter—energy released during chemical reactions between seawater

(a)   (b)

BOX FIGURE 7-3

[See color section] Abundant and bizarre life thriving under the harsh conditions of the deep-sea floor, in the vicinity of hydrothermal venting. (a) A black smoker chimney is shown spewing out sulfide-rich solutions that provide the energy source for this food chain. (From Dudley Foster, Woods Hole Oceanographic Institution.) (b) Tube worms, crabs, and other organisms can be seen. (From Richard A. Lutz and Michael J. Kennish, *Reviews of Geophysics*, 31, p. 210, August 1993. Copyright © 1993 by the American Geophysical Union.)

and hydrogen sulfide. Chemosynthetic bacteria do not use the energy of sunlight. Feeding off these bacteria are unusual species of clams, crabs, and giant red and white tube worms (Box Figure 7-3b).

---

tail, one finds that the divergence of the African and Arabian plates has created a system of rifts that radiate from a central point. This point is referred to as a *triple junction*. The East African rift valleys are at an earlier stage of rifting than the Red Sea and Gulf of Aden, where spreading has progressed to the point that new ocean basins have been formed.

Convergent Margins

Convergent margins are regions where two lithospheric plates are forced together. Although there has been some spirited controversy in the past, most geologists now are convinced that Earth is not increasing in size. (Like Wegener, those who have suggested that Earth changes its size have no physical explanation for how this could happen. Wegener, though, turned out to be right!) But if new sea floor is produced at mid-ocean ridges, then what happens to the old sea floor? The mapping efforts that followed World War II revealed deep basins in addition to mid-ocean ridges. Deep-sea trenches are long, narrow, very deep basins that are especially common along the margins of the Pacific Ocean (Figure 7-18). The discovery of these trenches provided the answer: The sea floor is consumed at deep-sea trenches about as fast as it is being produced at mid-ocean ridges.

Deep-sea trenches form at two of the three types of convergent plate boundaries: those that involve two oceanic plates and those that involve an oceanic and a continental plate. A third type of convergent margin

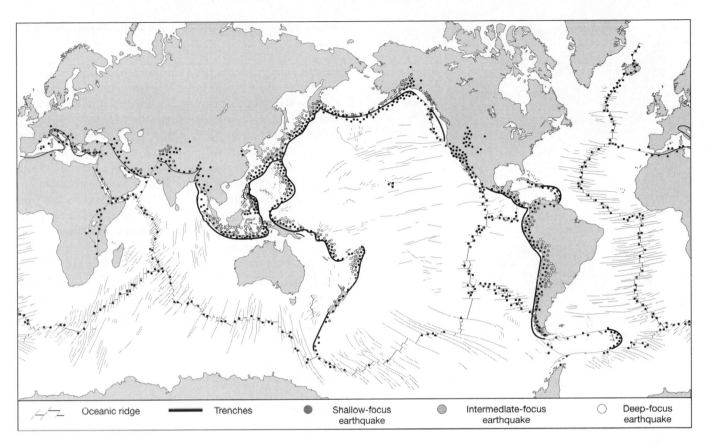

| ⚡ Oceanic ridge | ▬▬▬ Trenches | ● Shallow-focus earthquake | ◑ Intermediate-focus earthquake | ○ Deep-focus earthquake |

FIGURE 7-15

[See color section] Distribution of earthquakes of shallow, intermediate, or deep focus. Deep-focus earthquakes occur only at subduction zones. (From W.K. Hamblin and E.H. Christiansen, *Earth's Dynamic Systems*, 8/e, 1998. Reprinted by permission of Prentice Hall, Upper Saddle River, N.J.)

forms the world's highest mountains rather than the deepest trenches; these involve the collision of two continental plates. Let us explore each of these cases.

Oceanic/Continental and Oceanic/Oceanic Convergent Margins. Recall that the upper portion of a lithospheric plate (the crustal section) is brittle. When oceanic plates collide, the leading edge of one plate sinks entirely beneath the other. When the leading edge of one of the plates at a convergent margin is oceanic lithosphere and the leading edge of the other plate is continental lithosphere, the denser oceanic lithosphere sinks beneath the less-dense continental plate (Figure 7-19a). The sinking of an oceanic plate at a convergent margin is called **subduction,** and the entire region is called a *subduction zone*. The downgoing plate, called a **slab,** subsides into the mantle. In so doing, the plate bends, creating deep, linear depressions at the surface—deep-sea trenches. These trenches are the deepest parts of the oceans. Friction between the downgoing plate and the overriding plate generates a substantial amount of seismic activity near the surface (in the upper 60–100 km); other forces generate earthquakes with-

in the subducting slab at greater depths. The earthquake foci deepen as the distance from the trench toward the continent increases (Figure 7-15). Inland from the trench, melting at the top of the subducting slab of lithosphere produces igneous activity at the surface. This activity forms a range of volcanic mountains called a *volcanic arc*.

When two oceanic plates collide at a convergent margin, one subducts beneath the other (Figure 7-19b). Similar to what occurs in an oceanic–continental collision, a range of volcanic mountains forms to one side of the trench, but in this case the volcanoes rise up along the sea floor rather than on land. If they reach the ocean surface, they produce volcanic *island arcs*. The Marianas Islands, off the coast of the Philippine Islands, formed in this way. The Marianas Trench is the deepest trench of all, more than 10.5 km below sea level.

As the sea floor spreads from its place of origin at a mid-ocean ridge to its place of destruction at a subduction zone, sediment settling through the overlying water accumulates on the sea floor. Like a conveyor belt, the convergent motion of the plates in the subduction zone carries this sediment toward the trench. There the sedi-

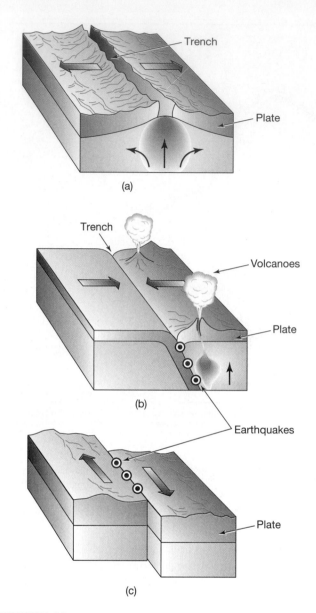

FIGURE 7-16

The three types of plate boundaries: (a) divergent; (b) convergent; and (c) transform fault. (From J.P. Davidson, W.E. Reed, and P.M. Davis, *Exploring Earth: An Introduction to Physical Geology*, 1997. Reprinted by permission of Prentice Hall, Upper Saddle River, N.J.)

ment may be scraped off by the opposing plate, forming wedges of deformed sediment. The rest of the sediment remains attached to the oceanic plate. The fate of this sediment is an area of active research. Part of it appears to become *underplated*—that is, attached to the base of the overlying plate—whereas some of it is carried into the asthenosphere. In the asthenosphere, it undergoes *dehydration* (loss of water) and *decarbonation* (loss of carbon) to the surrounding mantle, as well as a host of mineralogical transformations. These reactions prove to be very important to the global recycling of elements such as carbon, because volcanoes that form in such subduction

zones derive carbon dioxide gas from this sedimentary source.

Continental/Continental Convergent Margins. When two continental plates meet at a convergent margin, the continents collide abruptly. Because continental crust is too buoyant to be subducted, continental collision results in the separation of the crustal portion of the lithospheric plate from the mantle portion below. Subduction of the mantle portion of one plate may occur while the continental crust on both plates becomes compressed and crumpled. As a consequence, tall mountain belts and high plateaus form (Figure 7-19c). Around 50 million years ago, India collided with Asia. The collision between these two segments of continental lithosphere led to massive deformation and uplift of the continents. The tall peaks of the Himalayas and the uniformly high Tibetan Plateau bear firm testament to the awesome energetics of this collision. Older mountain belts (such as the Appalachians) are the products of collisional tectonics that occurred hundreds of millions of years ago. Subsequent erosion has reduced what were once majestic mountains into the more modest ridges we observe today.

Transform Margins

When the relative motion along a plate boundary is parallel to the boundary, lithosphere is neither created (extruded) nor destroyed (subducted); the plates merely slip past one another at a fault. Faults that form boundary-parallel margins are known as **transform faults.** The San Andreas Fault of California, which marks a segment of the boundary between the North American and Pacific plates, is a transform fault (Figure 7-20). Baja California and southern California (including Los Angeles) are moving slowly northward relative to the rest of California. In 50 million years or so, San Francisco and Los Angeles will be side by side, and beyond that time Los Angeles will actually be north of San Francisco. Geologists on the West Coast joke that California politics will at this time become completely reversed, with the northern part of the state being more conservative than the south.

Transform plate boundaries occur in oceanic settings as well. The jagged shape of parts of the mid-ocean ridge system is caused by offsets between ridge segments created by transform faulting (Figure 7-18). The Mid-Atlantic Ridge shows this type of behavior near the equator.

Overview of Plate Interactions

Figure 7-21 provides an overview of the types of plate interactions and the surface features they generate. The production of new oceanic lithosphere at mid-ocean ridges is matched by the destruction of older oceanic lithosphere at subduction zones, which are manifested at the surface by deep-sea trenches. Part of the sedimentary

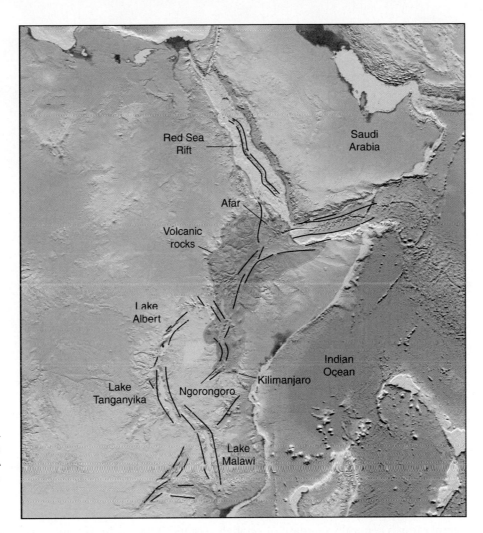

FIGURE 7-17

The rift valleys of East Africa, where Africa is being uparched and pulled apart. If spreading continues, the rift system may evolve into an elongate sea like the Red Sea to the north. (From W.K. Hamblin and E.H. Christiansen, *Earth's Dynamic Systems*, 8/e, 2001. Reprinted by permission of Prentice Hall, Upper Saddle River, N.J.)

layer riding atop the oceanic plate is incorporated into the overriding plate (oceanic or continental), and the rest is subducted into the mantle.

As Figure 7-21 shows, some lithospheric plates consist of oceanic lithosphere welded to continental lithosphere. The ocean–continent boundary is not a plate boundary at all. The North American plate, for example, is made up of the North American continent and oceanic lithosphere to the east. The North American plate continues beyond the continent–ocean lithospheric boundary to the axis of the Mid-Atlantic Ridge (the mid-ocean ridge down the center of the Atlantic Ocean). In such situations, the ocean widens as the continents drift away from the mid-ocean ridge; the continental margin is referred to as passive. *Passive continental margins* consist of a broad, gently seaward-dipping continental shelf that gives way to the more steeply dipping continental slope. The continental lithosphere is welded to the oceanic lithosphere beneath the continental shelf. In other cases, such as along much of the Pacific Ocean, the ocean–continent boundary is a convergent margin and is therefore a site of

subduction; such margins are said to be active. On *active continental margins,* including that off the Pacific coast of the northwestern United States, the continental shelf is narrow.

The oceanic ridge is offset in numerous places by transform faults. The relative motion across these faults is parallel to the plate boundary. Transform faults generate considerable amounts of earthquake activity. Around the globe, convergent and divergent margins are connected by transform faults.

The Physiology of the Solid Earth: What Drives Plate Tectonics?

Heat from the Deep

In the simplest sense, plate tectonics is the surface expression of the mechanism by which heat escapes from Earth's interior. Although there are spatial variations,

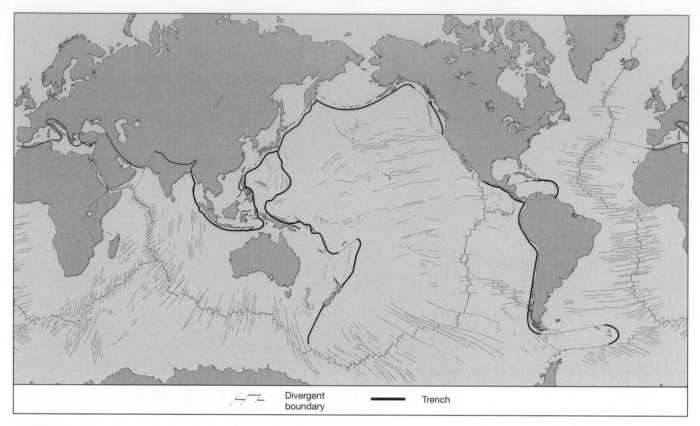

FIGURE 7-18

Distribution of oceanic trenches and mid-ocean ridges. (From W.K. Hamblin and E.H. Christiansen, *Earth's Dynamic Systems*, 8/e, 1998. Reprinted by permission of Prentice Hall, Upper Saddle River, N.J.)

temperatures generally increase through the mantle. This heat is transported to the surface, where it escapes to the atmosphere. (The average *geothermal heat flux,* or heat transported to the surface, is 0.06 W/m^2, which is trivially small compared to the net absorbed solar flux of about 240 W/m^2, the energy budget discussed in Chapter 3. It is the only heat available, though, in Earth's interior.) Heat is transported by convection in the mantle to the base of the lithosphere, and then by conduction through the lithosphere to the surface.

What is the origin of this heat in Earth's interior? It comes from two major sources: (1) radioactive decay and (2) residual heat from Earth's formation. A third source, the growth of the inner core (discussed above), drives convection of the outer core but is only a small contributor to the energy budget of Earth's interior.

Radioactive Decay. We discussed the fundamentals of radioactive decay back in Chapter 5. There, we applied these concepts to the decay of carbon-14. The important radioactive elements in the solid Earth are potassium, uranium, and thorium. Their *half-lives* are on the order of hundreds of millions to billions of years (whereas carbon-

14, if you recall, has a half-life of only 5,730 years; see Chapter 5). Thus, the decay rates of these isotopes are quite low. However, the crust and mantle contain significant concentrations of these elements, so their radioactive decay generates a considerable amount of heat. Because radioactive decay leads to a continuous loss of radioactive materials from Earth's interior, the abundance of these materials must have been greater in the past than it is now. Similarly, the rate of heat production (and of heat loss) must have been much higher in the past. On the basis of the abundance of potassium, uranium, and thorium in the crust and mantle, we can calculate that the amount of radioactive heat production has decreased by about a factor of 5 since Earth formed 4.6 billion years ago.

Other Heat Sources. Other sources of heat are residual; they are associated with heating events during Earth's formation. As we mentioned briefly earlier, Earth (and the other planets) was formed by the accretion of larger and larger clumps of matter into moon-sized objects called *planetesimals*. The planetesimals collided and merged, forming a large, primitive planet. A tremendous amount

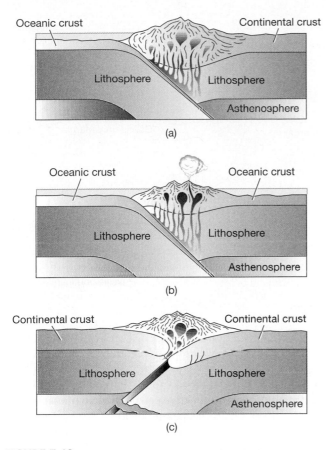

FIGURE 7-19

Three types of convergent plate boundaries: (a) oceanic–continental; (b) oceanic–oceanic; and (c) continental–continental. (From W.K. Hamblin and E.H. Christiansen, *Earth's Dynamic Systems*, 8/e, 1998. Reprinted by permission of Prentice Hall, Upper Saddle River, N.J.)

of energy was transferred to Earth during the accretion of the planet by collisions with planetesimals. The larger collisions probably caused widespread melting of Earth's upper mantle. The segregation of Earth into a less-dense mantle and crust and a denser core released gravitational energy in the form of heat. Convection of the outer core and mantle has been transferring this heat to Earth's surface ever since.

Convection in the Mantle

But how can a solid convect? Convection generally is thought of as a process that affects fluids. Yet solids need not be rigid; witness the flow of glaciers or the ductility of plastics. Rocks are ductile at the temperatures and pressures that occur within the mantle. When heated locally, these materials expand, become less dense, and rise buoyantly, although very slowly. Cooler, denser material sinks and replaces the buoyant material. In this way, mantle rocks can convect. Upon rising to the base of the lithosphere, the material cools as heat is transferred conductively to the lithosphere. As the material continues to cool, it travels laterally. It cools so much that it eventually becomes denser than the underlying lithosphere and descends. Thus, the material sinks back into the asthenosphere. The cycle continues as the material is again heated and becomes buoyant.

The size of mantle convection cells is unknown. Smaller cells may be generated separately within the upper mantle and within the lower mantle, or the whole mantle below the lithosphere may be involved. The nature of the mantle transition zone is the distinguishing factor between the two convection mechanisms. If this

FIGURE 7-20

Transform faults. Sea floor generated at the Juan de Fuca ridge moves southeastward, past the Pacific plate and beneath the North American plate, at the Mendocino transform fault. This fault connects a divergent boundary to a subduction zone. The San Andreas fault, another transform fault, forms the connection between two spreading centers: the Juan de Fuca ridge and a divergent zone in the Gulf of California. (From J.P. Davidson, W.E. Reed, and P.M. Davis, *Exploring Earth: An Introduction to Physical Geology*, 1997. Reprinted by permission of Prentice Hall, Upper Saddle River, N.J.)

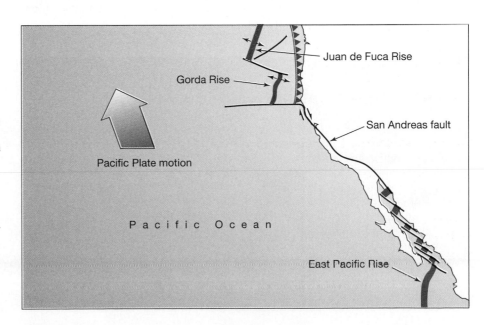

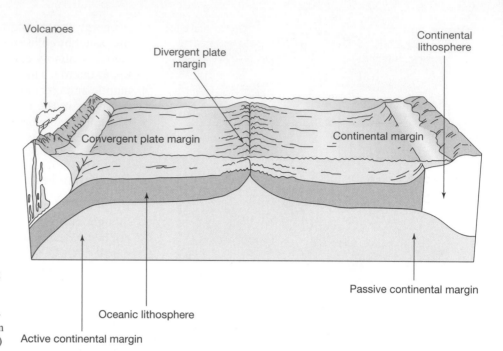

FIGURE 7-21

Schematic view of the relationships among the types of plate boundaries. (From J.P. Davidson, W.E. Reed, and P.M. Davis, *Exploring Earth: An Introduction to Physical Geology*, 1997. Reprinted by permission of Prentice Hall, Upper Saddle River, N.J.)

zone marks a change in chemical composition, then convection cells do not cross it because if they did, the compositional distinctions would be lost. In this case, there are likely to be separate convective cells in the upper and lower mantle (Figure 7-22a). Conversely, if the transition zone is the result of mineralogical rather than chemical changes, and if these changes take place quickly relative to the rate of convection, whole-mantle convection is possible (Figure 7-22b).

The lithosphere is an integral part of the mantle convection system. In a sense, the lithosphere is the cool upper boundary of the convective cell. However, the subduction of cool oceanic lithosphere undoubtedly perturbs the internal thermal structure of the upper mantle and the distribution of convective cells. Oceanic lithosphere is so dense that a slab of oceanic lithosphere at a subduction zone may sink to great depths within the mantle, become detached from the surface portion of the plate, and actually cool regions of the mantle from below. Such regions would become thermally stable, preventing convection locally. Lateral movements of the plates might also induce lateral movements of the underlying asthenosphere. And finally, the separation of the lithosphere at sea-floor spreading centers might drive the mantle to upwell. As the asthenosphere rises, it expands and melts, which further enhances the upwelling of more

A CLOSER LOOK
Radiometric Dating of Geological Materials

Suppose that we have a rock sample that contains both a radioisotope and its decay product, and we know exactly how much of that isotope existed when the material formed. If we also know the half-life of that isotope, then we can calculate the age of the sample. This method, called **radiometric dating**, has proved extremely useful in providing absolute dates for the geological time scale and for other specific events in Earth's history. The accuracy of radiometric dating drops rapidly after eight or nine half-lives. For example, the radioactive isotope of carbon, ^{14}C or *radiocarbon*, has a half-life of just 5,730 years. Thus, *radiocarbon dating* is accurately applied only to samples less than a few tens of thousands of years old. The long-lived radioisotopes of potassium and uranium, however, are useful for determining the ages of the oldest rocks on Earth, of lunar material, and of meteorites. Especially useful are the radioactive isotopes of uranium, ^{238}U (half-life of 4.5 billion years) and ^{235}U (half-life of 0.713 billion years), which decay through a series of intermediate steps to stable lead isotopes ^{206}Pb and ^{207}Pb, respectively. The radioactivity of the uranium in these ancient rocks is not high enough to be measured directly, but we can use an instrument called a *mass spectrometer* to determine the relative amounts of the lead isotopes. From these ratios and the known half-lives, we can accurately date the rocks.

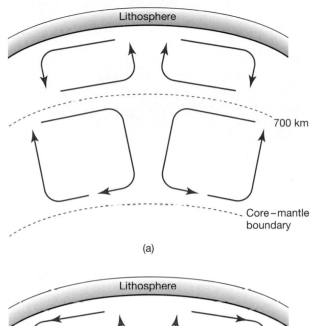

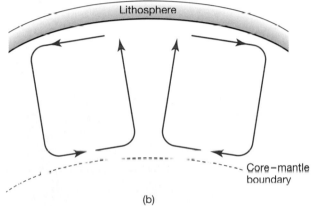

FIGURE 7-22
Mantle convection may (a) separate into upper and lower mantle convective cells or (b) involve the whole mantle. (From S. Judson and S.M. Richardson, *Earth: An Introduction to Geologic Change*, 1995. Reprinted by permission of Prentice Hall, Upper Saddle River, N.J.)

magma. Thus there are a number of lithosphere–asthenosphere interactions that affect the nature of convection in the mantle.

Forces Acting on Plates

The most important lithosphere–asthenosphere interaction is that which drives the motion of the plates. When plate tectonics was first proposed, *mantle drag,* or friction between the convecting asthenosphere and the overlying rigid lithosphere, was considered to be the cause of plate motions (force F_1 in Figure 7-23). Now geologists recognize a number of other forces that act on plates. These forces include the gravitational "push" generated by the high topography of a mid-ocean ridge on the rest of the oceanic plate (F_2); the increasing density of the

oceanic lithosphere as it cools, which pulls the opposite end of the plate into a subduction zone (F_3); the elastic resistance of the oceanic plate to being bent into a subduction zone (F_4); the tendency for the overriding plate to be drawn toward a subduction zone as the subducting slab bends (which otherwise would move the trench away from the overriding plate (F_5); friction between the subducting slab and the overlying lithosphere (F_6); and a tendency for the oceanic plate to sink as it cools and becomes denser (F_7). The overall motion of a given plate is the result of the balance of all these forces. Analysis of plate motions today argues for a predominant role played by the push of the ridges and the pull of the subduction zone, but in the past, mantle drag may have played a more important role in the rifting apart of supercontinents (see below).

Recycling of the Lithosphere: The Rock Cycle

All rocks of the lithosphere ultimately derive from igneous rocks. Igneous rocks of the oceanic lithosphere are born when extruded volcanically at mid-ocean ridges, and they die on average some 80 million years later when subducted and incorporated into the asthenosphere. (The oldest oceanic crust is about 200 million years old.) In contrast, the oldest-known continental rocks formed nearly 4 billion years ago. These record-setters reside in the old, previously active but now tectonically dormant regions of the continental interiors known as **cratons.** These very old continental blocks form the nucleus on which subsequent plate collisions have plastered on new material over the past 3–4 billion years, leading to the growth of the continents (Figure 7-24).

Weathering and Erosion

Once formed, igneous rocks are subject to a variety of processes that can alter their chemical composition and weaken their structural integrity, typically leading to complete disintegration or *dissolution* (dissolving away) of the original rock. On the sea floor, oceanic lithosphere is altered as hydrothermal solutions circulate the oceanic crust (see above), but these processes tend not to destroy the rocks. In contrast, igneous rocks that are exposed on land are subject to a variety of physical, biological, and chemical forces that tend to degrade the rock. These are referred to as *weathering* processes. Weathering transforms solid rock into small particles (sediments) and dissolved material.

A number of processes contribute to physical weathering. Rocks expand and fracture as the weight of over-

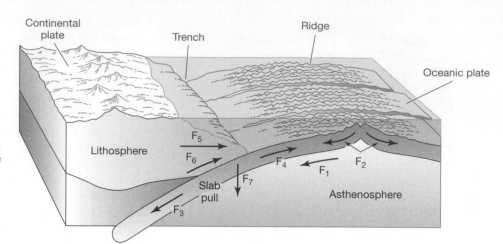

FIGURE 7-23

The various forces (labeled F) acting on plates at their leading and trailing edges. The motion of the plates responds to the sum of these forces. See the text for a discussion of the origin of each force. (After W.K. Hamblin and E.H. Christiansen, *Earth's Dynamic Systems*, 8/e, 1998. Reprinted by permission of Prentice Hall, Upper Saddle River, N.J.)

lying material is removed through erosion. In temperate latitudes, water seeps into these fractures in spring, summer, and fall. It then expands when it freezes during the winter, cracking the rock. Rocks exposed at the surface in high latitudes or at high altitudes are ground up as glaciers advance and retreat. Finally, there are biophysical mechanisms of weathering, including the action of plant roots that penetrate along these fractures, wedging the rock apart.

Chemical weathering results from the tendency for minerals to dissolve when exposed to rainwater and acidic soil waters generated by bacteria, fungi, or plant root discharges. The products of these chemical reactions include dissolved materials and relatively insoluble clays that form in the soils.

The transport of the products of weathering to basins where sediment accumulates is called **erosion.** In this process, crustal materials, decomposed and loosened by weathering, and the clays that form in the soils are transported by winds, landslides, and streams to sites of deposition. These sites include lakes and flood plains on land and deltas and deeper basins in the ocean. In the process, landscapes are created: Valleys form as erodable material is removed, leaving peaks and ridges of material that are resistant to erosive forces.

Sediment Accumulation

The accumulation of sediments depends on two factors: the rate of supply of sediment and the amount of space available to accumulate the sediment. The great depth of water in deep marine basins allows for the thick accumulation of sediments. As the sediments accumulate, however, the sea floor rises toward sea level, and the amount of available space diminishes. Sedimentary deposits also form in shallow-water settings, though, such as the margins of the Gulf of Mexico, where subsidence of the sea floor (resulting from tectonic forces that stretch and thin

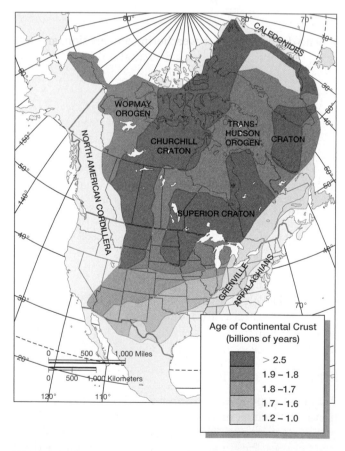

FIGURE 7-24

The ages of the components of the North American continent reveal that the continent has grown by the amalgamation of very old cratons, followed by accretion of younger material onto the periphery of the craton during plate collisions. (From J.P. Davidson, W.E. Reed, and P.M. Davis, *Exploring Earth: An Introduction to Physical Geology*, 1997. Reprinted by permission of Prentice Hall, Upper Saddle River, N.J.)

the lithosphere, or cooling and contraction) allows for continued accumulation of sediment.

As sediments continue to accumulate in these basins, the material is compacted by the increasing weight of overlying sediments. That weight can become so great that fluids trapped between the sediment grains are expelled. Eventually, as the burial process continues, sediments may become buried to a depth of several kilometers below the sea floor. At these depths, temperatures can exceed 200°C, and the pressures can be hundreds of times the atmospheric pressure. In addition, the fluids that circulate through these buried sediments are quite distinct chemically from surface waters. As a result of these environmental changes, sediments undergo further compaction, and the small voids that remain between sediment grains become filled with mineral cements precipitated from the subsurface fluids. In other words, the sediments lithify and become sedimentary rocks.

Uplift

If the continents were at sea level, there would be virtually no driving force for weathering and erosion, and thus no rock cycle. However, continental collisions crumple the crust, producing mountains and high plateaus, and oceanic collisions produce volcanic islands. These plate collisions generate the topography that is then destroyed by erosion. The elevation of a mountain range represents the competition between uplift and erosion. Young mountain belts like Taiwan or Papua New Guinea are undergoing rapid uplift. Steep mountain slopes develop that stimulate rapid erosion as well, but uplift exceeds erosion, and elevations grow. Old mountain belts such as the Appalachians of the eastern United States are not being rapidly uplifted. Erosion is dominating, and the high peaks of the geologic past are gone, having eroded to low ridges.

Erosion promotes the recycling of sedimentary rocks before metamorphism or melting occurs. Sedimentary deposits situated along active continental margins can be entrained into the convergent plate motions and uplifted onto the continents. As soon as the sedimentary rocks become exposed at the surface, weathering and erosion commence.

Metamorphism and Melting

If instead sedimentary (or igneous) rocks are subjected to high temperatures and pressures in Earth's interior, they metamorphose. If they ultimately melt, they form magma that can itself ascend and become igneous rock. Metamorphism occurs deep in sedimentary basins along passive margins, in the deformed regions of active margins, and in other regions of the crust where igneous activity has injected hot igneous rock into what was cooler crustal rock (igneous or sedimentary). The generation of

magma from crustal rocks typically occurs only where those rocks have been carried deep within Earth's interior, for example, in subduction zones.

The Rock Cycle

There are a number of alternative pathways in this process, but overall the complete regeneration of rock is called the **rock cycle** (Figure 7-25). The rock cycle is a consequence of plate tectonics. One complete cycle takes about 100 million years. However, the average lifetime of continental lithosphere as a whole is actually much longer (a few hundred million years), because the interiors of the continents are well insulated from the tectonic activity that occurs along their margins.

It is important to recognize that the rock cycle is not completely closed: New crustal material is produced through the emplacement of magmas derived from the mantle, and older crustal materials are taken back to the mantle at subduction zones. This slow exchange between the mantle and crust replenishes the crust, on average, every 2–3 billion years. In other words, a large proportion of the geologic record of the early Earth's crust has not only been through the rock cycle many times, but has been ingested (perhaps once and for all) into Earth's interior. Nevertheless, averages can be deceptive: The interior parts of the continents (the cratons) can be billions of years old and represent crustal material that has never been recycled, whereas the continental margins are recycled on a time scale much shorter than a billion years.

Plate Tectonics through Earth History

Evolution of the Driving Force

Earth has been losing heat throughout its 4.6 billion-year history. Although early in this history other mechanisms may have dominated, for at least the past 4 billion years heat loss has occurred by mantle convection. The rate of heat loss on the early Earth was several times the present value, presumably fueling higher relative rates of sea-floor production and subduction. But the style of subduction, and the balance of forces acting on the plates, may have been somewhat different on the early Earth than they are today. Presuming that plate velocities of a few centimeters per year have prevailed over the past 4 billion years or so, the continents have moved great distances during that time.

Wilson Cycles

A pattern seems to be emerging: Continents assemble into a supercontinent, which then breaks apart; these smaller continents eventually disperse and then reassem-

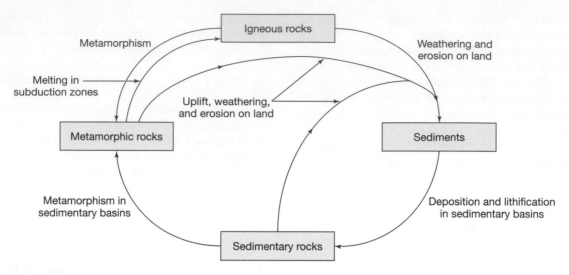

FIGURE 7-25
The rock cycle.

ble. This plate tectonic cycle has been dubbed the **Wilson cycle,** in honor of one of the pioneers of plate tectonics, Canadian geologist J. Tuzo Wilson. From paleogeographic reconstructions it appears that the cycle of supercontinent assembly and destruction takes about 500 million years. We can also approximate the duration of a Wilson cycle from the time it takes a plate to make its way halfway around Earth. Taking a typical plate speed of 4 cm/year (or 40 km/million years) and a half-circumference (at the equator) of around 20,000 km, we confirm that two continents rifting apart would meet each other again, on the far side of the planet, in around 500 million years. Accordingly, given that Pangea formed about 300 million years ago, the next supercontinent should be

formed in about 200 million years as the Pacific Ocean closes, swallowed up by the subduction zones that surround it.

Why do all the continents come together into a supercontinent rather than displaying a less organized, more random pattern of collision and rifting apart? Of course, drifting continents on a finite globe are bound to collide, so larger continents are likely to form. A somewhat controversial hypothesis argues instead that the continents are drawn toward cold regions of the asthenosphere (Figure 7-26). Once assembled, the thick supercontinent acts as an insulator, slowing the release of heat from the mantle. Mantle temperatures rise beneath the supercontinent, modifying the pattern of convection. The

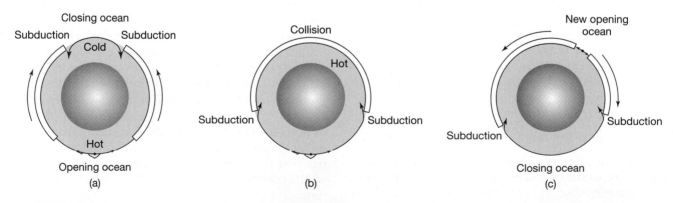

FIGURE 7-26

The Wilson cycle of supercontinent assembly and fragmentation. (a) The continents are drifting toward a region of cold asthenosphere. The closing ocean is lined by subduction zones and is contracting. The other ocean is opening, and the oceanic lithosphere is connected to the continental lithosphere at both margins. (b) The continental fragments have collided, forming a supercontinent. Subduction has begun along the margins of the formerly opening ocean. The insulating effects of the thick continental lithosphere lead to the buildup of heat and the initiation of rifting. (c) What once was an opening ocean has become a closing ocean, with cool asthenosphere beneath. One Wilson cycle is now complete. (After P. Kearey and F.J. Vine, *Global Tectonics*, 1990. Oxford: Blackwell Scientific)

resulting tension at the surface eventually rips the super-continent apart. The continents begin to move from this region of hot upwelling mantle to a site thousands of kilometers away, where the mantle has cooled and down-welling has commenced. This hypothesis is consistent with present-day plate speeds and seismic tomography, which together suggest that the continents, with the exception of

Africa, appear to be moving toward regions of cold mantle. Africa is situated above hot mantle, as evidenced by the East African Rift zone, a place where tension within the continental lithosphere is creating a rift in the continent. Africa has apparently moved little since the breakup of Pangea, and the underlying mantle still retains the heat built up during Pangea's existence.

Chapter Summary

1. The solid Earth is dynamic, not static. Wegener's idea of drifting continents proposed in the early 20th century has largely been substantiated.
   a. New sea floor is created at mid-ocean ridges and moves outward as ocean basins grow.
   b. Old sea floor is destroyed at deep-sea trenches in subduction zones. Earthquakes outline the surface of the slab of oceanic lithosphere being subducted beneath the continent; their foci along a continental margin become ever deeper away from the trench.
   c. Deeper probing of Earth's interior has revealed heterogeneity in composition and temperature that can be the result only of large-scale circulation in the mantle and outer core.
      i. This circulation is fueled by residual heat from the formation of the planet 4.6 billion years ago and by heat that continues to be produced as the result of the radioactive decay of potassium, uranium, and thorium in the mantle and crust.
      ii. Mantle circulation is the result of convection, not unlike that of the troposphere. Scientists continue to debate whether convective cells extend throughout the mantle or whether a dual system of upper and lower mantle convection is in operation.
2. The signature of plate tectonics is best seen at the surface, along the margins of lithospheric plates.
   a. At plate margins, divergent, convergent, and boundary-parallel plate motions generate impressive topographic features: mid-ocean ridges and continental mountain belts and volcanoes, deep-sea trenches, and transform faults, respectively. These features themselves change with time.
      i. Mountains grow through plate collision but shrink through weathering and erosion.
      ii. Sediments are transported to the oceans, where they fill in deep-sea trenches as well as basins generated

through subsidence.
      iii. Burial converts sediments to sedimentary rocks, but in time these rocks are likely to become re-exposed as the result of plate convergence and uplift, or metamorphosed and perhaps ultimately melted.
   b. This cycle of rock production and destruction (the rock cycle), driven largely by plate tectonic processes, continuously resurfaces the planet and recycles material between the crust and mantle.
3. The movement of lithospheric plates is driven by forces at plate margins and, at the base of the lithosphere, by friction with the convecting asthenosphere.
   a. Magnetic stripes displaying mirror-image patterns across the mid-ocean ridges document the creation of sea floor at the ridges.
   b. Plate motions are slow on human time scales (on the order of centimeters per year), but over geological time they can lead to the complete redistribution of continents on the globe.
      i. These movements appear to be organized, following a pattern called Wilson cycles. These cycles consist of the alternating assembly of supercontinents (perhaps at the position of mantle downwelling) and their subsequent breakup (as the insulating effects of the supercontinent lead to sublithospheric heating and mantle upwelling).
      ii. The most recent supercontinent, Pangea, formed more than 300 million years ago. Its breakup, beginning some 200 million years ago, led to the creation of the Atlantic Ocean and the separation of North America and South America from Europe and Africa.
      iii. The Pacific Ocean is currently shrinking, and it is estimated that in another 200 million years or so the continents will again reassemble into another supercontinent.

Key Terms

asthenosphere	earthquake	lithosphere
basalt	erosion	magma
body waves	granite	magnetic dynamo
chemosynthesis	half-life	metamorphic rocks
continental drift	igneous rock	mid-ocean ridge
craton	lithification	mineral

Moho	rock cycle	slab
Pangea	sea-floor spreading	subduction
plate tectonics	sediment	surface waves
polarity	sedimentary rock	S wave
P wave	seismic wave	transform fault
radioactive decay	silicate mineral	Wilson cycle
radiometric dating		

Review Questions

1. Why was the theory of continental drift not immediately embraced by the scientific community in the 1920s?
2. What is the Moho?
3. What are the bases for the two major divisions of Earth's interior—one that distinguishes crust, mantle, and core and the other that distinguishes lithosphere and asthenosphere?
4. Compare and contrast P and S seismic waves.
5. Why are earthquakes focused along plate margins?
6. What are the sources of heat in Earth's interior?

7. What is magnetic polarity? What role did it play in the generation of ideas regarding sea-floor spreading?
8. What are the three types of plate boundaries, and what surface features are characteristic of each?
9. What is erosion?
10. How can radioactivity be used to determine the age of a rock?
11. What are the driving forces for plate movement?
12. What is hypothesized to drive the Wilson cycle of plate fragmentation and reassembly?

Critical-Thinking Problems

1. We have seen that cooling of the oceanic lithosphere causes contraction, leading to subsidence of the sea floor away from the axis of spreading. The depth d of the ocean floor, measured in meters, increases with age t, measured in millions of years from the present, according to the following equation (valid for sea floor younger than 80 million years old):

    Graph a cross-section of a mid-ocean ridge that is spreading symmetrically in both directions at a rate of 1 cm/yr (10 km/million yr). The age of the oldest sea floor shown should be 80 million years.
2. Duplicate Figure 7-10 (see color version) and answer the following questions.

a. Draw a line from the tip of Florida horizontally across the Atlantic to northwest Africa, a distance of about 6400 km. Now, graph the age of the sea floor (on the y-axis) against the distance from the ridge axis (on the x-axis). From this graph, determine the spreading rate for each geologic interval represented, averaging the two values determined for eastward and westward spreading. Graph these values (y-axis) as a function of time (in million years, on the x-axis).
b. How has the Atlantic spreading rate varied over the last 200 million years?

Further Reading

General

Tarbuck, E. J., and Lutgens, F. K. 1998. *Earth: An Introduction to Physical Geology*, 6/e. Upper Saddle River, NJ: Prentice Hall.
"The Dynamic Earth," 1983, *Scientific American*.
Wegener, A. 1924. *The Origins of the Continents and Ocean Basins.* London: Metheune.

Advanced

Kearey, P., and Vine, F. J. 1996. *Global Tectonics*, 2/e. Oxford: Blackwell Scientific.

Recycling of the Elements: The Carbon Cycle

Key Questions

- What determines how reservoirs (such as the atmospheric CO_2 reservoir) respond to imbalances in the flow of material to and from them?
- Which reservoirs and processes are important to the recycling of carbon and other essential nutrients in the Earth system?
- Do feedback mechanisms regulate the amount of atmospheric CO_2?

Chapter Overview

The recycling of the elements among the components of the Earth system is key to the continued functioning of Earth as a living planet. Although many elements are critical, none is more central to the workings of the Earth system than carbon. All life is based on carbon; gaseous carbon dioxide is an important greenhouse gas; the acidity of the ocean is regulated by carbon compounds; and the maintenance of an oxygen-rich atmosphere depends on the transfer of carbon to sedimentary rocks. To perform these functions, the *carbon cycle* involves a hierarchy of subcycles that operate on different time scales, ranging from decades (for the replenishment of CO_2 in the atmosphere) to hundreds of millions of years (for the recycling of carbon through sedimentary rocks and for the exchange of carbon within Earth's

interior). Both biological and physical processes are involved in the recycling of carbon, and they are so closely intertwined that it becomes difficult to separate the two. In this chapter we trace the movement of carbon as it cycles through the Earth system and we develop additional systems theory notions of *steady state* and *residence time* along the way. We also discuss the *nutrient elements*, because the rate of carbon recycling depends strongly on the rate of nutrient recycling.

Systems Approach to the Carbon Cycle

Why is Earth the only planet in our solar system that supports life? The direct answer is that Earth is the only planet that has liquid water at its surface. (Jupiter's moon Europa may have water only a few kilometers beneath its icy surface.) But part of the reason Earth is able to maintain liquid water is that our planet has natural recycling systems for the elements essential for life, including carbon, nitrogen, phosphorus, and sulfur. These recycling systems are ultimately linked with the global process of plate tectonics, discussed in Chapter 6. The link between tectonic activity and the carbon cycle is important to the regulation of atmospheric CO_2 concentrations and thus to climate as well.

The winds and ocean currents discussed in Chapters 4 and 5 and the moving lithospheric plates discussed in Chapter 7 make up Earth's circulatory system: They transport energy and material to different parts of the Earth system where they are utilized in biological and physical processes. This mixing of Earth's fluid and solid parts also helps accomplish an important task: the recycling of the elements. Essential elements are released to the **biosphere** (the part of Earth that supports life, including the oceans, atmosphere, land surface, and soils) as rocks weather, volcanoes erupt, and nitrogen is made available from the atmosphere by chemical transformations stimulated by lightning discharge. Compared with the rates of utilization by the biota, these releases are very slow; they would support only very low rates of biological activity were there not highly efficient nutrient recycling mechanisms. **Nutrients** are substances, normally obtained in the diet, that are essential to organisms. Nutrient elements are incorporated into living tissue during growth and rapidly returned to the soil or ocean on death. This cycle is repeated many times before the elements are lost from the biosphere, mostly as constituents of sedimentary rocks. The situation is much like our recycling of aluminum cans: Recycling substantially reduces our dependence on the extraction of aluminum from Earth and allows us to produce aluminum products much more rapidly. Similarly,

element recycling within the biosphere allows for much higher rates of biological productivity.

A number of important recycling systems operate on Earth. We've already discussed the water cycle in Chapter 4. The cycles of the nutrient elements nitrogen and phosphorus are discussed later in this chapter. The recycling of carbon is especially important: As a major constituent of the greenhouse gases carbon dioxide and methane, it affects not only biological productivity but Earth's climate as well. We focus on the carbon cycle in this chapter because of its overarching importance to the Earth system.

A Journey through the Terrestrial Organic Carbon Cycle

As an introduction to just one part of the global carbon cycle, imagine that we could follow the carbon atom of a CO_2 molecule as it cycled through the terrestrial (land-based) part of the cycle (Figure 8-1). The carbon in CO_2 is **inorganic carbon**—it is not associated with compounds formed by living organisms and it does not contain carbon–carbon or carbon–hydrogen bonds. After spending nearly a decade moving with the winds in the troposphere,

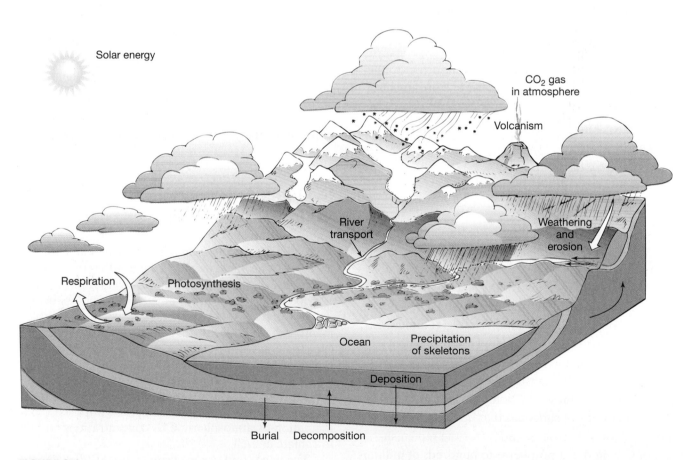

FIGURE 8-1

The global carbon cycle. (From J.P. Davidson, W.E. Reed, and P.M. Davis, *Exploring Earth: An Introduction to Physical Geology*, 1997. Reprinted by permission of Prentice Hall, Upper Saddle River, N.J.)

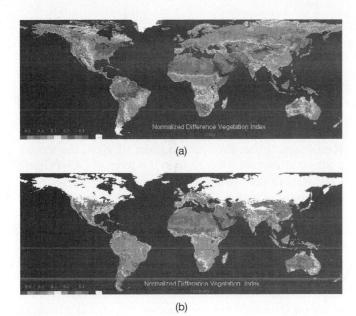

FIGURE 8-2

[See color section] Satellite image of the vegetation coverage of the land surface in the Northern Hemisphere in (a) summer and (b) winter. (From Felix Kogan, NOAA/NESDIS/ORA, Climate Research and Application Division.)

the gaseous CO_2 molecule will have visited both the Northern and Southern Hemispheres several times. Then one spring, during the annual greening of the Northern Hemisphere (Figure 8-2), the CO_2 molecule passes through a small opening in a leaf, the photosynthetic apparatus of a plant. Through a frenzy of collisions with other molecules and atoms, the oxygen atoms are ripped from the molecule while hydrogen, nitrogen, and other carbon atoms become attached. Our carbon atom, as part of the leaf, is now **organic carbon**. Some leaves are consumed and digested by animals. The carbon in these leaves is then released back to the atmosphere by the animals' respiration as CO_2.

Summer passes, fall arrives, and the leaf that contains the carbon atom has not been eaten. The nourishing substances and water that the leaf has received from the tree have ceased to flow. The leaf is released from the branch and settles to the ground. Other leaves fall on top, burying it in a thick mat of decaying matter. The carbon atom is part of the soil, where it will remain for about the next 50 years. By the end of that time, bacteria and fungi will have decomposed the organic matter that contains the atom. The chemical reactions that result transform the carbon atom once again into a gaseous CO_2 molecule, which escapes back to the atmosphere.

This life cycle of a carbon atom is repeated nearly 500 times on average before a "leak" occurs. Once in a while, before the organic matter that contains the carbon atom decomposes, the soil erodes and is transported by rivers to the oceans. There it settles with the other particles to the sea floor and is buried by subsequent sediments or carried with its underlying oceanic plate deep into a subduction zone. Under elevated temperatures and pressures the carbon atom may be converted into gaseous carbon atoms and escape to the surface, or be converted into a component of sedimentary or metamorphic rock.

The carbon atom may spend millions of years in the sedimentary/metamorphic rock reservoir, as mountain belts form, thrusting deeply buried rocks to Earth's surface and beyond to great elevations. Eventually our carbon atom will be transferred from its burial place within Earth's interior to the surface. Here environmental forces, both biological and physical, will cause the sedimentary rock containing the carbon atom to disintegrate during the process of weathering. In this process, the organic carbon reacts with oxygen from the atmosphere and forms (inorganic) CO_2, which escapes as a gas to the atmosphere. The weathering process, then, is the connecting link in the long path this particular carbon atom has taken—from the atmosphere, to the plant, to the soil, to the sediment, to the sedimentary rock, and back to the atmosphere.

The path the carbon atom has taken encompasses the *terrestrial organic carbon cycle*, operating on time scales that are short (years to decades) and those that are long (centuries and millennia to multimillion years). If instead of being incorporated into a plant leaf the carbon atom had entered the ocean and been converted to organic carbon by marine algae, it would have become part of the *marine organic carbon cycle*. There are also a host of processes not involving organic carbon that compose the *inorganic carbon cycle*. These various parts of the carbon cycle are discussed below.

Carbon Reservoir Dynamics

Carbon resides in many reservoirs at or near Earth's surface (Figure 8-3), ranging in size from the relatively tiny amount of carbon in atmospheric methane to the tremendous amount of carbon stored in sedimentary rocks. One of our ultimate goals is to understand how this system of reservoirs responds to perturbations. We will use the response to the release of carbon dioxide from the burning of fossil fuels as an example of the dynamics of the carbon cycle and of material recycling systems in general.

Figure 8-4 shows the seasonal fluctuations in the atmospheric CO_2 level for three years (1999–2001) measured from atop Mauna Loa, Hawaii. We saw a similar graph in Figure 1-2. In Figure 8-4, however, we focus on the natural seasonal cycle rather than the gradual increase in CO_2 from fossil-fuel burning and deforestation. The CO_2 content falls during the Northern Hemisphere summer, when photosynthesis (and the growth of leaves) surpasses respiration and decomposition. It then rises during

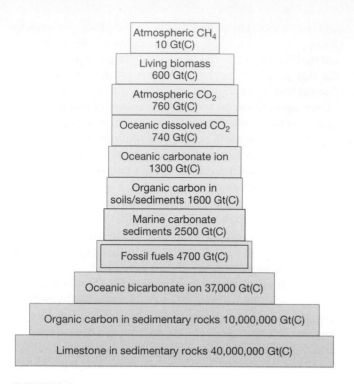

FIGURE 8-3

Reservoirs of carbon at or near Earth's surface.

the late fall to early spring, when respiration and decomposition of the previous season's crop of leaves exceeds photosynthesis. Because Hawaii is in the Northern Hemisphere (17° N), measurements made at Mauna Loa reflect the annual cycle in that hemisphere (see Chapter 1 for further discussion). Let us cast these observations in terms of systems theory.

Reservoirs. From the systems point of view, the atmosphere is a *reservoir* of carbon in the form of CO_2 (Figure 8-5). Reservoirs are typically characterized in terms of the amount of material they are holding at any particular time. Their sizes are commonly expressed either in mass units or volume units. (The units are typically *moles;* refer to the Box "The Concept of the Mole" if you need a refresher.) In Figure 8-5, the amount of carbon is expressed in gigatons (Gton) of carbon, or Gton(C). A gigaton is 1 billion metric tons, and 1 metric ton is 1000 kg. With the notation Gton(C), we are keeping track of the mass of only the carbon atoms, not the other atoms to which they are attached.

Reservoirs are temporary repositories for material that flows through them, and their sizes vary in response to imbalances between inflow and outflow, typically expressed in unit mass, unit volume, or moles per unit time. The *inflow* to the atmospheric CO_2 reservoir is the combination of the processes of respiration and decomposi-

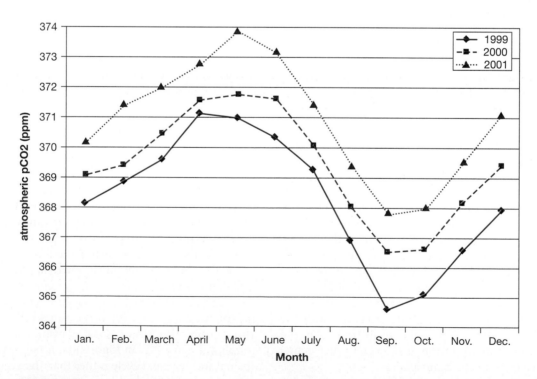

FIGURE 8-4

Seasonal fluctuations in atmospheric CO_2 from the Mauna Loa Observatory for 1999–2001. The gradual increase due to fossil-fuel burning and deforestation accounts for the offset from year to year. (Data courtesy Oak Ridge National Laboratory: http://cdiac.esd.ornl.gov/trends/co2/sio-mlo.htm)

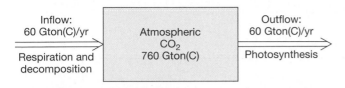

FIGURE 8-5

The atmospheric carbon (i.e., CO_2) reservoir, showing inflows and outflows.

tion. This inflow can be expressed in units of gigatons of carbon per year. The *outflow* from this reservoir is photosynthesis, and the rate of outflow is also expressed as gigatons of carbon per year.

Steady State. If the rates of inflow and outflow were such that the atmospheric CO_2 level remained at a constant value with time, we would say that *steady state* had been achieved. **Steady state** is a condition in which the state of a system component is constant with time. Steady state could be achieved if no inflow and no outflow existed, that is, if both processes ceased. A constant level could also be maintained if the rate of inflow of CO_2 into the atmosphere equaled the rate of outflow.

Any imbalance in these rates leads to a change in the level of atmospheric CO_2. When the inflow exceeds the outflow, the atmospheric CO_2 level rises. (This situation is analogous to the Northern Hemisphere winter condition.) When the outflow exceeds the inflow, the level

falls (analogous to the Northern Hemisphere summer condition).

In the record of atmospheric CO_2 variations in Figure 8-4, we see that one maximum and one minimum is reached each year. At these times the fluxes are in balance. The system is not really in steady state at these times, however, because the reservoir size is unchanging only for an instant. Averaged over longer times, though, the *natural* cycle of CO_2 is thought to be close to steady state, despite seasonal imbalances. Because of anthropogenic disturbances, the atmospheric CO_2 level is not currently at steady state, as demonstrated by the steady rise in CO_2 over the past several decades.

Steady state can be maintained over time only if the rates of inflow and/or outflow are sensitive to changes in the size of the reservoir. In systems terminology, this means that there must be couplings that link the reservoir size to the processes that govern inflow and outflow (see Chapter 2). Consider what would happen if a reservoir in steady state were perturbed by an addition of material. If the coupling governing inflow was negative or if that governing outflow was positive (i.e., inflow decreased or outflow increased), the reservoir would return to its original state. One such negative feedback loop exists between the photosynthetic rate of plants and atmospheric CO_2. As CO_2 levels go up, plants photosynthesize more rapidly; this effect has been called *CO_2 fertilization* (Figure 8-6). But as they do, CO_2 levels tend to fall, because CO_2 is consumed by plants during photosynthesis.

USEFUL CONCEPTS
The Concept of the Mole

Atoms and molecules are typically measured in units called moles. A *mole* (abbreviated mol) of a substance is defined as the amount of that substance that contains the same number of atoms or molecules (or any other particle) as the number of atoms in 12 g of ^{12}C. There are 6.02×10^{23} atoms in 12 g of ^{12}C; this number is called *Avogadro's number*, after the Italian chemist Amedeo Avogadro. A mole of any type of particle contains Avogadro's number of these particles. In concept, a mole is no different than, say, a dozen; it simply converts a number that would be cumbersome into one that is more practical. Even when we use moles, global scale reservoirs such as that of atmospheric CO_2 are huge. There are

presently about 6×10^{16} moles of CO_2 in the atmosphere.

One mole of ^{12}C weighs exactly 12 g. However, 1 mol of H weighs only approximately 1 g, and 1 mol of ^{16}O weighs approximately 16 g. These weights are defined as the mass numbers of the given isotopes. The mass number of an isotope is the total number of protons plus neutrons in the nucleus. Equivalently, atomic weight is the weight of 1 mol of a particular isotope. For an element with more than one stable isotope, the precise atomic weight (as opposed to the mass number) of the element is determined from the relative amounts of the various isotopes.

Expressing quantities of substances in moles rather than in mass units (e.g.

grams) can be useful when we are studying chemical reactions. Consider the chemical reaction for the formation of salt (sodium chloride, NaCl):

$$Na^+ + Cl^- \rightarrow NaCl$$

This equation shows that one atom of sodium (Na) will react with one atom of chlorine (Cl) to form one molecule of NaCl. Therefore, 1 mol of Na will react with 1 mol of Cl to form 1 mol of NaCl. To express this equation in mass units, we would need to use the atomic weights of sodium and chlorine. If we did, we would find that 22.99 g of Na reacts with 35.45 g of Cl to produce 58.44 g of NaCl—a more cumbersome calculation.

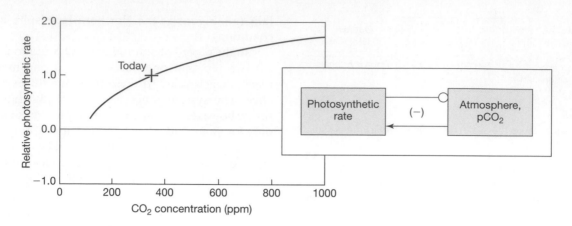

FIGURE 8-6

Effect of changes in CO_2 concentration on the photosynthetic rate of typical plants. Photosynthetic rates are relative to the value for today's atmospheric CO_2 level. Inset represents the negative feedback loop that results from this dependence.

Thus, the terrestrial biota tend to stabilize atmospheric CO_2 levels.

Residence Time. To help us monitor the recycling of elements through the Earth system, the concept of residence time can be useful. **Residence time** is defined as the average length of time a substance spends in a given reservoir that is at steady state. We can calculate the residence time by dividing the reservoir size at steady state by the inflow or outflow rate:

$$\text{residence time} = \frac{\text{reservoir size at steady state}}{\text{inflow rate or outflow rate}}.$$

Reservoir size has units of mass, and the inflow or outflow rate is in mass per unit time. Thus the quotient has units of time.

We can determine the residence time of the atmospheric carbon reservoir from the size of that reservoir, which Figure 8-3 gives as 760 Gton(C), and from the rate of respiration and decomposition (the inflow rate) or the rate of photosynthesis (the outflow rate), both of which are given as 60 Gton(C)/yr. Thus, if we assume steady state, the residence time of carbon in the atmosphere with respect to these processes is 760 Gton(C)/60 Gton(C)/yr = 12.7 yr. This means that carbon in the atmosphere is replenished about once per decade.

We can also think about residence time as an indicator of how long a reservoir takes to respond measurably to large imbalances in inflow or outflow. In our atmospheric carbon example, if photosynthesis were to cease but respiration and decomposition were to continue at their current rate, the atmospheric CO_2 level would double in about a decade. Thus, the residence time, defined

at steady state, becomes the **characteristic response time** when a system is *not* at steady state. The concept of a characteristic response time is similar to that of a half-life in radioactive decay (see Chapter 5). Stated formally, a disturbance from steady state (in a system where the rate of removal is proportional to the amount of disturbance) diminishes to 1/e (about 38%) of its original size in one characteristic response time.

Oxidized and Reduced Carbon

The many identities that carbon assumes in the Earth system can be lumped into two general categories: oxidized carbon and reduced carbon. **Oxidized carbon** is carbon that is combined with oxygen. Examples of oxidized carbon include the carbon in the skeletons of some organisms and in atmospheric CO_2. **Reduced carbon** is carbon that is combined mainly with other carbon atoms, hydrogen, or nitrogen. Organic carbon is a form of reduced carbon. Perhaps a more familiar pair of reduced and oxidized substances is metallic iron and its oxidation product, rust (iron oxide). In the presence of oxygen gas at Earth's surface, reduced substances, such as organic carbon and metallic iron, tend to be highly chemically reactive. Oxidized substances, such as CO_2 and rust, tend to be more inert.

In the next few sections, we explore the organic and inorganic carbon cycles. We begin with the organic carbon cycle as it operates on land—the terrestrial organic carbon cycle, which we already introduced in our journey through the carbon cycle. We then move to the oceans, where both oxidized and reduced carbon recycling are important on short time scales. Finally, we consider the longer-time-scale cycles, which involve geological processes.

The Short-Term Organic Carbon Cycle

The short-term organic carbon cycle involves processes ranging from those we can observe and appreciate on a daily to seasonal time scale (see Figure 8-2), such as the processes of photosynthesis and respiration, to processes of decomposition that are somewhat slower (Figure 8-7). The key step in this cycle is the conversion of inorganic carbon (atmospheric CO_2) to organic carbon by the process of **photosynthesis**. We are not so much interested in the process itself but rather in its impact on the global cycle, which is generally expressed as primary productivity. **Primary productivity** is the amount of organic matter produced by photosynthesis in a unit time over a unit area of Earth's surface. That amount depends on the population size of **primary producers**—that is, plants (or other types of photosynthesizers or even chemosynthesizers) that provide energy other organisms can use. The relationship is not simple, however, because some primary producer species are much more productive than others. In its simplest representation, primary production involves a chemical reaction between CO_2 and water to form organic matter and oxygen:

Photosynthesis: $CO_2 + H_2O \rightarrow CH_2O + O_2.$

| (Primary Production) | carbon dioxide | water | carbohydrate | oxygen gas |

Here, organic matter is represented by CH_2O, the simplest *carbohydrate*, or compound of carbon, hydrogen, and oxygen. In reality the molecules making up organic matter are much larger than this simple carbohydrate, and they contain small amounts of many other elements, including nitrogen and phosphorus. For our purposes, however, this simpler representation suffices.

Photosynthesis does not occur spontaneously at Earth's surface but instead requires an input of energy from the Sun. Plants, algae, and bacteria have evolved pigments that are able to capture the energy of sunlight and convert it to chemical energy, part of which is stored

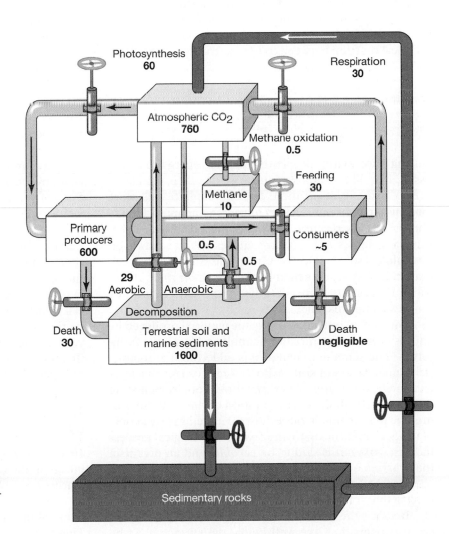

FIGURE 8-7

The short-term, terrestrial organic carbon cycle, showing reservoir sizes, inflows, and outflows. Dark shaded area represents the long-term cycle (see Figure 8-12). Reservoir sizes in Gt(C), fluxes in Gt(C)/yr.

in living tissues. This chemical energy is then utilized by other organisms that cannot utilize solar energy directly. Such organisms, including animals, are called **consumers**.

Most of the photosynthesis that occurs each year leads to the formation of tissue that is recycled rapidly, including the leaves of trees. This recycling is the cause of the seasonal variations observed in the CO_2 record of Figure 8-4. However, most of the organic carbon that makes up plant tissues has a residence time of many decades. That is because the bulk of the organic carbon in terrestrial plants is contained not in the leaves but in the roots and trunks of slow-growing trees. In other words, most of the biomass of primary producers on Earth is contained in tree roots and trunks. **Biomass** is the total mass of organic matter in living organisms in a particular reservoir. In terms of carbon, the total living biomass—the combined biomasses of all primary producers and consumers—is about equal to the atmospheric carbon reservoir (see Figure 8-3).

Consumer biomass is a small percentage (only about 1%) of the biomass of the producers. Consumers derive their metabolic energy from the chemical energy stored in plant tissues by ingesting the tissues and respiring. *Respiration* is the reverse of photosynthesis: It is the chemical reaction between oxygen and organic tissue that yields CO_2 and water:

$$\text{Respiration:} \quad \underset{\text{carbohydrate}}{CH_2O} + \underset{\substack{\text{oxygen} \\ \text{gas}}}{O_2} \longrightarrow \underset{\substack{\text{carbon} \\ \text{dioxide}}}{CO_2} + \underset{\text{water}}{H_2O}.$$

During photosynthesis, plants use solar energy to create tissue. But, like animals, plants produce their metabolic energy through respiration. Hence, respiration is more than just the "breathing" performed by animals. Unlike photosynthesis, respiration would proceed abiotically because it releases rather than requires energy. However, it would do so very slowly. Organisms are able to accelerate this chemical reaction by the use of *enzymes*, chemical compounds (specifically, proteins) that they synthesize for this purpose.

On land, about half the organic material produced by photosynthesis is respired by animals and by plants themselves. The remaining material is added to the organic-rich upper layers of soil. A host of microscopic bacteria and fungi live in soil. Their metabolic requirements are satisfied through the decomposition of the large store of organic matter that is buried there, fueled by the supply of oxygen from the overlying layers. A biological process that uses oxygen is said to be *aerobic*, and an organism that carries out aerobic metabolism is an *aerobe*. The chemical reaction for this aerobic decomposition is identical to that for respiration.

Because the only source of oxygen is the air, the microorganisms that live well below the surface must be adapted to environments that are devoid of oxygen. A biological process that occurs in the absence of oxygen is said to be *anaerobic*, and an organism that carries out anaerobic metabolism is an *anaerobe*. In the O_2-free environment of deep soil live anaerobic bacteria that decompose organic matter by an overall process known as methanogenesis. **Methanogenesis** is an anaerobic form of metabolism that involves multiple steps, carried out by different bacteria. One step involves fermentation of complex organic materials into simpler forms, including hydrogen (H_2) and acetate that methanogens can then use to form both oxidized carbon (in CO_2) and reduced carbon (methane). The overall process can be represented as:

$$\text{Methanogenesis:} \quad \underset{\text{carbohydrate}}{2CH_2O} \longrightarrow \underset{\substack{\text{carbon} \\ \text{dioxide}}}{CO_2} + \underset{\text{methane}}{CH_4}.$$

(Other, more complex compounds can also be used in and produced during anaerobic metabolism, but for the purposes of studying the carbon cycle, this representation is adequate.) The gases CO_2 and CH_4 can escape to the atmosphere. Once there, the CO_2 continues the short-term cycling path described earlier. Methane, however, is chemically unstable in our O_2-rich atmosphere and is destroyed by a series of oxidation reactions. The carbon contained in CH_4 combines with O_2 to form CO_2. With an atmospheric reservoir size of 5 Gton(C) and a supply rate from fermentation of 0.5 Gton(C)/yr (see Figure 8-7). the residence time of CH_4 in the atmosphere is approximately 10 years.

The land surface is continuously stripped of its soil cover by the action of winds and water. On average, about 5 cm of soil is eroded from the land surface every 1,000 years and transported by rivers to the oceans. Although river systems have a substantial capacity for storing sediment in flood plains and deltas, eventually most of the sediment makes its way to the oceans and is deposited on the sea floor. These sediments contain whatever organic matter has survived the trip from land to sea. Thus, there is a transfer of organic carbon from the terrestrial to the marine realm, one that amounts to about 0.1 Gton(C) per year. However, this transfer is small compared with the flux through the oceanic water column of organic matter produced in the ocean.

The Marine Organic Carbon Cycle on Short Time Scales

Producers and Consumers. The dominant primary producers in the ocean are the free-floating, photosynthetic marine microorganisms referred to as *phytoplankton*. (In more general terms, *plankton* are organisms with any type of metabolism that float freely in aquatic environments.) These organisms—primarily *diatoms* (Figure 8-8a) and

A CLOSER LOOK
Oxygen Minimum Zone

The decomposition of organic matter settling through the water column consumes oxygen, leading to oxygen depletion. Decomposition also releases nutrients to the water, so nutrient concentrations increase. The lowest oxygen concentrations are achieved at intermediate depths, about 1 km below the surface. This region is known as the *oxygen minimum zone* (Box Figure 8-1). In this zone, dissolved oxygen concentrations reach a minimum as a result of high oxygen demand by aerobic decomposers and low oxygen supply from the surface ocean or from below.

In the waters below the oxygen minimum zone, oxygen content increases with depth. Why is this so, if the source of oxygen is exchange with the atmosphere and production by phytoplankton in the surface ocean?

Recall from Chapter 5 that the circulation of the ocean resembles a giant conveyor belt that brings waters from the surface in the high-latitude North Atlantic and Antarctic oceans to the deep sea.

This water is enriched in oxygen as a consequence of having originated at the surface in the high latitudes. Like most gases, oxygen is more soluble in cold water than in warm water. So, gas exchange with the atmosphere leads to higher gas concentrations in high-latitude surface waters than in low-latitude surface waters. High-latitude surface waters are injected beneath the intermediate waters of the world's oceans by thermohaline circulation. Deep water therefore ends up with more dissolved O_2 than does the water above it. Thus, the oxygen minimum zone owes its existence to the combined operation of the biological pump and the global conveyor-belt circulation.

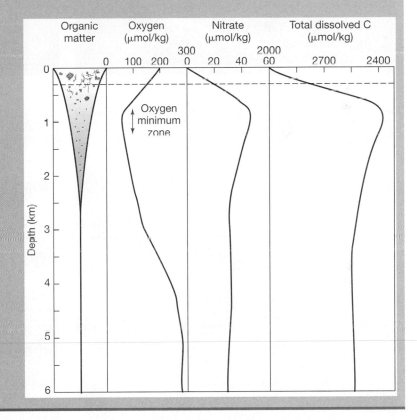

BOX FIGURE 8-1

The effect of the biological pump and thermohaline circulation on the chemical composition of the ocean. Typical vertical profiles of the amount of organic matter, dissolved oxygen, nitrate (a nutrient), and inorganic carbon are shown. Concentrations are in micromoles (10^{-6} mol) per kilogram of seawater.

---

other algae, such as *coccolithophorids* (Fig. 8-8b)—live in the photic zone. The **photic zone** is the uppermost part of the oceanic water column where there is sufficient light for photosynthesis: about the upper 100 m of the water column in the open ocean, and in shallower waters near shore, where water clarity is reduced. It roughly corresponds to the "surface ocean," which, as we saw in Chapter 5, is the upper part of the ocean mixed by the winds.

Phytoplankton consume CO_2 and produce O_2 through photosynthesis in much the same way as do land-based plants. Although the gases phytoplankton use and produce are dissolved in seawater, there is continuous gas exchange between the atmosphere and the ocean. Thus,

the activities of phytoplankton affect the atmosphere as well as the ocean.

Much of the organic matter produced in the surface ocean by phytoplankton is consumed by zooplankton. **Zooplankton** are free-floating marine consumers, including small invertebrates and microorganisms such as *foraminifera* (Figure 8-8c) and *radiolarians* (Figure 8-8d), that cannot photosynthesize. Zooplankton produce fecal pellets that, together with other large particles of decaying organic matter, settle through the water column to great depths. In contrast to the flux of organic matter from treetops to the ground, though, only about 1% of this material survives the trip to the sea floor. Even then, the material is subject to efficient recycling by aerobic and

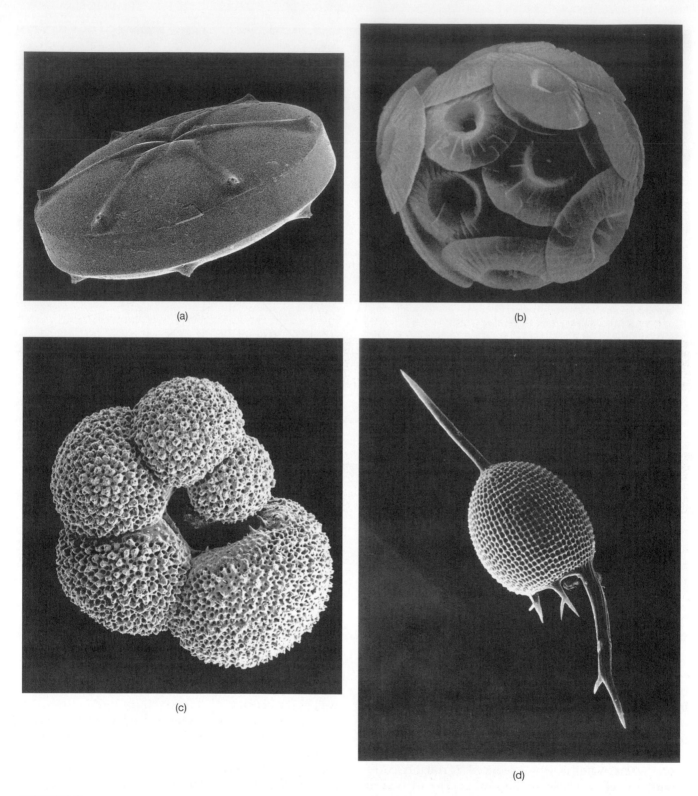

FIGURE 8-8

[See color section] Shells of typical phytoplankton: (a) diatom (SiO_2; approximately 50 μm wide) and (b) coccolithophorid ($CaCO_3$; about 10 μm in diameter). Typical zooplankton: (c) foraminifer ($CaCO_3$; approximately 600 μm in diameter) and (d) radiolarian (SiO_2; approximately 50 μm wide). (Courtesy of R. Bernstein, University of South Florida.)

anaerobic decomposers that live on the sea floor or in the uppermost layers of sediment. As a result, only about 0.1% of the organic matter that settles from the surface ocean is preserved in marine sediments.

Most marine organic matter is decomposed by animals and microbes as it settles through the water column. This decomposition releases CO_2 (the product of both oxygen-breathing animals and microbial respiration) and nutrients to the oceanic deep waters. For marine organisms, nutrients lead to high rates of primary productivity if they are available in the appropriate concentrations. Thus, it is critical to the productivity of the marine biota to get these nutrients back to the surface.

The Biological Pump.

The overall effect of photosynthesis in shallow waters, of the settling of organic matter, and of decomposition in deep waters is the transfer of CO_2 and nutrients from the surface waters to the deep ocean. This process is known as the **biological pump** (Figure 8-9). It is balanced by the conveyor-belt-like thermohaline circulation of the ocean (Chapter 5), which brings nutrients and carbon-rich waters back to the surface, replenishing the nutrients and carbon removed by the biological pump.

The biological pump has a profound effect on ocean chemistry. As a result of its operation, surface waters are measurably depleted in carbon and severely depleted in phosphate and nitrate (the major nutrient elements) relative to deep waters. If the biological pump were to cease—for example, as the result of a mass extinction—the ocean would assume a more uniform composition in a few thousand years, as the thermohaline circulation homogenized the ocean.

Nutrient Limitation.

Some elements are classified as nutrients for marine phytoplankton because they are essential for growth and exist at suboptimal concentrations; an increase in their concentration leads to higher rates of primary productivity. Other elements are *toxic* to marine life—that is, they are poisonous to certain organisms—because their concentrations in seawater are above the optimum for growth. (Even nutrient elements can be toxic if their concentrations become too high.) There is great variability in the concentration of many elements throughout the world's oceans. Each element, depending on its concentration, can be either a nutrient or a toxic substance. The situation is analogous to the parabolic growth curves for daisies as discussed in Chapter 2 (Figure 2-9a), except that element concentration substitutes for temperature. For each element, there is some optimum concentration that favors biological productivity.

Marine phytoplankton incorporate many nutrient elements into their tissues in ratios that appear to be nearly identical in all species. These ratios are called **Redfield ratios**, in honor of Alfred C. Redfield, the oceanographer who first described this phenomenon. Even more remarkably, the ratios of many of these elements in seawater is nearly identical to that in phytoplankton. For example, the elemental ratio of carbon:nitrogen:phosphorus is very nearly 106: 16: 1 (Table 8-1). A chicken-or-egg question arises: Does the composition of seawater determine the composition of organisms that live in the sea, or does the composition of marine organisms determine the composition of seawater? To answer this question, we must consider the distribution of primary productivity in the oceans.

Where are rates of primary productivity greatest in the oceans? We now have a sophisticated way of answering this question, by satellite. With a color scanner, researchers can quantify the color of seawater from space (Figure 8-10). The color of the ocean surface is strongly influenced by the density of phytoplankton, which contain photosynthetic pigments. In the early 1960s and 1970s, oceanographers began to study the relationship between the concentration of these pigments and the abundance and productivity of near-surface-dwelling phytoplankton. The researchers found that regions of the ocean that have low concentrations of *chlorophyll* (a

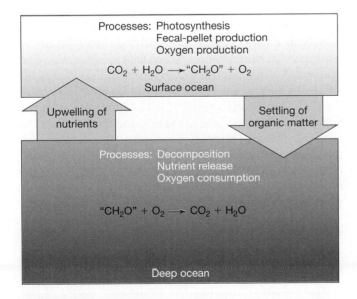

FIGURE 8-9

The marine biological pump.

TABLE 8-1

Redfield Ratios	
Element	Relative number of atoms in living phytoplankton
Carbon	106
Nitrogen	16
Phosphorus	1
Iron	0.01

FIGURE 8-10

[See color section] The concentration of photosynthetic pigments as determined by the Coastal Zone Color Scanner (CZCS) on the Nimbus 7 satellite. Pigment concentrations are indirect indicators of rates of primary production. (From Gene Feldman, NASA/GSFC/Science Photo Library, Photo Researchers, Inc.)

green pigment that is the dominant pigment in algae) are biological deserts, with low abundances of organisms and low nutrient concentrations. In contrast, more productive waters tend to have high chlorophyll concentrations.

Although we might expect that the warm, sunny low-latitude oceans would be most conducive to phytoplankton growth, the satellite patterns reveal that the waters with the highest productivities are the cold waters of the high-latitude Atlantic, Pacific, and Southern oceans. The reason is that the thermocline is weak or nonexistent at these latitudes because surface waters are just as cold as the deep waters there. Thus, wind-driven mixing of nutrient-rich deep waters up to the surface occurs much more readily at high latitudes than at low latitudes, where mixing is inhibited by the strong temperature and density gradients (Chapter 5). Apparently the inhibitory effects of cold temperatures and low light availability experienced by organisms that live at high latitudes is more than compensated for by an enhanced nutrient supply.

The satellite images also indicate high productivity in regions of upwelling. As we saw in Chapter 5, many of these regions occur along the western continental margins. There the wind-driven surface currents cause off-shore transport, which allows nutrient-rich waters from intermediate depths to well up to the surface. Other upwelling regions occur where surface currents diverge (for example, along the equator, due to the action of the trade winds and the Coriolis effect). Surface divergence also allows intermediate-depth waters to rise and replace the water transported away at the surface. Finally, the coastal regions of the continental shelves tend to be highly pro-

ductive because of nutrient inputs from rivers (often strongly enhanced by anthropogenic inputs of nitrogen and phosphorus) and localized upwelling driven by the topography of the sea floor.

Nutrient supply therefore seems to be a major limitation on the productivity of the surface ocean. Except in upwelling regions, phosphate and nitrate concentrations in surface waters are driven essentially to zero as a result of intense uptake by phytoplankton. Deep waters begin as surface waters, depleted in nutrients, at high latitudes. These waters increase in nutrient concentration as nutrients are released by decomposing organic matter that rains down from the overlying water column (the biological pump). In the North Atlantic, the effect of this "rain" on the composition of the water is small, because the water mass is young and has not had time to build up a large supply of organic matter. Farther along the conveyor belt's path, the aging water mass begins to show its increasing influence of the biological pump. Accordingly, the nutrient contents of deep North Pacific waters (the oldest waters in the ocean) are much greater (Figure 8-11a), and the O_2 contents much lower (Figure 8-11b), than those of the North Atlantic. Thus, the explanation for the observed similarity in nutrient elemental ratios between seawater and marine organisms—the Redfield ratios—is that the nutrient composition of the world's oceans is dominated by the production and decomposition of organic matter. In other words, the composition of marine organisms determines the composition of seawater.

The Long-Term Organic Carbon Cycle

The processes we have discussed thus far affect the atmospheric CO_2 balance on time scales shorter than a century. On longer time scales, these processes must be very closely in balance: Because the fluxes involved are so large, persistent imbalances would lead to intolerable fluctuations in atmospheric CO_2. Geological processes become the important controls on atmospheric CO_2 on longer time scales (Figure 8-12). The fluxes of carbon involved in these processes are small, and the reservoirs involved are large. Together these two adjustments in scale mean that the importance of the geological processes affecting the sedimentary reservoir and the atmosphere are influential only on long time scales (on the order of 1,000 years to 1 million years).

Carbon Burial in Sedimentary Rocks

The flux of land-derived and marine sediments to the sea floor fills sedimentary basins, many of which flank the margins of the continents. The continuous supply of sediment to these basins leads to the burial of previously deposited material. Eventually, as the process continues,

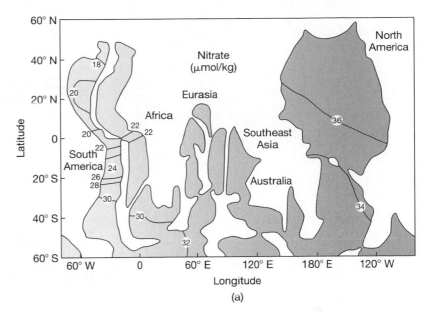

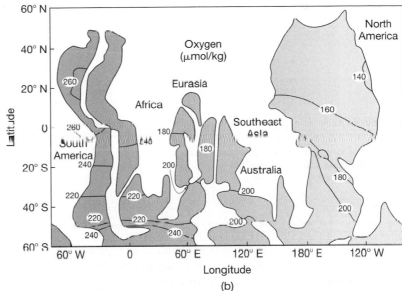

FIGURE 8-11

Concentrations of dissolved (a) nitrate and (b) O_2 in the deep ocean, in micromoles per kilogram of seawater. Maps are drawn for a water depth of 4000 m. At this depth the mid-ocean ridges appear as mountain ranges, and the continental outlines are substantially modified. (After W. S. Broecker and T.-H. Peng. 1982. *Tracers in the Sea*. New York: Eldigio Press, Columbia University, p. 31.)

sediments become buried to a depth of a few kilometers below the sea floor and become lithified. The organic carbon associated with these sediments is then entombed in sedimentary rock until weathering liberates the material to the biosphere.

Carbon Leaks and Oxygen Replenishment. This organic carbon burial represents a leak of material from the short-term organic carbon cycle (see Figure 8-12). It is this leak, rather than photosynthesis alone (which is nearly balanced by respiration and decay), that maintains the O_2 content of the atmosphere. Oxygen is continuously removed from the atmosphere by chemical reactions with reduced materials (especially organic matter) that are preserved in rocks exposed at Earth's surface and with re-

duced volcanic gases such as hydrogen, sulfur dioxide, and carbon monoxide. The loss of O_2 is very slow, but the O_2 concentration would reach zero in a few million years if that gas were not supplied from other sources. Oxygen is replenished by the leak of organic matter into the sedimentary rock reservoir. For every carbon atom that enters this reservoir as organic carbon, one oxygen molecule is left behind. This is because the O_2 liberated during the photosynthesis of that carbon was not utilized during respiration or decomposition, and thus the gas remains in the atmosphere.

Formation of Fossil Fuels. During burial, the organic material in the sediment also undergoes significant changes in its structure and chemistry, and fossil fuels may

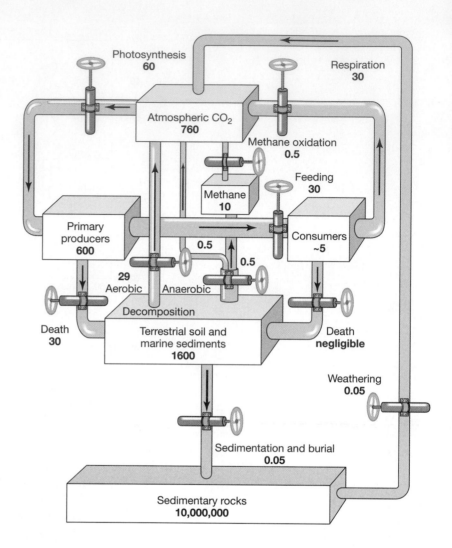

FIGURE 8-12

The combined short-term and long-term organic carbon cycles, showing the geological processes of sedimentation, burial, and weathering. Reservoir sizes in
Gt(C), fluxes in Gt(C)/yr.

form. Sediments derived from land plants (especially *peat*,
which forms in swamps and bogs) accumulate on land or
in basins near shore. If the concentrations of terrestrial
organic matter are high in these basins, burial processes
under high pressures and temperatures can lead to the
formation of **coal**. Similarly, high concentrations of organic matter in marine sediments can produce sedimentary rocks that, under high-pressure and high-temperature
conditions of burial, serve as sources of **petroleum**. This
material is fluid and tends to migrate through the basin
until it becomes trapped and accumulates. If the accumulation is sufficient, the petroleum will be an economical source of fossil fuel. More commonly, however, the
concentration of organic matter is 1% or less of the total
sediment material. As a result, most sedimentary rocks
do not represent economically viable energy sources.

The Sedimentary Organic Carbon Reservoir. Sedimentary rocks contain by far the greatest quantity of organic carbon on Earth: approximately 10^8 Gton(C) (see
Figure 8-3). Most of the organic carbon is found in *shales*,
which are fine-grained sedimentary rocks formed by the

lithification of muds. The residence time for the sedimentary organic carbon reservoir is about 200 million
years.

Weathering of Organic Carbon in Sedimentary Rocks

Weathering of the organic carbon in sedimentary rocks
is an oxidation process requiring atmospheric O_2, either
by direct exposure to the atmosphere or by exposure to
groundwaters containing dissolved O_2. The oxidation of
this material can be represented by the same schematic
chemical reaction that we used previously for respiration
and aerobic decomposition. The organic matter reacts
with oxygen, releasing carbon dioxide to the atmosphere
or groundwater.

In this sense, the mining, pumping, and combustion of
fossil fuels represent merely an acceleration of the weathering process. The rocks from which humans have removed these fuels would likely have become exposed at
the surface and undergone oxidation to form CO_2 some
time in the distant future. Human intervention, however,

has speeded up this process by a factor of a million or more for these fossil-fuel deposits. The release of organic matter from sedimentary rocks is occurring much faster than it can be replaced. Hence fossil fuels represent only a short-term energy source.

Summary of the Organic Carbon Cycle

We have now explored the entire global organic carbon cycle. Pathways exist to recycle carbon from all reservoirs (see Figure 8-12). Every reservoir in this cycle is directly connected to the atmosphere. Thus, the CO_2 concentration of the atmosphere changes continuously in response to changes in the flux of carbon to and from these reservoirs. The responses to these changes are rapid for the large fluxes associated with the biota, soils, and marine sediments but slow for the small fluxes from the sedimentary rock reservoir. This observation will prove to be important in later chapters when we consider the fate of CO_2 added to the atmosphere from the burning of fossil fuels.

The Inorganic Carbon Cycle

The photosynthesis of CO_2 to reduced carbon in organic matter, and its subsequent reoxidation to CO_2 through respiration, decomposition, and weathering, is central to the organic carbon cycle. But there are other sources and sinks for atmospheric CO_2. Carbon dioxide readily dissolves in rainwater and seawater and then undergoes rapid chemical reactions to other ionic forms of inorganic carbon. The oxidized carbon in these waters is chemically reactive and becomes involved in a number of chemical processes. Because they do not involve organic carbon directly, these processes are together referred to as the **inorganic carbon** *cycle*.

The important reservoirs of inorganic carbon are the atmosphere, which we have discussed at length; the ocean; sediments; and sedimentary rocks (see Figure 8-3). The sediment and sedimentary-rock carbon reservoirs consist primarily of limestone. **Limestone** is a rock composed largely of *calcium carbonate* ($CaCO_3$), generally in the form of the mineral *calcite*. The magnesium-rich carbonate mineral *dolomite*, $CaMg(CO_3)_2$, is abundant in older sedimentary rocks.

Carbon Exchange between Ocean and Atmosphere

Carbon dioxide is continuously exchanged between the atmosphere and ocean. The distribution of sources and sinks of CO_2 is tied to the circulation and productivity patterns of the oceans (Figure 8-13). In regions of the ocean where high rates of primary productivity have created surface waters with low CO_2, CO_2 *diffuses* from the atmosphere to the ocean. In other words, the net flow of CO_2 is down the concentration gradient—from regions of higher CO_2 concentration (in this case, the atmosphere) to regions of lower concentration (the ocean). Conversely, upwelling regions, such as the equatorial Pacific surface waters, have high CO_2 concentrations because deep wa-

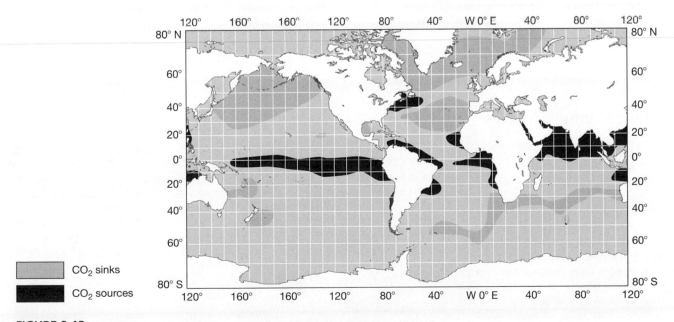

FIGURE 8-13
Oceanic sources (darker shading) and sinks (lighter shading) of atmospheric CO_2. Sources have CO_2 concentrations larger than those in equilibrium with the atmosphere, whereas sinks have lower-than-equilibrium CO_2 concentrations. (After T. Takahashi. 1989. *Oceanus*, 32, pp. 22–29.)

ters rich in CO_2 (as the result of decomposition associated with the biological pump) have risen to the surface there. In such regions, CO_2 flows from the ocean to the atmosphere. Thus, the oceans serve as both a source and a sink for atmospheric CO_2.

Before the carbon cycle was disturbed by human activity, the flux of CO_2 from oceanic CO_2 source areas was probably closely balanced by the flux to oceanic CO_2 sinks. This pattern is changing in response to the burning of fossil fuels. As we have seen, the atmospheric CO_2 concentration has been increasing. Regions of the ocean that were previously weak sources have now become sinks, and the ocean as a whole has become a sink for CO_2.

The Chemistry of Inorganic Carbon in Water

In the oceans, inorganic carbon exists in a number of dissolved forms. When CO_2 dissolves in water (whether it be freshwater or seawater), carbonic acid is generated:

$$CO_2 + H_2O \longleftrightarrow H_2CO_3. \qquad (1)$$

carbon dioxide   water   carbonic acid

The double arrows in this reaction indicate that the reaction can proceed both forward and backward. The rates of the forward and reverse reactions depend on the concentrations of the *reactants* (on the left-hand side) and of the *products* (on the right-hand side), respectively. If the concentration of the reactants is high, the forward reaction proceeds fast, depleting the concentration of the reactants and enhancing that of the products until the forward rate matches the reverse rate. Conversely, if the concentration of the products is high, the reverse reaction proceeds more rapidly, and the products become depleted (and reactants enriched) until the rates balance. In other words, *chemical equilibrium* is rapidly achieved in such reactions.

Like many substances that dissolve in water, carbonic acid molecules break apart, or *dissociate*, into ions. Ions are charged atoms or molecules. Ions with negative charges are referred to as **anions**. Ions with positive charges are referred to as **cations**. When carbonic acid dissociates, carbon-bearing anions and hydrogen cations are formed. The relative abundance (concentration) of the carbon anions is thus linked to the *pH* of seawater (see Box, "Useful Concepts: pH"), which is a measure of the concentration of hydrogen ions in solution.

Carbonic Acid, Bicarbonate, and Carbonate Ion Equilibrium. Returning to the discussion of the forms of inorganic carbon in water, we see that the dissociation of carbonic acid involves the release of one or both of its hydrogen atoms to yield carbon–oxygen anions. When the first hydrogen atom is lost, *bicarbonate ion* (HCO_3^-) is formed:

$$H_2CO_3 \longleftrightarrow H^+ + HCO_3^-. \qquad (2)$$

carbonic acid   hydrogen ion   bicarbonate ion

If the H^+ concentration were to decrease (i.e., the pH increased), more carbonic acid would dissociate to balance the equilibrium between the two forms (the reaction would proceed to the right). If, instead, seawater were to become acidic, this equilibrium would shift to the left, forming carbonic acid at the expense of bicarbonate ion.

A CLOSER LOOK
Other Element Cycles

Our focus in this chapter and throughout the book is on the carbon cycle because of its paramount importance to the functioning of the Earth system. However, most elements on the periodic table are recycled along with carbon on a variety of time scales. Rocks exposed to weathering contain a host of elements that are carried by rivers to the ocean, taken up by marine organisms or scavenged onto settling particles, removed to sediments and sedimentary rocks, and entrained in the rock cycle.

Two elements are of particular interest: nitrogen (N) and phosphorus (P). These two elements are required by all living organisms for the synthesis of such essential compounds as proteins and ATP (adenosine triphosphate, an important molecule in metabolism). Generally they are not provided in the necessary proportions for growth, and so one becomes limiting: typically P on land and N in the sea. Organisms have evolved special enzymes and metabolic pathways to facilitate the extraction of these nutrient elements from otherwise unavailable forms (e.g., *alkaline phosphatase* to release P from organic compounds, and *nitrogenase* to convert N_2 to nutrient ammonia), thus speeding the recycling of the elements.

Biological processes tend to accelerate the recycling of nutrient elements relative to their slow recycling with the rock cycle. This biological acceleration can be as high as a factor of 1000 for phosphorus. As a result, the productivity of the world's biota is many times greater than it would be without biological acceleration.

The release of the second hydrogen ion converts bicarbonate ion in the previous reaction to *carbonate ion* (CO_3^{2-}):

$$\underset{\substack{\text{bicarbonate} \\ \text{ion}}}{HCO_3^-} \longleftrightarrow \underset{\substack{\text{hydrogen} \\ \text{ion}}}{H^+} + \underset{\substack{\text{carbonate} \\ \text{ion}}}{CO_3^{2-}}. \qquad \textbf{(3)}$$

(Note that we have previously used the term "carbonate" to refer to calcium and magnesium carbonate minerals.) For a given hydrogen ion concentration (pH), the relative amounts of bicarbonate and carbonate ions are adjusted until equilibrium is achieved. A pH above neutral favors the carbonate ion, whereas a pH below neutral favors the bicarbonate ion.

USEFUL CONCEPTS
pH

The hydrogen ion H^+ is the smallest of all cations. The small size and ionic charge make hydrogen ions extremely reactive: They tend to infiltrate solids, breaking bonds and causing the molecules that make up the solids to dissolve. Solutions (liquids with dissolved material) with high concentrations of hydrogen ions are called **acids**. Vinegar and hydrochloric acid are common acids. Solutions with low concentrations of hydrogen ions are called **bases** (or alkalis). Baking soda and lye dissolved in water are common bases.

Strong acids are solutions that completely dissociate: When dissolved in water, the anions separate completely from the hydrogen cations. This dissociation leads to high hydrogen ion concentrations. For example, the strong acid hydrochloric acid (HCl) dissociates to form hydrogen ions and chloride ions:

$$\underset{\substack{\text{hydrochloric} \\ \text{acid}}}{HCl} \longrightarrow \underset{\substack{\text{hydrogen} \\ \text{ion}}}{H^+} + \underset{\substack{\text{chloride} \\ \text{ion}}}{Cl^-}.$$

Other strong acids include nitric acid (HNO_3) and sulfuric acid (H_2SO_4). Weak acids, such as boric acid (H_3BO_3) and carbonic acid (H_2CO_3), only partially dissociate when dissolved in water.

The concentration of dissociated hydrogen ions (expressed in moles per liter of solution) determines the acidity of the solution. Acidity is commonly measured by the pH scale, which is a logarithmic scale. The pH of a solution is a close approximation of the negative of the logarithm (to the base 10) of the hydrogen ion concentration, $[H^+]$ (in moles per liter):

$$pH = -\log [H^+].$$

(We say "approximation" because chemists define pH in terms of the activity of hydrogen, which is the concentration of hydrogen ions available for chemical reaction. A small fraction of the hydrogen ions are involved in electrostatic interactions with other ions.) For example, a solution with a hydrogen ion concentration of 10^{-4} mol per liter would have a pH of 4, because the logarithm of 10^{-4} is -4. At room temperature, pure water is defined as having a neutral pH, or a pH of exactly 7. Solutions with a pH less than 7 are acids, whereas solutions with a pH greater than 7 are bases (Box Figure 8-2). The pH of the surface and deep oceans is slightly basic—about 8 and 7.5, respectively. Rainwater that is in equilibrium with atmospheric CO_2 has a slightly acidic pH, between 5 and 6. Most lakes, rivers, and streams range in pH from about 6 to 9, that is, from slightly acidic to mildly basic.

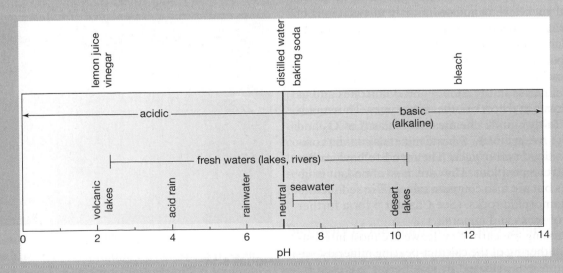

BOX FIGURE 8-2
The pH scale.

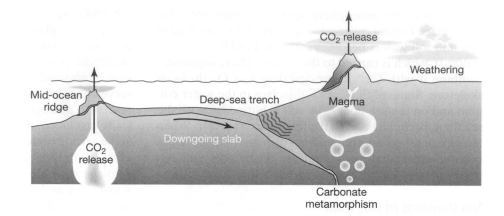

FIGURE 8-17

Pictorial representation of the carbonate–silicate geochemical cycle.

that transform the sediments into metamorphic rock. Among these reactions is the reaction between sedimentary carbonate minerals and silica-rich sediments that forms silicate minerals and releases CO_2:

Carbonate metamorphism:

$$CaCO_3 + SiO_2 \longrightarrow CaSiO_3 + CO_2.$$
calcite   silica   wollastonite   carbon dioxide

This process is termed **carbonate metamorphism**. As before, we use the mineral wollastonite to represent the more complex silicate minerals that are typically generated by this process.

If sufficiently high temperatures are reached at depth during carbonate metamorphism, magmas are generated. These magmas may erupt in volcanoes at the surface, releasing CO_2 to the atmosphere. The CO_2 in these volcanoes probably includes some mantle-derived CO_2 and some CO_2 from the subducted crust and sediments, but scientists do not yet know the relative proportions of these two sources. Under metamorphic conditions the CO_2 produced can migrate as a fluid toward the surface. Although a substantial fraction of it reacts with minerals along the way, some CO_2 is released through springs and seeps to the atmosphere.

Together, silicate weathering, carbonate precipitation, and ocean-atmosphere exchange are the reverse of carbonate metamorphism. (Compare the equation for the net result of these processes, given in the previous section, with the carbonate metamorphism equation.) Without a fairly close balance between the inflows and outflows of carbon, the supply of this important greenhouse gas to the atmosphere and ocean would, on geological time scales, quickly be depleted. Earth would soon become a frozen ball of ice.

That may have been the fate of Mars, which appears to once have had flowing water on its surface and, perhaps, a CO_2-rich atmosphere with a stronger greenhouse

effect. In contrast, on Earth the release of CO_2 after carbonate metamorphism and volcanism has essentially balanced the consumption of CO_2 during silicate weathering over the history of the planet. What has ensured this balance? We cannot call on simple chemical equilibrium, because the reactions involved are representative of a whole host of processes rather than a single chemical reaction. Rather, we must look for feedback loops that, according to the amount of CO_2 in the atmosphere, adjust the rates of CO_2 input by volcanism or of CO_2 removal by silicate weathering and thereby keep the reservoir at steady state.

Long-Term Feedbacks in the Carbonate–Silicate Cycle

Because it is driven largely by heat flow from the Earth's interior, the rate of volcanism is probably not very sensi-

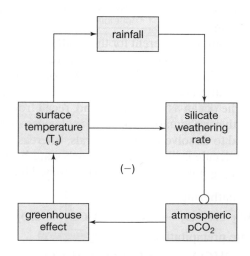

FIGURE 8-18

Systems diagram showing the negative feedback loop that results from the climate dependence of silicate–mineral chemical weathering and its effect on atmospheric CO_2. This feedback loop is thought to be the major factor regulating atmospheric CO_2 concentrations and climate on long time scales.

tive either to the amount of CO_2 in the atmosphere or to the climate of Earth's surface. In contrast, many climatic factors affect the rate of chemical weathering. The regulation of atmospheric CO_2 on long time scales (millions of years) likely is the consequence of the feedback between climate and rates of silicate weathering. The climatic factors that help regulate the chemical weathering rates of silicate rocks include the following:

- Temperature: rates of reactions, including chemical weathering, tend to increase as temperature increases.
- Net rainfall: weathering requires water as a medium both for the dissolution of minerals and for the transport of the dissolved material to the oceans, and thus weathering rates rise as precipitation increases.

These environmental factors are responsive to atmospheric CO_2 levels. Recall from Chapter 3 that, as a result of the greenhouse effect, global temperatures rise as the atmospheric content of CO_2 increases and that rates of evaporation increase with increasing temperature. We know from Chapter 4 that water that evaporates from the ocean must fall as precipitation. For these reasons, we would expect that a warmer world would be a wetter world: Net precipitation should increase as temperature increases. Thus, the silicate weathering rate should increase as the atmospheric CO_2 level rises. Figure 8-18 shows a feedback loop for these processes. On the other side of the feedback loop, increased silicate weathering rates tend to reduce atmospheric CO_2 levels because silicate weathering uses up carbonic acid.

A CLOSER LOOK
Biological Enhancement of Chemical Weathering

A particularly active and controversial area of geological research centers on the extent to which terrestrial biota may enhance rates of chemical dissolution of minerals exposed at Earth's surface. Several factors are involved, including the following:

- Microbial decomposition and respiration by plant roots in soils generate large amounts of carbon dioxide. The CO_2 concentration in soils is generally 10 to 100 times that in the atmosphere.
- Microbial decomposition also releases organic acids that cause mineral dissolution. By increasing weathering rates, these biological processes enhance the release rate of essential mineral nutrients to the biota.
- Plant root development leads to the stabilization of soils on steep slopes, where erosion would otherwise strip soils away. This process creates a more stable weathering environment where mineral dissolution can take place for longer times before erosion takes place.
- Roots penetrating through fractured rock tend to enlarge the fractures, causing further disintegration of the rock and allowing water and soil acids to penetrate deeper into the rock.

Field experiments to test the hypothesis of biological enhancement of chemical weathering span environments that have huge contrasts in climate (Iceland and Hawaii) but similar types of exposed rocks (basalts). In Iceland, scientists have found that streams draining areas of plant cover have significantly higher dissolved ion concentrations than do those draining areas of bare rock.

In Hawaii, scientists are looking at rock surfaces covered with lichens (Box Figure 8-3) and comparing them with bare surface exposures. Underneath the lichen there generally is a thin layer of material that is leached of the more soluble elements; bare surfaces have no such layer. These

leached layers appear to be the result of mineral dissolution aided by the production of organic acids, although some scientists argue that the leached layers actually are wind-blown dust trapped by the lichens. Further research is needed to resolve this controversy.

Calculations of the extent of biological enhancement from various field studies are somewhat wide-ranging but indicate that biological activity accelerates rates of chemical weathering by at least a factor of 2. Some lichens studies have concluded that that factor is several hundred times. In any case, biological processes are very important components of the feedback loop that regulates global chemical weathering rates, and thus atmospheric CO_2 levels, on geological time scales.

BOX FIGURE 8-3

[See color section] The lichen *Xanthoria parietina* and other lichens on a seashore rock, England. (*Source:* Biophoto Associates/Photo Researchers Inc.)

The overall feedback loop, as shown in Figure 8-18, is negative. The feedback tends to stabilize Earth's climate against perturbations, as we will see later.

Links between the Organic and Inorganic Carbon Cycle

Although we have differentiated between the terrestrial and marine organic carbon cycles and between the organic and inorganic carbon cycles, all these cycles are inextricably linked as parts of the global carbon cycle. Changes that occur on land rapidly affect the oceans through the transport of carbon and nutrients by rivers and through variations in atmospheric CO_2. Changes in the recycling of organic matter affect atmospheric and oceanic CO_2 and in turn the whole of the carbon cycle. Thus, the cycle divisions we have made are artificial, but they help us represent a complicated system in simple terms.

Chapter Summary

1. The global carbon cycle involves processes that occur on land and in the oceans and involve both biological and non-biological chemical reactions.
2. The terrestrial and marine organic carbon cycles operate on a variety of time scales. On the short time scale (tens to hundreds of years),
   a. carbon dioxide is removed from the atmosphere during photosynthesis on land and returned during respiration and decomposition; methane is released to the atmosphere from soils, where anaerobic metabolism is taking place.
   b. a small amount of terrestrial organic carbon survives respiration and decomposition and is buried in sedimentary basins on land or is transported to the sea.
   c. in the oceans, phytoplankton produce organic matter that is consumed by zooplankton and decomposed by aerobic and anaerobic bacteria.
   d. a small fraction of the organic matter settling through the water column is not decomposed and is instead buried in marine sediments.
3. On longer time scales (millions of years), the organic matter buried in sediments undergoes lithification with the sediments. Most of these sediments are muds, and the rocks formed are shales.
   a. When concentrations of organic matter are very high in the sediment, fossil fuels may form during burial and lithification.
   b. The sedimentary rocks may eventually undergo uplift through tectonic processes, exposure, and weathering. During weathering, organic matter undergoes oxidation, producing CO_2, which escapes to the atmosphere.
4. An inorganic carbon cycle, involving oxidized forms of carbon, is important on both short and long time scales (millions of years).
   a. Atmospheric CO_2 exchanges with CO_2 dissolved in the surface ocean on a time scale of decades. The uptake of CO_2 is enhanced by reactions among the forms of dissolved inorganic carbon in seawater.
   b. Atmospheric CO_2 dissolves into rainwater, creating an acidic solution. When the rain falls on the land surface, reactions with carbonate and silicate minerals convert carbonic acid to bicarbonate ion. The bicarbonate ion is carried by rivers to the oceans.
   c. In the oceans, carbonate-secreting organisms use the bicarbonate ion in the construction of their shells or skeletons. This material may dissolve in transit or it may become part of the sediment that covers the sea floor.
5. The regulation of atmospheric CO_2 on long time scales (millions of years) is the consequence of the feedback between climatic factors and rates of chemical weathering of silicate rocks, as part of the long-term inorganic carbon cycle, referred to as the carbonate–silicate geochemical cycle.
   a. Any disturbance in the amount of atmospheric CO_2 affects climate through the greenhouse effect. Changes in climate affect silicate weathering rates and thus the rates of CO_2 consumption.
   b. The overall feedback loop is negative, implying that on long time scales the climate system is stable against a wide range of perturbations.
6. The terrestrial and marine organic carbon cycles, the organic and inorganic carbon cycles, and the carbonate–silicate geochemical cycle are inextricably linked as parts of the global carbon cycle.
   a. Changes on land are "communicated" rapidly to the oceans by riverine transport of carbon and nutrients and through variations in atmospheric CO_2.
   b. Changes in the recycling of organic matter affect atmospheric and oceanic CO_2 and in turn the whole of the carbon cycle.

Key Terms

acid	biomass	characteristic response time
anion	biosphere	coal
base	carbonate metamorphism	consumers
biological pump	cation	inorganic carbon

limestone
nutrient
organic carbon
oxidized carbon
petroleum

pH
photic zone
photosynthesis
primary producer
primary productivity

Redfield ratio
reduced carbon
residence time
zooplankton

Review Questions

1. Which of the following carbon reservoirs has the longest residence time: plants, the oceans, or sedimentary limestone?

2. One or more of the following processes involves organic carbon; identify it (them): the precipitation of a calcite skeleton, the exchange of carbon between the oceans and the atmosphere, dissolution at the sea floor, or oxidation during weathering?

3. Describe the biological pump.

4. Why is plate tectonics critical to the maintenance of an atmosphere–ocean reservoir rich in carbon?

5. Limestone (carbonate) weathering does not lead to the net removal of carbon dioxide from the atmosphere. Why not?

Critical-Thinking Problems

1. The key to stability is feedback between the reservoir and the fluxes into and/or out of the reservoir. Assume that the rate of outflow from a reservoir depends on the size of the reservoir according to the following relationship: outflow rate = k • (size of reservoir), where k is a constant.

   a. A reservoir of water has a volume of 5000 liters, and the rate of outflow at steady state is 25 liters per minute. What is k? (Give both the numerical value and its units.) What is the residence time? What is the relationship between k and the residence time?

   b. The inflow rate is 25 liters per minute. Describe graphically and in words how the reservoir size would change with time, beginning with a reservoir size of zero and continuing until the reservoir reaches steady state.

2. Use Figure 8-4 to answer the following questions:

   a. During which months is the rate of photosynthesis greatest, relative to the combined rate of respiration and decomposition, and during which months is it smallest? Explain your reasoning. Why aren't these coincident with the minimum and maximum CO_2 levels for the year, respectively?

   b. On the basis of your answer to part (a), estimate, for each year, the maximum net rate of photosynthesis and the maximum net rate of respiration/decomposition for each of the three years shown.

   c. Are there significant differences in these rates from year to year? If so, propose an explanation for them.

3. A giant meteor crashes into Earth, causing devastating environmental changes that kill off all life in the oceans.

   a. Describe how the vertical distribution of dissolved oxygen, carbon, and nutrients would respond.

   b. Would the temperature and salinity of the ocean be affected by the loss of the biological pump? Why or why not?

   c. If global warming from CO_2 released to the atmosphere from the meteor impact site caused, instead of the com-

plete loss of marine life, the sudden cessation of thermohaline circulation in the oceans, what would be the effect on the vertical and spatial distribution of dissolved nutrients, carbon, and oxygen in the world's oceans?

4. a. The atmosphere consists of 78% N_2, 21% O_2, 1% Ar, and about 0.036% (360 ppm) CO_2. What is the mean molecular weight of air? Round your answer to three significant figures, and use the following table of atomic weights:

Element	Atomic Weight
C (Carbon)	12.011
N (Nitrogen)	14.0067
O (Oxygen)	15.9994
Ar (Argon)	39.948

   b. The total mass of the atmosphere is about 5×10^{18} kg. How many moles each of air, O_2, and CO_2 are present in the atmosphere? (Note: Calculate the latter two answers from the first one rather than by computing the masses of O_2 and CO_2. The values listed in part (a) for the various gases are abundances by volume, not by mass. This fact, and the fact that a mole of any gas takes up the same volume at a given pressure and temperature, mean that you need to work in moles.)

   c. Forests contain about 600 Gton(C) in the form of wood and leaves. Suppose that all the world's forests were to burn down instantaneously. By how much would atmospheric CO_2 increase? By how much would O_2 decrease? Express your answers in percentages. Assume that the equation for burning is the same as that for respiration (given earlier in this chapter).

5. Explain why lakes and rivers have slightly basic pH values, whereas rainwater (the ultimate source of water for lakes and rivers) is slightly acidic.

Further Reading

General

Mackenzie, F. T. 1995. "Biogeochemistry." In *Encyclopedia of Environmental Biology*. New York: Academic Press, vol. 1, p. 249.

Advanced

Post, W. M., Peng, T.-H., and Emanuel, W. R. 1990. "The Global Carbon Cycle." *American Scientist,* 78(4):310.

Wigley, T. M. L. and Schimel, D. S. 2000. *The Carbon Cycle.* Cambridge: Cambridge University Press.

Hanson, R. B., Ducklow, H. W. and Field, J. G. 2000. *The Changing Ocean Carbon Cycle.* Cambridge: Cambridge University Press.

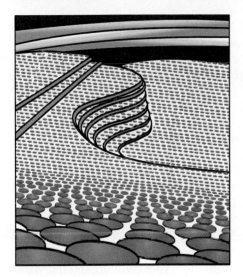

Focus on the Biota: Metabolism, Ecosystems, and Biodiversity

Key Questions

- What are the characteristics of life on Earth that allow it to interact with physical processes at the global scale in such a significant way that it creates a habitable planet?
- How is the biosphere structured?
- How is energy transferred within the biosphere?
- What is an ecosystem?
- What is biodiversity and how is it measured?
- How is the diversity of interactions between the biota and the physical world related to the stability of the Earth system?

Chapter Overview

In this chapter we highlight the role that life plays in the operation of the Earth system. We begin with a general discussion of life and its unique characteristics, and then explore the varied metabolic pathways different forms of life take to grow and reproduce. Organisms interact at a variety of scales, so we find that populations of organisms group into communities, which at a larger scale interact among themselves and with their environment in ecosystems. The level of diversity of ecosystems can be expressed in a variety of ways. We can simply count the number of species or we can take into account the more complex diversity of interactions that take place

between species and between organisms and their environment. This diversity of interactions, a defining characteristic of life on Earth, is important in our understanding of the feedbacks between the biota and the physical world that create a habitable planet, and helps us further understand the complexity of the Earth system.

Life on Earth

Characteristics of Life

Earth is unique among the planets in our solar system in that it apparently is the only one to support life. Earth more than supports life, it flaunts it. Life is involved in almost every process occurring at the surface of the planet. Some fundamental characteristics of life allow it to have such an influence.

- *Life spreads exponentially.* The rate of population growth depends on the number of individuals reproducing at a particular time. This characteristic leads to the phenomenon of *exponential growth.* If left unchecked, 2 individuals become 4 in one generation, 4 become 8 in two generations, 8 become 16 in three generations, and 16 become 32 in four generations. In nature, however, exponential growth ceases as resources become limiting.
- *Life needs energy.* Photosynthesizers use solar energy, *chemosynthesizers* use chemical energy, and most other

organisms utilize the chemical energy that is packaged into the material produced by photosynthesizers and chemosynthesizers.

- *Life pollutes.* Every organism needs to metabolize, and when it does so, it releases waste products. These waste products can be of use to other organisms, and they may affect the environment (e.g., the release of the greenhouse gases CO_2 and CH_4 through respiration and decomposition).

- *Life is versatile.* There is considerable versatility in how organisms interact with each other and with the environment. Plants and animals exist in a variety of forms and express various behaviors. But their versatility is modest compared to that of microbes. Microbes express a wide array of metabolic activities that have tremendous impact on the environment and allow them to occupy a wider range of environments than eukaryotes.

All these characteristics of life allow it to interact with the physical processes that occur on the planet in such a way that Earth is a habitable planet. Let's explore this in more detail by developing a classification scheme for life that is based on metabolic rather than genetic similarities, and is structured around the flow of energy through the ecosystem.

Autotrophs and Heterotrophs

Although life can be categorized *taxonomically* (according to species, genera, families, etc.), a classification system that focuses on the ways in which organisms obtain energy and metabolize it is more useful from an Earth systems point of view (Table 9-1). As we saw in Chapter 8, the most fundamental distinction is between those organisms that grow using a source of energy to reduce carbon dioxide to organic carbon (primary producers or **autotrophs**) and those that require organic matter to grow (consumers or **heterotrophs**). Autotrophs include plants, algae, and a host of microbes that can photosynthesize (e.g., cyanobacteria, purple and green sulfur bacteria) or chemosynthesize (e.g., colorless sulfur bacteria). These autotrophic organisms produce organic matter from inorganic carbon sources, a process that requires energy (i.e., the chemical reactions do not occur spontaneously in nature). In the case of photosynthesis, the sun provides the necessary energy. In chemosynthesis, energy-releasing inorganic chemical reactions (those that occur even without the involvement of organisms *because* they release energy), often involving oxygen and reduced compounds (see Chapter 8), are the energy source. Chemosynthesis is the mechanism of primary production of the midocean-ridge hydrothermal vent communities that exist at great depths in the ocean where sunlight does not penetrate (see Chapter 7). The organic material that autotrophs produce is a storehouse of energy, and will decompose *abiotically* (without the intervention of organisms), albeit at a slow rate, releasing that energy as heat. Heterotrophs simply accelerate these chemical reactions that would otherwise proceed at a slower pace abiotically, and in doing so, gain the energy they need to grow and reproduce.

The heterotrophic pathway that releases the most energy is aerobic respiration, which uses molecular oxygen to decompose organic matter through the process of oxidation, converting the organic carbon to carbon dioxide.

TABLE 9-1

Metabolic Pathways for Life[1]				
General Method for Acquiring Energy	*Specific Pathway*	*Subcategory*	*Reactants*	*Byproducts*
Autotrophy				
	Photosynthesis		Solar energy, CO_2	
		Oxygenic	H_2O	Molecular oxygen (O_2)
		Anoxygenic	Molecular hydrogen (H_2), reduced sulfur or reduced iron	Oxidized sulfur (native sulfur or sulfate), iron oxide (solid)
	Chemosynthesis		H_2, reduced forms of sulfur, nitrogen, iron or manganese	Oxidized sulfur, nitrate, iron and manganese oxides (solids)
Heterotrophy			Organic matter	
	Aerobic Respiration		O_2	CO_2, H_2O
	Anaerobic Respiration		Nitrate, sulfate, iron and manganese oxides	CO_2 and molecular nitrogen, ammonia, hydrogen sulfide, reduced and dissolved iron and manganese
	Fermentation		Complex organic molecules	Simple organic molecules

[1] After K.H. Nealson and D.A. Stahl, Chapter 1 in *Geomicrobiology, Interactions between Microbes and Minerals*, Rev. in Mineral., **35,** 5–34.

In environments where oxygen isn't present (e.g., in muds on the seafloor and lake bottoms, and in the guts of animals), anaerobic heterotrophs, especially bacteria, substitute other oxidized inorganic compounds in lieu of oxygen to decompose the organic matter. Bacteria use such oxidants as dissolved nitrate (through a process known as *denitrification*) or sulfate (sulfate reduction) or particulate metal oxides of iron and manganese. Other heterotrophic organisms (certain fungi such as yeasts and some bacteria) perform fermentation, an important process that breaks down large, complex organic compounds into simpler ones that can be used by other heterotrophs. Fermenters do not oxidize organic matter, but they are able to utilize the energy that is released when complex organic materials are broken apart.

Methanogenic bacteria are an important group of organisms for our consideration of the Earth system, in particular because they may have been very significant in biogeochemical cycling on the early Earth and because they produce an especially effective greenhouse gas, methane (CH_4), through their metabolism. In fact, the word *methanogenic* means methane-producing. Methanogens can be either autotrophic or heterotrophic:

Autotrophic methanogenesis: $CO_2 + 4H_2 \rightarrow CH_4 + 2H_2O$
Heterotrophic methanogenesis: $CH_3COOH \rightarrow CH_4 + CO_2$

Autotrophic methanogenesis takes advantage of the energy yield of the chemical reaction between carbon dioxide and molecular hydrogen (H_2) when H_2 concentrations are relatively high, a situation that currently occurs in organic-rich muds and may have also occurred on the early Earth surface. Heterotrophic methanogens utilize the simpler carbohydrates (such as acetic acid, CH_3COOH, shown above) produced through fermentation. Both pathways produce methane, and heterotrophic methanogenesis produces both methane and carbon dioxide. Given their global abundance, it is clear that these bacteria can have a significant impact on the greenhouse effect.

Moreover, as James Lovelock (see Chapter 2) pointed out long ago, the combined activity of methanogens (such as methanogenic bacteria) and oxygenic photosynthesizers (such as plants), which produce the organic matter that the fermenters convert to acetic acid, releases both oxygen and methane to the atmosphere. We can represent this chemically:

Oxygenic photosynthesis:
$$2CO_2 + 2H_2O \rightarrow 2\text{ ``}CH_2O\text{'' } + 2O_2$$
Fermentation:
$$2\text{ ``}CH_2O\text{''} \rightarrow CH_3COOH$$
Heterotrophic methanogenesis:
$$CH_3COOH \rightarrow CH_4 + CO_2$$

NET: $\qquad\qquad CO_2 + 2H_2O \rightarrow CH_4 + O_2$

The net effect of these coupled processes creates an unstable, highly reactive, far-from-chemical-equilibrium atmosphere that is as strong a signature of life on our planet as any other. Equally amazing is the dynamic stability of this reactive atmosphere. As you'll see in later chapters, aerobic life has persisted on Earth for hundreds of millions of years, indicating that the atmosphere has remained oxygen-rich through this interval of Earth history. Strong feedbacks must exist to maintain atmospheric compositions over geologic time intervals. As you will see in the following discussion of ecosystems, the constant and complex interaction between all living things on Earth contributes to the atmospheric conditions that are key to the stability of the Earth system.

Structure of the Biosphere

The metabolic processes we have just described represent the main ways in which organisms interact with other species and with their environment. These interactions are not random. Rather, they make up higher levels of organization that we can recognize and study. Recall from Chapter 1 that the *biosphere* comprises that part of Earth inhabited by organisms; it includes both living and non-living components. A simple hierarchy has been developed that subdivides the biosphere (Figure 9-1). The smallest subunit is the *species*, which consists of all closely related organisms that can potentially interbreed. (Note that this definition only applies to species that reproduce sexually.) All the members of a single species that live in a given area make up a **population**. In any area you will tend to find a characteristic assemblage of two or more groups of interacting species, known as a **community**. A community may include any combination of animals, plants, fungi, and microbes. A region with a characteristic plant community (such as a desert or tropical rainforest) is called a **biome**. A community of animals, plants, fungi, and microbes, together with the physical environment that supports it, is referred to as an **ecosystem**. All the ecosystems on Earth in turn make up the biosphere. Although it is usual to discuss biodiversity and extinction (a topic of a later chapter) in terms of species, it is important to recognize that no one species exists independent of the other species around it. Species coexist and interact with a specific assemblage of other species and with their environment in ecosystems.

Ecosystems

What are Ecosystems?

As we have said, ecosystems are subsets of the (global) biosphere, assemblages of animal, plant, fungal, and microbial species that interact with each other and their

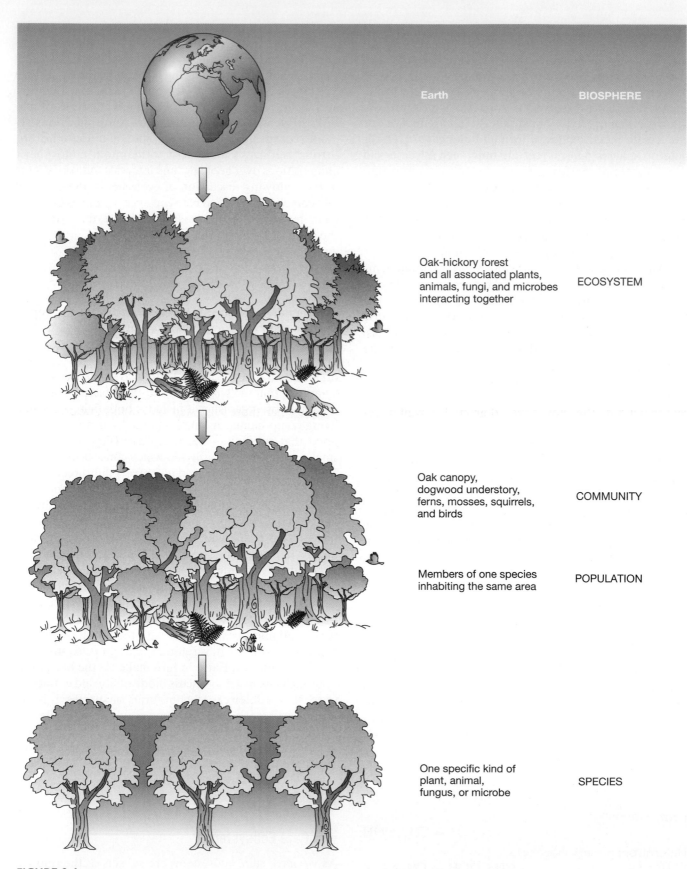

Earth   **BIOSPHERE**

Oak-hickory forest
and all associated plants,
animals, fungi, and microbes
interacting together   ECOSYSTEM

Oak canopy,
dogwood understory,
ferns, mosses, squirrels,
and birds   COMMUNITY

Members of one species
inhabiting the same area   POPULATION

One specific kind of
plant, animal,
fungus, or microbe   SPECIES

FIGURE 9-1

Division of the biosphere into ecosystems, communities, populations, and species. (From Nebel and Wright, *Environmental Science, 4/e*, 1993. Reprinted by permission of Prentice Hall, Upper Saddle River, N.J.)

surrounding environment (see "Species Interactions," below). For terrestrial ecosystems the environment includes the topography, soils, atmosphere, and climate. For aquatic ecosystems the environment includes the physical and chemical characteristics of the particular freshwater or marine water body concerned. Since each ecosystem is located in a slightly different physical environment, there is a tendency to think of the environment as determining the type of ecosystem that develops. This is true to some degree, but it is not the whole story: The organisms within a particular ecosystem interact with their environment. For example, the type of soil present may help determine which plants will grow, but the plants themselves add organic matter to the soils that changes the soil chemistry, possibly allowing different species to grow there. Certain species might tolerate only particular temperature and precipitation regimes, but those species may be capable of altering local climate, thereby promoting their own growth or the growth of other species.

We indicated this possibility with James Lovelock's Daisyworld example in Chapter 2. We also illustrate it here with an example of his from the boreal forests of North America and Asia. The coniferous trees in the boreal forest hide the snow-covered ground in winter, reducing the albedo of the forested region. The reduced albedo should result in higher winter temperatures than would occur if the forest were not present. By modifying its local environment to be warmer at its northern edge, the forest may be able to push farther north beyond the point where temperatures would otherwise be too cold to allow the trees to grow. Is there any indication that this might, in fact, be happening?

Scientists at the National Center for Atmospheric Research in Boulder, Colorado, conducted a general circulation model (see Chapter 6 for more information on GCMs) experiment in which they changed all of the forest north of 45°N to bare ground. This is equivalent to moving the border between the boreal forest and the treeless tundra southward. The effect of this change was to produce a large increase in the wintertime albedo, because the white snow cover was revealed by the removal of the forest. The increased albedo caused a large drop in air temperatures, the greatest change in the month of April—up to 12°C (21.6°F) over the land surface. The colder winter temperatures increased the sea-ice cover, and the higher albedo further enhanced the cooling effect. The colder temperatures were maintained through the summer, with July being as much as 5°C (9°F) colder than before the removal of the forest.

These processes are illustrated in the systems diagram in Figure 9-2. The solid lines indicate the interactions simulated by the model, including the ice–albedo feedback discussed in several earlier chapters. The dashed line completes another positive feedback loop implied by the model results. The forest cover is not an interactive part of the model, so we cannot see the feedback from the change in temperature to the change in forest cover. However, the model does show that the July 18°C (64.4°F) isotherm (which correlates well with the present northern limit of the forest) shifts southward far enough to prevent forest regrowth. Although the interactions are more complex than those suggested by Lovelock, we do see that the forest has a significant impact on the climate system. By keeping high-latitude temperatures from being as cold as they would otherwise be, the forest helps perpetuate an environment conducive to its own growth.

It is apparent, therefore, that ecosystems are not divorced from their environment; the environment is part of the ecosystem. As the environment changes, the types of organisms in the ecosystem and the interactions among them change, and as they do, the local environment may change. The obvious conclusion is that ecosystems are not static. Changes in climate can cause ecosystems to move gradually to new locations, such as arctic tundra and its biologic community spreading equatorward during glacial

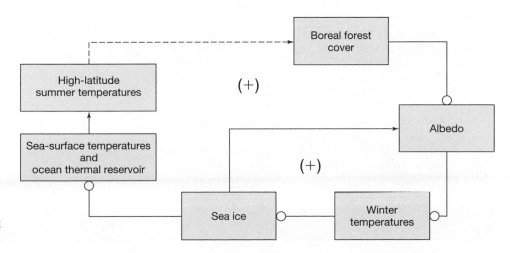

FIGURE 9-2

Possible feedbacks between the boreal forest and climate.

periods and retreating poleward during interglacials. This is in part a consequence of the physiological requirements of the individual organisms that make up the ecosystem's biota; each species has minimum, optimal, and maximum conditions for growth. As the environment changes, organisms may find themselves in less than optimal conditions (see "A Closer Look: Physiological versus Ecological Optima for Growth"). More interestingly, new

environmental conditions could give rise to a totally new assemblage of species: an ecosystem that has not been seen before. In fact, because ecosystems and the environment interact, it may be possible for a new ecosystem to evolve without any large-scale change in the environment. In this case, the initial environmental conditions support one ecosystem but the interactions change the local environment, so the ecosystem evolves into something new.

A CLOSER LOOK
Physiological versus Ecological Optima for Growth

When studied under controlled, laboratory conditions, the growth rate of most organisms responds to environmental change in a fashion similar to that proposed for the daisies of Daisyworld in Chapter 2: there are minimal, optimal, and maximal conditions for growth. This relationship can be clearly expressed for the response of photosynthetic rates of plants grown in greenhouses to changes in temperature (Box Figure 9-1a). This figure also shows a distinction between C_3 and C_4 plants. C_3 autotrophs comprise all the trees, most of the other plants, the cyanobacteria, and all algae; they are called C_3 because an important sugar produced during photosynthesis has three carbon atoms. C_4 plants are relative newcomers to the Earth system, evolving in the last 10–20 million years in response to either lower atmospheric CO_2 levels or drier climates. They include many grasses, corn plants, and pineapples, to name a few. They are called C_4 because they produce a 4-carbon sugar during their photosynthetic cycle. C_3 plants can grow at lower temperatures, but have lower maximum temperatures for growth than do C_4 plants. C_4 plants, however, are able to very efficiently scavenge CO_2 from the atmosphere (Box Figure 9-1b), allowing them to spend less time with their stomata (pores) open. (Plants typically obtain CO_2 by opening their stomata.) This ability is a great advantage in arid environments because open stomata also release water vapor to the atmosphere, causing water stress in plants. It has also proven advantageous from the perspective of Earth history: Atmospheric CO_2 levels have fallen over the last several million years (see Chapter 12), falling ever closer to the break-even point for C_3 plants (~30–70 ppm) where photorespiration (respiration by plants) equals photosynthesis. As atmospheric CO_2 levels have fallen, plants that could more efficiently photosynthesize under these atmospheric conditions have presumably thrived. However, when temperatures drop, as Box Figure 9-1a shows, the C_3 pathway becomes favorable.

Of course, organisms have many environmental requirements, each of which may exhibit a parabolic relationship under otherwise optimal conditions (as in Box Figure 9-1), but nature does not provide such ideal conditions. It is important that we understand these relationships both in the laboratory and in nature so that we can establish the coupling and feedback that govern environmental change (as we

did in Chapter 8 when considering the controls on atmospheric CO_2). Many of the factors that affect growth are interdependent, and can create apparent paradoxes that can only be understood when considered simultaneously. Since

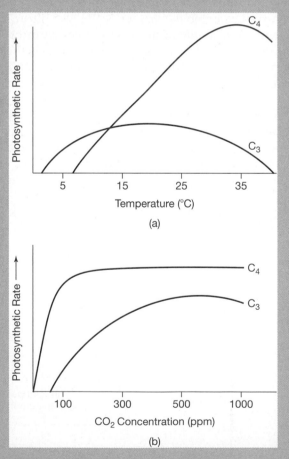

BOX FIGURE 9-1

Typical physiological responses of plants grown under greenhouse conditions to changes in some environmental conditions. (a) The response of C_3 and C_4 plants to changes in temperature. (b) The response of C_3 and C_4 plants to changes in the partial pressure of CO_2 in the atmosphere. Note that the scale is nonlinear.

Although Figure 9-1 implies an ordered hierarchy, the levels are not discrete and ecosystems themselves are not discrete units. Each level in the hierarchy interacts with all other levels, and ecosystems overlap each other. One ecosystem gradually merges into another geographically at a diffuse boundary called a transitional ecosystem or **ecotone** (Figure 9-3). An ecotone may include an entirely different assemblage of species that do not match those in the ecosystems on either side of the boundary (Figure 9-4). There is also considerable overlap between ecosystems in a structural sense: Several ecosystems may share many common physical attributes. Based on the various plant communities ecosystems support, we can identify distinct types of terrestrial biomes. The most important are shown in Figure 9-5 (see color plate as well).

interspecies interactions are discussed in the main body of the text, let's focus here on interdependencies of *environmental* factors. Recall our discussion of the marine algae in Chapter 8. In laboratory culture, algae exhibit optimal *physiological* growth rate at temperatures in the range of 20–25°C. Thus, one might predict that maximum rates of oceanic primary production would be in the tropical to subtropical ocean. Instead, what one finds are high rates of photosynthesis at high latitudes and in coastal zones irrespective of latitude, as reflected in satellite images of ocean chlorophyll concentration (Figure 8-10). This paradox is reconciled if we think back to Chapter 5. In that chapter we learned that the supply of nutrients to marine ecosystems is generally dependent upon upwelling of nutrient-rich deep waters to the surface. Upwelling is prevalent along west-facing coastlines (because of *Ekman pumping*) and at high latitudes, where the lack of a strong thermocline (pycnocline) allows for deep wind-driven mixing. Thus, the *ecological optimum* for algal growth is closer to 8°C, a compromise between the detrimental effects of colder water and the beneficial effects of enhanced nutrient supply. This fact will prove important to our consideration of the causes of glaciation in Earth history (Chapter 14).

Another paradox is the extremely high productivity of tropical rainforests. Perhaps contrary to expectation, tropical soils have severely depleted nutrient concentrations compared to temperate forest soils. How can they sustain such high productivities? The answer is that nutrients are very efficiently recycled in tropical forest ecosystems. Most of the nutrients are stored in the trees themselves. When a tree dies, it falls to the forest floor, which is warm and damp, the ideal conditions for decomposers (fungi and bacteria). Breakdown and release of mineral nutrients is thus quite quick. Moreover, trees in tropical forests have extensive, shallow root systems that rapidly extract nutrients as they are released by the decomposers. As a result, the residence time of nutrients in tropical soils is extremely short (Box Table 9-1); a typical molecule of nutrient phosphate is retained less than two years in a tropical soil, whereas its lifetime in a temperate forest is closer to six years, and in a boreal forest, is over 300 years. The rapid recycling of nutrients in tropical soils draws down the steady-state nutrient concentration but sustains high rates of productivity. However, if the trees are removed, so too are the nutrients, and the soil left behind is infertile. We'll find in Chapter 18 that this is one of the serious consequences of deforestation of the tropical rainforest.

Recognition of the distinction between physiological and ecological optima for growth is an important step in developing a deeper understanding of the Earth system. Life, including humans, is influenced by a variety of interacting factors. The overall optimal growth condition thus may be suboptimal for many if not all factors that influence growth. Thus, an environmental change that should increase primary productivity (say, warming of the high-latitude ocean in response to buildup of atmospheric CO_2 levels) may in fact diminish it. The systems approach provides the answers to these seeming paradoxes.

BOX TABLE 9-1

The mean residence time (in years) of organic matter and nutrients and the net primary productivity (NPP) of four biomes.

Biome	Organic matter	Nitrogen	Phosphorus	Potassium	Calcium	Magnesium	NPP (g C/m²/yr)
Boreal forest	353	230	324	94	149	455	360
Temperate forest	4	5.5	5.8	1.3	3.0	3.4	540
Chaparral	3.8	4.2	3.6	1.4	5.0	2.8	270
Tropical rain forest	0.4	2.0	1.6	0.7	1.5	1.1	900

(From Bush, M.B. *Ecology of a Changing Planet, 3/e*, 2003. Reprinted by permission of Prentice Hall, Upper Saddle River, N.J.)

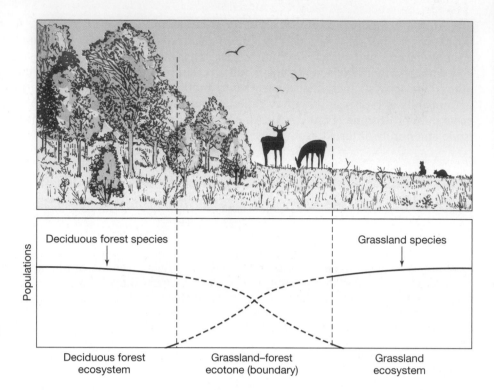

FIGURE 9-3

Ecosystems and ecotones. (From Nebel and Wright, *Environmental Science, 6/e,* 1998. Reprinted by permission of Prentice Hall, Upper Saddle River, N.J.)

Species Interactions

Although ecosystems may appear to be very different from one another, they all exhibit a common biotic structure. For example, all ecosystems include autotrophs and heterotrophs. In this type of organization, primary consumers (e.g., zooplankton in the sea or rodents on land) live off the producers (algae, plants), secondary and high-er-order consumers (fish, hawks) feed on lower-order and primary consumers (zooplankton, rodents), and the decomposers (bacteria and fungi) and detritus feeders feed on dead organic matter of both producers and consumers. In assisting the chemical breakdown of organic matter, the decomposers and detritus feeders return nutrients to the system that are then reused by the producers.

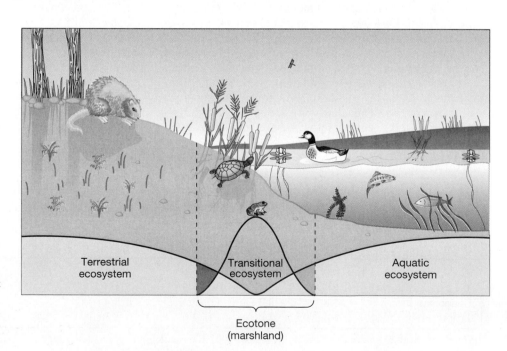

FIGURE 9-4

An ecotone may create a unique habitat of its own. (From Nebel and Wright, *Environmental Science, 6/e,* 1998. Reprinted by permission of Prentice Hall, Upper Saddle River, N.J.)

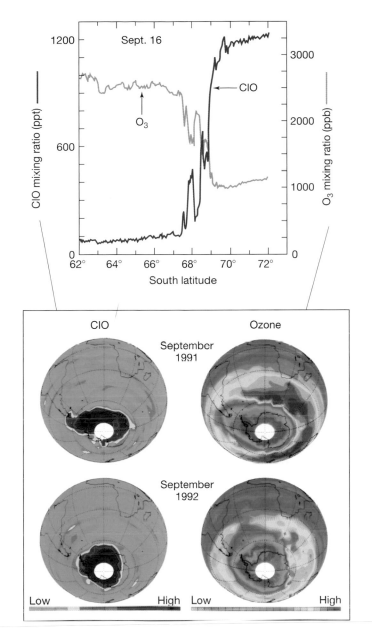

FIGURE 1-6

(Top) Simultaneous measurements of ozone (O_3) and chlorine monoxide (ClO) made from a NASA aircraft as it flew into the Antarctic ozone hole in September 1987. The hole was entered at a latitude of about 68° S. The units ppt and ppb stand for "parts per trillion" and "parts per billion," respectively. (Bottom) Contour plots of ClO and O_3 concentrations obtained from spacecraft measurements. These data also show that ozone is low where ClO is high. (From R.W. Christopherson, *Geosystems: An Introduction to Physical Geography, 3/e,* 1997. Reprinted by permission of Prentice Hall, Upper Saddle River, N.J.)

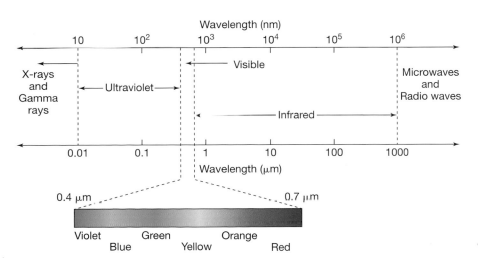

FIGURE 3-3

The electromagnetic spectrum.

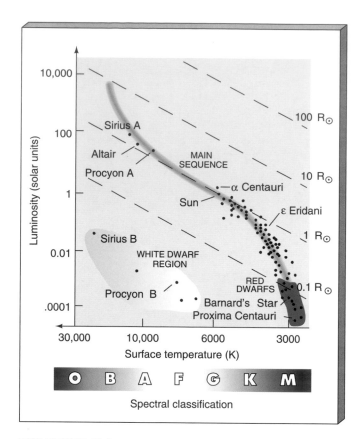
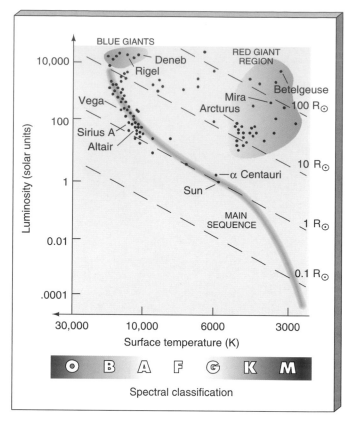

BOX FIGURE 10-2

Hertzsprung–Russell diagram showing different classes of stars. (From E. Chaisson and S. McMillan. *Astronomy: A Beginner's Guide to the Universe,* 2/e, 1998. Reprinted by permission of Prentice-Hall, Upper Saddle River, N.J.)

FIGURE 10-6

Picture of a black smoker. (Ken MacDonald/Science Photo Library.)

FIGURE 11-2

Shark Bay in western Australia. These "living stromatolites" are formed by communities of microbes and may be an analog to early microbial life. (From P. Cloud. *Oasis in Space: Earth History From the Beginning.* W. W. Norton and Co., New York, 1988.)

FIGURE 11-4

Saturn's moon, Titan, showing the orangish organic haze. (Photo courtesy of NASA.)

FIGURE 11-21
A Carboniferous dragonfly with a wingspan of 60 cm (2 ft.). (From © The Field Museum, Chicago. Photographer: John Weinstein.)

(a)

(b)

FIGURE 13-7
Shocked quartz and (b) microspherules from K–T boundary clays. (From Dr. David Kring/Science Photo Library, Photo Researchers, Inc.)

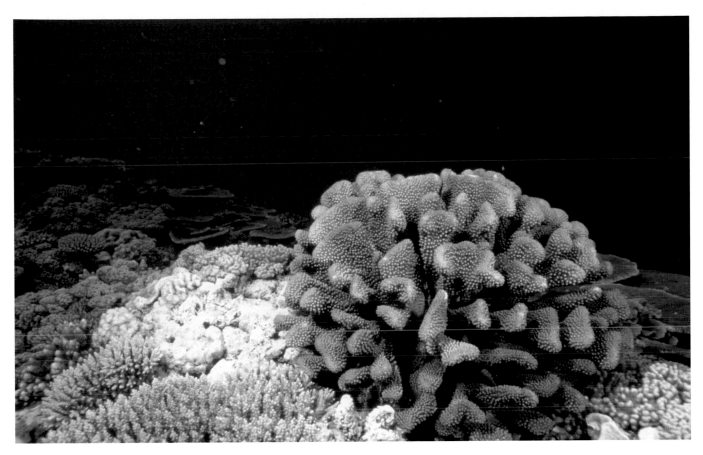

FIGURE 14-14

Coral reefs may be responsible for the changes in atmospheric carbon dioxide concentrations that occur between glacials and interglacials. (From T. McKnight. *Physical Geography: A Landscape Appreciation*, 5/e, 1996. Reprinted by permission of Prentice Hall, Upper Saddle River, N.J.)

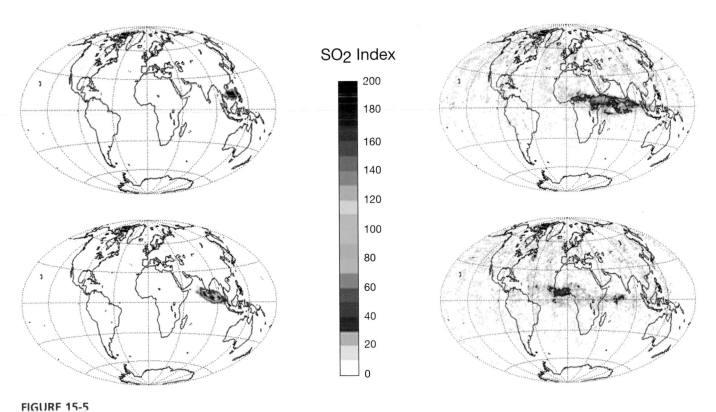

FIGURE 15-5

Satellite observations of the Mount Pinatubo aerosol cloud, 1991. (a) June 16. (b) June 18. (c) June 23. (d) June 30. (After G.J.S. Bluth, et al. Global Tracking of the SO_2 Clouds from the June 1991 Mount Pinatubo Eruption. *Geophysical Research Letters,* 19:151–154, 1992.)

15

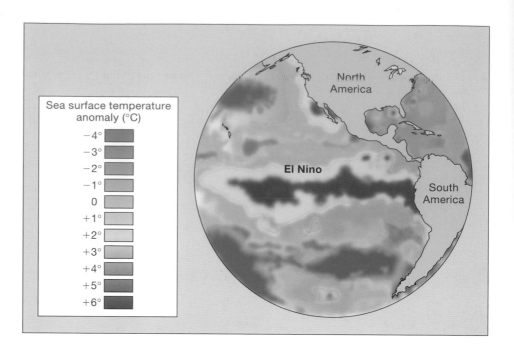

FIGURE 15-15

Sea-surface temperature anomalies in the central and eastern Pacific during the 1997–1998 ENSO event.

Sea surface temperature anomaly (°C)

-4°
-3°
-2°
-1°
0
+1°
+2°
+3°
+4°
+5°
+6°

El Nino

North America

South America

FIGURE 19-3

The surface of Venus, as observed by radar from the Magellan spacecraft. (From Jet Propulsion Laboratory, NASA Headquarters.)

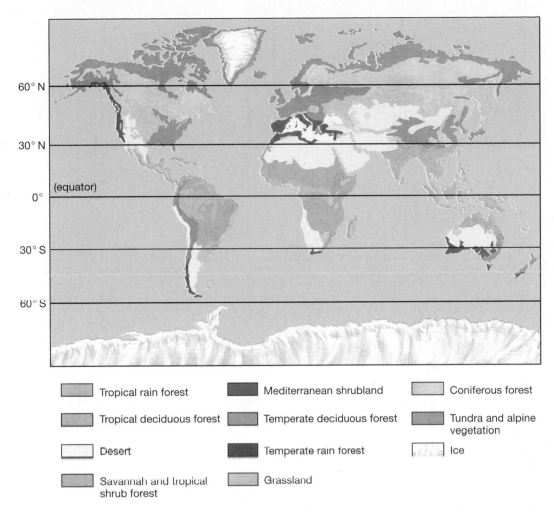

Tropical rain forest

Tropical deciduous forest

Desert

Savannah and tropical shrub forest

Mediterranean shrubland

Temperate deciduous forest

Temperate rain forest

Grassland

Coniferous forest

Tundra and alpine vegetation

Ice

FIGURE 9-5

[See color section] World distribution of the major terrestrial biomes. (From Audesirk and Audesirk, *Biology: Life on Earth, 5/e*, 1999. Reprinted by permission of Prentice Hall, Upper Saddle River, N.J.)

We can represent which organisms feed on which by means of a **food chain** that links particular organisms in an ecosystem. Because one organism may feed on several other types or may be eaten by several different types of organisms, food chains are usually interconnected into **food webs** (Figure 9-6). Despite the potential complexity of these webs, the overall structure is very simple: Each web consists of a series of feeding levels called *trophic levels.* For example, the following trophic levels range from the bottom up: producers > primary consumers > secondary consumers > higher-order consumers. There are normally no more than four trophic levels in any system.

How many organisms are there at each level? Rather than looking at the number of individuals, we can simplify the comparison between levels if we talk in terms of **biomass**. Biomass is the total combined weight of organic material in each trophic level. Each trophic level, except for the producers, ingests its food (organic matter) from the level below. The organisms utilize that organic matter for growth and to produce energy. As we move from lower-order to higher-order trophic levels, much organic matter is lost when it is converted to energy. In terrestrial ecosystems, the biomass is decreased by 90 to 99% at each higher level. An alternative way to think about trophic interactions is in terms of **exploitation efficiency**. Of 100 carbon units of net primary productivity, approximately 20 units are exploited by herbivores and 80 units are "wasted," expended by the herbivores without translation into biomass or unutilized and transferred to the soil ecosystem where decomposers take over. In turn, carnivores are able to exploit only about 0.2 units of the original 100 for growth, with the rest being expended in their metabolism.

Ecosystems are not organized entirely according to which species is feeding on whom; other forms of interaction are also found. These may include mutually supportive relationships, such as the relationship between

Third
trophic
level:
all
primary
carnivores

Second
trophic
level:
all
herbivores

First
trophic
level:
all
producers

FIGURE 9-6

A simple food web. Food (energy and nutrients) is transferred from one organism to another along these pathways. (From Nebel and Wright, *Environmental Science, 6/e,* 1998. Reprinted by permission of Prentice Hall, Upper Saddle River, N.J.)

flowering plants and insects: Insects feed off of the nectar or pollen from the flowers, which are then pollinated in the process. The ultimate example of this mutualism is **symbiosis**, a relationship in which two species benefit from living together in intimate contact. The relationship between corals and plant-like protists known as *dinoflagellates* is a good example of a symbiosis. This symbiotic relationship involves a beneficial cooperation between the coral animal (*polyp*) and a dinoflagellate that lives within the digestive cavity of the coral polyp. The coral provides protection, inorganic nutrients through excretion, and carbon dioxide for the dinoflagellate. In turn, the dinoflagellate provides nutrition (*photosynthate*), helps the coral synthesize some organic compounds (*lipids*), and removes carbon dioxide, making it easier for the coral to precipitate its $CaCO_3$ skeleton. The dinoflagellate in this case is the coral's *symbiont*. Under stress—for example, when sea temperatures rise during El Niño climate events (see Chapter 15)—the dinoflagellates may be expelled from the coral polyp. This is called a "bleaching" event, because the corals turn white; their beautiful coloration comes from the pigments of the dinoflagellate symbiont. If deleterious conditions are sustained, the coral can die as a result of the lack of its symbiont.

At the other extreme are species that coexist in a competitive relationship, although this competition tends to occur less frequently than we might suppose. Most species tend to adapt to a particular habitat and, even where potentially competitive species occupy the same habitat, each tends to develop its own particular *niche*. An animal's niche describes not only the food it eats, but also where and when it eats, where it lives, where it nests, and so on. Specialization to this degree, in which different species occupy the same geographic location but have different living habits, reduces potential contact and helps reduce competition among species at the same trophic level.

The species in an ecosystem interact with their environment as well as with each other, and different species thrive under different physical conditions. There is usually some optimal range of conditions over which each species is best adapted. The species comes under stress as the environment moves away from that range—for example, as it becomes wetter or drier, warmer or colder, shadier or sunnier, or more or less acidic. At some point, the stress may be great enough that the organism reaches its limit of tolerance for those conditions, and death occurs.

When we throw human beings into the mix, the level of complexity and the nature of the interactions increase substantially. We then must take into account the social, political, and economic interactions among different human societies and also the dramatic effects that these societies can have on the physical environment.

Ecosystem Disturbance and Succession

Natural or human disturbances of an ecosystem that seriously disrupts the existing ecosystem structure—for example, wildfire or deforestation—initiate a response, often of rebuilding, that in some cases follows a predictable pattern called **succession**. The job of rebuilding can be quite extensive because initial disturbances such as wildfires or deforestation can promote subsequent effects like soil erosion, nutrient loss, or microclimate changes such as aridification from the loss of evapotranspirative pumping of water into the atmosphere. The resulting arid, nutrient-poor environment is not necessarily conducive to the regrowth of the preexisting community. The first species to reinvade a disturbed environment are called *opportunists* or *pioneer* species. They tend to be fast growing, rapidly reproducing, environmentally tolerant species that can spread across the disturbed area quickly. In colonizing a previously disturbed area, these organisms tend to begin the process of repair, improving the soil or modifying the local climate in ways that can result in their replacement by other, slower growing organisms that ultimately have competitive advantages. The establishment of a mature forest can take decades to hundreds of years after disturbance; the forests of New England are still undergoing succession 200 years or more after the original logging that occurred during colonization of America. In general, succession patterns are predictable, at least in terms of the types of plants that will become prevalent at various stages of succession. If many species are equally well suited to the environmental conditions at a particular stage of succession, however, an element of unpredictability may be introduced into the succession pattern. Thus a diversity of outcomes is possible, and this diversity may be reflected in the *biodiversity* of a region subject to disturbance. Indeed, we will see in the next section that a modest level of disturbance may be required for high-diversity ecosystems to become established.

Succession is one indication that the biosphere has the capacity to "heal," that it is resilient to perturbation. In what follows we will ask a related question: is resilience a general characteristic of diverse ecosystems? Before we do, though, we need to come to a better understanding of what we mean by the term *biodiversity*.

Biodiversity

How do we measure the "health" of the biosphere? By analogy to living systems, a healthy planet should actively transport nutrients from where they are not needed to where they are and should eliminate wastes (as demonstrated in Chapter 8). Its important environmental variables (temperature and atmospheric and oceanic compositions) should not fluctuate wildly. And it should be capable of responding to natural and anthropogenic disturbances, such as volcanic eruptions, meteorite impacts, deforestation, and pollution, in such a way as to minimize their consequences.

These characteristics of a stable Earth system are ones that we normally associate with living organisms. Indeed, it is these very characteristics that are most suggestive of an important role for the biota in the regulation of the Earth system. The biota have affected Earth's long-term climate evolution by modifying its greenhouse gas content much in the same way that daisies modified the climate history of Daisyworld (Chapter 2). They have also created an oxygen-rich atmosphere; in Chapter 11 we'll find that our oxygen-rich atmosphere is a direct consequence of oxygenic photosynthesis. The evolution of this metabolic pathway has been called the greatest pollution event of Earth history. As a result of the prevalence of oxygenic photosynthesis, anaerobes, previously able to inhabit a diversity of habitats, survived only in environments such as seafloor sediments where oxygen does not penetrate. In these cases, the perturbations to which the biota has responded have been gradually imposed, perhaps over millions of years. But how has the Earth system responded to *rapid* environmental change? Is the resilience revealed over the long term also a characteristic in the short term? The answer, as we will see in later chapters, is yes. Earth has been subjected to insults the magnitude of which we are unable to imagine: Meteorites broader than the ocean is deep have struck Earth many times during its 3.5 billion years of inhabitation by organisms. The robustness of the planetary system is revealed by a fossil record displaced but not interrupted and by a geological and geochemical record that suggests that the long-term environmental consequences of these sudden disturbances were small.

How is the diversity of life forms on the planet—its **biodiversity**, the number of species in an area—related to the health of the Earth system? At the local scale, if we measure "health" in terms of biological productivity, we might conclude low diversity ecosystems are the healthiest. For example, highly productive lakes that have been impacted by fertilizer additions in runoff tend to be dominated by a very few species that are highly productive under high nutrient loadings. However, we don't normally consider contaminated lakes "healthy." We might instead propose that Earth's health can be measured by the number of species it supports. This assumption is implicit in the concern over the loss of biodiversity Earth is currently experiencing as a result of deforestation and loss of habitat. But do the abilities we associate with a healthy planet depend on its biodiversity? Is global biodiversity an indicator of the functional status of the Earth system?

Measures of Biodiversity

Biodiversity can be determined in a number of different ways. Perhaps the simplest measure of biodiversity is the number of species present in a community. A community with five species is much less diverse than one with 100 species. There are some problems with this simple definition, however. One problem has to do with heterogeneity. Suppose there are two communities, both with two species of organisms, as shown in Table 9-2. According to the simple definition, the two communities are equally diverse. Community I, however, has 99 individuals of species A and one individual of species B, whereas community II has 50 individuals of species A and 50 individuals of species B (Figure 9-7). The chance of encountering species B in community I is quite remote, only 1 in 100; this community is quite homogeneous. In contrast, community II seems more diverse, because there is an equal likelihood of encountering an individual of species A and of species B. Community II is thus more heterogeneous.

To capture the importance of heterogeneity, measures of diversity other than simply the number of species have been proposed. The *Simpson's diversity index* measures the likelihood that two individuals drawn from the same community will be of different species. This likelihood is expressed quantitatively as follows:

$$\text{Simpson's diversity} = 1 - [(\text{proportion of species A})^2 + (\text{proportion of species B})^2 + \ldots].$$

The proportion of each species in the community is identical to the probability that an individual chosen at random will be of that species. The probability of choosing two individuals of that species in a row is in the square of the proportion, just as the probability of throwing two "heads" in a row during a coin toss is $(0.5)^2 = 0.25$, or 1/4. The value of this index for our two simple communities I and II are $1 - (0.99^2 + 0.01^2) = 0.02$ for the homogeneous Community I and $1 - (0.5^2 + 0.5^2) = 0.5$ for the heterogeneous Community II, as shown in Table 9-2. When the number of species is large and the composition heterogeneous, the maximum Simpson's diversity approaches 1.0.

The Simpson's diversity index is clearly superior to a simple species count in expressing biodiversity. Nevertheless, in most discussions of global biodiversity of the

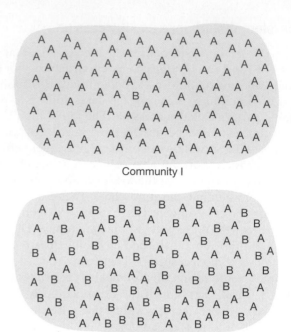

FIGURE 9-7

Two communities comprising two species (A and B) each. Although equal numbers of species are represented, Community II appears more diverse than Community I.

past, present, and future (to some extent, this book included), only the species count is used.

Diversity and Stability

A long-standing debate exists among ecologists about the relationship between diversity and stability. For most communities, diversity increases from the poles to the tropics: Most of the highly diverse communities exist within 10 to 20 degrees of the equator. Why are the tropics so diverse? Tropical climates tend to be stable over a range of time scales. In the short term, the lack of large seasonal variations in solar insolation leads to only small monthly contrasts in temperature and rainfall. Moreover, investigation of the geological record reveals that on long time scales, tropical temperatures have changed relatively little. Even during the Ice Ages, tropical temperatures fell only slightly while temperate to polar climates cooled substantially. The high diversity and climatic monotony of the tropics have been taken to indicate that environmental stability leads to high diversity—a premise called the *time stability hypothesis*. The persistence of uniform environmental conditions in the tropics presumably allows evolution to proceed without disruption, leading to higher diversity.

In contrast, the *intermediate disturbance hypothesis* states that the high diversity of tropical ecosystems is the result of disturbances that occur with intermediate frequency and intensity. This hypothesis is in direct contra-

TABLE 9-2

Diversity of Two Simple Communities			
	Number of Individuals, Species A	*Number of Individuals, Species B*	*Simpson's Diversity Index*
Community I	99	1	0.01
Community II	50	50	0.50

diction with the time stability hypothesis. Highly diverse tropical rainforests tend to have some species with few or no young trees (indicating that they are dying out) and other species with a very high proportion of young trees (indicating that they are increasing in abundance). This turnover is presumed to be the result of a fairly recent natural disturbance. Regions of rainforest that are known to have been relatively undisturbed over historical time tend to have lower diversities. Also, highly diverse coral reef ecosystems tend to occur at the outer edge of barrier reefs, where these ecosystems are periodically confronted with the damaging effects of waves and storms.

Both hypotheses link the diversity of life on Earth to the stability (or instability) of the environment. As ecologists continue to debate this issue, Earth systems scientists begin to wonder whether environmental stability might, in turn, depend on biodiversity.

Diversity of Interactions

Even the more elaborate measures of diversity, such as the Simpson's diversity index, fail to account for a characteristic of ecosystems that is important to our understanding of the feedbacks between the biota and the physical world: the diversity of interactions. A community consisting of 500 species of ants, with relatively uniform populations of each, along with a few species of plants and predators, is highly diverse according to this index. However, in terms of the diversity of roles played by these organisms, the community is extremely homogeneous.

Here the infancy of Earth system science is clearly a limitation: No diversity index has been proposed that captures the degree of interaction between biological and physical components of the Earth system. Such a diversity index should increase as the number of couplings among the biota and between the biota and the physical world increases. It should also incorporate the attribute of redundancy; the Earth system is more resilient if there are alternative ways of performing important functions, such as photosynthesis or decomposition. If one of these pathways is lost (e.g., through extinction), the others can compensate. An ecosystem with 10 interactions and only 20 species is then not as diverse as one with 40 species interacting in these 10 ways. The final attribute to incorporate into a systems diversity index is potential diversity. Species in small abundance today may come to dominate after a disturbance. In doing so, they will ensure that some vital function of the Earth system continues with little interruption or modification. If biodiversity is defined in this way, it seems clear that a more biologically diverse world is a more stable, resilient world, that biodiversity does indeed enhance environmental stability at the global scale.

In later chapters we explore how biodiversity has varied over Earth history, and how human activities today are affecting the diversity of life on the planet. In Chapter 13, we'll see that the biosphere has suffered from unimaginable catastrophes that reduced the species diversity by up to 95%, yet recovered. We also discuss how current practices of monoculture and genetic engineering may make us susceptible to the sort of widespread blight experienced by the Irish people when their monoculture of potatoes succumbed to a fungal infection in the 19th century.

Chapter Summary

1. Some of the characteristics of life that allow it to play an important role in the Earth system are its tendency toward exponential growth, its need for energy, its tendency to pollute, and its versatility.
2. Organisms can be placed into broad groups according to whether they are producers (autotrophs) or consumers (heterotrophs).
   a. Autotrophs include those that use solar energy (photosynthesizers) and those that use chemical energy (chemosynthesizers).
   b. Heterotrophs, including aerobes (use oxygen), anaerobes (use other oxidants), and fermenters (who do not oxidize organic matter), get energy from the food they consume.
3. Populations of organisms live in communities with other organisms that interact among themselves and their environment in ecosystems. Boundaries between ecosystems are typically gradational ecotones rather than sharply contrasting adjacent ecosystems.
4. The flow of energy (food) through ecosystems is often displayed as a food chain from producers to consumers and decomposers.

   a. Closer inspection of natural communities indicates that the relationships form more of a web than a chain.
   b. Exploitation efficiency is quite low; much of the food (energy) available to higher levels in the food chain is not used for growth but rather expended during metabolism.
5. Species also interact in other ways, including some that are competitive but others that are mutually beneficial (symbiosis).
6. After a disturbance, an ecosystem often responds with a predictable succession of organisms, from opportunistic, fast-growing species to slower-growing but ultimately more competitive species.
7. The diversity of life on Earth is a function not only of the number of species, but also of the degree to which the populations of those species are nonuniformly distributed (heterogeneous).
8. Environmental stability seems to lead to high biodiversity in some instances; however, modest disturbance enhances diversity in others.

Key Terms

autotrophs

biodiversity

biomass

biomes

community

ecosystem

ecotone

exploitation efficiency

food chain

food web

heterotrophs

population

symbiosis

Review Questions

1. What are the characteristics of life that allow it to influence the environment at a global scale?

2. What are the two fundamental groupings of organisms based on their metabolisms?

3. Describe the two mechanisms of autotrophy. Where on Earth might you expect to find one or the other of these two pathways to dominate?

4. Describe the three mechanisms of heterotrophy. Where on Earth might you expect to find one or the other of these two pathways to dominate?

5. Why is a food web often a better description than a food chain of the way in which energy (food) is passed through an ecosystem?

6. How does symbiosis differ from other forms of species interactions?

7. Describe a typical successional sequence following a disturbance of an ecosystem. What are the characteristics of opportunistic species that allow them to rapidly repopulate a disturbed area?

8. What is the advantage of the Simpson's Diversity Index over a simple census of the number of species in quantifying the diversity of an ecosystem?

9. What do we mean by "diversity of interactions?"

Critical-Thinking Problems

1. Figure 9-2 presents a systems diagram of the feedbacks involving boreal forest cover, albedo, temperatures, sea ice, and the oceans. We used this diagram to show that it is possible for the northern boreal forest to have a significant impact on the larger-scale climate. Using the information you now have about the possible impacts of anthropogenically induced greenhouse climate change, expand on this diagram and discuss the implications in terms of climate and forest cover.

2. In the final section of this chapter we presented some thoughts about what a diversity index useful for Earth sys-

tem scientists might include. Using these thoughts or those of your own, develop a quantitative index, similar to the Simpson's Diversity Index, that reflects the diversity of interactions at the global scale.

3. Using the information from Table 9-1, design a layered microbial ecosystem that could be self-sustaining with the exception of the import of solar energy from above. All of the inorganic compounds listed in the table are available for your use in building this ecosystem.

Further Reading

General

Bradbury, I. K. 1998. *The Biosphere, 2/e.* New York: Wiley.

Wilson, E. O. 1992. *The Diversity of Life.* New York: W. W. Norton.

Advanced

Volk, T. 1998. *Gaia's Body.* New York: Copernicus (Springer-Verlag).

Westbroek, P. 1991. *Life as a Geological Force: Dynamics of the Earth.* New York: W.W. Norton.

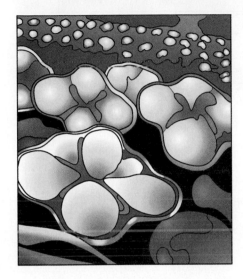

CHAPTER 10

Origin of the Earth and of Life

Key Questions

- How old is Earth?
- How did the solar system form?
- How did the atmosphere and ocean form?
- What was the composition of the atmosphere early in Earth's history?
- When and how did life originate?
- Why did the earliest organisms show a preference for hot environments?

Chapter Overview

Earth formed some 4.6 billion years (b.y.) ago by the accretion of solid particles from a cloud of gas and dust surrounding the young Sun. Earth's atmosphere and ocean started forming as the planet itself was being built as a consequence of the release of volatile materials during impacts. The atmosphere and ocean continued to grow during the "heavy bombardment period" between 4.6 and 3.8 b.y. ago. The composition of the atmosphere is unknown because little or no rock record survives from that time, but it probably consisted mostly of N_2 and CO_2. Life may have originated during the heavy bombardment period from reactions between organic chemicals created in Earth's surface environment or imported from space. This may explain why the last common ancestor of all extant life appears to have come from a hot environment.

Introduction

Earth is a smallish planet that orbits an ordinary star, our Sun. Earth is special, however, because it is the only planet in the universe that is known to harbor life. How was Earth formed, and how did it come to be habitable? These are questions we need to understand if we are to assess the possibility that life might exist elsewhere. We must also try to understand how life itself originated. Was it a chance occurrence, or was it a phenomenon that was almost unavoidable on a young, habitable planet like Earth? We don't know the answers to these questions yet, but scientists have made progress over the last few decades in determining how both our planet and ourselves came to exist.

Introduction to Geologic Time

One of the most important points that any geology professor makes to an introductory class is the immense amount of time represented in the geologic record. For reasons outlined below, scientists believe that Earth and the rest of the solar system formed about 4.6 b.y. ago. The universe itself has existed for roughly 12 billion years, based on estimates of its current rate of expansion. Both the age of the Earth and the age of the universe are almost inconceivably longer than a typical human lifetime of about 80 years or even the total amount of time that human-like species have been in existence, about 4 million years (m.y.). An analogy that is sometimes made is to scale Earth's age down to a single calendar year beginning at 12:01 A.M. on January 1. In that case humans would have first appeared at about 4:40 P.M. on December 31, and the oldest human (about 120 years) would have been born less than one second before midnight later that evening.

In Chapter 1, we introduced the geologic time scale and discussed two events, the Pleistocene glaciations and the K–T mass extinctions, that occurred relatively recently in Earth history. In this chapter, we focus on events that occurred billions of years ago. The major **eons** into which geologic time is divided are the Hadean (4.6–3.8 b.y. ago), Archean (3.8–2.5 b.y. ago), Proterozoic (2.5–0.54 b.y. ago), and Phanerozoic (0.54 b.y. ago–present). The first three of these time intervals are often collectively termed the Precambrian because they come before the Cambrian Period, when **shelly fossils** (fossilized remains of shelled organisms) became abundant in the rock record. Figure 10-1 shows some of the other major events that have occurred during Earth's history.

Formation of the Solar System

How did Earth and the rest of the solar system form? This question has fascinated astronomers for hundreds of years. It is interesting to Earth system scientists as well because it sets the boundary conditions for the rest of Earth's subsequent evolution.

Formation of the Solar Nebula

The Sun is thought to have formed from a collapsing cloud of interstellar dust and gas. Such **interstellar clouds** are observed today by both optical and radio telescopes (Figure 10-2). The one shown in Figure 10-2 is a particularly spectacular one that happens to be backlit by some bright blue stars. It is a portion of the Eagle Nebula, which is sometimes referred to as the "Pillars of Creation." Interstellar clouds are concentrated in the spiral arms of our Milky Way galaxy, where the density of material is highest. If such a cloud is dense enough and cold enough (about 10 K), it will collapse under its own self-gravity and the process of star formation can begin. It may seem counterintuitive that a cold cloud is required to form a hot star, but this is indeed the case. The reason is that, according to the ideal gas law (Chapter 4), a cold gas exerts less pressure than a warm gas of the same density. A cold interstellar cloud has less internal pressure to counteract the force of gravity and is therefore more likely to collapse. Once it does so, the cloud immediately warms up because the infall of material releases gravitational energy, which is converted into heat. The innermost part of

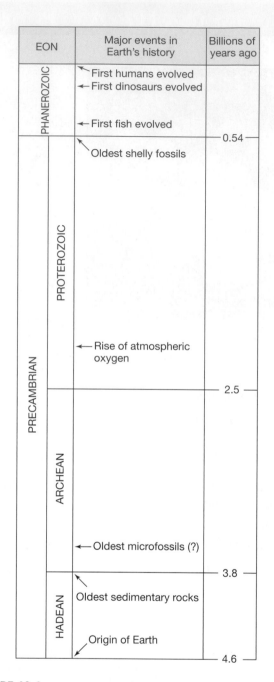

FIGURE 10-1

The geologic time scale, showing major events in Earth's history. (From R.W. Christopherson. *Geosystems: An Introduction to Physical Geography,* 3/e, 1997. Reprinted by permission of Prentice Hall, Upper Saddle River, N.J.)

the cloud becomes hot enough for thermonuclear fusion reactions to begin, and a new star is born.

The clouds that are observed in interstellar space are typically many thousands of times the mass of the Sun. As they contract, however, they produce smaller fragments that can themselves contract to form one or more stars. Whether the collapse results in a single star, or a multiple star system, depends largely on how fast the

A CLOSER LOOK
Determining the Age of the Earth

The problem of determining Earth's age is somewhat complicated even though the basic principles of radiometric dating (Chapters 5 and 7) are straightforward . The reason is that standard parent–daughter age-dating techniques, like the uranium–lead system, yield only the crystallization ages of the minerals to which they are applied, not the age of the material itself. But, as far as geologists know, none of the rocks and minerals that composed Earth's original crust have been preserved. The oldest minerals that can be dated by standard techniques are a handful of **zircons** (a zirconium silicate mineral) that yield ages of about 4.1 b.y. How do we deduce that Earth is actually half a billion years older than this?

To find Earth's age, one must first answer a related question: How old are meteorites? **Meteorites** are pieces of rock and/or metal that are thought to have drifted around the solar system for billions of years before hitting Earth. They have been collected from all over the planet but are found most readily in Antarctica, where they are easily spotted on the ice. The gradual flow and melting of the ice concentrates meteorites in a few specific localities, making them especially easy to find. The most primitive meteorites (because they have not been altered by melting) are called **chondrites**. Dating of chondrites provides an upper limit on Earth's age because these objects are thought to have formed at the same time as the solar system as a whole.

Meteorites can be dated using standard parent–daughter techniques, but the most accurate method—**lead–lead dating**—involves the use of multiple lead isotopes. The reason this is useful is that the U/Pb ratio differs from one meteorite to another and even among different minerals within the same meteorite, so it is difficult to determine what the initial U/Pb ratio must have been. By using multiple isotopes, we can avoid this problem. The three isotopes used are ^{204}Pb, ^{206}Pb, and ^{207}Pb. The lead isotopes ^{206}Pb and ^{207}Pb are radiogenic isotopes that derive, respectively, from the decay of ^{238}U and ^{235}U. The half-lives of these decay processes are 4.5 b.y. and 0.713 b.y. The third lead isotope, ^{204}Pb, is a nonradiogenic isotope that is used for comparison in the measurements. By measuring the abundances of these isotopes in different meteorites and then plotting the $^{206}Pb/^{204}Pb$ ratio on one axis and the $^{207}Pb/^{204}Pb$ ratio on the other, we can construct what is known as an **isochron diagram** (Box Figure 10-1). If all the samples being analyzed have the same age, as is true for chondritic meteorites, then the data should (and do) all fall on a straight line. The slope of this line tells us the age of the collection of meteorites, which is accurately determined as 4.55 b.y.

Moon rocks, which were brought back by the Apollo spacecraft missions of the late 1960s and early 1970s, can be dated in a similar manner. The oldest Moon rocks are about 4.44 b.y. old, suggesting that the Moon formed soon after the solar system itself. The Moon could be even older than this, as there is no guarantee that we have found the oldest Moon rock.

Earth's age is more difficult to obtain because the oldest rocks are all gone. They were probably recycled back into the interior by plate tectonics. However, Earth's age can be deduced indirectly by examining lead isotope ratios in rocks containing lead minerals, that is, minerals that initially contained lead, but little or no uranium. (U–Pb dating, in contrast, is performed on minerals that initially contained uranium, but little or no lead.) The isotopic composition of lead minerals is the same as that of the magma from which they formed. Because Earth's mantle contains uranium as well as lead, the abundances of ^{206}Pb and ^{207}Pb in the mantle have increased with time, as have those of the magmas derived from it. If we plot the lead isotope ratios from rocks of different ages and then analyze the resulting curve mathematically, we can show that, 4.5 to 4.6 b.y. ago, Earth's mantle should have had the same $^{206}Pb/^{207}Pb$ ratio as meteorites. This, in turn, implies that Earth formed at the same time as the rest of the solar system, about 4.55 b.y. ago. This radiometric age scale is the fundamental underpinning to most of our theories about how Earth formed and evolved.

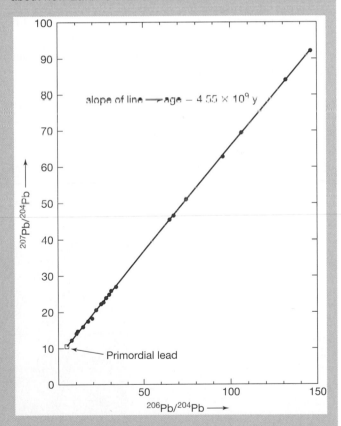

BOX FIGURE 10-1

An isochron diagram, showing lead isotope ratios from a collection of chondritic meteorites. The fact that all the data fall on a straight line shows that all the meteorites have the same age. The age of the meteorites, 4.55 b.y., is determined from the slope of the line. (From K. K. Turekian. *Global Environmental Change: Past, Present, and Future*, 1996. Reprinted by permission of Prentice-Hall, Upper Saddle River, N.J.)

FIGURE 10-2

[See color section] The Eagle Nebula viewed from the Hubble Space Telescope. (NASA Headquarters, *http:// hubblesite.org/gallery/ showcase/nebulae/n6.shtml*)

cloud fragment is rotating: the faster it rotates, the more likely it is to form two or more stars. In the case of our own Sun, a single star formed. This is fortunate for us, because a multiple star system would probably be a very difficult place to form a habitable planet like Earth. (It is difficult to identify stable orbits in such systems and probably even more difficult to form a planet in just the right place.) The cloud fragment did have a certain amount of rotation, however, and this caused some of its material to spread out into a disk. The gas and dust that made up the disk are referred to as the **solar nebula**. Astronomers have now been able to see similar disks around other Sun-like stars. Figure 10-3 shows the disk around the star Beta Pictoris, which was the first such disk to be discovered.

Formation of Planets

Once the solar nebula was in place, the process of planetary formation would have begun. The nebula itself would have been heated by the emerging Sun—the **proto-Sun**— so its interior would have warmed. At the same time, small particles of solid material would have begun to condense from the gas. In the hot, inner parts of the nebula, these grains consisted mainly of rocky materials such as iron and silicate minerals that can condense at temperatures as high as 2000 K. That is why Earth and the other **terrestrial planets** (Mercury, Venus, and Mars) are composed principally of rock. In the cooler, outer parts of the nebula, icy materials such as water (H_2O), methane (CH_4), and ammonia (NH_3) would also have condensed. Thus, the giant planets (Jupiter, Saturn, Uranus, and Neptune) contain large amounts of these more volatile compounds. **Volatile compounds** are substances that have low boiling points. The particles formed by condensation were gravitationally attracted to the mid-plane of the nebula, where they clumped together to form **planetesimals**, small proto-planets. These planetesimals collided with each other, often sticking together to form larger bodies in a process called **accretion**. This process is shown schematically in Figure 10-4. Over tens to hundreds of million of years, the planetesimals grew to form the planets that we see today.

While Earth was growing by accretion, its core should have started to form. Recall that the innermost parts of Earth are its solid inner core and liquid outer core, both composed mainly of iron and nickel. Core formation was once thought to have been triggered by radioactive heating after Earth was fully formed. But it is now believed that core formation occurred as the planet itself was form-

FIGURE 10-3

[See color section] The disk around the star Beta Pictoris, as seen from the Hubble Space Telescope. Top panel: Visible light image. Bottom panel: False color image created by image processing to highlight features in the disk structure. (NASA Headquarters, *http://hubblesite. org/newscenter/archive/ 1996/02/image/a*)

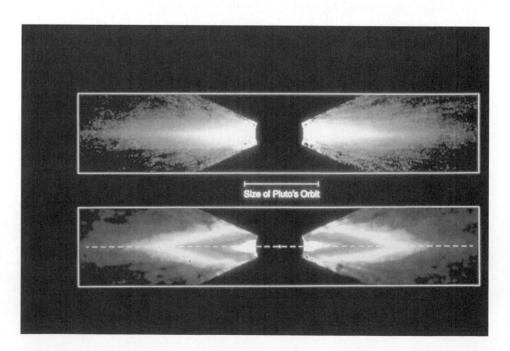

Size of Pluto's Orbit

A CLOSER LOOK
Main-Sequence Stars and the Hertzsprung–Russell Diagram

The **Hertzsprung–Russell diagram**, or H-R diagram for short, is a standard means of categorizing different types of stars (Box Figure 10-2). The horizontal axis represents the effective radiating temperature of the star, as determined by Wien's law (Chapter 3) or by some equivalent method of analyzing the star's spectrum. For historical reasons, temperature is always shown increasing to the left. The vertical axis represents the luminosity of the star relative to that of the Sun.

As Box Figure 10-2 shows, most stars fall along a well-defined band that runs from the upper left of the diagram to the lower right. This band is referred to as the **main sequence**. It consists of "normal" stars—that is, stars that are in the slowly evolving, middle phase of their existence. Stars are further grouped into seven classes on the basis of their spectra as O, B, A, F, G, K, or M. The brightest, bluest, and most massive (O and B) stars are referred to as early-type stars; the dimmest, reddest, and least massive (K and M) stars are called late-type stars. Within these categories, stars are further assigned numbers ranging from 0 (early) to 9 (late).

Our Sun is an unremarkable G2 star that occupies a spot near the middle of the main sequence. In about 5 b.y., the Sun will evolve off the main sequence and become first a red giant (at the upper right in the H-R diagram) and eventually a white dwarf (at the lower left). The Sun is drifting slightly upward on the H-R diagram during its main-sequence lifetime as a consequence of the conversion of hydrogen to helium. This effect is small compared with the changes that occur before and afterward, but it has important effects on planetary climates. It is this slow, main sequence evolution that gives rise to the faint young Sun paradox and that may limit the lifetime of Earth's biota in the future.

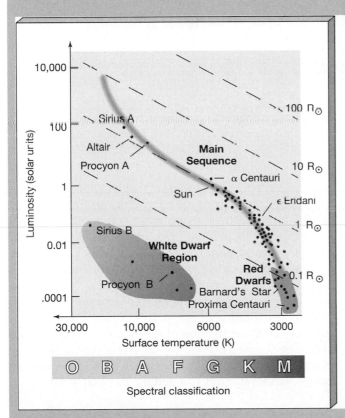

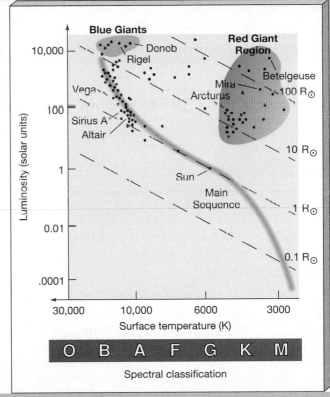

BOX FIGURE 10-2

[See color section] Hertzsprung–Russell diagram showing different classes of stars. (From E. Chaisson and S. McMillan. *Astronomy: A Beginner's Guide to the Universe*, 2/e, 1998. Reprinted by permission of Prentice-Hall, Upper Saddle River, N.J.)

(a) A slowly rotating portion of a large nebula becomes a distinct globule as a mostly gaseous cloud collapses by gravitational attraction.

(b) Rotation of the cloud prevents collapse of the equatorial disk while a dense central mass forms.

(c) A protostar "ignites" and warms the inner part of the nebula, possibly vaporizating preexisting dust. As the nebula cools, condensation produces solid grains that settle to the central plane of the nebula.

(d) The dusty nebula clears by dust aggregation into planetesimals or by ejection during a T-Tauri stage of the star's evolution. A star and a system of cold bodies remains. Gravitational accretion of these small bodies leads to the development of a small number of major planets.

FIGURE 10-4

The process of planetary accretion. By colliding with each other, (a) small planetesimals (b) grow gradually into (c) large planets. (From W.K. Hamblin and E.H. Christiansen. *Earth's Dynamic Systems,* 8/e, 1998. Reprinted by permission of Prentice Hall, Upper Saddle River, N.J.)

ing. Some of the planetesimals that collided with Earth during accretion were so large that they melted large portions of the crust and upper mantle. This allowed the iron and nickel to separate out and flow down to form the core.

Formation of Jupiter

Two specific events that occurred during the process of planetary accretion have special significance for surface conditions on Earth. The first was the formation of the giant planet Jupiter. Jupiter has over 300 times the mass of Earth and more than three times the mass of the next largest planet, Saturn. The reason Jupiter is so large, astronomers believe, is that its core accreted early enough to capture large amounts of hydrogen and helium from the solar nebula before the nebula dissipated. Accretion would have been rapid at Jupiter's orbit because volatiles could condense in addition to metal and silicates. To capture hydrogen efficiently, Jupiter's core must have grown rapidly to a mass several times that of Earth. Observations of young stars indicate that nebular gas and dust persist for at most a few million years, after which time they are either incorporated into planets or they spiral back into the star.

Jupiter affects surface conditions on Earth by perturbing asteroids from the asteroid belt into Earth-crossing orbits and by preventing most comets from reaching the inner solar system. The first effect makes Earth a more dangerous place to live, whereas the second one tends to make it safer. As mentioned in Chapter 1, impacts of comets and asteroids are thought to have played a major role in the evolution of life. A large asteroid impact may have caused the extinction of the dinosaurs, and this may in turn have paved the way for the rise of mammals. Thus, biological evolution might have taken an entirely different course had Jupiter not attained the size that it did.

Formation of the Moon

Another celestial event that had a profound influence on Earth's subsequent evolution was the formation of the Moon. Most of us know that the Moon's gravitational pull affects ocean tides. However, few people are aware that the Moon affects climate as well. It does so by stabilizing Earth's obliquity. Recall from Chapter 4 that it is Earth's *obliquity,* or tilt (currently 23.5°), that gives rise to the normal progression of the seasons at middle to high latitudes.

We will see in Chapter 14 that Earth's obliquity varies slightly and that these variations have influenced the glacial–interglacial cycles of the past 3 million years. Computer modeling studies have shown that, without the Moon, Earth's obliquity would vary by much larger amounts, occasionally reaching values as high as 85°. This would wreak havoc with climate because the seasonal cycles would be extremely large over much of Earth's surface. Thus, from the standpoint of planetary habitability, the formation of the Moon may be one of the most important events to occur during the formation of the solar system.

How exactly did it happen? Many theories have been advanced, including co-accretion (accreting in Earth's orbit at the same time as did Earth), fission (splitting apart of a rapidly rotating Earth), and capture (gravitational capture of a body that originated elsewhere in the solar system). Most of these theories, however, involve steps that are either physically implausible or that would produce a lunar composition different from that observed in Moon rocks. From the samples collected by the Apollo astronauts, we know that the Moon is depleted in volatile elements compared to the bulk Earth and that its oxygen isotopic composition is similar to that of Earth's mantle. Its density is substantially lower than that of Earth, indicating that the Moon is also depleted in iron. Furthermore, the Moon appears to have had a completely molten surface, or **magma ocean**, shortly after it formed.

One theory that is consistent with all of the available evidence is the **giant impact hypothesis** (Figure 10-5). According to this hypothesis, Earth received a glancing blow from a Mars-sized planetesimal during the latter stages of accretion, and the debris from the impact reassembled in orbit around Earth to form the Moon. This type of cosmic accident is statistically unlikely, but not so unlikely as to be implausible. We know that it is not a commonplace occurrence because none of the other terrestrial planets have large moons. (Mars has two tiny moons, Phobos and Deimos, but these are thought to be captured asteroids.) Current models of accretion suggest that giant impacts themselves should occur fairly frequently, but that most would not have the right geometry to form a large moon. Our own Moon resulted from one giant impact that did, which is fortunate for us, because the Moon has made Earth's climate much more stable than it might otherwise have been.

Formation of the Atmosphere and Ocean

Even after Earth had recovered from the effects of the Moon-forming impact, its surface would still have been an active and hostile environment. Continued bombardment of the surface by smaller planetesimals should have released large amounts of water and other volatile compounds directly into the atmosphere. This phenomenon is referred to as **impact degassing**. The process has been studied in the laboratory by firing high-speed bullets into targets of volatile-compound–rich material, such as carbonate rock. The shock from the bullet's impact causes the carbonate rock to release gaseous CO_2. Similar shock-induced degassing should occur when volatile-compound–rich planetesimals from the asteroid belt region or from the outer solar system collided with Earth's surface.

The energy released by the impacts, combined with the greenhouse effect of the gases given off, may have kept Earth's surface so hot that all the water would have remained in the atmosphere as steam. Alternatively, Earth's oceans may have periodically condensed and been re-evaporated many times by impacts. Any incoming object larger than about 450 km in diameter would have had sufficient energy to evaporate today's oceans. In either case, both the atmosphere and the ocean should have begun forming as the planet itself formed. This modern conception of planetary formation contrasts with older theories in which Earth was thought to have accreted as a cold, airless body, and in which the atmosphere was thought to have formed later from gases given off by volcanos. Volcanos undoubtedly contributed material to the surface, but the bulk of the atmosphere and ocean were probably formed directly by impacts.

The main period of accretion is believed to have lasted for only about 100 million years. After this time, Earth's surface would have become somewhat more quiescent. Big impacts probably continued to occur sporadically, however, until about 3.8 b.y. ago. The evidence for this **heavy bombardment period** comes from the Moon, Mars, and Mercury. All three bodies appear to have been heavily cratered during their early histories. (Venus does not provide a record of this period because, like Earth, during its history it has been resurfaced many times by volcanism.) The lunar cratering record is the best understood because some of the craters have been dated by using rocks collected near their rims. If the Moon and the other terrestrial planets were being pelted by large objects, it is almost certain that Earth was getting hit, too. The bombardment may have made it difficult for life to have originated before about 4 to 4.2 b.y. ago. We will return to this issue later in the chapter. It may also have brought additional water and other volatile compounds to Earth, particularly if the impactors were comets or volatile-compound–rich asteroids from the outer asteroid belt. So, the atmosphere and oceans may have continued to grow throughout this period.

Composition of the Early Atmosphere

What would have been the composition and surface pressure of the atmosphere during the heavy bombardment period? No one knows for sure, but we can make some

A) A Mars-sized body, 0.1 to 0.2 Earth masses, approaches the proto-Earth at an oblique angle.

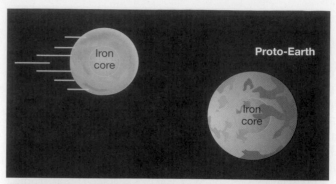

B) The two bodies collide. The ejecta from the impact fly off at an angle and the proto-Earth starts spinning rapidly.

C) The ejecta from the impact form a disk around the proto-Earth. The debris in the disk collide with each other and accrete to form the Moon. The iron core of the impactor collides again with the proto-Earth and becomes part of its core.

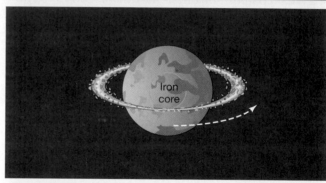

D) The Moon is initially only a few Earth radii (~20,000 km) away from the now nearly fully-formed Earth. Earth spins rapidly as a result of the collision. The daylength is 5-6 hours. Because they are so close together, the Moon and the Earth generate huge tides in each other that dissipate energy and transfer angular momentum from the Earth to the Moon.

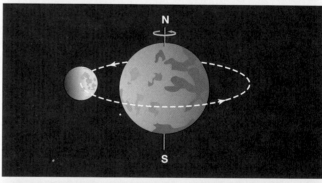

E) Over time, Earth's rotation slows, while the Moon retreats to its present orbital distance of ~60 Earth radii (384,400 km). The Moon's current rate of recession is ~1 cm/yr.

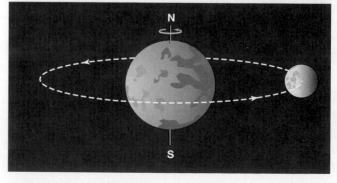

FIGURE 10-5

Schematic diagram illustrating the formation of the Moon by a giant impact.

educated guesses. Free oxygen, O_2, which makes up about 21% of today's atmosphere, should have been virtually nonexistent, because life—and therefore photosynthesis, the source of free oxygen—had probably not yet arisen. Nitrogen, N_2, does not participate very actively in geochemical cycles; hence, most of it should have been in the atmosphere, as it is today, at about 78% of the total. The present partial pressure of N_2 is about 0.8 bar. Nitrogen would have formed from nitrogen-rich organic compounds and ammonia ice (NH_3) in incoming planetesimals. The shock of their impact on Earth's surface should have converted much of this nitrogen to N_2.

Estimating the CO_2 partial pressure of the primitive atmosphere is much more difficult. On one hand, we know that Earth's total inventory of carbon is huge—the equivalent of 60–80 bars if it were all oxidized to CO_2. As discussed in the previous chapter, most of this carbon is presently stored on the continents in the form of carbonate rocks such as limestone and dolomite. It was originally delivered to Earth as organic carbon in incoming asteroids and comets. As discussed above, much of this carbon would have been immediately released into the atmosphere by the process of impact degassing. The chemical form of the carbon is difficult to calculate: some models predict that it would have been released as CO (carbon monoxide), while other models suggest that it would have been released as a mixture of CH_4 and CO_2. In either case, most of the carbon would have ended up as CO_2 because it would have been oxidized by photochemical reactions involving water vapor. (See the Box "A Closer Look: Oxidation of the Atmosphere by Escape of Hydrogen.")

Exactly how much CO_2 would have been present in the atmosphere during the heavy bombardment period is difficult to determine. Almost no rocks have been preserved from this time interval, and those that do exist tell us little or nothing about how much CO_2 was present. Hence, we are forced to rely on theoretical models to try to estimate the CO_2 partial pressure at that time. Unfortunately, different theoreticians get different answers depending on what they think was most important.

If, for example, the continents were originally much smaller, as some geologists believe, the process of silicate weathering on land may have been much slower than today. Because silicate weathering is the long-term loss process for CO_2 (Chapter 9), this would have tended to make the atmospheric CO_2 level higher. Some geologists have predicted that Earth could have had a 10-bar CO_2 atmosphere for the first several hundred million years of its history, until the continents began to grow. In that case, the surface temperature could have been quite hot (80–90°C), in spite of the 30% decreased luminosity of the young Sun.

On the other hand, other geologists point out that CO_2 should have reacted rapidly with the fresh seafloor and with the finely powdered **ejecta** produced by impacts. In that case, atmospheric CO_2 levels could have been quite low, and early Earth would have been very cold. We shall return to the question of the temperature of early Earth in Chapter 12. For now, though, we simply acknowledge that we do not know whether the atmosphere was thick or thin during the heavy bombardment period, and we are equally uncertain whether the climate was warm or cold. This might not matter much, either, except that life may have originated during this era, and theories of life's origin depend strongly on the ambient temperature. Perhaps, as our theoretical understanding of planetary formation increases, we will learn more about this earliest period of Earth's history.

The Origin of Life

The question of how life on Earth originated has been a topic for both religious and scientific speculation. Nearly all religions have their own creation "myths." While many of these stories have considerable moral and intellectual value, most are directly contradicted by the geologic record on Earth and by the sheer immensity of geologic time. For example, a literal reading of the Bible implies that God created the Earth and all forms of life over a space of 7 days only a few thousand years ago. We saw earlier in this chapter that radiometric dating places the age of Earth at over 4.5 billion years. Unless the laws governing radioactive decay change with time themselves, an unlikely possibility, the Biblical creation story cannot be true in a literal sense. A Creator or Supreme Being could indeed have played a role in the creation of both the universe and life, but if so, both events must have happened a very long time ago.

The modern scientific theory of life's origin was first formulated in the 1920s by Russian scientist Alexander Oparin and independently by British scientist J. B. S. Haldane. The *Oparin–Haldane hypothesis,* as it came to be called, postulated that life arose from chemical reactions that were initiated in a strongly reduced early atmosphere and came to completion in the early oceans. Recall that reduced carbon is carbon that is bonded to other carbon atoms or to hydrogen. A **strongly reduced atmosphere** is one that is rich in hydrogen-containing gases, such as methane (CH_4) and ammonia (NH_3). Oparin and Haldane proposed that energy sources such as sunlight and lightning caused these gases to react with each other to form organic compounds in a process termed **chemical evolution.** Ultimately, and in a manner that admittedly is still not understood today, these organic compounds as-

A CLOSER LOOK
Oxidation of the Atmosphere by Escape of Hydrogen

The compounds that formed Earth's primitive atmosphere were initially highly reduced. Recall from Chapter 9 that reduced carbon is carbon that is bonded to other carbon atoms, hydrogen, or nitrogen. Most of the carbon in meteorites is in the form of reduced or organic carbon, and this was presumably true of the planetesimals from which Earth formed as well. When these planetesimals impacted the young planet, much of the carbon would have been released as the reduced gases CO and CH_4. It would not have remained in those forms very long, however. In the absence of oxygen and ozone, ultraviolet radiation from the Sun would have **photolyzed** (split apart) water molecules, creating hydrogen atoms (H) and **hydroxyl radicals** (OH):

$$H_2O + \text{UV photon} \rightarrow H + OH$$

(A *radical* is a molecule that is highly reactive because it has an unpaired electron in its outer shell.) OH radicals play an important role in today's atmosphere, where they are responsible for oxidizing various gases, including CH_4 and CO. In the case of CO, the reaction is:

$$CO + OH \rightarrow CO_2 + H.$$

For CH_4, the reaction sequence is more complicated but the result is the same: the carbon ultimately ends up as CO_2.

What makes these reactions important is the fact that they are essentially irreversible. The hydrogen atoms that are produced are light enough to escape from Earth's atmosphere. As this happens, both CO and CH_4 are converted irreversibly to CO_2. Hence, Earth's atmosphere tends to become more oxidized with time, simply because hydrogen is always being lost to space. Note, however, that this process does *not* produce free oxygen, O_2. A very small amount of O_2 can be produced by other reactions, as described later in the chapter, but almost all of the O_2 in our present atmosphere was produced by photosynthesis.

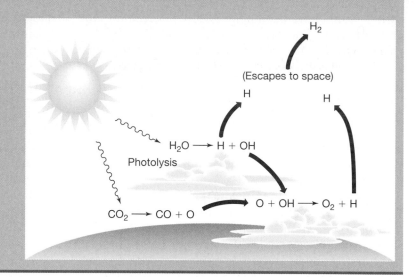

BOX FIGURE 10-3
Diagram illustrating escape of hydrogen to space and resultant oxidation of the early atmosphere.

sumed the characteristics of living systems. (See the Box "A Closer Look: What Does It Mean to be Alive?")

The Miller–Urey Experiment

The Oparin–Haldane theory of life's origin received a gigantic boost from a series of laboratory experiments performed in 1953 by a graduate student at the University of Chicago, Stanley Miller, working under the guidance of famous geochemist Harold Urey. Miller and Urey filled flasks with mixtures of gases that were considered at that time to be representative of Earth's primitive atmosphere. (These gases had just recently been discovered in Jupiter's atmosphere. Because Jupiter does not lose hydrogen, its

atmosphere was considered by Urey to be "unevolved." Urey reasoned, not quite correctly, that Earth's atmosphere would have been similar in composition before hydrogen had had time to escape.) The flasks contained methane and ammonia, along with water vapor and molecular hydrogen. The researchers then sparked the flasks with powerful electric discharges, simulating lightning in the prebiotic atmosphere.

After several minutes of electrification, the walls of the flasks became coated with a sticky, brownish material. When analyzed, this material was found to contain an assortment of organic compounds, including amino acids. **Amino acids** are compounds—containing an amino group (NH_2) and a carboxyl group (COOH)—that are impor-

A CLOSER LOOK
Prebiotic O_2 Concentrations

How much O_2 would have been present in the atmosphere prior to the origin of life? We already know that most of Earth's O_2 was produced by photosynthesis. This, of course, is a biological process. We would like to know the prebiotic atmospheric O_2 level for two reasons: (1) free O_2 would have poisoned the chemical reactions leading to the origin of life, as discussed in the text; and (2) if prebiotic O_2 levels were low on Earth, then the presence of O_2 in another planet's atmosphere may be a useful indicator that life is present. As we will see in Chapter 19, scientists hope to eventually look for the presence of O_2 in the atmospheres of planets around other stars to determine whether such planets might be inhabited.

How could O_2 have been produced in the absence of photosynthesis? Probably the most important mechanism for producing O_2 abiotically is the following. First, water vapor is photolyzed by a UV photon, producing atomic hydrogen (H) and a hydroxyl radical (OH):

$$H_2O + UV\ photon \rightarrow H + OH.$$

Then, another UV photon splits a CO_2 molecule, producing carbon monoxide (CO) and atomic oxygen (O):

$$CO_2 + UV\ photon \rightarrow CO + O.$$

The OH and O radicals then combine to form O_2:

$$O + OH \rightarrow O_2 + H$$

If the hydrogen atoms produced in these reactions escape to space, which they can do because they are light, then a net production of O_2 has occurred.

This does *not* necessarily imply that O_2 will build up in the atmosphere of an abiotic planet, however, because there are also loss processes for O_2 that tend to remove oxygen as fast as it is produced. The most important of these is oxidation of reduced volcanic gases, such as H_2 and CO. These gases do not react directly with O_2 at room temperature (unless one provides a flame to get them started!). However, they can react with oxygen indirectly by way of reactions that involve the by-products of water vapor photolysis. The net result is:

$$2\ H_2 + O_2 \rightarrow 2\ H_2O$$
$$2\ CO + O_2 \rightarrow 2\ CO_2$$

Thus, the oxygen reacts with these reduced gases to form H_2O and CO_2. When one studies these processes with detailed models, one finds that such reactions will quickly use up almost all of the O_2 produced by the photolysis of H_2O and CO_2, followed by the escape of hydrogen to space. The net result is a **weakly reduced atmosphere** with a composition similar to that shown in Box Figure 10-4. (Evidently, the amount of free O_2 generated by photochemical reactions in an early Earth-type atmosphere is extremely low. Hence, the presence of O_2 in a planet's atmosphere is a strong indication of the presence of life.)

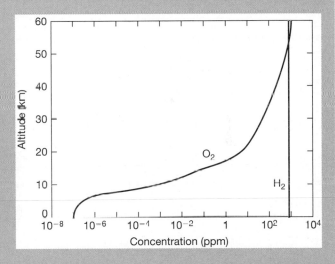

BOX FIGURE 10-4
Vertical profiles of H_2 and O_2 in a weakly reduced primitive atmosphere.

---

tant building blocks for proteins (see the Box "A Closer Look: The Compounds of Life"). **Proteins**, composed of one or more chains of amino acids, are key molecules in organisms. Proteins may be *enzymes* (guiding chemical reactions), structural components, *hormones* (maintaining constant body conditions), or transport molecules.

The Miller–Urey experiment, as it is now called, made headlines around the world. It was indeed a revelation to discover that many of the basic compounds on which life depends can be synthesized by a straightforward process that could have occurred in nature. Many scientists working in this field thought that we might be close to understanding how life itself began.

Today, the Miller–Urey experiment is still held in high regard scientifically, but researchers are less certain that it represents a critical step in the origin of life. One reason is that just mentioned: Current theories of the early atmosphere suggest that it was not as strongly reducing as the gas mixtures in Stanley Miller's flasks. Methane and ammonia, if they were present at all, would probably have been held to relatively low concentrations because they are photolyzed by solar UV radiation. The hydrogen then

A CLOSER LOOK
What Does It Mean to be Alive?

Almost anywhere we go on Earth today, we can see things that we know are alive: plants, animals, insects, people, and other **organisms**. How do we know that these various, extremely different things are alive?

It is surprisingly difficult to come up with a good definition of life. The ability to move, for example, is not an identifying characteristic of life because while animals move, plants do not. Also, many inanimate objects, cars for example, move under their own power. At a chemical level, all organisms that we know of consist of organic (carbon-based) compounds and are reliant on the presence of liquid water during at least part of their life cycles. However, we cannot be certain that this is true of life in general. Carbon is particularly well suited for making long chains and big, complex molecules, so it may well be that all life is carbon based. However, biologists still seek some more general definition.

The definition of life that is generally quoted by biologists is based on Charles Darwin's theory of evolution. According to this theory, organisms evolve by way of the combined processes of **replication**, **mutation**, and **natural selection** (Box Figure 10-5). As the name suggests, *replication* is the process by which an organism reproduces itself. Given that all organisms have a finite (and relatively short) life span, life obviously could not exist for very long if organisms were unable to replicate. *Mutation* simply means that the replication process is not exact, that is, the organisms that are produced can in some instances be different from the original organisms. If not for mutation, organisms could not evolve into different and more complex forms. *Natural selection* refers to the process by which certain, better-adapted organisms survive in greater numbers than do others. Thus, a favorable mutation can lead to a new type of organism that may gradually replace the old organism or, alternatively, to an organism that is capable of living in some different environment. Through this combined process of replication, mutation, and natural selection, life has evolved into a myriad of different forms that have successfully colonized nearly all parts of Earth's surface.

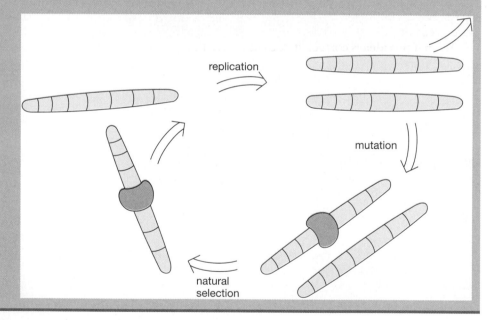

BOX FIGURE 10-5

Cartoon illustrating the processes of replication, mutation, and natural selection, by which life is defined.

---

escapes to space, and the carbon and nitrogen atoms left behind are converted into CO_2 and N_2. Furthermore, modern volcanos do not emit much methane or ammonia. Early volcanic gases might have been more highly reducing; however, even this was probably not enough to produce a Miller–Urey-type atmosphere.

The RNA World

A second way in which our ideas about the origin of life have changed is that most biologists now believe that proteins were not among the earliest structural elements of life. Rather, life might have relied exclusively on RNA or some simpler variant of RNA. (See Box "A Closer Look: The Compounds of Life" for a description of what RNA, DNA, and proteins consist of chemically.)

The evidence that RNA preceded proteins was discovered in the mid-1980s by Thomas Cech of the University of Colorado and Sydney Altman of Yale University, and it earned them the 1989 Nobel Prize in chemistry. Cech and Altman discovered that one particular type of RNA molecule was capable of cleaving (cutting) itself into smaller pieces. This capability meant that it was theoretically possible for an RNA molecule to replicate (duplicate) itself without help from any other molecule. DNA-based organisms cannot do this. The DNA molecule can replicate only with the aid of complex enzymes made of proteins. An **enzyme** is a biological molecule that speeds up, or *catalyzes,* a particular biochemical process.

Life depends on a complex interaction among DNA, RNA, and proteins. The DNA carries the basic genetic information (the blueprint for the organism); the RNA is used to transfer this information to other parts of the cell, where proteins are made; and the proteins perform many different cell functions, including the replication of DNA and RNA.

A primitive, RNA-based organism could have been much simpler than today's organisms. RNA itself could have been the molecule in which the genetic information was stored. RNA is less stable than DNA, but it carries essentially the same information. And because it can cleave itself, RNA could have reproduced without the aid of enzymes. The elegance and simplicity of this idea has led biologists to suggest that DNA-based life was preceded by an **RNA World**, in which only RNA-based organisms were present.

Even if this idea is correct, it does not solve the problem of life's origin. We still need to make the basic compounds of which RNA is composed and then assemble them into a self-replicating molecule. The required phosphate molecules were probably present in the primitive ocean as a result of weathering of rocks, but ribose and the four nitrogen-containing bases have more complicated structures that may or may not have been easy to form.

Prebiotic Synthesis of Organic Compounds

In the Miller–Urey experiment, the investigators found that they could synthesize amino acids by starting from gaseous mixtures that contain methane and ammonia. If RNA-based organisms came first, the first requirement for originating life would have been to synthesize ribose and the four bases adenine, guanine, cytosine, and uracil. Alternatively, the earliest organisms could have used simpler compounds that later evolved into RNA, but this scenario presents the difficult problem of how life could have switched from one molecular basis to another. (Switching from RNA to DNA is not considered difficult from an evolutionary standpoint because the molecules are so similar. Indeed, the first step in synthesizing DNA within a cell is to form the corresponding RNA molecule.)

Could ribose and the bases have formed from compounds present in the primitive atmosphere and oceans? We can get a clue by looking at chemical formulas. The molecular formula for ribose is $C_5H_{10}O_5$. (The corre-

sponding structural formula, which shows how the atoms are arranged, is shown in Box Figure 10-7b.) Simple division shows that ribose can be formed from five molecules of H_2CO—the compound *formaldehyde,* which is commonly used to preserve dead animals. (If you dissected a frog in high school biology class, you may recall its distinctive smell.) The molecule H_2CO should not be confused with the shorthand notation for organic carbon, CH_2O, that we have used elsewhere in this book. Whereas H_2CO is an actual molecule, CH_2O merely represents complex hydrocarbons that have approximately the same relative ratios of C, H, and O atoms.

Thus, the first step in forming ribose is to synthesize formaldehyde. This step, it turns out, is easily accomplished. Photochemical reactions in weakly reducing, CO_2-rich atmospheres are predicted to produce large quantities of formaldehyde. Because formaldehyde is soluble in water, much of it should have dissolved in rainwater and been transported into the early ocean. Converting formaldehyde into ribose is also not difficult, because formaldehyde spontaneously reacts to form sugars in water solution. The problem for life's origin, which we will not discuss here, is that lots of other sugars form in addition to ribose and that these molecules might have interfered with the synthesis of RNA. What we can say is that the necessary starting material for forming ribose should have been available.

What about the four bases—could they have been formed on the prebiotic Earth? Let us take the same approach and begin with chemical formulas. For simplicity, we consider only the simplest base, adenine. Its molecular formula is $C_5H_5N_5$. Evidently, adenine can be formed from five molecules of HCN, *hydrogen cyanide.* Hydrogen cyanide is an extremely deadly poison to most higher organisms. To prebiotic chemists, however, it is considered an essential building block for life.

Forming hydrogen cyanide in the prebiotic atmosphere is more difficult than forming formaldehyde. In a Miller–Urey-type atmosphere containing methane and ammonia, HCN would have been generated by lightning. In a weakly reduced, CO_2–N_2 atmosphere, lightning would not have sufficed because the carbon and nitrogen atoms produced by the lightning would have combined with oxygen atoms from the CO_2. However, the primitive atmosphere might have contained a few tens of parts per million of methane, CH_4. Methane photolysis in the stratosphere produces molecular fragments that can combine with N atoms that flow down from the ionosphere, forming HCN. The N atoms are produced from the breaking apart of the ion N_2^+ when it recombines with an electron.

The key to this mechanism is to identify a source for atmospheric methane. The methane in today's atmosphere is almost entirely of biological origin. However, some abiotically generated methane is released in fluids

A CLOSER LOOK
The Compounds of Life

Life depends on a complex array of organic (carbon-containing) compounds that organisms use to perform different tasks. Perhaps the most fundamental of these compounds are amino acids and nucleic acids. Amino acids are the building blocks of proteins, which are essential to many different cell functions. Nucleic acids, which include both ribonucleic acid (RNA) and deoxyribonucleic acid (DNA), are the carriers of genetic information. DNA stores this information, and RNA transfers the information to different parts of the cell and makes proteins and other compounds.

Chemically, an amino acid is an organic compound that contains an amino group (NH_2) and a carboxyl group (COOH). The two simplest ones are glycine and alanine (Box Figure 10-6). These compounds are similar except for the side chain: The hydrogen atom, H, in

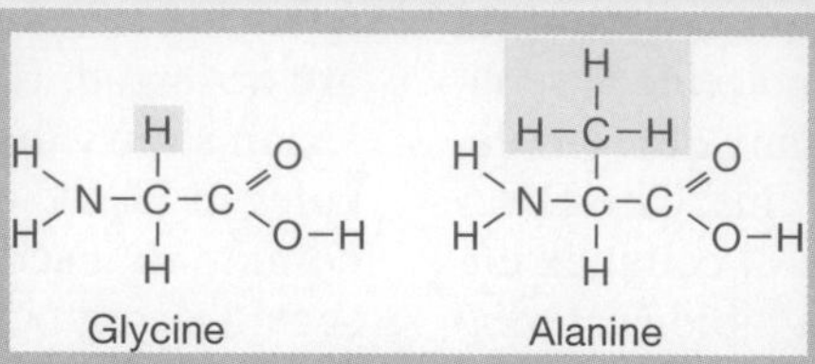

BOX FIGURE 10-6

Two of the simpler amino acids found in proteins.

glycine is replaced by a methyl group, CH_3, in alanine. Other amino acids have more complicated side chains containing oxygen, nitrogen, or sulfur. Twenty different amino acids are found in naturally occurring proteins, but many other amino acids are chemically possible.

RNA and DNA are more complicated compounds that consist of chains of molecules called **nucleotides**. Each nucleotide consists of three parts. In RNA, these include a phosphate molecule, a

ribose (sugar) molecule, and a nitrogen-containing base. The base can be any of four molecules: adenine (A), guanine (G), cytosine (C), or uracil (U) (Box Figure 10-7). The nucleotides in RNA are linked together in a long, single strand.

A DNA nucleotide is similar to an RNA nucleotide except that the sugar molecule is deoxyribose instead of ribose and that one of the four bases, uracil, is replaced by thymine (T). The DNA nucleotides are strung together in two chains that form a double helix. This double-stranded, twisting structure was discovered in 1953 by James Watson and Francis Crick. The sequence of bases in DNA carries information in the form of the genetic code. Groups of three individual nucleotides code for specific amino acids. For example, the sequence CCA codes for glycine and CGA codes for alanine.

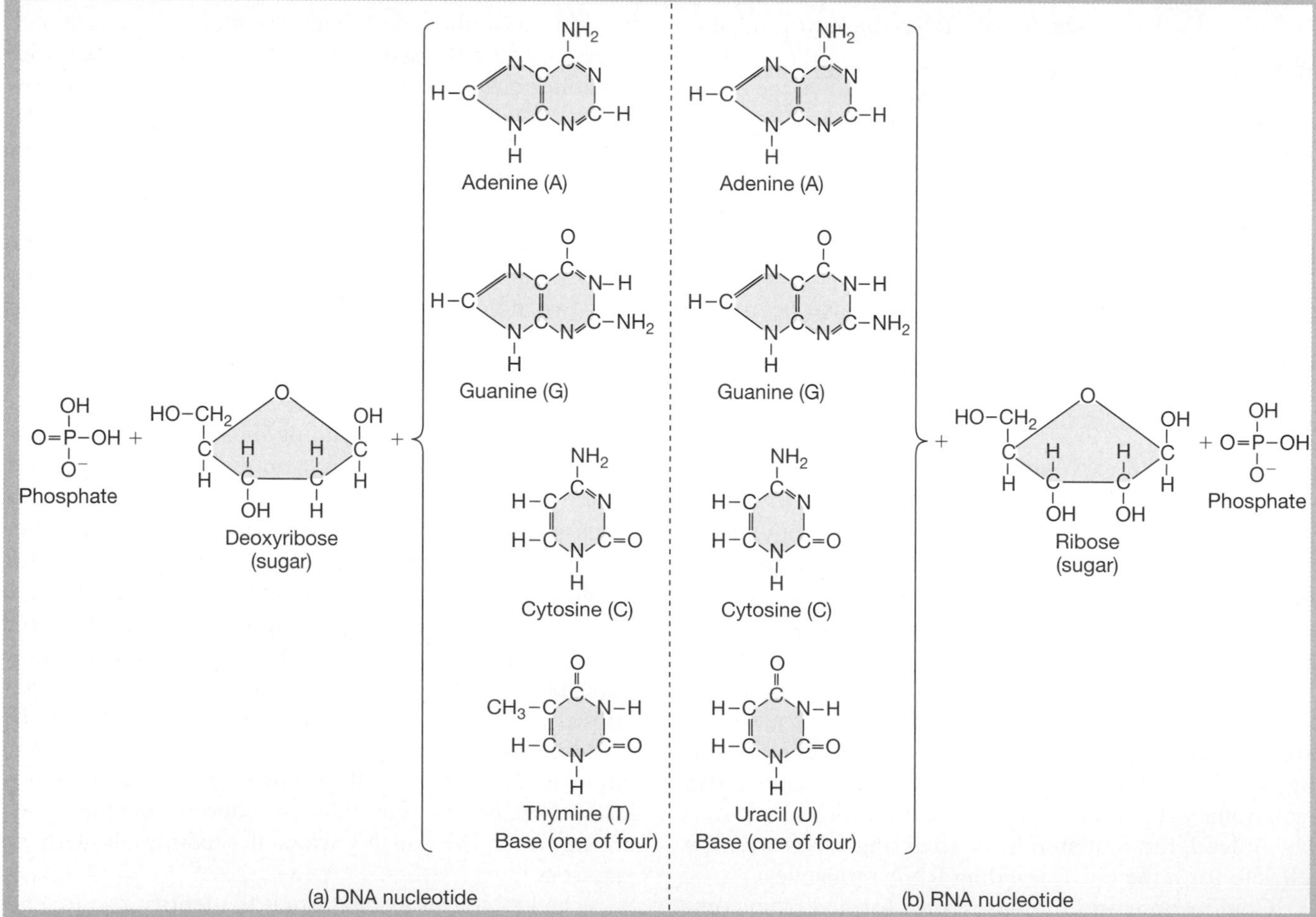

BOX FIGURE 10-7

Structural diagrams of the components of (a) DNA and (b) RNA. (From G. S. Kutter. *The Universe and Life,* 1987. Jones and Bartlett Pub., Boston.)

coming from hydrothermal vents at mid-ocean ridges. These vents, which are described in more detail below, are places where hot, mineral-laden water flows into the deep ocean. The water contains dissolved carbon compounds, including both CO_2 and CH_4. The fluids emanating from the hottest vents contain mostly CO_2, but the cooler, off-axis vents on certain, slow-spreading ridges (e.g., the Mid-Atlantic Ridge) are rich in CH_4 and H_2. These gases are thought to be produced by a process called **serpentinization**, a chemical reaction in which seawater reacts with **ultramafic rocks** (rocks rich in magnesium and iron) to form compounds called *serpentine minerals*. We will return to this topic in the next chapter because it may bear on the rise of atmospheric oxygen. For now, though, we simply point out that some abiotically produced methane should have been available on the early Earth even if the highly reduced, Miller–Urey-type atmosphere never existed.

Other Theories of Life's Origin

Some researchers remain skeptical that life could have formed on Earth's surface or in its oceans. Although the fundamental building blocks of life, H_2CO and HCN, were probably available, the chance that they would have been concentrated sufficiently to allow further reactions to occur might have been small. And the more complex organic compounds that might have formed in this way would not have lasted long in the surface-ocean environment, because they would have been destroyed by photochemical and thermal (heat-driven) reactions. Therefore, researchers have sought alternative ways of forming complex organic compounds.

One possibility is that the relevant organic compounds were formed in space and brought to Earth by asteroids or comets or as tiny dust particles. **Interplanetary dust particles (IDPs)** are small particles recovered from the stratosphere that are known to be of extraterrestrial origin. We know that organic compounds, including amino acids, exist in IDPs as well as in some meteorites. Indeed, amino acids and many other complex organic compounds have now been identified in interstellar dust clouds (Figure 10-2). They are believed to form from reactions between ions and neutral molecules that occur at very low temperatures. Typical temperatures in interstellar dust clouds are on the order of 10 K, not much above absolute zero. It may seem surprising that organic chemistry could occur in this environment, but it is precisely the extremely low temperatures involved that allow complex organic molecules to exist. The organic molecules form from reactions between other molecules and ions (charged particles), and then they live for long times because temperatures are too cold to allow them to decompose.

Some of the molecules formed in the interstellar environment are thought to have survived the collapse of the cloud that formed our own Sun and solar nebula. They would have been incorporated into solid materials that condensed out of the nebula and accreted to form asteroids and comets. Such materials might have been delivered to Earth in great quantities during the heavy bombardment period of solar system history, between 4.5 and 3.8 b.y. ago.

A third theory of life's origin is that it took place in or around **hydrothermal vents** in the mid-ocean spreading ridges. Recall from Chapter 7 that mid-ocean ridges are places where new sea floor is being created. The ridges are cooled by seawater that flows a kilometer or more down through cracks in the rock, is heated, and then rises rapidly back to the surface. In the process, the water picks up reduced substances such as hydrogen (H_2), hydrogen sulfide (H_2S), and dissolved ferrous iron (Fe^{2+}). *Ferrous iron* is a reduced form of iron that is soluble in seawater. When it hits the cold water, the hot (350°C) vent water produces a dark plume of precipitating material called a **black smoker** (Figure 10-6). The majority of the dark material is iron sulfide, FeS, produced by the reaction between ferrous iron and hydrogen sulfide.

Are submarine hydrothermal vents a likely place for life to have originated? The vent systems are rich in the types of reduced materials from which organic molecules can be synthesized. They contain liquid–solid interfaces that some researchers think are needed to organize organic molecules into specific patterns. One model suggests that life originated on the surface of *pyrite* (FeS_2) mineral grains, which are abundant in hydrothermal vent systems. However, complex organic molecules are not stable at the high (350°C) temperatures observed in vents located directly on the ridge axes. If life did originate at the mid-ocean ridges, it probably did so in cooler, off-axis

FIGURE 10-6

[See color section] Picture of a black smoker. (Ken MacDonald/Science Photo Library.)

vents. Some researchers argue that even the off-axis vents are too warm and that the best place for life to have originated would be in some near-freezing surface environment. The debate as to whether life originated in a hot or cold environment is likely to continue until we have a better idea of how the process actually occurred.

When Did Life Arise?

Thus far, we have discussed how life may have arisen but we have not talked about when this event occurred. The question of when life originated is currently the topic of much debate. Until about the middle of the last century, **paleontologists** (geologists who study fossils) believed that life originated only about 540 m.y. ago at the dawn of the Cambrian Period. Fossils of this age or of more recent time periods are easy to find because the organisms that formed them were large enough to see with the naked eye and because many of them formed shells of silica or calcium carbonate that became preserved within sedimentary rocks. In the 1940s, though, paleontologists such as Elso Barghoorn at Harvard University began to discover **microfossils**. As their name implies, microfossils are the fossilized remains of tiny, single-celled organisms (or, in some case, organisms formed of chains of individual cells). Unlike **macrofossils** (the remains of multicellular organisms), microfossils are difficult to find and even more difficult to classify. And it is easy to be confused between bonafide microfossils and structures that look like microfossils but are formed abiotically. As this chapter is being written, a vigorous debate is occurring over what had been thought to be the world's oldest microfossils (see Figure 10-7). These specimens were collected from the Apex Chert, which is part of the 3.5 b.y.-old Warrawoona Formation in Australia by paleontologist J. William Schopf from the University of California, Los Angeles, and they do indeed look remarkably like some modern bacteria. (See further discussion in Chapter 12.) But British paleontologist Martin Brasier and his colleagues believe that these structures are not biological at all. Rather, they think that these are bits or chains of organic carbon that formed abiotically within fluids emanating from a hydrothermal vent. The jury is still out as to which protagonist in the debate is correct.

Regardless of who turns out to be right about the Apex Chert microfossils, it does appear that life had originated by 3.5 b.y. ago or slightly thereafter. Many more structures that are plausible microfossils have been found in slightly younger rocks, along with macroscopic structures called *stromatolites,* which we will discuss in Chapter 12. More interesting is the question of whether life had originated even earlier. *Isotopically light* organic carbon has been found in 3.85 b.y.-old rocks from Isua,

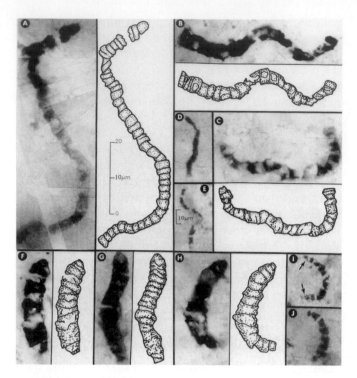

FIGURE 10-7

Apex Chert "microfossils." These structures are from the 3.5 billion year old Warrawoona Formation in Australia. A debate is currently raging as to whether they are biogenic or not. (J. W. Schopf. *Science* 260, 640–646, 1993).

West Greenland, and from nearby Akilia Island. By "isotopically light," we mean that the organic carbon is depleted in the heavier ^{13}C isotope compared to the normal ^{12}C isotope. As we will discuss in more detail in the next chapter, metabolic processes such as photosynthesis tend to discriminate against the heavier carbon isotope, so "light" organic carbon is usually considered to be evidence for biological activity. However, some abiotic processes can also favor one isotope over another. Furthermore, some (but not all) of the rocks in which this organic carbon has been found have recently been reclassified as being of igneous, rather than sedimentary, origin. It is difficult to imagine how biologically generated carbon could end up trapped within a rock that formed from a molten magma. So, the jury is still out on this question as well. But it is certainly possible, even likely, that life was already around by 3.8–3.9 b.y. ago.

The Universal Tree of Life

Some of the most useful information concerning the origin of life comes from studying modern organisms. Over the past two decades, molecular biologists have learned to sequence both RNA and DNA. Sequencing a molecule of nucleic acid means determining the order of the individual nucleotides. Recall that a nucleotide consists of one of four bases attached to a ribose molecule for RNA, or deoxyribose, for DNA, which in turn is connected to other nucleotides by phosphate linkages. (See the Box, "A Closer Look: The Compounds of Life.") By using powerful new techniques such as **PCR**, the **polymerase chain reaction**, biologists have been able to unravel the genetic code of all sorts of different organisms, including humans. Both DNA and RNA are extremely large and complicated molecules that contain information about every facet of an organism. Particular parts of these molecules can be used to look way back into early evolutionary history. The particular molecule that has been found to be most useful is the RNA found within **ribosomes**. The ribosome is a part of the cell in which proteins are manufactured. All organisms have ribosomes and ribosomal RNA, which makes this molecule useful for making comparisons. Protein manufacture is also an extremely ancient and slowly evolving metabolic capability, so sequencing ribosomal RNA provides a way of looking deeply into evolutionary history. Because PCR acts on DNA, not RNA, this is actually done in practice by sequencing the part of the DNA molecule that codes for ribosomal RNA, but it is essentially the RNA sequences that are being compared.

The results of comparing sequences of ribosomal RNA from various organisms can be used to draw a "tree" (Figure 10-8). This tree was first constructed by Carl Woese at the University of Illinois and his graduate student, George Fox. It shows that organisms can be divided into three main categories, or **domains**: **Bacteria**, **Archaea**, and **Eukarya**. The Bacteria and Archaea are both composed entirely of single-celled organisms that we commonly refer to as *bacteria*. The Eukarya contain some single-celled organisms as well, such as the intestinal parasite *Giardia*, but they also include all higher plants and animals, including humans. As Figure 10-8 shows, humans (*Homo*) and corn (*Zea*) are closely related to each other by comparison to the very deep divisions that occur between the three different domains of life.

A Hyperthermophilic Last Common Ancestor?

All sorts of interesting information can be derived from this Tree of Life, and we shall take advantage of some of it in the next chapter when we discuss the rise of oxygen. For now, though, let us concentrate on the shaded branches near the point at which the Archaea and Eukarya split from the Bacteria. This point is thought to lie close to the common ancestor of all life.

The shaded branches represent **hyperthermophilic bacteria**—organisms that live at temperatures above 80°C. Surprisingly, nearly all of the organisms near the "root" of the tree are hyperthermophiles. Today, by contrast, hyperthermophiles are restricted to a few unusual environments, such as the mid-ocean ridge vent systems and geothermal hot spots like Yellowstone National Park.

What is the Tree of Life telling us? Does this imply that life originated in a hot environment like the mid-ocean ridge hydrothermal vents? Maybe. But there are other possibilities as well. Some biologists believe that this is merely an artifact of the fact that guanine-cytosine (G-C) bonds in DNA are slightly more stable at high temperatures than are adenine-thymine (A-T) bonds. The DNA of hyperthermophiles has therefore evolved to be rich in G-C bonds, so these organisms all tend to cluster together and to look artificially "ancient." But it might also be telling us something very interesting. Remember, at least some scientists believe that life originated prior to 3.8 b.y. ago. This would have been within the heavy bombardment period discussed earlier in this chapter. During that time, Earth was still being pummelled by large comets or asteroids. Some of these impacts may have been large enough to vaporize the uppermost layers of the ocean and to completely sterilize the land surface. Suppose that life originated in some cool surface environment and then proceeded to colonize most of Earth's surface, including the mid-ocean ridge vents. Suppose further that, after this had occurred, a giant impactor hit Earth and destroyed all of the surface-dwelling organisms. The mid-ocean ridge dwellers would have been protected from all but the very largest impacts because they were sheltered underneath 2–3 kilometers of ocean. Once the effects of the collision on Earth's surface had died down, roughly 1,000–2,000 years following the impact, organisms from the vent systems could have begun the process of recolonizing. So, it may be that the last common ancestor of all modern organisms was indeed a hyperthermophile, even though the origin of life itself took place in some cooler environment. The information in the Tree of Life does not allow us to distinguish between these two possibilities.

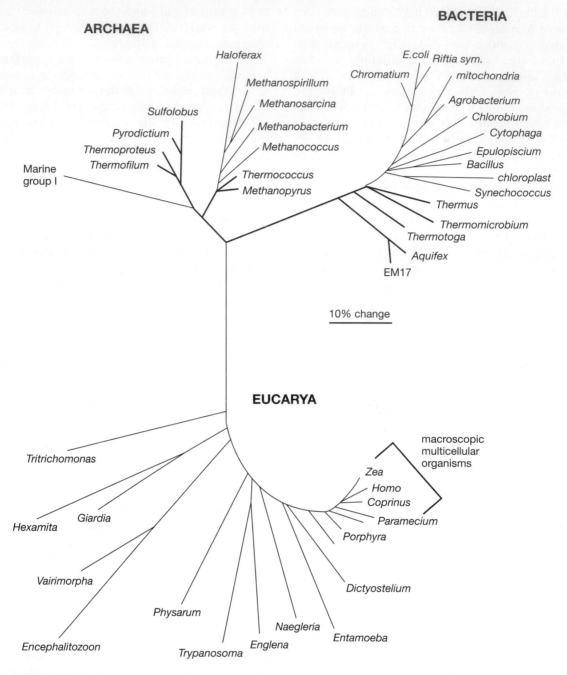

FIGURE 10-8

The Universal Tree of Life derived by sequencing ribosomal RNA. (Courtesy of Norman Pace, University of Colorado.)

Chapter Summary

1. Earth formed from accretion of solid materials that condensed out of the solar nebula soon after the Sun itself formed.
   a. The age of Earth is identical to that of meteorites, 4.55 b.y., as determined by radiometric age dating.
   b. Earth's core probably formed as the Earth itself was forming as a result of heating and stirring by large impacts.
2. The Moon is thought to have formed as a consequence of a glancing impact by a Mars-sized planetesimal and, hence, is something of a cosmic accident.
3. Earth's atmosphere and ocean formed along with the planet from impact degassing of incoming planetesimals and from volcanic outgassing. The resulting atmosphere was probably rich in N_2, and possibly CO_2, but contained little O_2 prior to the origin of life and the evolution of photosynthesis.
4. Life originated on Earth by a process termed *chemical evolution*.
   a. This process may have occurred on Earth's surface using chemicals formed by energetic processes within the atmosphere.
   b. Alternatively, it may have formed from chemicals synthesized in space and imported in interplanetary dust particles or from chemicals synthesized in hydrothermal vents at mid-ocean ridges.
5. Much of what we know about the early evolution of life comes from analyzing ribosomal RNA, or the DNA equivalent thereof.
   a. Strong evidence shows that RNA preceded DNA as an informational molecule.
   b. Weaker evidence indicates that the common ancestor of all extant organisms lived in a hot environment. This could be explained either by a high temperature origin of life or by extinction of nonhyperthermophilic organisms by a giant impact. The latter hypothesis is consistent with life having originated during the heavy bombardment period prior to 3.8 b.y. ago.

Key Terms

accretion	hyperthermophilic bacteria	photolyzed
amino acids	IDPs	planetesimals
Archaea	impact degassing	polymerase chain reaction
Bacteria	interplanetary dust particles	proteins
black smoker	interstellar clouds	proto-Sun
chemical evolution	isochron diagram	replication
chondrites	lead–lead dating	ribosomes
domains	macrofossils	RNA world
ejecta	magma ocean	serpentinization
enzyme	main sequence	shelly fossils
eon	meteorites	solar nebula
Eukarya	microfossils	strongly reduced atmosphere
giant impact hypothesis	mutation	terrestrial planets
heavy bombardment period	organisms	ultramafic rocks
Hertzsprung–Russell diagram	natural selection	volatile compounds
hydrothermal vents	paleontologists	weakly reduced atmosphere
hydroxyl radicals	PCR	zircon

Review Questions

1. How old is the solar system, and how is this age determined?
2. How is the age of Earth determined if no rocks older than 4.1 billion years have been preserved?
3. How do Jupiter and the Moon affect the habitability of Earth?
4. How and when did the atmosphere and ocean form? Which gases are thought to have been present in the early atmosphere?
5. In what types of environments might life have originated?
6. Why is RNA thought to have preceded DNA in evolution?
7. How is the Universal Tree of Life constructed?
8. Into what three different domains are modern organisms divided?
9. List two possible reasons why organisms near the base of the Tree of Life are hyperthermophilic.

Critical-Thinking Problems

Write a 1 to 2-page, typed essay on the following question:

1. What do you feel is the best theory for how life originated? Do you think that life might exist elsewhere besides Earth?

Further Reading

General

Lovelock, J. E. 1988. *The Ages of Gaia: A Biography of our Living Earth.* New York: W. W. Norton and Co.

Lunine, J. 1999. *Earth, Evolution of a Habitable World.* Cambridge: Cambridge University Press.

Advanced

Brack, A. 1998. *The Chemical Origins of Life: Assembling the Pieces of the Puzzle.* Cambridge: Cambridge University Press.

Effect of Life on the Atmosphere: The Rise of Oxygen and Ozone

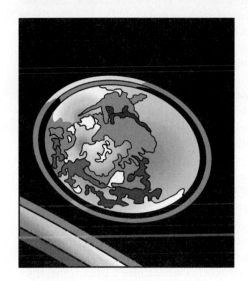

Key Questions

- What were the earliest forms of life, and how did they affect atmospheric composition?
- When and why did atmospheric O_2 become abundant?
- When did the ozone layer form, and how did its formation affect Earth's surface environment and the evolution of the biota?
- By how much has atmospheric O_2 varied over the last 540 million years?
- What controls the atmospheric O_2 concentration today?

Chapter Overview

Earth's present atmosphere is rich in molecular oxygen and has a well-developed ozone layer that shields the planet's surface from harmful solar ultraviolet radiation. This oxygen is produced by photosynthesis and, hence, would not have been present prior to the origin of photosynthetic life. The first organisms to evolve were probably not photosynthesizers. Rather, they lived in other ways, for example, by converting carbon dioxide and hydrogen into methane. Photosynthesis originated at or before 2.3 b.y. ago, the time when atmospheric O_2 levels first rose. Indeed, photosynthesis was probably occurring for several hundred million years prior to this time, but the initial rise of O_2 was de-

layed for reasons that may be related to the gradual oxidation of Earth's mantle. Atmospheric O_2 has varied by modest amounts for the past few hundred million years because of changes in the rate of organic carbon burial. The fluctuations in O_2 are small, however, because of a negative feedback mechanism that appears to involve the oxygenation of the deep oceans and the availability of dissolved phosphorus.

Introduction

In the last chapter, we saw that the prebiotic Earth probably had an atmosphere dominated by carbon dioxide and molecular nitrogen (N_2). But we also saw that life probably originated very early—within the first 700 m.y. of Earth's history—even though some of the evidence in favor of this idea has been questioned. As soon as life had evolved, it began to be a force that could change the composition of the atmosphere and eventually the surface as well. What was the nature of these earliest organisms, and how did they alter atmospheric composition? We can gain some insight into these questions by studying the structure of the ribosomal RNA tree and deducing which modern organisms look most "primitive."

An even bigger change in atmospheric composition occurred when organisms evolved that were capable of oxygenic photosynthesis. The production of oxygen by such organisms eventually led to the establishment of our modern, O_2-rich atmosphere. But this change from reduced to oxidized atmospheric conditions appears to have occurred well after the invention of photosynthesis, for reasons that are poorly understood. What other nonbiological changes in the planet were needed in order to allow atmospheric O_2 to accumulate?

A related change in atmospheric composition that has affected both microbial and advanced life has been the development of a protective ozone layer. We have already seen that ozone is important because it blocks out harmful solar ultraviolet (UV) radiation. But, if the concentration of atmospheric O_2 was initially very low, there must have been a time when the ozone layer did not exist. How did life cope with the solar UV flux at that time? When was an effective solar UV screen established? These are also questions that we address in this chapter.

Effect of Life on the Early Atmosphere

As we saw in the previous chapter, we still do not know how life originated on Earth. We also do not know exactly when it originated because both the 3.5 b.y.-old Apex Chert microfossils and the 3.85 b.y.-old carbon isotopic evidence for life have been called into question. Starting at about 3.5 b.y. ago, geologists have also found layered structures called **stromatolites** that are believed to be the fossilized remains of bacteria that form layered, sedimentary mats (Figure 11-1). The organisms that formed them are thought to have been similar to communities of photosynthetic bacteria that inhabit the shallow, salty waters of Shark Bay on the coast of Western Australia (Figure 11-2).

How did the presence of life affect the composition of the atmosphere? The organisms that formed stromatolites were almost certainly photosynthetic—the fact that they lived in layers implies that they needed sunlight. However, this does not necessarily imply that they were producing oxygen. "Normal," **oxygenic photosynthesis** is carried out by higher plants and algae today. But some bacteria make their living by a related process called anoxygenic photosynthesis. In **anoxygenic photosynthesis**, H_2S or H_2 is used instead of H_2O to reduce CO_2 to organic carbon; this process does not yield O_2. Finally, there are certain bacteria called **cyanobacteria**, formerly referred to as blue-green algae, that can perform both oxygenic and anoxygenic photosyntheses. In practice, some cyanobacteria switch back and forth between these two metabolic processes in response to changes in their local environment. When hydrogen sulfide, H_2S, is present, these cyanobacteria photosynthesize anoxygenically; when H_2S is absent, they generate O_2.

Until just recently, it was widely believed—based on their relatively large sizes—that the Apex Chert microfossils (Figure 10-7) were the remains of cyanobacteria. The new evidence from Brasier's group (see Chapter 10), however, indicates that the environment in which they were living was similar to the modern, hydrothermal vents in the deep ocean. No light penetrates to these depths, so if these fossilized structures were indeed organisms, they are unlikely to have been photosynthetic. Instead, they probably survived on the energy produced from chemical reactions. Such organisms are termed **chemosynthetic**. Today, many organisms that live near the hydrothermal vents derive energy from reacting H_2S from the vent fluids with O_2 dissolved in the surrounding ocean water. On the early Earth, free O_2 is unlikely to have been present, for reasons discussed in the previous chapter. How might these organisms have been making their living?

Production of Methane

To get some idea of what types of organisms may have been living on the primitive Earth, we can turn once again to the "Universal" Tree of Life derived from sequencing of ribosomal RNA. In the last chapter, we saw that many of the organisms near the root of the tree appear to be *hyperthermophiles*—organisms that live at high temperatures. Only slightly further away from the root, along one branch of the Archaea, are the methanogenic bacteria (Figure 11-3). **Methanogenic bacteria**, or **methanogens** for short, produce energy from chemical reactions that generate methane. The simplest of these is

$$CO_2 + 4\,H_2 \rightarrow CH_4 + 2\,H_2O$$

Both CO_2 and H_2 are abundant in modern vent fluids and in surface volcanic gases. On the early Earth, they would have been abundant in the atmosphere as well. CO_2 may have built up to very high levels and probably played an important role in keeping the Earth warm, despite reduced solar luminosity. (See Chapter 12.) The abundance of H_2 can be estimated by balancing the rate of escape of hydrogen to space with the rate at which hydrogen and other reduced gases were outgassed from volcanos. Such a calculation predicts that hundreds to thousands of ppm

FIGURE 11-1
Stromatolites from the 3.5 billion-year-old Warrawoona formation in Australia. (From J. W. Schopf, ed. *Earth's Earliest Biosphere.* Princeton, NJ: Princeton University Press, 1983.)

FIGURE 11-2

[See color section] Shark Bay in western Australia. These "living stromatolites" are formed by communities of microbes and may be an analog to early microbial life. (From P. Cloud. *Oasis in Space: Earth History From the Beginning.* W. W. Norton and Co., New York, 1988.)

of H_2 should have been present (possibly even more if hydrogen escape was slower than it is today because the upper atmosphere was colder). Many methanogens can survive at these H_2 concentrations, although most require about 1% H_2 (0.01 bar) to reproduce. Methanogens can also metabolize other reduced substances, such as formate ion ($HCOO^-$), which is formed when carbon monoxide (CO) dissolves in water. CO, like H_2, is thought to have been an important constituent of the early atmosphere. Thus, there is good reason to believe that methanogens were widespread on the early Earth. If primitive methanogens produced methane at the same rate that it is produced biologically today, the atmospheric methane concentration could have exceeded 1000 ppm, more than 600 times its present level (1.6 ppm). The higher abundance results from the fact that methane would have been destroyed less rapidly in a low-O_2 atmosphere.

The presence of high concentrations of methane may have caused the early atmosphere to look quite different from today's atmosphere. We can get some idea of what

it might have looked like by observing Saturn's moon, Titan. Titan has a dense (1.5-bar) atmosphere that consists of about 98% N_2 and 2% CH_4. But the most striking feature of Titan is that it is enveloped in an orangish haze layer that completely obscures the surface (Figure 11-4). This haze layer is thought to consist of hydrocarbon aerosols formed from photolysis and charged particle bombardment of atmospheric CH_4. Although Titan is not a perfect analog for early Earth—its surface temperature is only 94 K—computer models predict that the same type of haze could have formed in Earth's early atmosphere if CH_4 was more abundant than CO_2. In the next chapter, we will show that this condition may indeed have been satisfied.

Although this discussion is highly speculative, the geologic record provides indirect evidence that something like this actually happened. In particular, some of the sedimentary organic carbon from the Late Archean (around 2.7 b.y. ago) is highly depleted in ^{13}C (Figure 11-5)—more so than can be explained by photosynthesis alone. This depletion is thought to be an indication of widespread

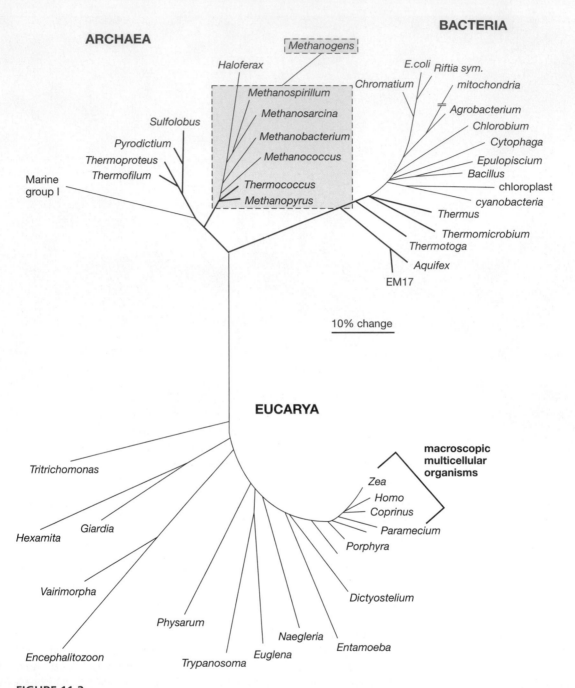

FIGURE 11-3

The ribosomal RNA tree, showing where methanogens and cyanobacteria fit into the picture. (Courtesy of Norman Pace, University of Colorado.)

methanogenic activity. The methane produced by methanogens is depleted in ^{13}C relative to ^{12}C. If this methane was taken up by other organisms, its low-^{13}C carbon may have made its way into sediments and could account for the low-^{13}C carbon preserved in the geologic record. Alternatively, some of the organic carbon in ancient sediments may have come directly from the atmosphere. The haze particles formed from methane photolysis are predicted to be extremely ^{13}C depleted.

Perhaps Figure 11-5 is telling us that such a haze actually existed during the time period around 2.7–2.8 b.y. ago.

Cycling of Atmospheric Nitrogen

A second way in which organisms might have affected the atmosphere is by cycling nitrogen through the atmosphere–ocean system. Organisms need nitrogen for making proteins and nucleic acids as well as for other biochemical

FIGURE 11-4

[See color section] Saturn's moon, Titan, showing the orangish organic haze. (Photo courtesy of NASA.)

functions. However, most of them cannot use nitrogen in its normal molecular form, N_2. Instead, they require **fixed nitrogen**, in which nitrogen atoms are bonded to other types of atoms. Ammonia (NH_3) and nitrate ion (NO_3^-) are two examples of fixed-nitrogen compounds.

Modern marine organisms acquire fixed nitrogen in two ways. One of these is from lightning. In the high-temperature region surrounding a lightning discharge, nitrogen and oxygen react to form nitric oxide, NO:

$$N_2 + O_2 \rightarrow 2\,NO.$$

Nitric oxide is a radical that plays an important role in ozone photochemistry. In today's atmosphere it is eventually oxidized to nitric acid, HNO_3. Nitric acid is soluble in water and, thus, is quickly rained out of the troposphere. In solution, it **dissociates** (comes apart) to form hydrogen ions and nitrate ions:

$$HNO_3 \leftrightarrow H^+ + NO_3^-$$

The resulting nitrate ion can be directly used by organisms as a source of fixed nitrogen.

Some marine organisms can make their own fixed nitrogen by a process termed **nitrogen fixation**. Most of the biologically available nitrogen in the oceans is fixed by cyanobacteria, the same type of organisms that we believe were responsible for the initial rise in O_2. However, the first nitrogen-fixers were probably not cyanobacteria. The ability to fix nitrogen is widespread among **prokaryotes**, which are single celled organisms that lack *cell nuclei*—structures that house genetic material (Figure 11-6a). (In prokaryotes, the genetic material is not concentrated within a specific structure.) The prokaryotes include both the Bacteria and the Archaea. Methanogens are prokaryotes, and many of them can fix nitrogen. Most **eukaryotes** (organisms with cell nuclei) cannot fix nitrogen (Figure 11-6b). Prokaryotes are thought to be more primitive than eukaryotes because their cell structure is simpler. They are also more resistant to UV radiation, which is consistent with the idea that they evolved under a low-O_2 atmosphere that lacked a protective ozone layer.

How did the very earliest organisms acquire fixed nitrogen? Lightning is the most likely answer, although the reaction between N_2 and O_2 could not have been the source because there would have been little O_2 available. However, the analogous reaction

$$N_2 + 2\,CO_2 \rightarrow 2\,NO + 2\,CO$$

could have provided fixed nitrogen at a modest rate.

If nitrogen fixation were not balanced by some reverse process, N_2 would be completely removed from the atmosphere in only about 20 million years. This does not happen, however, because atmospheric nitrogen is recycled by several different processes. Today, the dominant

FIGURE 11-5

Carbon isotope values from organic carbon (kerogen) in ancient rocks. Organic carbon produced by oxygenic photosynthesis typically has $\delta^{13}C$ values near $-25‰$ to $-30‰$. The extremely "light" ($-50‰$ to $-60‰$) values around 2.7–2.8 b.y. ago probably require cycling of carbon through methane. (From A. A. Pavlov et al. *Geology* 29, 1003–1006, 2001.)

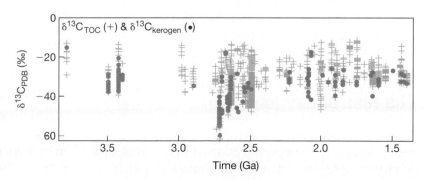

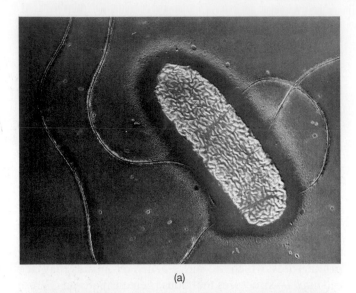

(a)

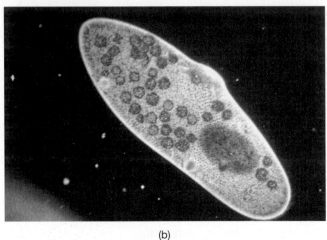

(b)

FIGURE 11-6

[See color section] (a) Prokaryotes have no nucleus, and the DNA is dispersed within the cell. (From Eric V. Grave, Photo Researchers, Inc.) (b) Eukaryotes have their DNA enclosed within a cell nucleus. (From A.B. Dowsett/Science Photo Library, Photo Researchers, Inc.)

recycling mechanism is bacterial **denitrification**. Some organisms can derive energy by reacting nitrate with organic matter. In the process, nitrogen is released as either N_2 or as nitrous oxide, N_2O. Most of the nitrous oxide is converted back into N_2 by photolysis, although some of it reacts in the stratosphere to form NO. Denitrification is particularly rapid in **anoxic** (oxygen-free) regions of the ocean and in anoxic soils.

The Rise of Oxygen

At some point in Earth's history, a historic biological event occurred. An organism evolved that was capable of producing O_2 through oxygenic photosynthesis:

$$CO_2 + H_2O \rightarrow CH_2O + O_2.$$

We have already seen that these first photosynthetic organisms were the *cyanobacteria.* These organisms are still important components of the ecosystems in modern oceans, lakes, and rivers. They have many different forms: some are round, or *coccoid,* and live as individual cells, while others grow in long, multicellular filaments (Figure 11-7). As mentioned earlier, many of them are able to fix nitrogen from the atmosphere. This is quite an extraordinary feat for cyanobacteria because the enzyme used to reduce N_2, *nitrogenase,* is poisoned by O_2. Thus, cyanobacteria have been forced to find ways to protect their nitrogenase from the O_2 that they produce photosynthetically. Some types of filamentous cyanobacteria do this by developing special cells called **heterocysts** that are devoted to fixing N_2 (see Figure 11-7c). No O_2 is produced within a heterocyst, so the inside of the cell can be kept virtually oxygen-free. Other types of cyanobacteria photosynthesize during the day and fix nitrogen at night, so they protect their nitrogenase in a different way. Still others, for example the abundant tropical marine species *Trichodesmium,* fix nitrogen in the morning and photosynthesize in the afternoon. Clearly, even though they are "simple" prokaryotes, cyanobacteria are very advanced organisms in a metabolic sense.

Cyanobacteria are important for yet another reason. Most photosynthesis today is carried out by eukaryotic algae or by higher plants. But we are virtually certain that these organisms did not reinvent photosynthesis on their own. A glance at the universal, ribosomal RNA tree (Figure 11-3) shows why. As the diagram shows, cyanobacteria are closely related to the chloroplasts in higher plants. **Chloroplasts** are the parts of plant cells in which oxygenic photosynthesis takes place. Chloroplasts contain their own DNA, which is why they can be placed on the Tree of Life along with free-living organisms. As biologist Lynn Margulis of the University of Massachutsetts pointed out more than 20 years ago, this shows conclusively that all higher plants (including algae) acquired their ability to produce oxygen by way of an **endosymbiotic** event: some eukaryotic organism ingested, or enveloped, a prokaryotic cyanobacterium without killing it. After that, the two organisms lived together in a mutually beneficial arrangement. The eukaryotic host cell provided nutrients to the cyanobacterium, and the cyanobacterium in turn provided O_2 to the host. This O_2 could then be used as an energy source (via respiration) by the host cell.

When Did Cyanobacteria Evolve?

A critical question from the standpoint of understanding Earth history is: when did the cyanobacteria evolve? Or, to be more precise, the question could be phrased: When did cyanobacteria evolve the capability of producing O_2?

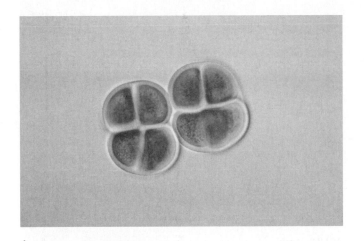

A

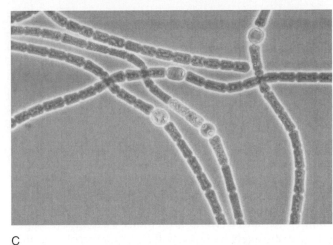

C

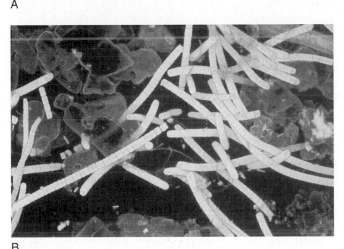

B

FIGURE 11-7

[See color section] Three different types of cyanobacteria: (a) *Chroococcus* (coccoid), (b) *Oscillatoria* (filamentous), (c) *Nostoc* (heterocystic). (From: N. A. Campbell. *Biology*, 2nd ed. Benjamin/Cummings, Redwood City, CA, 1990.)

We have seen (Figure 10-7) that organisms, or at least structures, resembling modern cyanobacteria were already present by 3.5 b.y. ago. But, even if they were indeed alive, these organisms could not have been producing oxygen by photosynthesis because they lived in a deep ocean, hydrothermal vent environment where light levels were extremely low.

The first real evidence for the existence of cyanobacteria comes from organic chemicals in 2.7 b.y.-old sedimentary rocks from the Fortescue Group in western Australia. (Western Australia is a haven for Precambrian geologists because it contains very old rocks *and* it is extremely dry, so that these rocks are not covered up by vegetation.) These rocks contain organic carbon that has not been as highly degraded as most organic material of that age. In essence, they contain 2.7 b.y.-old oil. In this oil are compounds called *2α-methylhopanes* (Figure 11-8). These compounds are thought to derive from the breakdown of *lipid* molecules that are present in the cell walls

of modern cyanobacteria but are not found in other organisms. If the oil in these rocks formed at the same time that the rocks were deposited, rather than migrating in at some later date, then cyanobacteria were probably around by this time. That said, this geochemical evidence does not by itself prove that those cyanobacteria were pro-

2α(Me), 17α(H), 21β(H)-hopane

FIGURE 11-8

A 2α-methyl hopane molecule, thought to be an indicator for cyanobacteria. (Courtesy of Jen Eigenbrode, Penn State University.)

ducing oxygen. Like the Apex Chert microfossils discussed earlier, these organisms may have resembled modern cyanobacteria—in this case by having a similar cell wall—but they did not necessarily have the same metabolism. We mention this because the first evidence for the presence of free O_2 in the atmosphere does not come until almost 400 m.y. later. Thus, if O_2 was being produced photosynthetically 2.7 b.y. ago, it must have been entirely consumed by reactions with reduced substances.

Additional evidence that O_2 was being produced at this time comes from the presence of other organic compounds called **steranes** in these same 2.7-b.y.-old sediments. Steranes derive from the breakdown of organic compounds such as **cholesterol** that are thought to be produced exclusively by eukaryotes. (Cholesterol is familiar to us as the fatty substance that builds up in one's arteries as one gets older and that can lead to heart attacks.) Most eukaryotes use O_2 for respiration and, hence, require at least 1 percent of present dissolved oxygen in the water in which they are living. So, the presence of steranes implies that O_2 was being produced within the water column, presumably by cyanobacteria. The presence of O_2 in surface water does *not* necessarily imply the presence of O_2 in the atmosphere because the rate at which oxygen (or any gas) can flow between the surface ocean and atmosphere is limited by diffusion through the gas-liquid interface.

To determine when O_2 first rose to appreciable concentrations in the atmosphere, we must turn once again to the geologic record. Until recently, the question of when O_2 first rose was hotly debated. The reason is that most of the geologic evidence bearing on the O_2 rise is difficult to interpret. As we will see, new evidence from sulfur isotopes may have finally resolved this long-standing question. Let us begin, however, by discussing the various types of geologic evidence that have been used to try to track atmospheric O_2.

Banded Iron-Formations

One type of geologic evidence that bears on the rise of oxygen is the occurrence of banded iron-formations. **Banded iron-formations (BIFs)** are laminated sedimentary rocks that consist of alternating, millimeter-thick layers of iron-rich minerals and chert (Figure 11-9). Such minerals include *magnetite* (Fe_3O_4) or *hematite* (Fe_2O_3), and *chert* (SiO_2). They are of enormous economic importance today. Much of the iron that is used in making steel and automobiles come from BIFs in Canada and Australia. But these deposits do not form today; nor have they done so at any time in the recent past. Radiometric age dating shows that almost all BIFs formed prior to 1.9 b.y. ago. The only exceptions are a few BIFs that formed in the Late Proterozoic, about 0.6–0.8 b.y. ago. As we will discuss further in the next chapter, the Late Proterozoic

FIGURE 11-9

[See color section] A banded iron-formation, or BIF. (From J.W. Schopf.)

is unusual for several reasons, the most intriguing of which is the evidence of low-latitude glaciation. Some researchers have suggested that the two phenomena are linked: extensive ice cover inhibited oxygen transfer between the atmosphere and ocean, and this led to the reappearance of BIFs. This is one piece of evidence that the so-called "Snowball Earth" events actually occurred.

The reason that BIFs are useful as oxygen indicators is that iron can exist in more than one oxidation state. The **oxidation state** of an atom, molecule, or compound is its degree of oxidation. Substances with a low oxidation state have a large number of available electrons; substances with a high oxidation state do not (see the Box "Useful Concepts: Oxidation States of Iron"). We have already mentioned one oxidation state of iron, **ferrous iron** (Fe^{2+}). A second, more oxidized state is termed **ferric iron** (Fe^{3+}). Iron ions in these two oxidation states have very different chemical properties: Fe^{2+} is soluble in seawater, whereas Fe^{3+} is not. Because iron switches from Fe^{2+} to Fe^{3+} when oxygen is present, it can provide indirect information about past O_2 levels.

To make use of this information, we need to understand the process by which BIFs formed. Although no consensus has been reached regarding the precise mechanism, there is general agreement on some parts of the story. To explain the voluminous quantities of iron deposited in BIFs, large portions of the deep oceans must have been anoxic. This condition would have allowed iron to be transported as dissolved Fe^{2+}. The iron was probably supplied originally from continental weathering and from mid-ocean ridge hydrothermal vent fluids. Trace element patterns in BIFs, especially those of the **rare earth elements** (atomic numbers 57–71 in the periodic table), show that at least some of the iron must have come from the vents. What happened next is not well understood. We know that BIFs did not form on the floor of the deep

USEFUL CONCEPTS
Oxidation States of Iron

Iron occurs in three oxidation states in nature: elemental (or metallic), ferrous, and ferric. The elemental form is located mostly in Earth's core, which is composed largely of iron–nickel alloy. Ferrous and ferric iron occur in the mantle and crust. They exist in dissolved form as Fe^{2+} and Fe^{3+} ions, respectively. The positive charges arise because these ions are missing electrons: Fe^{2+} is missing two electrons, and Fe^{3+} is missing three. We say that these two ions are in the 2+ and 3+ oxidation states. Elemental iron, by comparison, has an oxidation state of zero.

When iron reacts with oxygen, the oxidation state of the iron atom determines the number of oxygen atoms with which it combines. Oxygen, when it reacts, can be thought of as having a valence of −2. Thus, to produce a neutral molecule, one Fe^{2+} ion combines with one O_2^{-2} ion to form the mineral FeO (wüstite). Two Fe^{3+} ions combine with three oxygen ions to form Fe_2O_3 (hematite).

Most of the iron in BIFs consists of magnetite, Fe_3O_4. This is equivalent to one molecule of FeO bonded to one molecule of Fe_2O_3. So, BIFs contain a 1:2 mixture of ferrous iron and ferric iron.

---

ocean. If they had, they would have been destroyed when the sea floor was subducted. Instead, the dissolved iron must have been transported to the margins of the continents, where it was deposited on stable continental shelves. To be precipitated, the iron must first have been oxidized from Fe^{2+} to Fe^{3+}.

Exactly how this occurred is still a subject of debate. One suggestion is that the iron was brought to the surface by wind-induced upwelling of the type that occurs in some modern coastal settings. (Recall from Chapter 5 that this upwelling is produced by Eckman pumping.) The fine, millimeter-scale banding observed in the BIFs could have been caused by seasonal changes in winds that produced upwelling at certain times of the year but not at others. Iron-rich sediments would have formed when the upwelling was strong; silica-rich sediments would have formed when upwelling was weaker. Once the dissolved ferrous iron was brought to the surface, it could have been oxidized by dissolved O_2 produced by photosynthetic cyanobacteria. Alternatively, the iron could have been oxidized abiotically. Laboratory experiments have shown that dissolved Fe^{2+} can be oxidized to Fe^{3+} by UV radiation. (In the process, H_2O is reduced to H_2.) A third possibility is that Fe^{2+} could have been oxidized by **phototrophic** bacteria (bacteria that use sunlight) that did not produce O_2. Whether the iron in BIFs was oxidized by UV radiation or by biologically generated O_2 is still unresolved.

Because the oxygen in BIFs could have come from several sources, we cannot use them directly to infer the O_2 content of the atmosphere. We can, however, safely conclude that the deep oceans must have been anoxic prior to 1.9 b.y. ago, during the time that BIFs formed. This in turn implies that the atmospheric O_2 concentration was lower than it is today. So, BIFs are a strong indicator that the atmosphere has changed during recorded geologic history.

Detrital Uraninite and Pyrite

More direct information on atmospheric O_2 levels can be obtained from other types of geologic indicators. Elements other than iron are also capable of changing their oxidation state. Uranium, for example, has two common oxidation states: U^{4+} and U^{6+}. As with iron, these two ions differ in their solubilities. In this case, however, it is the oxidized (6+) form that is soluble and the more reduced (4+) form that is not.

The U^{4+} ion combines with oxygen to form **uraninite**, UO_2. This mineral occurs in rocks today, but it is normally oxidized to the soluble, 6+, state during weathering. (Notable exceptions to this rule occur in some of the rivers that drain the Himalayas, where the eroded sediments are redeposited very quickly, before the uranium can be oxidized.) Dissolved uranium is transported to the oceans, where it diffuses into anoxic sediments, is reduced, and precipitates as UO_2.

This modern cycle of uranium weathering and deposition does not seem to have operated early in Earth's history. Sedimentary rocks older than about 2.2 b.y. contain uraninite in *detrital* form (Figure 11-10a). A **detrital mineral** is a mineral that survived the weathering process and was transported to the site of deposition as a solid particle rather than in solution. Such a mineral can often be identified by its appearance, which closely resembles the texture of the source rock. The presence of the detrital form implies that, at the time when the source rock was weathered, the atmospheric O_2 content was too low to oxidize uraninite. Quantitative analysis suggests that the O_2 concentration during this time was less than 10^{-3} bars, or 0.005 PAL. (*PAL* means "times the present atmospheric level.") Unfortunately, this is only an upper bound. The actual O_2 concentration prior to 2.2 b.y. ago could have been much lower than this. Indeed, it might have been essentially the same as on the prebiotic Earth, where

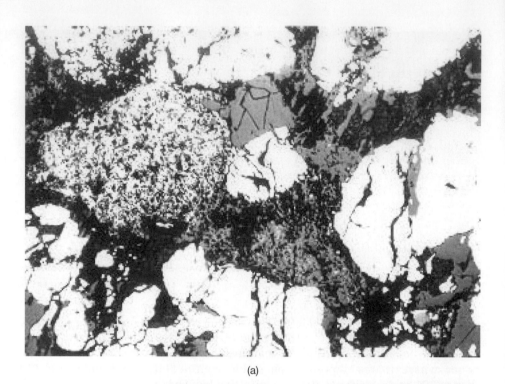

(a)

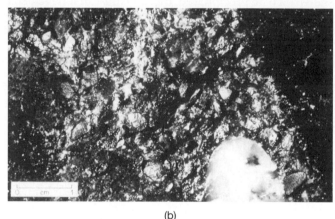

(b)

FIGURE 11-10

[See color section] Samples of the detrital form of (a) uraninite and (b) pyrite. (From J.W. Schopf.)

the surface O_2 concentration is thought to have been on the order of 10^{-13} PAL (see the Box "A Closer Look: Prebiotic O_2 Concentrations" in the previous chapter).

A second mineral that was deposited in detrital form prior to about 2.2 b.y. ago, but not since then, is **pyrite**, FeS_2 (Figure 11-10b). It tells essentially the same story as uraninite. When pyrite is weathered under today's oxidizing atmospheric conditions, the sulfur is oxidized to sulfate, SO_4^{2-}, and the iron is oxidized to Fe^{3+}. But this oxidation apparently did not happen on the early Earth. Eroded grains of pyrite were transported long distances by streams and rivers, as evidenced by the rounded appearance of some samples, and were deposited in chemically unaltered form. This evidence is another indication that atmospheric O_2 was low during the Archean Eon.

Paleosols and Redbeds

Other geologic indicators that have been used to study the rise of atmospheric O_2 include **paleosols** (ancient soils) and **redbeds** (reddish-colored sandy and silty sediments). Heinrich Holland of Harvard University has conducted chemical analyses of numerous Precambrian paleosols. He has found that most paleosols older than about 2.2 b.y. have lost significant amounts of iron, whereas paleosols younger than 1.9 b.y. have retained it. This finding is consistent with the story outlined above. Prior to 2.2 b.y. ago, atmospheric O_2 was low, so the iron released during weathering remained as soluble Fe^{2+} and was carried away by groundwater. The resultant paleosols are iron-poor. After 1.9 b.y. ago, atmospheric O_2 was relatively abundant, so the

FIGURE 11-11

A sequence near Uranvan, Colorado, showing Triassic redbeds overlying Upper Jurassic sandstones. (From P. Cloud. *Oasis in Space.* Norton, New York, 1988.)

iron released by weathering was oxidized to insoluble Fe^{3+} and was retained in the soil. Holland's detailed calculations predict that atmospheric O_2 was less than 0.01 PAL before 2.2 b.y. ago and greater than 0.15 PAL after 1.9 b.y. ago.

Redbeds are sandy sediments that were deposited on land by rivers or as windblown dust (Figure 11-11). They form today in arid regions, such as the American Southwest. The reddish color of these deposits comes from a thin layer of **hematite**, Fe_2O_3, that coats the surfaces of the sediment grains. The iron in hematite is of the oxidized, Fe^{3+}, variety, so redbeds indicate oxidizing atmospheric conditions at the time of their formation. The earliest confirmed redbeds are thought to have formed about 2.2 b.y. ago, which is consistent with the other evidence for a rise in atmospheric O_2 at about this time.

Mass-Independent Sulfur Isotope Ratios

As mentioned earlier, new evidence from sulfur isotopes may have clinched the question of when atmospheric O_2 first rose. Sulfur is unusual in that it has four stable isotopes that occur naturally: ^{32}S, ^{33}S, ^{34}S, and ^{36}S. These isotopes can be separated, or **fractionated**, by a variety of physical and chemical processes. In such processes, the lighter sulfur isotopes typically react faster than do the heavier isotopes. For example, **biological sulfate reduction**, in which certain bacteria use sulfate to oxidize organic matter, discriminates strongly against the heavier isotopes. Scientists usually measure just the two most abundant isotopes, ^{32}S and ^{34}S. The reduced sulfur that is formed from this reaction is preserved as pyrite (FeS_2), which we have just discussed. The pyrite produced by bacterial sulfate reduction is strongly depleted in ^{34}S relative to ^{32}S. Indeed, such "isotopically light" pyrite is considered evidence for biological activity. (We say that the

pyrite is "light" because it is enriched in the light isotope of sulfur compared to the heavier one.)

If one examines modern, sulfur-bearing rocks, one observes that the four sulfur isotopes are distributed in a highly predictable manner. ^{33}S is fractionated relative to ^{32}S by about half as much as is ^{34}S, and ^{36}S is fractionated by about twice as much. This is because the mass difference between ^{33}S and ^{32}S (1 atomic mass unit) is half that between ^{34}S and ^{32}S (2 atomic mass units), whereas the mass difference between ^{36}S and ^{32}S (4 atomic mass units) is twice as much. We say that all of the sulfur isotopes fall along the normal *mass fractionation line (MFL)*. But if one examines sulfur isotopes in Archean sediments, the results are quite different (Figure 11-12). In Figure 11-12, the pyrite sediments have more ^{33}S than would be expected and the barite ($BaSO_4$) sediments have less.

The units used in Figure 11-12 deserve some explanation. Isotopic abundances are measured in parts per thousand (‰), also called "parts per mil." The standard way of expressing isotope abundances is to use *delta* (δ) *notation*. For ^{34}S, for example, we write

$$\delta^{34}S(\text{‰}) = \left[\frac{(^{34}S/^{32}S)_{\text{sample}} - (^{34}S/^{32}S)_{\text{standard}}}{(^{34}S/^{32}S)_{\text{standard}}} \right] \times 1000$$

Negative $\delta^{34}S$ values mean that the sample is depleted in ^{34}S relative to the standard. The standard employed is FeS (the mineral *troilite*) from the Canyon Diablo meteorite. Hence, sulfur isotopes are said to be measured on the *CDT* scale.

Figure 11-12 by itself does not tell us when atmospheric O_2 first rose. However, if one replots these data in terms of the ages of the various samples, a very clear result appears (Figure 11-13). In Figure 11-13, $\Delta^{33}S$ represents the deviation of the measured $\delta^{33}S$ values from the

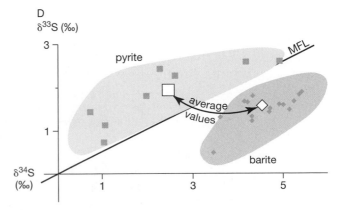

FIGURE 11-12

Diagram showing sulfur isotope concentrations measured in Archean rocks. Barite is $BaSO_4$; pyrite is FeS_2. (From J. Farquhar et al. *J. Geophys. Res.* 106, 1–11, 2001.)

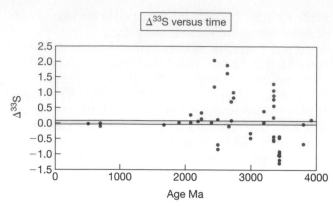

$\Delta^{33}S$ versus time

(Farquhar et al., *Science*, 2000)

FIGURE 11-13

Diagram illustrating the deviation of $\delta^{33}S$ from the normal mass fractionation line. Phanerozoic samples fall close to zero on this scale. (From J. Farquhar et al. *Science* 289, 756–758, 2000.)

normal mass fractionation line. The solid bar through $\Delta^{33}S = 0$ represents 73 samples of Phanerozoic age, that is, younger than 540 m.y. old. The data show that so-called "mass-independent fractionation" of sulfur isotopes is observed in sedimentary rocks older than about 2.3 b.y. of age but not in younger rocks. This marked change in the sulfur isotope distribution occurs at exactly the same time that other, conventional types of geologic evidence indicate that oxygen levels first rose.

How are sulfur isotopes in sediments affected by atmospheric O_2? Laboratory experiments show that isotopes can be fractionated in a mass-independent manner by **photochemical reactions** occurring in the atmosphere. In particular, photolysis of sulfur dioxide (SO_2) fractionates sulfur isotopes in this unusual way. According to the theory of quantum mechanics, only gas phase reactions can do this. Reactions occurring in liquids or solids cannot do this (this includes all biological reactions, which take place in liquid water solution). Today, SO_2 is not photolyzed in the atmosphere because the ultraviolet radiation needed to cause this reaction is absorbed by O_2. Much lower O_2 levels would be needed in order to allow this reaction to occur. Even more importantly, today all the sulfur that enters the atmosphere is eventually oxidized to sulfur dioxide or to sulfuric acid (H_2SO_4). These gases are soluble in water; hence, they are removed by rainout and end up in the ocean as dissolved sulfate ($SO_4^=$). Even if some photochemical reaction within the atmosphere did cause mass-independent fractionation of sulfur isotopes, the effects would disappear because all of the byproducts of the reactions would be recombined as oceanic sulfate before they entered into sediments.

By contrast, in a low-O_2 atmosphere, sulfur photochemistry is much more complex (Figure 11-14). Here, the horizontal scale represents the oxidation state of sulfur, which ranges from -2 for hydrogen sulfide (H_2S) to $+6$ for sulfuric acid. Because of the absence of O_2, sulfur

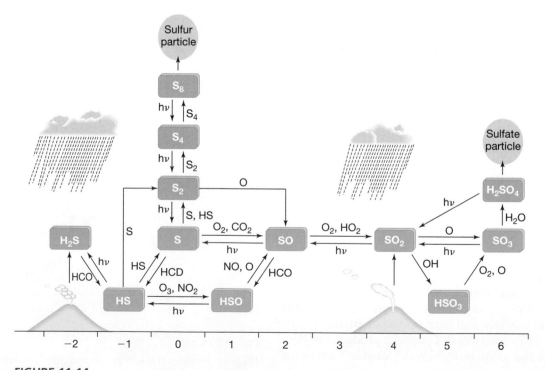

FIGURE 11-14

The atmospheric sulfur cycle in an anoxic, Archean atmosphere. The horizontal scale represents the oxidation state of sulfur. (From J. F. Kasting et al. *Science* 293, 819–820, 2001.)

photochemistry can proceed in both directions, left and right in the figure. Sulfur outgassed as SO_2 can be reduced to elemental sulfur (S_8) or to H_2S. Conversely, sulfur outgassed as H_2S can be oxidized all the way to H_2SO_4. The important point is that sulfur can leave the atmosphere in a variety of different oxidation states, and it does not all end up as dissolved sulfate in the ocean. Hence, mass-independent fractionation patterns produced in the atmosphere can, in theory, be preserved in sediments. Detailed photochemical modeling shows that atmospheric O_2 concentrations must have been at least 10^5 times lower than today in order for this type of chemistry to occur. Hence, the new data on sulfur isotopes is strong evidence that the Archean atmosphere was essentially anoxic.

What Delayed the Rise of Oxygen?

When combined with the other geologic evidence bearing on the rise of atmospheric oxygen, the sulfur isotope data tell us fairly conclusively that O_2 first rose to appreciable levels some time close to 2.3 b.y. ago. This creates a puzzle, though, because the organic biomarker evidence mentioned earlier indicates that cyanobacteria had evolved at least 400 m.y. before this time. Hence, a question that researchers are still working on is this: What delayed the rise of atmospheric O_2 by almost half a billion years? One commonly held idea is that it simply took this long for photosynthetic O_2 to remove all the reduced ferrous iron that was initially present in the oceans. This idea is probably incorrect, however, because if oxygen was produced at today's rate, it would have oxidized all the dissolved iron in the oceans in only a few thousand years. Something else must have been suppressing O_2. Several possibilities have been suggested. Perhaps the organisms that generated the 2.7 b.y.-old 2α-methylhopanes were merely the precursors to cyanobacteria and were not themselves generating oxygen. Or, perhaps they were generating oxygen, but they had not yet learned to fix nitrogen simultaneously, and so their production of O_2 was ex-

tremely limited. (We saw earlier that modern cyanobacteria have evolved complicated mechanisms for protecting their nitrogenase from the O_2 that they produce.) A third idea is that volcanic gases were more reduced prior to 2.3 b.y. ago, and that this additional sink for oxygen kept atmospheric O_2 levels suppressed. In this view, volcanic gases became more oxidized with time as a consequence of escape of hydrogen to space and corresponding oxidation of the upper mantle. (The escaping hydrogen would have come from H_2O initially; thus, oxygen would have been left behind when hydrogen escaped.) We will not dwell on this problem any further, except to point out that many questions regarding the rise of atmospheric oxygen remain to be investigated. Scientists, as always, find new questions when old questions are answered.

The Rise of Ozone

The rise of atmospheric oxygen would have been accompanied by an increase in stratospheric ozone. As we mentioned previously, ozone is critical for life because it shields out harmful solar ultraviolet radiation. The principal wavelength region in which ozone shielding is important is between about 200 and 300 nm (Figure 11-15). (Recall that 1 nm [nanometer] = 10^{-9} m = 10^{-3} μm.) Ultraviolet radiation with wavelengths shorter than 200 nm is also harmful to organisms, but it is effectively absorbed by both O_2 and CO_2. Thus, it should not have been a problem even in an anoxic atmosphere. Ultraviolet radiation between 300 and 400 nm is much less harmful to organisms and is not considered to be a threat, even though a substantial flux of such radiation reaches Earth's surface today.

As we mentioned in Chapter 1, the amount of UV radiation absorbed by ozone depends on the ozone *column depth*, that is, the total amount of ozone between the surface and the top of the atmosphere. The column

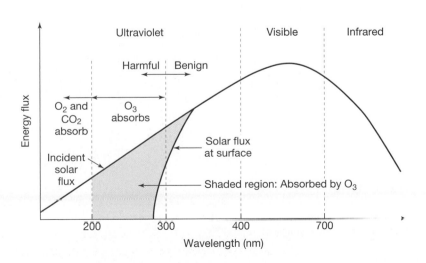

FIGURE 11-15
The wavelength region where absorption by ozone is important.

depth is usually measured in *Dobson units*. One Dobson unit (DU) is equivalent to a layer of pure ozone 0.01 mm thick at the ground. The average ozone column depth today is about 320 DU, equivalent to a 0.32-cm-thick layer of pure ozone at 1 atm pressure.

How much ozone would have been present if the atmospheric O_2 concentration were lower than it is today? **Photochemical models** can be used to address this question: From a model calculation that is capable of reproducing today's average ozone column depth, O_2 is gradually removed from the model atmosphere. The results of one such calculation are shown in Figure 11-16. The calculation shows that the ozone column depth increases nonlinearly with atmospheric O_2 level: Even a small amount of O_2 produces a substantial ozone column depth. The reasons have to do with the details of ozone photochemistry, which we will save for Chapter 17. Exactly when a biologically effective UV screen would have been established is not clear because organisms differ greatly in their tolerances for UV radiation. However, examination of the calculated UV fluxes shows that most of the harmful UV radiation would already have been absorbed once the ozone column depth exceeded about 100 DU, or roughly one-third of today's value. As Figure 11-16 demonstrates, this column depth would have been reached at an O_2 level of only 0.01 PAL. For comparison, the paleosol evidence discussed earlier indicates that atmospheric O_2 was greater than 0.15 PAL by 1.9 b.y. ago. So, we can infer that a reasonably effective UV screen was already established by this time.

Variations in Atmospheric O_2 Over the Last 2 Billion Years

The initial rise of atmospheric oxygen and ozone occurred over 2 b.y. ago. This does not necessarily imply, however, that O_2 concentrations immediately jumped up to the modern value of 21% by volume. Indeed, although the available evidence is far from conclusive, we have reason to believe that O_2 concentrations remained well below the present level until shortly before the dawn of the Cambrian Period. One line of reasoning is the following. Recall that eukaryotes (organisms whose cells have nuclei) had already evolved by 2.7 b.y. ago, according to organic biomarker evidence. Recall also that the same Archean "oil" that contained 2α-methylhopanes from cyanobacteria contains *steranes*. These compounds are formed by the breakdown of cholesterol, which is thought to be manufactured only by eukaryotes. Most eukaryotes are aerobes, that is, they require oxygen, so their presence at 2.7 b.y. ago implies that free O_2 was present in their immediate environment, most likely the shallow surface ocean or lakes. Free O_2 was not present in the atmosphere, however, because this is conclusively ruled out by the geologic evidence just described.

The Ediacaran Fauna

The curious thing is this: multicellular organisms did not appear in the fossil record until about 560 m.y. during a time period referred to as the Vendian. Until then,

FIGURE 11-16

Ozone column depth at different atmospheric O_2 levels, as calculated by a photochemical model. "PAL" stands for "times the present atmospheric level."

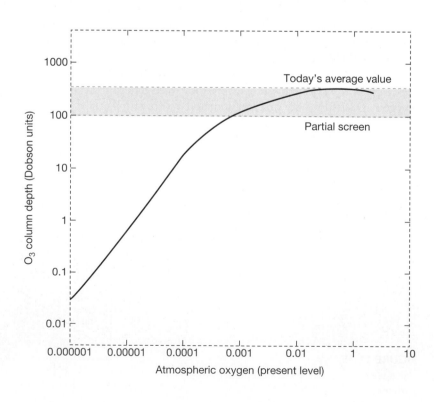

all eukaryotic organisms were single-celled. Why did the evolution of multicellular organisms take so long if eukaryotes were already present by 2.7 b.y. ago? One possible answer is that atmospheric O$_2$ levels were too low to support them. A study performed more than 30 years ago by Donald Rhoads and John Morse showed that animal life disappears below a few tens of meters depth in the anoxic Gulf of California. Their results indicate that modern multicellular organisms need at least 10–20% of present dissolved oxygen in order to survive. The evolution of multicelled animals could have been delayed because atmospheric O$_2$ concentrations and, hence, dissolved O$_2$ concentrations as well were below this level prior to the Vendian.

Another clue that atmospheric oxygen levels were still low in the Vendian period comes from the nature of the multicellular organisms that did evolve at that time. The animals of the Vendian period, termed the *Ediacaran fauna* because they were first discovered in the Ediacaran hills of Australia, have flattened bodies that may have been designed to maximize surface uptake of oxygen. For

example, the Ediacaran fossil *Dickinsonia* (Figure 11-17) was about 30 cm in length but only 1 or 2 cm in thickness. Bruce Runnegar of the University of California, Los Angeles, has argued that *Dickinsonia* lacked a circulatory system, so that it had to acquire all of its oxygen through its skin. His analysis of this fossil organism indicates that O$_2$ concentrations had to be above 0.1 times the present level in order for *Dickinsonia* to survive. However, oxygen levels were probably not much higher than this, if O$_2$ limitation was the reason for this animal's flattened shape.

Variations in Atmospheric O$_2$ During the Phanerozoic

The Phanerozoic Eon—the time during which advanced, multicellular life has thrived—began 540 m.y. ago with what is often termed the *Cambrian explosion*. At about this time, organisms acquired the capability of making hard shells. As a result of this invention, the fossil record becomes much more detailed after this time. Figure 11-18 shows a sample of the early Cambrian fauna collected

FIGURE 11-17

The Ediacaran organism, *Dickinsonia*. (Simon Conway Morris, University of Cambridge, Cambridge, United Kingdom.)

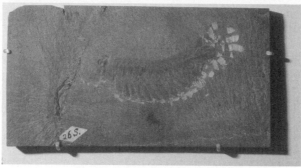

FIGURE 11-18

Early Cambrian fauna from the Burgess shale. (From Chip Clark Archives/Smithsonian Insitution Photo Services.)

from the famous Burgess Shale in western Canada. These organisms almost certainly required high levels of atmospheric oxygen, perhaps approaching those of the modern Earth. Indeed, numerous researchers, beginning with Lloyd Berkner and William Marshall back in the 1960s, have suggested that the Cambrian explosion was triggered by an increase in atmospheric O_2 This speculation remains unproven, but it remains a leading hypothesis for why multicellular life diversified so rapidly beginning at that time.

The record of atmospheric O_2 variations since that time is almost as poor as it is during the Proterozoic. However, we can make some estimates of how much O_2 has varied by looking at carbon isotopes. Carbon, like sulfur, has more than one stable isotope, and these isotopes behave differently in chemical reactions. For carbon, the two isotopes of interest are ^{12}C and ^{13}C. During photosynthesis, plants consume CO_2 that contains both of these carbon isotopes. The lighter isotope, ^{12}C, reacts faster, however; thus, the organic matter that is formed is depleted in ^{13}C. For typical rates of photosynthesis, the depletion in ^{13}C is about 25 parts per mil. (Carbon isotope concentrations are expressed using "delta" notation as well. For carbon, the standard is carbonate from the Peedee belemnite, a fossil cuttlefish which is related to modern squid. Hence, the carbon isotope scale is referred to as the *PDB* scale.)

Carbon isotopes can be used to estimate the rates of O_2 production in the following way. As we saw in Chap-

ter 9, it is burial of photosynthetically produced organic carbon that results in net production of O_2. The organic carbon that is buried is isotopically light, that is, it is depleted in ^{13}C. Hence, the carbon that remains in the atmosphere and in the ocean as dissolved bicarbonate and carbonate ion becomes isotopically heavy. The carbonate sediments that are formed are in equilibrium with dissolved carbonate and bicarbonate; thus, they are isotopically heavy as well compared to the buried organic carbon. And the faster that organic carbon is buried, the heavier the carbonates become.

Figure 11-19 shows the carbon isotope composition of carbonate sediments during the last 540 m.y. Today, carbonate sediments are at about 0 per mil on the PDB scale. It can be shown from mass balance arguments (see the Box "Thinking Quantitatively: Carbon Isotopes and Organic Carbon Burial") that this implies that about 20% of the CO_2 entering the system from volcanic outgassing and weathering of carbonate and organic carbon-containing rocks on land is buried as organic carbon, while the other 80% is buried as carbonates. Figure 11-19 shows, however, that this has not always been the case. Carbonates were isotopically heavy during the Carboniferous and early Permian Periods, 360–250 m.y. ago, and again during the Cretaceous Period, 144–65 m.y. ago. By the logic that we just went through, this implies lots of organic carbon burial, and hence lots of O_2 production dur-

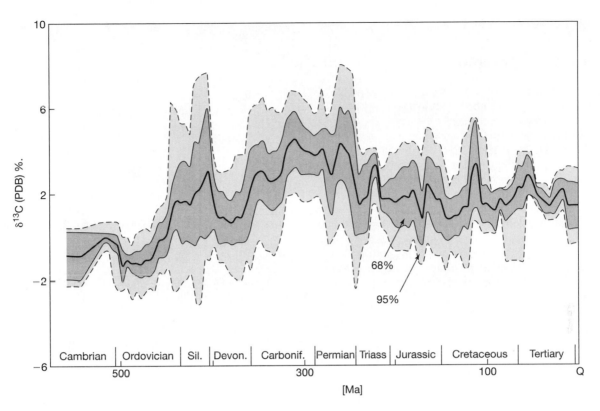

FIGURE 11-19

The carbon isotope record from carbonate rocks deposited during the Phanerozoic. Solid curve is the running mean. Shaded areas include 68% and 95% of all the data. (From J. Veizer et al. *Chemical Geology,* 161, 59–88, 1999.)

ing both time periods. The biggest excursion in the carbon isotope record and, hence, the biggest increase in organic carbon burial, was during the mid-Carboniferous Period. Carbonate sediments were about 5‰ heavier at that time, indicating that organic carbon was being buried about twice as fast as today. It is not hard to understand why if one considers what was happening at that time. The Carboniferous Period gets its name from the extensive coal deposits that were formed during this time interval. The peak of this long-lived period of coal formation was during the Pennsylvanian Epoch, 325–286 m.y. ago. The coal formed during that time has been the basis for the coal mining industry of western Pennsylvania and West Virginia.

Robert Berner and Donald Canfield of Yale University put these carbon isotope data into a model of the global carbon cycle and used them to estimate atmospheric O$_2$ levels. Their results are shown in Figure 11-20. Atmospheric O$_2$ concentrations were probably higher than today during both the Carboniferous and the Cretaceous Periods. The predicted variations in O$_2$ are not as large as one might anticipate from Figure 11-19 because the actual situation is more complicated than just described. Some of the excess oxygen that was produced

was taken up by the sulfur cycle in oxidizing sulfide to sulfate. One can see evidence for this from sulfur isotopes, which covary with carbon isotopes in just the manner one would expect if oxygen was shuttling back and forth between the two reservoirs. However, Berner and Canfield's prediction is probably at least qualitatively correct. Other researchers have picked up on this idea and have suggested that the giant dragonflies of the Carboniferous (Figure 11-21) and the dinosaurs of the Mesozoic may both have been breathing an enriched blend of air. Like the Ediacaran organism *Dickinsonia* mentioned earlier, insects take in O$_2$ through their "skin"; thus, some paleontologists have suggested that it was the higher O$_2$ levels during the Carboniferous that allowed them to achieve their gigantic size. On the other hand, other researchers have pointed out that the level of biological competition was not as great then as it is today. One wonders how long 2-foot dragonflies would have survived in the presence of modern eagles or falcons. The gigantic pterosaurs of the Mesozoic would probably have made short work of them as well. So, the giant dragonflies of the Carboniferous could have existed simply because there was no one around to eat them. If this was true, they are not very reliable oxygen indicators.

THINKING QUANTITATIVELY
Carbon Isotopes and Organic Carbon Burial

Carbon isotopes provide a useful way of analyzing the behavior of the organic carbon cycle in the distant past. The reason is that organisms fractionate carbon isotopes when they convert CO_2 into organic matter during photosynthesis. As described in the text, ^{13}C is taken up faster during photosynthesis than is ^{12}C. By measuring the $^{13}C/^{12}C$ ratio in carbonate sediments, we can determine how fast organic carbon was being buried and, hence, how fast O_2 was being produced.

We can analyze the carbon isotope record mathematically by keeping track of the total amount of ^{12}C and ^{13}C flowing through the system. For the purposes of this analysis we can treat the atmosphere and oceans as one combined reservoir. Carbon enters the atmosphere–ocean system in two different ways: (1) outgassing of CO_2 from volcanoes, and (2) weathering of carbonate rocks and sedimentary organic carbon (**kerogen**) on land. Carbon leaves the system in the form of sedimentary organic carbon and carbonates. Let

F_{in} = flux of carbon into the atmosphere– ocean system

F_{carb} = burial rate of carbonates

F_{org} = burial rate of organic carbon

δ_{in} = isotopic composition of carbon entering the system

δ_{carb} = isotopic composition of buried carbonate carbon

δ_{org} = isotopic composition of buried organic carbon

The total amount of carbon (including both ^{12}C and ^{13}C) leaving the system must be equal to the amount of carbon entering it. Thus

$$(1) \qquad F_{in} = F_{carb} + F_{org}$$

The amount of ^{13}C leaving the system must also equal the amount that enters. Because the ratio of ^{13}C to ^{12}C is small, we can express this mathematically by the equation

$$(2) \qquad F_{in}\,\delta_{in} = F_{carb}\,\delta_{carb} + F_{org}\,\delta_{org}$$

To make use of this system of two equations, we need one additional piece of information. During photosynthesis, most plants (and photosynthetic algae) fractionate carbon isotopes by about 25‰. Or, to say it another way, the organic matter that is formed from photosynthesis is depleted in ^{13}C by this amount. Let

$$\Delta_B = \delta_{carb} - \delta_{org} \cong 25‰$$

To simplify things still further, let

$f_{carb} = F_{carb}/F_{in}$ = fraction of carbon entering the system that is buried as carbonate

$f_{org} = F_{org}/F_{in}$ = fraction of carbon entering the system that is buried as organic carbon

Dividing equation (1) by F_{in} yields the relationship

$$(3) \qquad 1 = f_{carb} + f_{org}$$

This just tells us an obvious result: all of the carbon entering the system must leave either as carbonate carbon or organic carbon. Meanwhile, dividing equation (2) by F_{in} yields

$$(4) \qquad \delta_{in} = f_{carb}\delta_{carb} + f_{org}\delta_{org}$$

Now, we can make use of equation (3) to write: $f_{carb} = 1 - f_{org}$. Thus,

$$(5) \qquad \delta_{in} = (1 - f_{org})\,\delta_{carb} + f_{org}\delta_{org}$$
$$= \delta_{carb} - f_{org}(\delta_{carb} - \delta_{org})$$

But the quantity $(\delta_{carb} - \delta_{org})$ is just Δ_B, the average fractionation produced during photosynthesis. Thus

$$(6) \qquad \delta_{in} = \delta_{carb} - f_{org}\,\Delta_B$$

or, rearranging terms and solving for f_{org}

$$(7) \qquad f_{org} = (\delta_{carb} - \delta_{in})/\Delta_B$$

Finally, we assume that the average isotopic composition of carbon entering the atmosphere–ocean system is the same as the carbon isotopic composition of the mantle, about −5‰. Using this value for δ_{in} and setting $\Delta_B = 25‰$ gives

$$(8) \qquad f_{org} = (\delta_{carb} + 5)/25$$

When $\delta_{carb} = 0‰$, as it does today for marine carbonates, equation (8) predicts that $f_{org} = 0.2$. This is consistent with the value given in the text: about 20% of the CO_2 entering the atmosphere–ocean system is reduced to organic carbon and leaves the system in that form. Figure 11-19 shows that δ_{carb} was greater than +5‰ around 300 m.y. ago. According to equation (8), this implies that the fraction of carbon leaving the system as organic carbon was >0.4, or more than twice the modern value. If the carbon input rate from weathering and outgassing was the same as today, then the rate of O_2 production was more than twice as much. So, it would not be at all surprising if the atmospheric O_2 concentration was higher at that time as well.

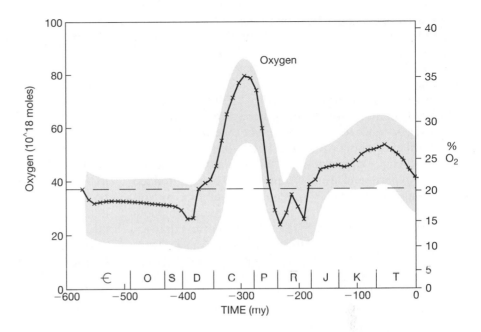

FIGURE 11-20

Calculated variation in atmospheric O_2 during the Phanerozoic. (From R. A. Berner and D. Canfield. *Amer. J. Sci.* 289, 333–361, 1989.)

Modern Controls on Atmospheric O_2

Let us now turn our attention to the modern system. We have offered some suggestions for why atmospheric O_2 concentrations may have increased when they did. A related question, which we might hope would be somewhat easier, is this: What controls the atmospheric O_2 concentration today?

The answer, surprisingly, is that we do not know for sure, although researchers do have a number of ideas. Whatever the oxygen control mechanism is, it appears to be very efficient. The modern atmospheric O_2 level is 21%

by volume, or 0.21 bar. It seems unlikely that the O_2 concentration has strayed from this level by more than ±50% since the late Devonian Period, about 360 million years ago. The evidence is that forests have existed since that time and, while they have always been able to burn, they have never disappeared entirely.

Forest Fires and Atmospheric Oxygen

Fires burn more intensely when the oxygen content of the air is increased. A graphic, and tragic, illustration of the phenomenon occurred in the mid-1960s, when an *Apollo* space capsule burned up on its test pad in Houston with

FIGURE 11-21

[See color section] A Carboniferous dragonfly with a wingspan of 60 cm (2 ft.). (From © The Field Museum, Chicago. Photographer: John Weinstein.)

three astronauts aboard. At that time, NASA was using pure O_2 in its spacecraft to minimize launch weight. After the fire, the space agency returned to using normal air because they realized, belatedly, that pure O_2 was too flammable.

A similar problem could occur globally if the atmospheric O_2 content got too high. Forest fires, ignited by lightning or other mechanisms, might rage out of control and burn everything within reach. It is difficult to determine the exact O_2 level at which this would occur, but laboratory experiments using wet matchsticks and shredded paper suggest that an O_2 concentration of 35% by volume would be enough to destroy most of the global biota (Figure 11-22). A catastrophe of this magnitude may actually have taken place in the aftermath of the K–T impact event, 65 m.y. ago. But, as we will discuss in Chapter 13, this was a very special event in which fires may have been ignited all over the globe by finely dispersed ejecta re-entering Earth's atmosphere. Other than that scenario, there is no evidence that such widespread forest fires have occurred. We can infer that the atmospheric O_2 concentration has probably remained below 35% by volume ever since forests first appeared.

Forest fires can also be used to place a lower bound on atmospheric O_2 levels. Fires will not ignite when the O_2 concentration falls below about 13% by volume. This limit is somewhat firmer than the upper bound on O_2 because it depends on simple physics: At O_2 concentrations below 13%, a flame loses heat by convection more rapidly than it gains heat by combustion. Sedimentary rocks preserve a more-or-less continuous record of charcoal since the late Devonian period. Charcoal is produced from the in-complete combustion of organic matter by fire, so forest fires must have burned on and off throughout this period. Hence, we conclude that the atmospheric O_2 concentration has not fallen below 13% by volume during the last 360 m.y.

What mechanism could have stabilized atmospheric O_2 within the "fire window" during the past few hundred million years? The main loss processes for O_2, surface weathering and the oxidation of reduced volcanic gases, are thought to be independent of O_2 concentration within this range. The control mechanism must therefore act on the O_2 source—namely, photosynthesis followed by burial of organic carbon. Let us consider the factors that affect the organic carbon burial rate.

Oxygenation of the Deep Ocean

Most of the organic matter being buried today is deposited in marine sediments rather than on land. The same seems to have been true during most of the past few hundred million years, except during the Carboniferous Period, when the great coal beds were forming. Thus, the control of organic carbon burial appears to lie in the ocean.

For many years, researchers have believed that the feedback mechanism that stabilizes atmospheric O_2 involves the deep oceans. The deep oceans today contain relatively high concentrations of dissolved O_2. Indeed, over much of the globe, the dissolved O_2 content of deep water is higher than that of surface water (Figure 11-23. See also the box on p. 155 in Chapter 8). The reason is that the water in the deep ocean originates at the surface at high latitudes, where temperatures are very cold. Like most other gases, O_2 increases in solubility as the temperature goes down. Thus, cold, high-latitude surface water contains more dissolved O_2 than does surface water at lower latitudes. This high O_2 content is passed on to

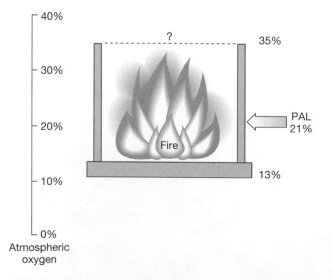

FIGURE 11-22

The oxygen "fire window," showing proposed minimum and maximum atmospheric O_2 levels consistent with the continuous record of charcoal deposition since the Devonian period. (From T. P. Jones and W. G. Chaloner. *Global and Planetary Change* 97, 39, 1991.)

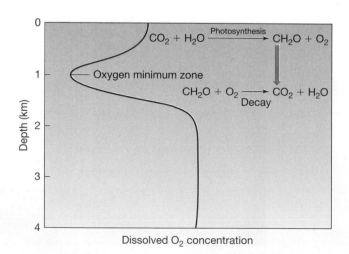

FIGURE 11-23

The vertical profile of dissolved O_2 in the (low-latitude) ocean.

the deep ocean when the cold surface water sinks as part of the global thermohaline circulation.

Oxygen concentrations in the deep ocean would drop dramatically, however, if the atmospheric O_2 concentration were lowered. Even today, over much of the ocean a pronounced **oxygen minimum zone** exists at a depth of about 1 km (see Figure 11-23). The low O_2 concentrations at this depth are caused by the decay of organic matter that falls from the surface ocean. In some regions—the Black Sea, for example—the rate of vertical mixing is slow enough to cause the deep ocean itself to become anoxic. If the atmospheric O_2 concentration were to decrease from its present level, the area covered by such **anoxic basins** would expand. The lower O_2 content of deep water could, in principle, allow more organic carbon to be buried on the ocean floor because the organic matter would decay less quickly. This, in turn, could provide a negative feedback that might stabilize atmospheric O_2. Unfortunately, however, measurements do not support this idea. There is no direct correlation between the concentration of dissolved O_2 and the organic carbon content of marine sediments. Evidently, organic carbon in sediments can be oxidized quite efficiently by bacteria that utilize either sulfate or nitrate rather than O_2.

Dissolved Oxygen and Sedimentary C:P Ratios

The actual O_2 control mechanism appears to be slightly more complicated than that outlined above. Recall from Chapter 9 that marine productivity is usually limited by the availability of key nutrients, especially nitrogen (N) and phosporus (P). Of these, P is thought to be the most critical, because N can be fixed by organisms such as cyanobacteria. The concentration of P in seawater is controlled by the rate of supply of P from the weathering of rocks on the continents and by the loss of P due to incorporation in sediments. The C:P ratio in sediments is related to the dissolved oxygen concentration in seawater. Sediments overlain by well-oxygenated water have lower C:P ratios (i.e., they contain more phosphorus) than do sediments overlaid by anoxic water. The reasons are complex: Apparently, bacteria in sediments store phosphorus when oxygen is available and use this stored material as an energy source when oxygen levels become too low. Phosphorus also tends to be bound up with iron compounds in sediments when oxygen is present. Both processes tend to create higher C:P ratios in sediments deposited under anoxic conditions.

The existence of such a relationship means that the dissolved O_2 content of deep water can affect the burial rate of organic carbon indirectly by altering the availability of phosphorus. This effect creates a negative feedback loop that may stabilize atmospheric oxygen (Figure 11-24). If atmospheric O_2 were to decrease for some reason, deep ocean O_2 would decrease, and the C:P ratio in

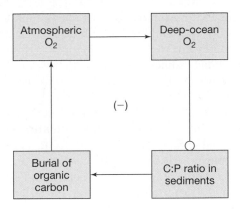

FIGURE 11-24

A likely feedback loop for controlling atmospheric O_2, involving dissolved O_2 concentrations and the C:P ratio in marine sediments.

sediments would increase. This change would allow more organic carbon to be buried without removing any additional phosphorus. Increased burial of organic carbon would result in increased O_2 production, which would help restore atmospheric O_2 to its original level.

Although this mechanism may help stabilize atmospheric O_2, it does not imply that O_2 levels have remained constant. Atmospheric O_2 levels can be affected by changes in the rate of organic carbon burial on land, such as those that occurred during the Carboniferous Period. Furthermore, the degree of deep-ocean oxygenation may depend on the details of ocean circulation. Periods such as the Mesozoic, when the poles were much warmer, are unlikely to have had the same type of thermohaline circulation that operates today. Data from oxygen isotopes indicate that during most of this time deep water was 10–15°C warmer than it is today. Warmer deep water would have contained less dissolved O_2. Thus, anoxia should have tended to be more prevalent if other factors had remained the same.

If the oxygen control mechanism suggested above is correct, however, the atmospheric O_2 concentration should have responded to this change. Larger areas of anoxia would have promoted the deposition of sediments with high C:P ratios. This deposition in turn should have led to increased oxygen production and higher atmospheric O_2 levels. So, as suggested earlier, the dinosaurs may have breathed a somewhat enriched mixture of air. The possible connection between deep ocean circulation and atmospheric O_2 concentration illustrates yet another interesting linkage in the highly intertwined Earth system.

Forest Fires and C:P Ratios in Sedimentary Organic Matter

Although the control mechanism described above is attractive, it leaves one important question unanswered: If the control of atmospheric O_2 lies in the oceans, why have O_2 levels always remained within the fire window for forests? Either this is just a coincidence, or forests themselves must have something to do with regulating O_2. But forests are not part of the control mechanism outlined above.

The answer to this question may also involve sedimentary C:P ratios. The C:P ratios of marine sediments and terrestrial sediments are very different. The C:P ratio of typical marine organisms is 105:1, whereas organic matter derived from trees has a characteristic C:P ratio of about 1,000:1. Thus, the burial of terrestrial organic carbon removes much less phosphorus than does the burial of an equivalent amount of marine organic matter. And, although the burial rate of organic matter on land is relatively low, some of the organic matter produced there is carried away by rivers and buried in river deltas.

Could this mechanism create a feedback loop that keeps atmospheric O_2 within the fire window? Suppose that atmospheric O_2 levels were to become so high that the forests all burned down. The phosphorus that was being utilized by terrestrial vegetation would go directly into rivers and, thence, to the oceans, where it would be utilized to make low C:P marine organic matter. The burial of this marine organic matter would generate less oxygen than would the burial of the same amount of terrestrial organic matter. Oxygen production would go down, causing atmospheric O_2 levels to decrease. The net result would be a negative feedback cycle, which could conceivably help to keep atmospheric O_2 within the limits acceptable to terrestrial vegetation (Figure 11-25).

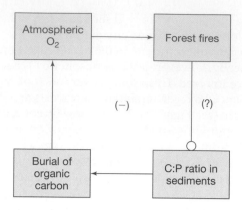

FIGURE 11-25

A possible feedback loop by which forests might help control atmospheric O_2. The link between forest fires and the C:P ratios in near-shore marine sediments is not well established. Some such control mechanism may be required to keep atmospheric O_2 levels within the "fire window."

For reasons that are not entirely understood, however, the C:P ratio of river delta sediments collected to date does not appear to be much higher than that of normal marine sediments. Either other plants and algae with low C:P ratios provide most of the terrestrial organic material, or else sufficient reprocessing of material occurs during sediment deposition to smooth out the C:P differences between terrestrial and marine organic matter. Most of the work done so far has been on passive margins, though, where reprocessing times are long. Current work on active margins, where terrestrial organic matter is rapidly buried, may reveal the site of high C/P organic matter burial.

Chapter Summary

1. Evidence from ribosomal RNA, combined with our knowledge of the early environment, suggests that methanogenic bacteria were among the earliest organisms. Once methanogens evolved, they changed the composition of the atmosphere by converting hydrogen into methane. The biota should also have increased the rate at which nitrogen was cycled between the atmosphere and ocean.

2. The first O_2-producing organisms, the cyanobacteria, are thought to have originated at or prior to 2.7 b.y. ago, based on the presence of organic biomarkers (2-methylhopanes) in ancient rocks. Nevertheless, the geologic record indicates that atmospheric O_2 did not become abundant until some time after 2.3 b.y. ago. Prior to that time, O_2 levels were suppressed by reaction with reduced volcanic gases. A variety of different types of geologic evidence, including mass independently fractionated sulfur isotopes, show when the initial rise in O_2 took place.

3. Accompanying the rise in atmospheric O_2 was a corresponding rise in the abundance of ozone. A reasonably efficient UV shield should have developed once the O_2 concentration exceeded about 1% of its present level. Based on the geologic evidence, this threshold should have been passed by 1.9 b.y. ago, or perhaps a few hundred million years earlier.

4. Atmospheric O_2 levels have probably fluctuated over the last 2 b.y. as a consequence of changes in the rate of organic carbon burial. O_2 concentrations may have increased markedly just before the beginning of the Cambrian Period, 540 m.y. ago, perhaps triggering the explosion of multicellular life at that time. O_2 levels were probably also higher during the Carboniferous and Cretaceous Periods, based on evidence from carbon isotopes. The formation of the Carboniferous coal beds created the largest positive excursion in atmospheric O_2 during the Phanerozoic Era.

5. Despite these changes in the rate of organic carbon burial, atmospheric O_2 concentrations have probably not changed by more than ±50% during the past 360 million years, as evidenced by the continuity of the fossil charcoal record. The most plausible control mechanism involves the degree of oxygenation of deep water and the C:P ratio of marine sediments. Less-oxygenated deep water promotes a higher sedimentary C:P ratio that, in turn, allows more organic carbon to be buried for a given burial rate of phosphorus. The burial of terrestrial organic matter in river deltas may also influence the C:P ratio of sediments, allowing forests to exert some control over ambient O_2 levels.

Key Terms

anoxic
anoxic basin
anoxygenic photosynthesis
banded iron-formation
BIFs
biological sulfate reduction
chemosynthetic
chloroplasts
cholesterol
cyanobacteria
denitrification
detrital mineral
dissociate

endosymbiotic
eukaryotes
ferric iron
ferrous iron
fixed nitrogen
fractionated
hematite
heterocysts
kerogen
methanogenic bacteria
methanogens
nitrogen fixation
oxidation state

oxygen minimum zone
oxygenic photosynthesis
paleosols
photochemical models
photochemical reactions
phototrophic
prokaryotes
pyrite
rare earth elements
redbeds
steranes
stromatolites
uraninite

Review Questions

1. To which domain of life do methanogens belong? Why are they thought to be evolutionarily ancient?
2. What is the difference between prokaryotes and eukaryotes? Which are seen first in the fossil record?
3. What types of organisms have heterocysts and what are they used for?
4. Which organisms were the first to produce O_2?
5. What types of geologic evidence are used to study the rise of atmospheric O_2?
6. How did plants and algae acquire their ability to photosynthesize?
7. When did the ozone layer become thick enough to provide an effective UV screen?
8. What do carbon isotopes tell us about the rise of atmospheric O_2?
9. What does the fossil charcoal record tell us about atmospheric O_2 levels?
10. How is the atmospheric O_2 content maintained at its current level?

Critical-Thinking Problems

1. a. The atmosphere consists of 78% N_2, 21% O_2, 1% ^{40}Ar, and about 350 ppm CO_2. What is the mean molecular weight of air? Round your answer to three significant figures. (Note: The atomic weights of N and O are 14 and 16, respectively.)
   b. The total mass of the atmosphere is about 5×10^{18} kg. How many moles of air does it contain? How many moles of O_2 and CO_2 are present? (Note: Calculate the latter two answers from the first one, not by computing the mass of O_2 and CO_2. The concentrations listed for the various gases are percentages by volume, not by mass, so you need to work in moles.)
   c. Forests and soils contain roughly 2160 Gton of carbon in the form of wood, leaves, and humus. (The actual value is not known this accurately, but this choice makes the

numbers work out well.) Suppose that all of the world's forests were to burn down instantaneously and that all of the soil carbon was oxidized as well. By how much would atmospheric CO_2 increase? By how much would O_2 decrease? Express your answers in percentages. Assume that the burning equation is the same as that for respiration and decay: $CH_2O + O_2 \rightarrow CO_2 + H_2O$.

2. The combined burial rate of organic carbon in marine sediments and in coal is approximately 0.05 Gton(C) per year. This burial is the net source of atmospheric O_2. In steady state, this source of oxygen is balanced by the weathering of reduced materials in rocks (kerogen, sulfides, and iron). If the weathering rate were to remain constant following the disaster in Problem 1c, and if all photosyntheses were shut off (in the oceans as well as on land), how long would it take for atmospheric O_2 to disappear?

Further Reading

General

Cloud, P. *Oasis in Space: Earth History From the Beginning.* New York: W. W. Norton and Co., 1988.

Advance

Kasting, J. F. Earth's early atmosphere. *Science,* 259, 920–926, 1993.

Long-Term Climate Regulation

Key Questions

- Why was Earth's climate warm despite reduced solar luminosity in the distant past?
- Was Earth's surface ever totally frozen?
- Has Earth's climate generally been warmer or colder than today's climate?
- Why was the climate warm during the time of the dinosaurs, and why has it cooled over the past few tens of millions of years?

Chapter Overview

Solar evolution models predict that the Sun was about 30% dimmer when it first formed and that its luminosity has increased more or less linearly since then. Nevertheless, Earth appears to have had liquid water at its surface for as far back as the geologic record extends, roughly 4.4 b.y. Warm temperatures on the early Earth were maintained by a combination of powerful greenhouse gases, especially CO_2 and CH_4. CH_4 concentrations fell abruptly around 2.3 b.y. ago when atmospheric O_2 levels rose, throwing the Earth into its first major glaciation. Indeed, this glaciation appears to have been the first of three "Snowball Earth" episodes, during which times Earth's surface was entirely covered with ice and snow. Earth recovered from this climatic catastrophe because volcanic CO_2 accumulated in the at-

mosphere, thereby increasing the greenhouse effect and eventually melting the ice. As the Sun gradually brightened, atmospheric CO_2 concentrations diminished, maintaining the climate within the limits favorable to life. This decrease in CO_2 was not accidental; rather, it was a natural consequence of a negative feedback in the carbonate–silicate cycle. Over the past few hundred million years, Earth's climate has fluctuated between warm and cold conditions, primarily as a result of changes in atmospheric CO_2 induced by plate tectonics and the carbonate–silicate cycle.

Introduction

Earth has been in existence for billions of years and has been inhabited by organisms for most of that time. All known organisms require water during at least part of their life cycles (although some can do without water for extended periods). This implies that Earth's surface temperature has remained warm enough to support liquid water throughout this entire time. But we have already seen that the Sun was approximately 30% less bright early in the solar system's history. Why did the climate remain relatively warm despite such a large change in solar heating?

Based on what we learned in Chapter 3, the solution to the "faint young Sun paradox" probably involves either a stronger atmospheric greenhouse effect or a lower planetary albedo, which is what is needed to balance the planetary energy budget. When we last looked at this problem, we had no good reason to believe that either of these factors should have changed in the needed direction. In Chapter 8, however, we studied another component of the Earth system, the carbon cycle, in some detail. In particular, we found that a strong negative feedback exists between atmospheric CO_2 concentrations and the rate of silicate weathering. This feedback has a tendency to stabilize climate over long time scales because it causes atmospheric CO_2 levels to increase when the climate gets too cold or to decrease when it becomes too warm.

This negative feedback in the carbonate–silicate cycle is a plausible solution to the faint young Sun paradox. But it cannot explain all the details of Earth's climate evolution. As we saw in the last chapter, there are good reasons to believe that methane was much more abundant prior to the rise of atmospheric O_2. Methane is a good greenhouse gas; hence, it may well have played a role in keeping the early Earth warm. Furthermore, if we examine Earth's climate history more closely, we find that Earth's surface temperature has experienced substantial swings: there have been periods (like the present) when the polar regions have been covered with ice, and there have been other, longer periods when polar ice appears to have been completely absent. There have even been periods, like the Neoproterozoic Era (about 700 million years ago), when continental ice sheets may have existed in the tropics! In this chapter, we examine the question of just how stable Earth's climate has been, and we look at how plate tectonics and biological evolution may have caused large climatic shifts.

The Faint Young Sun Paradox Revisited

Let us return now to the problem that we considered briefly at the end of Chapter 1: the faint young Sun paradox. We have already mentioned that the Sun gets brighter as it ages. The luminosity increase is a direct consequence of the density change caused by the conversion of hydrogen into helium. As such, it is considered to be a *robust* prediction of solar evolution models, meaning that it does not depend sensitively on the details of the model. Thus, even if physicists are wrong about precisely how fusion occurs in the Sun's core, the faint young Sun problem is not likely to go away.

How Well Do We Understand Solar Evolution?

Recently, our faith in solar evolution models was increased by an important new observation. Physicists had been bothered for years about the apparent underabundance of neutrinos emitted by the Sun. **Neutrinos** are nearly massless particles emitted during nuclear reactions.

The word "nearly" here is important. Until just recently, it was not known whether they had any mass at all. Like photons, they could have been totally massless. However, researchers at two new underground neutrino detectors, one in Sudbury, Canada, and another in northern Japan, have shown that this cannot be true. Neutrinos come in three different "flavors," only one of which was able to be measured by earlier detectors. These detectors measured only about one-third as many neutrinos as they should have, based on our theories of what is happening in the solar interior. However, if neutrinos have mass, then they can convert to the other two forms of neutrinos on their way from the Sun's core to Earth. The Super-Kamiokande detector in Japan measured all forms of neutrinos, whereas the Sudbury detector measured only one type. By comparing the numbers in the two experiments, the researchers were able to show that the total number of neutrinos emitted by the Sun agrees with the number that is predicted theoretically. Thus, solar physicists have renewed faith that they understand how the Sun produces energy.

There is one way in which the faint young Sun paradox could be avoided or, at least, modified. If the Sun was a few percent more massive in the past, it would have been brighter than it is now. But this would imply that it must have lost large quantities of material during its lifetime. The Sun does lose mass by way of the **solar wind**, an outflow of charged particles from the Sun's **corona** (the hot, outermost layer that is visible during a solar eclipse; see Figure 12-1). But the solar wind mass flux is about

FIGURE 12-1

The solar corona, as revealed during a solar eclipse. (Courtesy of NASA: *http://science.nasa.gov/ssl/pad/solar/images/ecl1991a.jpg*)

10,000 times too small to account for a 1% change in the Sun's mass over geologic time. Rapid mass loss does occur in stars that are still in the process of forming. The outflow from young stars is often referred to as a **T-Tauri wind**, named after a particular star in the constellation Taurus that is behaving in this way. However, T-Tauri winds are expected to last only about 1 million years after the star begins to form, so this process should have no bearing on Earth's climate hundreds of millions of years later. (Earth itself is thought to have taken at least 10 million years to form.) Thus, the faint young Sun paradox appears to require an Earth-bound explanation.

Defining the Faint Young Sun Problem Mathematically

Let us see if we are any closer to resolving this paradox now that we have examined the Earth system in more detail. If you worked out "Critical-Thinking" Problem 5 in Chapter 3, you should have determined that a 30% reduction in solar luminosity would have led to a 22° decrease in mean surface temperature, T_S, if Earth's albedo and greenhouse effect remained unchanged. As T_S is 15°C today, this would make the surface temperature at 4.5 b.y. ago equal to about −7°C, well below the freezing point of water. The actual surface temperature could have been even colder because of the positive feedback provided by water vapor and ice cover. A climate model calculation that includes the water vapor feedback (by assuming fixed relative humidity) is shown in Figure 12-2. The solar luminosity curve is the same as the one shown in Chapter 1. The lower dashed curve represents the effective radiating temperature, T_e, calculated using the principle of planetary energy balance from Chapter 3. The upper dashed curve is the mean surface temperature calculated by the climate model. The shaded region between the two curves represents the greenhouse effect, ΔT_g; this increases with time because of the water vapor feedback. A constant atmospheric CO_2 concentration of 340 ppm and a constant surface albedo are assumed.

Under these assumptions, T_S falls below the freezing point of water prior to 1.9 b.y. ago and reaches a chilly 255 K (−18°C) at 4.6 b.y. ago. This prediction is at odds with geologic evidence discussed in the previous two chapters, which shows that liquid water was present at 3.8 b.y. ago and that life has probably been present since that time as well.

Possible Solutions to the Problem

How can the faint young Sun problem be resolved? According to the planetary energy balance equation, three types of solutions exist. Either the planetary albedo must have been lower in the past, the greenhouse effect must have been larger, or additional heat sources besides the Sun must have been present.

One additional heat source that is sometimes suggested for warming the early Earth is **geothermal heat** from Earth's interior. As mentioned in Chapter 7, some of this heat is produced by the decay of radioactive elements in Earth's crust and mantle and some is left over from Earth's formation. Both heat sources should have been larger in the distant past, so it is not unreasonable to ask whether geothermal heat might have helped keep Earth warm.

The problem with this suggestion is that the geothermal heat flux is simply not big enough to supply the required energy. A deficit of 30% in the solar flux is equivalent to a loss of about 70 W/m^2 of heating, averaged over Earth's surface. The modern geothermal heat flux is only about 0.06 W/m^2. Theoretical models of Earth's interior evolution suggest that the geothermal heat flux at 4 b.y. ago was higher than the present value by a factor of perhaps 4 or 5, so the available heat flux should have been approximately 0.3 W/m^2. The release of this much heat at the sea floor would have prevented the oceans from freezing to the bottom, but they should still have been covered with an ice layer several hundred meters thick if no additional surface warming was present. Little or no light would have penetrated such an ice layer, so this would be hard to reconcile with the evidence for ancient photosynthetic life. (Recall from Chapter 11 that stromatolites provide evidence that some type of photosynthetic bacteria were already extant by 3.5 b.y. ago.) The early Earth was not a global iceball.

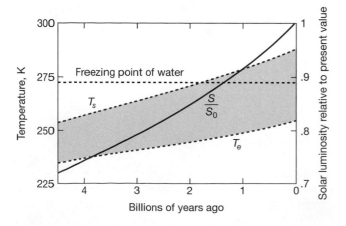

FIGURE 12-2

The faint young Sun paradox. The scale on the right applies to the solar luminosity curve, labeled S/S₀; the scale on the left applies to temperature curves. The shaded area represents the magnitude of the atmospheric greenhouse effect. (From J.F. Kasting et al. How Climate Evolved on the Terrestrial Planets. *Scientific American* 256(2):90–97, 1988. Used with permission, © George V. Kelvin/*Scientific American*.)

It is also difficult to solve the faint young Sun problem by simply changing Earth's albedo. The current planetary albedo is about 0.3, so the factor $(1-A)$ in the energy balance relation is equal to 0.7. If the solar luminosity, S, were 30% lower, A would have to be near zero to keep T_e constant. It is difficult to imagine how a water-covered planet could have an albedo near zero, because clouds, which are highly reflective, would almost certainly have been present. If the surface was actually as cold as predicted by Figure 12-2, sea ice would likely have been present also, even if continents and continental ice sheets were not. So, although Earth's albedo may have changed with time, it is unlikely that this by itself could have kept the planet warm.

The most likely solution to the faint young Sun problem is that Earth's greenhouse effect was larger in the past. If so, though, which greenhouse gases would have been more abundant? As we have seen previously, water vapor is the strongest greenhouse gas in the modern atmosphere. However, water vapor cannot solve the faint young Sun problem by itself because it is close to saturation and, hence, acts as a feedback on climate rather than as a forcing. Sagan and Mullen, who pointed out the faint young Sun problem in the first place, suggested that ammonia, NH_3, might be the solution. Ammonia is a good absorber of infrared radiation and it is also a **reduced gas**, that is, one that combines with oxygen to form stable compounds. As discussed in the previous chapter, reduced gases should have been much more abundant prior to the rise of atmospheric oxygen around 2.3 b.y. ago. That is why Sagan and Mullen thought that ammonia might be a good candidate for warming early Earth. After their paper was published, however, other researchers demonstrated that ammonia would have been rapidly destroyed by ultraviolet radiation. Hence, it is unlikely to have been abundant enough to have provided the necessary warming.

A CO_2-Rich Early Atmosphere?

One greenhouse gas that could have kept early Earth warm is carbon dioxide. There are several reasons for suspecting that atmospheric CO_2 levels were originally much higher. As discussed in Chapter 10, smaller continents would have reduced the amount of land available on which to weather silicate rocks and to store carbonate rocks; thus, the CO_2 sink should have been smaller. Impact degassing of late-arriving planetesimals, along with enhanced volcanism on the hot, young Earth, would have created a larger CO_2 source. All these changes would have favored higher atmospheric CO_2 concentrations. On the other hand, weathering of finely dispersed impact ejecta could have drawn down CO_2 levels by converting silicate minerals in carbonates. So, it is difficult to say whether atmospheric CO_2 concentrations would have been high or low during the first several hundred million years of Earth's history when the impact rate was high. There is also little geologic evidence with which to test one's theories. Climate during the heavy bombardment period remains largely a mystery.

Beginning about 3.8 b.y. ago, the geological record improves and one can draw inferences about climate with somewhat more confidence. As we have seen, the presence of liquid water and (possibly) of life suggests that additional greenhouse gases were present. CO_2 is among the most likely of these for the following reason. As discussed in Chapter 8, the carbonate–silicate cycle, which affects atmospheric CO_2 levels over long time scales, contains a strong negative feedback. If Earth's surface temperature were lower as a result of low solar luminosity, the rate of silicate weathering should have been slower, thereby lowering the CO_2 loss rate. CO_2 emitted from volcanos would have accumulated in the atmosphere until the global rate of silicate weathering balanced the volcanic outgassing rate. If Earth had ever become entirely ice-covered, silicate weathering would have ceased entirely and volcanic CO_2 should have accumulated in the atmosphere until the associated greenhouse effect became large enough to melt the ice. Thus, the Earth system has a natural way of recovering from global glaciation. Most of the time, the feedback is strong enough that global glaciation is avoided. When it does happen, however, the system apparently recovers in exactly the manner described.

How much atmospheric CO_2 would have been required to keep the early oceans from freezing? If CO_2 and H_2O were the only important greenhouse gases, then the minimum CO_2 level needed to compensate for a 30% reduction in solar luminosity is 0.3 bar—about 1,000 times the amount of CO_2 in the atmosphere today (Figure 12-3). Although this sounds like a lot, this amount of CO_2 is not large compared to the total amount of carbon available. The CO_2 stored in carbonate rocks today would produce a partial pressure of some 60 bars were it all present in the atmosphere. Only a small fraction (0.5%) of this CO_2 is needed to resolve the faint young Sun problem. Indeed, atmospheric CO_2 levels may initially have been much higher than this, as mentioned earlier. The upper limit of 10 bars shown in Figure 12-3 corresponds to the amount predicted on an ocean-covered early Earth. Climate model simulations show that such a CO_2 concentration would produce a global average temperature of 80–90°C. While we do not consider such a situation to be very likely, it is difficult to rule it out entirely. This would provide an alternative explanation for the prevalence of hyperthermophiles near the base of the evolutionary tree (Chapter 10, Figure 10-8).

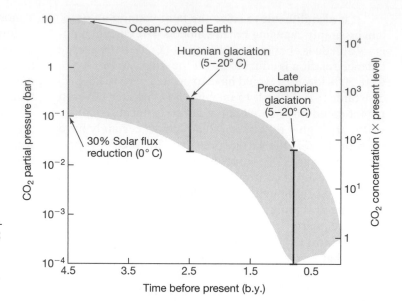

FIGURE 12-3

Atmospheric CO_2 concentrations needed to compensate for changing solar luminosity if CO_2 and H_2O were the only important greenhouse gases. The vertical bars at 2.5 b.y. ago and 0.65 b.y. ago show limits estimated from climate model calculations during glacial periods. (Reprinted with permission from J.F. Kasting. *Science* 25:920–926. Copyright 1993 American Association for the Advancement of Science.)

Effect of Methane on Archean Climate

CO_2 was probably not the only greenhouse gas that affected Earth's early climate. We saw in the previous chapter that CH_4 could also have been relatively abundant prior to 2.3 b.y. ago, when atmospheric O_2 levels were low. This CH_4 could have come from a variety of biological and abiotic sources. Prior to life's origin, CH_4 could have been produced by impacts and by *serpentinization* of rocks on the sea floor. As mentioned in Chapter 10, serpentinization is a process by which various *serpentine* minerals are formed from reaction of water with iron- and magnesium-rich basalts. In the process, ferrous iron is oxidized to magnetite (Fe_3O_4) and water is reduced to molecular hydrogen (H_2). If CO_2 is present in the water, CH_4 is formed instead. These processes could have introduced modest amounts (10–100 ppm) of CH_4 into the prebiotic atmosphere. We saw earlier that this may have helped in the synthesis of the key biological precursor molecule, HCN.

Once life evolved, the source of methane to the atmosphere should have increased greatly. In Chapter 10 we argued that methanogenic bacteria were probably among the earliest organisms. These bacteria would have converted much of the H_2 in the atmosphere into CH_4 by way of the reaction: $CO_2 + 4\,H_2 \rightarrow CH_4 + 2\,H_2O$. They could also have generated CH_4 from organic matter created by photosynthetic bacteria. Theoretical models suggest that atmospheric CH_4 concentrations of 1000 ppm or more are likely to have existed during the post-biological Archean and early Paleoproterozoic Eras, 3.8–2.3 b.y. ago.

If atmospheric CH_4 was indeed present at these concentrations, it would have had a strong warming effect on global surface temperatures, as seen in Figure 12-4. The horizontal axis of the figure represents atmospheric CO_2 partial pressure in bars (1 bar ≅ 1 atm). The labels on the

curves show the CH_4 *mixing ratio*. Mixing ratio is just fractional abundance, so a mixing ratio of 10^{-3} is equal to 1000 ppm. The solid curves show global average surface temperatures calculated with a one-dimensional radiative-convective climate model, or RCM. (Look back at Chapter 3 if you have forgotten what an RCM is.) The calculations were performed for an assumed solar luminosity of 80% of the present value, which is the value expected 2.8 b.y. ago.

What these curves demonstrate is the following: If CH_4 was *not* present in the atmosphere at that time, a CO_2 concentration of about 0.02 bar, or about 600 times the present level, would have been needed to keep Earth's surface from freezing at that time. If, however, CH_4 was

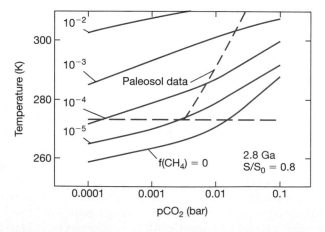

FIGURE 12-4

Average surface temperature as a function of atmospheric CO_2 and CH_4 concentrations. The dashed curves show the freezing point of water and the upper limit on Late Archean CO_2 derived from paleosols. (From A. A. Pavlov et al. *J. Geophys. Res.* 105, 11, 981–990, 2000.)

present at a mixing ratio of 10^{-3}, as expected on the basis of biological considerations, then the surface temperature could have remained above freezing even if CO_2 levels were no higher than today. This does *not* imply that the negative feedback involving CO_2 and climate did not operate. It merely says that high CO_2 levels were not needed if CH_4 was as abundant as suggested here. Indeed, high CH_4 concentrations would have led to lower CO_2 concentrations because the warming they produced would have resulted in faster silicate weathering.

There is an additional reason for liking this solution to the faint young Sun problem, which is indicated in Figure 12-4. The nearly vertical dashed curve represents an upper limit on atmospheric CO_2 levels derived from *paleosols* (ancient soils). Robert Rye and colleagues from Harvard University examined several paleosols of Late Archean age and noticed that none of them contained the mineral *siderite* ($FeCO_3$). Siderite is formed from the reaction between ferrous iron and carbonate ion ($CO_3^=$). Their analysis indicates that siderite should have formed in these soils if atmospheric CO_2 concentrations were higher than those indicated by the dashed curve. (Their limit is temperature-dependent because the chemical reactions involved in siderite formation depend on temperature.) Thus, their study is consistent with an Archean atmosphere in which CH_4 was an important greenhouse gas.

A Pink Sky During the Archean?

As mentioned in the previous chapter, high CH_4 concentrations could have made the Archean atmosphere appear very different from today's atmosphere. Today, the skies are blue as a consequence of scattering of sunlight by gas molecules (predominantly N_2 and O_2). **Scattering** is when a photon interacts with a particle (or molecule) and is sent off in a different direction. Gas molecules are smaller than the wavelength of visible light, so their interaction with sunlight is termed **Rayleigh scattering**. In Rayleigh scattering, the shorter wavelengths of light are scattered preferentially compared to longer wavelengths. Thus, if you are looking away from the Sun, the blue rays are scattered into your line of vision more effectively than are the red ones, giving the sky its bluish appearance. Conversely, when one looks directly at the Sun near sunrise or sunset, the blue rays are scattered out of your line of sight during their long pathlength through the atmosphere, and the Sun appears reddish-orange.

In the Archean, the interaction of sunlight with the atmosphere may have been quite different. When CH_4 becomes as abundant as CO_2 in an atmosphere, photochemical models predict that it can **polymerize** to form particles of higher hydrocarbons. Higher hydrocarbons are molecules in which carbon atoms are attached to each other in long chains. Planetary scientists believe that polymerization of CH_4 accounts for the orangish haze in

Titan's atmosphere (see Figure 11-4 in the previous chapter). The reason that Titan's atmosphere appears orange is that the particles are approximately the same size as the longer (red) wavelengths of solar radiation. Scattering by particles that are comparable in size to the wavelengths being scattered is called **Mie scattering**. Mie scattering also predominates in Mars' atmosphere because of the large amount of suspended dust. Thus, the martian atmosphere looks pinkish, particularly during or after a global dust storm.

Earth's Archean atmosphere may have looked pinkish as well if the methane greenhouse story is correct. Indeed, there is a positive feedback that may have pushed Earth's climate system into a state in which organic haze would have started to form. Most methanogens are either **thermophiles** or hyperthermophiles. Recall that hyperthermophiles are organisms whose optimum growth temperatures are 80°C or above. Thermophiles are organisms with optimum growth temperatures between 40°C and 80°C. Laboratory culture experiments have shown that the methanogens with higher optimum growth temperatures also have faster growth rates and shorter doubling times. Hence, a positive feedback loop exists whereby higher surface temperatures should have favored faster-growing methanogens (Figure 12-5). This, in turn, would have led to increased methane production, a bigger greenhouse effect, and hence still higher surface temperatures. Thus, the most likely solutions to the Archean climate problem fall into the upper left-hand quadrant of Figure 12-4, where CH_4 and CO_2 levels are comparable. So, it would not be at all surprising if Earth was obscured by a global organic haze layer at that time.

Climate Regulation by the "Anti-Greenhouse Effect"

The discussion above makes it appear as though the early Earth should have become hotter and hotter until temperatures became too high for even the hyperthermophilic methanogens. (The highest temperature at which

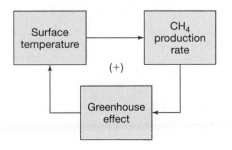

CH₄-Climate Feedback Loop

FIGURE 12-5

Feedback diagram showing the positive feedback loop between atmospheric CH_4 levels and surface temperature.

life has been found today is 113°C. Such temperatures can be reached without having the water boil if the overlying pressure is greater than one bar.) Well before this happened, however, another phenomenon would have occurred. Both methane and the haze that it produces are strong absorbers of visible radiation. Methane itself absorbs in the red part of the visible and in the near-infrared. (This is why Uranus and Neptune appear blue. The red wavelengths of the incident sunlight are absorbed by the 2% methane in their atmospheres, so only the blue light is reflected.) Absorption of sunlight by methane and organic haze could have produced an **anti-greenhouse effect**, as occurs on Titan today. Indeed, the term "anti-greenhouse" was coined by Christopher McKay of NASA Ames Research Center in a paper about Titan's climate. In the anti-greenhouse effect, sunlight is absorbed high in the atmosphere and is reradiated back to space as infrared energy without ever reaching the planet's surface. This cools the surface, which is why it is called the "anti-greenhouse" effect. On Earth, the organic haze layer could not have been too thick, or it would have cooled the surface below the freezing point of water. Once this happened, the methanogenic bacteria that were producing the CH_4 would have died off, and the haze layer would have thinned.

All of this suggests that Earth's climate at this time may have been regulated by a negative feedback loop between surface temperature, atmospheric CH_4 and CO_2, and the organic haze layer (Figure 12-6). Higher temperatures would have led to an increase in CH_4 (for reasons described above) and to a decrease in CO_2. The decrease in atmospheric CO_2 with surface temperature is because of the well-known feedback involving the silicate weathering rate, which we have just been studying. As CH_4 levels went up and CO_2 levels went down, methane would have begun to polymerize and an organic haze layer would have begun to form. As soon as it became thick

enough to block out sunlight, however, the anti-greenhouse effect would have set in and the surface would have begun to cool. The overall effect of this negative feedback loop should have been to stabilize Earth's surface temperature somewhere above the freezing point of water. Whether this would have kept the Archean Earth hot, or merely warm, depends on a variety of factors (e.g., continental size) that are difficult to determine. So, this hypothesis does not tell us what the Archean climate was like, but it does suggest why it was stable against glaciation for a very long time.

We note parenthetically that the Archean climate stabilization mechanism that we have just outlined is quite "Gaian" in nature—much more so than the present climate system. If methanogens truly were keeping the early climate warm, and if they themselves depended on this warmth in order to flourish, then they were clearly modifying climate in such a way as to benefit themselves. And, just as clearly, they would have been doing so without any intelligence or foresight, just as Lovelock and Margulis suggested that a Gaian feedback mechanism ought to do. So, Gaian control of a planetary climate system does indeed appear to be possible. But it is not *necessary*, at least in Earth's case, because the largely abiotic CO_2-weathering feedback can operate as well and would presumably have done so had methanogens not been present. Let us try to bear this in mind when we get to Chapter 19 and discuss the possibility of life on Earth-like planets around other stars.

The Long-Term Climate Record

Our discussion to this point has been focused on the very early Earth and on processes that may have contributed to climate stabilization. After all, the most significant characteristic of Earth's climate on long time scales is that it has remained conducive to the presence of life for something close to 4 b.y. If one examines the geologic record in more detail, however, one finds that climate is anything but stable. A variety of different geologic indicators, discussed below, show that **paleoclimate** (past climate) has actually varied in a complex manner, with long periods of warmth separated by shorter periods of intense cold. And there may have been several "Snowball Earth" episodes, when Earth's surface actually did freeze over. This suggests that other factors that we have not yet talked about affect long-term climate as well. Determining what those other factors might be occupies the last two sections of this chapter. Before discussing them, let us examine some of the different types of geologic evidence that are used to study paleoclimate and see if we can piece them together to determine the broad outlines of Earth's climate history.

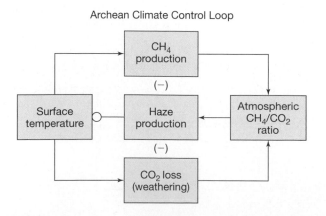

FIGURE 12-6

Feedback diagram showing a possible climate control mechanism that may have regulated Earth's surface temperature during the Archean Era.

Geological Indicators of Paleoclimate

The types of geological indicators that are useful for determining paleoclimate depend on the time scale being considered. For recent periods of Earth history, it is possible to obtain a reasonably accurate estimate of ocean temperatures by measuring oxygen isotopes in carbonate sediments obtained from *deep-sea cores*. This technique will be described in Chapter 14, where it is used to identify the glacial–interglacial cycles of the past 3 m.y. This method works only for time periods more recent than about 200 m.y. because most of the sea floor older than this has been subducted.

During the past 540 m.y. (the Phanerozoic Era), we can learn about paleoclimate by examining the fossil record. Species of plants and animals that are known to live in certain climates can be used to estimate surface temperatures in the localities where their fossils are found. In doing so, we must account for the fact that the continents have drifted over Earth's surface as a consequence of plate tectonics. This technique is of limited use in the Precambrian Era, however, because the single-celled organisms that were the only extant forms of life during most of this time could have survived under a wide range of climatic conditions.

Evidence of Past Glaciations

The best available evidence for climate change on billion year time scales comes from geologic deposits formed by glacial ice. Three such types of deposits are shown in Figure 12-7. **Tillites** (Figure 12-7a) are mixtures of cobbles, pebbles, sand, and mud that have been packed together to form rocks. They are formed from debris produced when glaciers grind up surface rocks. The debris is carried along by the glaciers as they move and is deposited in piles of rubble called **moraines** along the margins of the ice sheets. (Moraines mark the terminal points or flanks of the glaciers.) In association with tillites, geologists sometimes find rocks with long, parallel scratches, or **glacial striations**, formed when moving glaciers drag other rocks across their surfaces (Figure 12-7b).

A third signature of glaciation is **dropstones**, "misplaced" chunks of rock that occur in otherwise finely laminated marine sediments (Figure 12-7c). They form when rocks trapped in glacial ice are carried out to sea by icebergs, a process termed **ice-rafting.** When the iceberg melts, the trapped debris falls to the bottom of the ocean and becomes incorporated into sediments.

The Long-Term Glacial Record

Geologists who have studied the long-term glacial record have concluded that Earth's climate history has been marked by five main periods of glaciation (Figure 12-8). The first such period occurred at approximately 2.3 b.y.

(a)

(b)

(c)

FIGURE 12-7

Geological indicators of glaciation: (a) a tillite from the 2.4-b.y.-old Gowganda formation in Canada; (b) glacial striations from the 0.65-b.y.-old Smalfjord tillite in Norway; (c) a dropstone from the Gowganda formation. (From J.W. Schopf, *Earth's Earliest Biosphere: Its Origin and Evolution.* Copyright 1983 by Princeton University Press. Reprinted by permission of Princeton University Press.)

EON	GLACIATIONS	ERA	Duration in millions of years	Millions of years ago
PHANEROZOIC		CENOZOIC	65	—65—
		MESOZOIC	186	—251—
		PALEOZOIC	293	—544—
PRECAMBRIAN (PROTEROZOIC)	Late Proterozoic glaciations	Neoproterozoic	330	—900—
	WARM	Mesoproterozoic	700	—1600—
		Paleoproterozoic	900	
	Huronian glaciations			—2500—
PRECAMBRIAN (ARCHEAN)	WARM (?)	LATE	500	—3000—
		MIDDLE	400	—3400—
		EARLY	400	—3800—
(HADEAN)			800	—4600—

Era	Period	Epoch	Glaciations	Duration in millions of years	Millions of years ago
CENOZOIC	Quaternary	Holocene	Pleistocene glaciations	0.01	0.01
		Pleistocene		1.8	1.8
		Pliocene		3.5	5.3
	Tertiary	Miocene		18.5	23.8
		Oligocene		9.9	33.7
		Eocene		21.1	54.8
		Paleocene		10.2	65
MESOZOIC	Cretaceous		WARM	79	144
	Jurassic			62	206
	Triassic			45	251
PALEOZOIC	Permian		Permo-Carboniferous glaciations	35	286
	Carboniferous (Pennsylvanian)			39	325
	Carboniferous (Mississippian)		WARM	35	360
	Devonian			50	410
	Silurian			30	440
	Ordovician		Late Ordovician glaciations	65	505
	Cambrian		WARM	39	544
PRECAMBRIAN					

FIGURE 12-8

The major cold and warm periods during Earth's history. (From W.K. Hamblin and E.H. Christiansen. *Earth's Dynamic Systems*, 8/e, 1998. Reprinted by permission of Prentice Hall, Upper Saddle River, N.J.)

ago, the time from which the tillite and dropstone pictured in Figure 12-7 derive. Because these deposits were first identified near the banks of Lake Huron in North America, this early cold period is sometimes called the "Huronian glaciation." The Huronian glaciation was followed by over 1 billion years of ice-free conditions during the Late Paleoproterozoic and Mesoproterozoic.

Why did Earth experience a major glaciation in the midst of what was otherwise an extended, ice-free stretch of its history? The discussion we have just been through offers a convenient explanation. Suppose that methane *was* an important component of the atmospheric greenhouse during the Late Archean and early Proterozoic. The rise of atmospheric O_2 around 2.3 b.y. ago would have eliminated most of this methane, throwing climate into a temporary deep-freeze. And indeed, the geologic record is entirely consistent with this story. Figure 12-9 shows a stratigraphic section from the Huronian sequence of southern Canada. This particular section was deposited from 2.2–2.45 b.y. ago, based on radiometric age dating.

The three hatched layers represent glacial deposits (tillites), indicating that there were three separate episodes of glaciation. Beneath the lowermost glacial layer one finds rocks containing *detrital uraninite*. As discussed in the previous chapter, such deposits are indicative of a low-O_2 atmosphere. Above the uppermost glacial layer is the Lorraine *redbed* formation, which must have formed under a high-O_2 atmosphere. As pointed out over 30 years ago by Canadian geologist Stuart Roscoe, it appears as if the Huronian glaciation (or glaciations) is contemporaneous with the rise of O_2.

After this time, the climate became warm again, as evidenced by the complete absence of evidence for ice. Why, one might ask, should the climate have warmed back up when the methane that was keeping it warm originally had largely disappeared? The carbonate–silicate cycle provides one possible explanation. CO_2 should have been outgassed from volcanos at about the same rate after the rise of O_2 as it had been before. Hence, it had to have been removed by silicate weathering at that same rate.

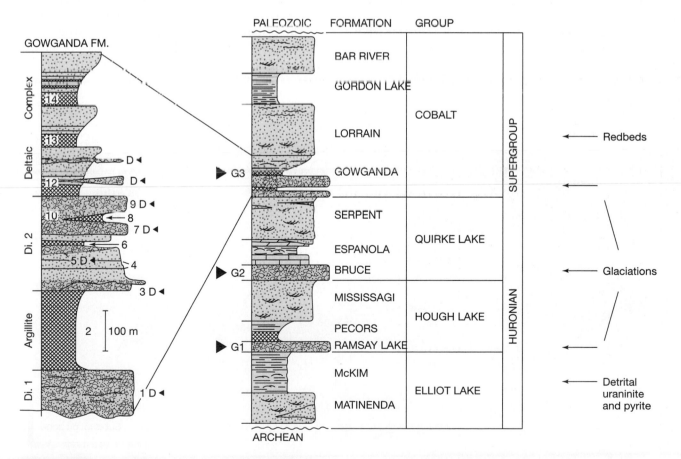

FIGURE 12-9

A stratigraphic section from the Huronian sequence in southern Canada. The hatched regions show the three glacial layers. (From G. M. Young. *Stratigraphy, Sedimentology, and Tectonic Setting of the Huronian Supergroup*. Field Trip B5 guidebook, joint meeting Geol. Assoc. Canada, Mineral. Assoc. Canada, Soc. Econ. Geol., Toronto, 1991.)

The silicate weathering rate is largely a function of surface temperature; thus, after the glaciation, the surface temperature had to recover to approximately its original value in order to keep the rate of silicate weathering the same as before. To accomplish this, atmospheric CO_2 could have jumped to a substantially higher concentration. This prediction remains speculative for the time being, but it might eventually be tested if more paleosol data become available to constrain past CO_2 concentrations.

There is another possibility for the extended warmth of the middle Proterozoic that climatologists have begun to think about only recently. It may be that after disappearing briefly just prior to the Huronian glaciation, atmospheric methane recovered and again reached levels much higher than today. Why, one might ask, should it have done so? The answer—which is admittedly still speculative—is that both atmospheric O_2 and dissolved oceanic sulfate levels may still have been much lower than today. This model for the mid-Proterozoic has been championed by Donald Canfield and his colleagues from the University of Southern Denmark. Today, most of the organic matter that reaches the sea floor is recycled back to CO_2 either by aerobic bacteria that cause decay or by sulfate-reducing bacteria. (This is the process of bacterial sulfate reduction that was mentioned in the previous chapter.) If neither O_2 nor sulfate was present in appreciable concentrations, then much more organic matter may have decayed by the combined processes of fermen-

tation and methanogenesis. These processes occur in modern marine sediments today, but only at depths below which the O_2 and sulfate are exhausted.

Low-Latitude Glaciation: the Snowball Earth

Eventually the climate became cool once again. Indeed, the Late Proterozoic, between 0.8 and 0.6. b.y. ago, was so cold that it is considered a great mystery. Evidence for glaciation during this time interval is found on six of the seven present-day continents. (The only exception is Antarctica. Antarctica may have been glaciated, too, but it is impossible to tell because it is still largely buried in ice.) The reconstruction shown in Figure 12-10 suggests that the continents were at that time grouped into two supercontinents, one of which was centered near the equator. Alternatively, the continents may have been grouped into a single supercontinent centered on the equator but extending a significant distance to the north and south. In all continental reconstructions of this time period, one feature remains constant: The continent of Australia is situated at or near the equator. This is considered remarkable because the geologic evidence indicates that at this time Australia was glaciated from one end to another. The evidence includes tillites, glacial striations, and dropstones, the latter demonstrating that ice sheets extended right to the margins of the paleocontinent. Today, in contrast, there *is* glaciation in the tropics but it is con-

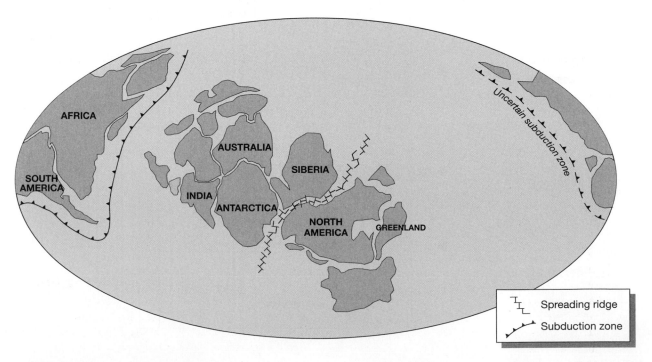

FIGURE 12-10

Possible continental reconstruction for the Neoproterozoic Era. All the continents appear to have been glaciated at that time. (After J. L. Kirshvink. *The Proterozic Biosphere: A Multidisciplinary Study*, J. W. Schopf and C. Klein, eds., Ch. 12.1, Cambridge University Press, Cambridge, 1992.)

fined to high mountain ranges like the Andes of South America. There, the glaciers never make it down below about 5 km elevation.

Why are geologists convinced that Australia was situated in the tropics at this time? Recall from Chapter 7 that past north–south continental positions are determined from paleomagnetic data. In Australia, glacial deposits are found mixed in with rocks in which the remnant magnetic field lines are parallel to the original bedding plane of the rock. If Earth's magnetic field was approximate to a dipole field, as it is today, this implies that these rocks formed near the equator. Many different geologists have looked at the paleomagnetic data from the Neoproterozoic rocks in Australia and have concluded that the evidence for low-latitude glaciation is real. The Huronian sequence described above also shows evidence for low-latitude glaciation; hence, it too may represent a "Snowball Earth" episode.

How could the Earth have possibly gotten cold enough to glaciate the tropics? In one sense, it's not that difficult. After all, the entire first half of this chapter is devoted to the question of how Earth *avoided* global glaciation during the earlier parts of its history. If the Sun was less bright in the distant past, then the atmospheric greenhouse effect had to be larger, or else global glaciation would be expected. Conversely, if CO_2 and CH_4 levels were low at some time in the past, then the lower solar luminosity would ensure that Earth's climate was very cold.

Let us take the Neoproterozoic glaciations as an example because they are the ones that have been best studied. Fieldwork in Namibia, West Africa, by Paul Hoffman and his colleagues at Harvard University has shown that there were two main episodes of glaciation: one at ~750 m.y. ago and a second at ~600 m.y. ago. Consider the more recent of these two episodes. In the recent past, solar luminosity has been increasing at a rate of about 1% every hundred million years. Thus, the Sun should have been about 6% less bright at the time of the second glaciation. Under these reduced luminosity conditions, GCM climate modeling by William Hyde and colleagues from Texas A&M University indicates that CO_2 concentrations would have to have been more than twice the preindustrial level (i.e., 2×280 ppm) in order to avoid global glaciation. Other GCMs yield slightly different critical CO_2 levels, but all of them agree that there is some CO_2 level below which global ice cover cannot be avoided.

How exactly would a Snowball Earth glaciation proceed? To understand in detail, see the Box "A Closer Look: How Did Life Survive the Snowball Earth." The general outline, however, is as follows. For one reason or another, atmospheric CO_2 concentrations were drawn down to relatively low values. (We will not consider the Huronian glaciation here. That one appears to have been caused by the rise of atmospheric oxygen.) Hoffman

and his colleagues originally suggested that increased organic carbon burial on newly created continental shelves was the cause, but it appears likely that other factors were important as well. Perhaps the most important of these is that a significant fraction of the continental area was situated in the tropics. This allowed silicate weathering to proceed even though the Earth was growing colder and colder, so that atmospheric CO_2 could continue to be drawn down. This first step in initiating global glaciation is counterintuitive because the conventional wisdom regarding glaciations has been that they occur when a continent drifts near or over one of the poles. Global glaciations are different. They probably require continents at low latitudes. This may also explain why they have occurred only at certain times in Earth's history. When the continents are not concentrated at low latitudes, the silicate weathering feedback prevents global glaciations from occurring.

The rest of the sequence may have gone like this: As atmospheric CO_2 levels declined, the polar ice sheets gradually crept down to lower latitudes. Once they reached approximately 30 degrees, however, something spectacular happened. All of a sudden, within a few decades perhaps, the oceans froze all the way down to the equator. The reason is that the positive feedback loop between ice albedo and surface temperature (Chapter 3, Figure 3-21) became so strong that it made the system unstable. This result can be demonstrated quantitatively using climate models. (See the Box "Thinking Quantitatively: Energy Balance Modeling of the Snowball Earth.") Once the ocean surface had frozen entirely, Earth's surface would have become extremely cold, $-50°C$ or lower, because the albedo would have been very high (>0.6, as compared to 0.3 today) and most of the incident sunlight would have been reflected back to space. However, as soon as the surface froze, silicate weathering on the continents would have virtually ceased, and volcanic CO_2 would have begun to accumulate in the atmosphere. In approximately 10 m.y., given modern volcanic outgassing rates, the atmospheric CO_2 partial pressure would have reached 0.1 bar (300 times the current level) and, all of a sudden, the ice would have begun to melt. The positive ice–albedo feedback loop would now have worked in the other direction: increased melting would have led to decreased albedo, which would in turn have led to increased surface temperature and more melting. Models predict that the ice cover would have disappeared entirely within a few thousand years. At this point, the Earth would have a dense, CO_2-rich atmosphere and a low albedo, and so it would have become very hot, with an average surface temperature as high as 50–60°C. Silicate weathering would now have proceeded rapidly, drawing down atmospheric CO_2 levels and eventually restoring the climate system to its original state.

THINKING QUANTITATIVELY
Energy Balance Modeling of the Snowball Earth

Although detailed modeling of the Snowball Earth climate requires a general circulation model, or GCM, much can be learned from simpler climate models. Indeed, much of the theoretical framework for understanding runaway glaciation was developed independently in the late 1960s by Soviet climatologist Michail Budyko and English climatologist William Sellers using what were later termed "energy-balance climate models," or EBMs. In a typical EBM, the Earth's surface was divided into 18 different latitude bands, each 10 degrees wide. In their simplest form, these models calculated the annually averaged solar heating in each latitude band, along with the average outgoing infrared radiation flux. Heat transport between different latitude bands was parameterized as *diffusion*. Diffusion of heat is usually called *conduction*. We know, of course, that the atmosphere does not really transfer heat in this way. Rather, it does so by the complex system of winds and ocean currents described in Chapters 4 and 5. Budyko and Sellers did a clever thing, however: they adjusted their diffusion coefficients so that their models matched the observed equator-to-pole temperature gradient. Thus, their models were capable of reproducing the average latitudinal distribution of temperature and, most importantly, they could estimate the size of the polar ice caps. This allows such models to be used to study the phenomenon of runaway glaciation.

Results from a more up-to-date version of an EBM are shown in Box Figure 12-1. The figure is somewhat complicated, so let us go through it carefully. The horizontal axis is the effective solar flux (S_{eff}), that is, the solar flux divided by the modern value. Thus, $S_{eff} = 1$ corresponds to the modern solar constant. The vertical scale is the sine of the ice-line latitude. This marks the extent of the polar ice caps. (The model is symmetric in each hemisphere.) Recall from trigonom-

etry that $\sin 30° = 0.5$, so a point halfway up the vertical scale corresponds to a latitude of 30°. Exactly half Earth's surface area is located poleward of 30°; the other half is located equatorward of this point.

The lines and curves in the figure represent the ice-line extent as calculated by the EBM climate model. Solid curves represent stable solutions; dashed curves represent unstable solutions. The three different curves shown correspond to three different atmospheric CO_2 levels. The curve furthest to the right is for a CO_2 partial pressure of 3×10^{-4} bar, which corresponds to a CO_2 concentration of 300 ppm, close to today's value. The points where the curves (or lines) intersect a vertical line at $S_{eff} = 1$ represent stable climate solutions for the modern Earth. Surprisingly, there are three different, stable solutions. The one that actually corresponds to the modern climate is the "small ice cap" solution. The sine of the ice-line latitude is ~0.95, which puts the boundary of the polar ice cap at about 72°. But there are two other stable solutions as well: an ice-free solution (no polar cap) and an ice-covered solution. The ice-covered solution corresponds to the "Snowball Earth." It is stable because the high albedo of the ice causes most of the incident sunlight to be reflected back to space.

The most interesting features of the figure, however, are the unstable solutions (dashed curves). As one can see, all solutions in which the ice line is equa-

torward of ~30° are unstable. If the polar ice caps ever reached this latitude, the ice albedo feedback would have become completely unstoppable: increases in ice cover past this point would cause more sunlight to be reflected back to space, which would result in decreased surface temperatures and further increases in ice cover. Very quickly, within a few decades, the ocean surface would have frozen all the way down to the equator. This is thought to have been how the climate system became trapped in the Snowball Earth.

The figure also shows how the system could have recovered from the Snowball Earth. Once Earth's surface was totally frozen, silicate weathering would have ceased and volcanic CO_2 would have accumulated in the atmosphere. One can see from the figure that when the CO_2 partial pressure reaches 0.12 bar (12,000 ppm), the unstable solution intersects the equator (sine ice-line latitude = 0) at $S_{eff} = 1$. Physically, this means that the ice-covered solution is no longer stable. Instead, the system would transition spontaneously (and rapidly) up to the ice-free solution. The climate would become extremely warm, 50–60°C, and would remain that way until silicate weathering was able to remove the excess CO_2 from the atmosphere. As described in the text, it looks as if this is exactly what happened during the Late Precambrian Snowball Earth episodes 600 m.y. and 750 m.y. ago.

BOX FIGURE 12-1

Energy-balance climate calculations for the Snowball Earth model. Solid curves represent stable solutions; dashed curves represent unstable solutions. Dots show equilibrium solutions for today's solar flux. (From K. Caldeira and J. F. Kasting. *Nature*, 359:226–228, 1992.)

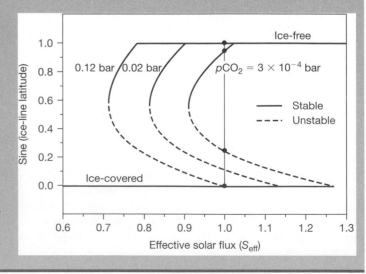

A CLOSER LOOK
How Did Life Survive the Snowball Earth?

Perhaps the most interesting question regarding the Snowball Earth is: How did life manage to survive through it? One can estimate the thickness of the ice cover over the oceans by means of a fairly simple calculation. (See "Critical Thinking" Problem 1 at the end of the chapter.) The thickness of the ice is limited by the geothermal heat flux that must be conducted upward through it. If you do the problem correctly, you should find that the sea ice during the Neoproterozoic glaciations was over a kilometer thick on average. This is far too thick to allow sunlight to penetrate. Very little sunlight makes it any deeper than 5 or 10 m depth even in very clear ice. And, yet, we are certain that photosynthetic life survived these two catastrophes, as well as the earlier Huronian glaciation, which may also have been a Snowball Earth episode. How can one resolve this apparent paradox?

One solution that has been suggested by William Hyde and his colleagues from Texas A&M University is that the tropical ocean may not have been entirely frozen. In their model (published in the journal *Nature* in May 2000), the tropical oceans remained ice-free, but at least some tropical continents were ice-covered. They suggested that the ice sheets formed at high elevations where the local temperatures were below freezing and that they flowed down to the continental margins before they melted. In this way, they could account for the occurrence of dropstones in Neoproterozoic marine sediments around Australia. Their model has been termed a "Slushball Earth" because the planet was never entirely ice-covered, if they are correct. However, their model has difficulty explaining the reoccurrence of BIFs and the presence of cap carbonates (see text, p. 244); hence, it does not appear to us that their solution is the right one.

Another idea is that life survived in geothermally heated environments in which liquid water remained even as the rest of the surface froze. The hydrothermal vents of the mid-ocean ridges are one obvious refuge. As most mid-ocean ridge vents are over 2.5 km deep, they would have been well beneath the ice layer, and organisms living within them would have been essentially unaffected by a global glaciation. Such organisms are not photosynthetic, however. Continental geothermal areas such as Yellowstone Park in the United States are another possibility. However, the water in Yellowstone's geysers and hot pools ultimately derives from rainwater. During a Snowball Earth event, the hydrological cycle would have virtually shut down and such areas should have dried up. A few geothermal areas that are well connected to the ocean, the island of Iceland for example, might have remained habitable for photosynthetic life. However, there are very few such areas in the world. If life had been restricted to only a few such locations, biologists believe that the Universal Tree of Life (Chapter 10, Figure 10-8) would show evidence for this. So, they think that photosynthetic life must have remained more widespread.

A third idea is that the ice may not have been as thick in all locations as models predict. Christopher McKay from NASA Ames Research Center near San Francisco has suggested that the ice could have been thin enough in the tropics to allow some sunlight to penetrate. McKay has spent several field seasons studying ice-covered lakes in the so-called "dry valleys" of Antarctica. The lakes there are covered with about 5 m of exceptionally clear ice, beneath which is found a thriving photosynthetic biota. The ice is clear because it forms very slowly and, hence, excludes air bubbles, which would otherwise give it a cloudy appearance. Box Figure 12-2 shows the fraction of light transmitted through such clear ice as a function of its thickness. If the ice in the tropics was as thick as predicted above, then obviously very little light would make it through. But, if the ice was only a few meters thick, then as much as 10% of the incident sunlight might penetrate it. This energy would have to get out by conduction through the ice, just like the geothermal heat from Earth's interior. If you work "Critical Thinking" Problem 2 at the end of the chapter, you should be able to show that a stable solution can exist with ice as thin as 2 m in the tropics. Approximately 10% of the incident sunlight could make it through this ice, according to Box Figure 12-2. Perhaps this "thin ice" model is the solution to the Snowball Earth paradox.

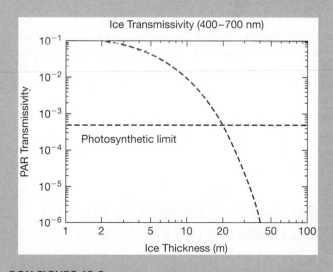

BOX FIGURE 12-2

Visible light transmissivity (400–700 nm) versus depth for clear ice. (From C. P. McKay. *Geophys. Res. Lett.* 27, 2153–2156, 2000.)

Additional Geological Evidence for the Snowball Earth: BIFs and Cap Carbonates

To some, this story sounds simply too extraordinary to be true. Is there any additional evidence that such a sequence of events actually occurred? The answer is yes. There are a number of other observations that are consistent with the Snowball Earth model. Of these, two features in particular stand out. The first is the reappearance of banded iron-formations (BIFs). Recall from the previous chapter that most BIFs were deposited prior to 1.8 b.y. ago and that their formation was linked to anoxic conditions in the deep ocean. Surprisingly, BIFs reappear briefly during the Neoproterozoic, at exactly the time of the glaciations. This reappearance is explained by the Snowball Earth hypothesis because the global ice cover would have cut off the ocean from the large reservoir of atmospheric O_2. The dissolved O_2 in the oceans, or at least in some ocean basins, was used up by oxidation of organic matter in sediments. Ferrous iron emanating from the hydrothermal vents in the mid-ocean ridges accumulated in the ocean and was ultimately upwelled on continental shelves (after the ice had melted) and deposited as BIFs.

The other piece of evidence is even more telling. Directly above each of the two Neoproterozoic glacial deposits in Namibia is a layer of carbonate rock approximately 400 m thick. These carbonate deposits are often termed **cap carbonates** because they "cap" the glacial layers. The bottom parts of these caps are fine-grained and show evidence of having been deposited very rapidly, as would be expected in the immediate aftermath of a Snowball Earth episode. This close association between glacial deposits and carbonates has puzzled geologists for many years because glaciers typically form at high latitudes, where it is cold, whereas carbonates typically form at low latitudes where carbonate minerals are less soluble. The Snowball Earth hypothesis explains this association naturally; indeed, the model *predicts* that such carbonate deposits should have been formed. Thus, there is compelling evidence to indicate that the Snowball Earth model is correct.

Variations in Atmospheric CO_2 and Climate During the Phanerozoic

Although the most spectacular extremes in climate appear to have occurred during the Precambrian, climate has varied over the past 540 m.y. as well. Three of the five glacial periods shown in Figure 12-8 have occurred during the Phanerozoic Eon: a brief one during the Late Ordovician Period (about 440 m.y. ago), a long series of glaciations near the boundary between the Permian and Carboniferous Periods (about 280 m.y. ago), and the most recent episode, which is called the Pleistocene glaciation.

This name is something of a misnomer because Antarctica began to be glaciated some 15–30 million years ago, well before the Pleistocene Epoch began, and because both Antarctica and Greenland continue to be ice-covered today. To qualify as a long-term "glacial period," it is only necessary that ice be present over part of Earth's surface for an extended period of time. This is different from common usage, whereby the term "glacial period" refers to one of several episodes of maximum ice extent during the last 3 m.y. These shorter-time-scale climate fluctuations, which will be discussed in Chapter 14, can be thought of as modulations of an overall climate that is basically "glacial."

CO_2 and Climate During the Paleozoic Era

The first part of the Phanerozoic Eon, which extended from about 545 m.y. ago until 250 m.y. ago, is called the Paleozoic Era. From this time on, considerable information is available to tell us about climate because the fossil record is much more detailed. By examining the types of plants and animals that existed and determining where they lived, scientists can deduce not only whether the climate was warm or cold, but also how wet it was and by how much it varied between the pole and equator. The general results of such studies are shown in Figure 12-11. Following the intense cold of the Neoproterozoic, the climate warmed. The Cambrian Period was decidedly warmer than today, and the ensuing three geological periods were mostly ice-free as well, except for a brief (and poorly understood) spike of glaciation during the Late Ordovician. Climate cooled markedly, however, during the Carboniferous Period, culminating in a series of glaciations that spanned almost 80 m.y. and are termed the "Permo-Carboniferous glaciations."

Why did climate get cold at this time? We have already seen that, on long time scales, climate is largely determined by a trade-off between solar luminosity and the atmospheric greenhouse effect. By this time, the atmosphere was well oxygenated, so we can assume that CH_4 concentrations were modest and that the greenhouse effect was mostly attributable to CO_2 and H_2O. Thus, the cooling in Earth's climate was probably caused by a decrease in atmospheric CO_2 levels.

Why should CO_2 levels have declined during the Late Paleozoic? On long time scales, atmospheric CO_2 is largely controlled by the carbonate–silicate cycle, but the organic carbon cycle cannot necessarily be neglected. Consider the carbonate–silicate cycle first. How might that have changed? As we saw in the previous section, the rate of silicate weathering is enhanced when continents move toward the equator. So, perhaps changes in continental positions were once again the key. But, biological innovations could have been important as well. As discussed in Chapter 8, plants and microorganisms are

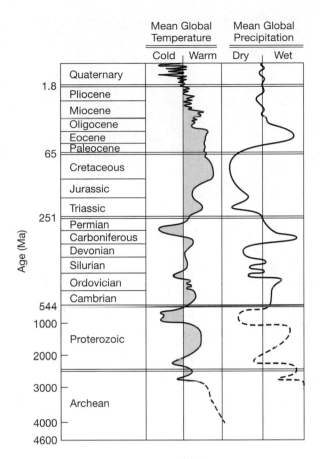

FIGURE 12-11

Estimated change in surface temperature during the Phanerozoic eon. (From K.C. Condie and R.E. Sloan. *Origin and Evolution of Earth: Principles of Historical Geology*, 1998. Reprinted by permission of Prentice Hall, Upper Saddle River, N.J.)

thought to accelerate weathering by increasing the CO_2 partial pressure in soils and by releasing organic acids that help dissolve rocks. **Vascular plants** have a well-developed stem or trunk for transporting water and nutrients from the ground up to their leaves, and typically a well-developed root system which takes up water and nutrients and helps support the plant. So, one might expect that the spread of vascular plants would have lowered atmospheric CO_2 (Figure 12-12). Unfortunately for this idea, the timing is not right. Vascular plants originated in the Late Silurian and spread widely during the Devonian Period, well before the climate began to cool. So, let us consider other factors that might have affected atmospheric CO_2.

A strong clue as to what might have happened is suggested by the carbon isotope record. As discussed in the previous chapter (see Figure 11-19), carbon isotopes indicate that the organic carbon burial rate nearly doubled during the Carboniferous as a result of the formation of large coal beds. In Chapter 11 we suggested that this may have led to an increase in atmospheric O_2. However, it

could have led to a decrease in atmospheric CO_2 as well. Robert Berner from Yale University has included this information in a model that tries to predict CO_2 levels during the Phanerozoic (Figure 12-13). As expected, the increased burial of organic carbon during the Carboniferous leads to a substantial drop in atmospheric CO_2 levels. As partial confirmation of Berner's results, the CO_2 concentrations predicted by his model are in approximate agreement with CO_2 levels estimated from paleosol data. This gives us some confidence that the basic idea is correct: during the latter half of its history, Earth's climate has been tightly coupled to atmospheric CO_2 levels.

The Warm Mesozoic Era

The Mesozoic and Cenozoic Eras span the past 251 m.y. of Earth history. The Mesozoic, the age of the dinosaurs, is thought to have been considerably warmer than today (Figure 12-11). Because large animals existed during this time, the fossil evidence relating to Mesozoic climate is quite extensive. For example, during the mid-Cretaceous Period, around 100 m.y. ago, lush ferns and alligators resided in what is now Siberia. Dinosaur skeletons have been recovered from north of the Arctic Circle in Alaska. These and other pieces of evidence indicate that the mid-Cretaceous climate was on the order of 2 to 6 degrees Celsius (3.6–11°F) warmer at the equator and 20 to 60 degrees Celsius (36–110°F) warmer at the poles (Figure 12-14).

A second type of evidence that indicates that the Mesozoic climate was warm comes from measurement of oxygen isotope ratios in carbonate sediments recovered from deep-sea cores. Although we will postpone detailed discussion of this topic until Chapter 14, oxygen isotopes provide information both on the temperature of the water in which they formed and on the amount of water stored in the polar ice caps. The measured isotope ratios tell us that Mesozoic ocean water was much warmer than today. This is especially true of the deep ocean, which currently has an average temperature of only 2°C. Deep-ocean temperatures during the Mesozoic Era were as high as 15°C. Furthermore, the polar ice caps, which today hold enough water to raise sea level by nearly 80 m, appear to have been absent throughout the entire Mesozoic and during the early Cenozoic as well.

What factor or combination of factors was responsible for the extreme warmth of the Mesozoic? In keeping with the discussion of the previous section, suspicion currently centers on higher atmospheric CO_2 levels. Climate models suggest that an increase in CO_2 by a factor of 4 from the present value could explain the warm climate of the mid-Cretaceous Period. This hypothesis is bolstered by paleomagnetic evidence that indicates that the sea floor was spreading faster at that time than it has in the more recent geologic past. Recall that the spreading rate

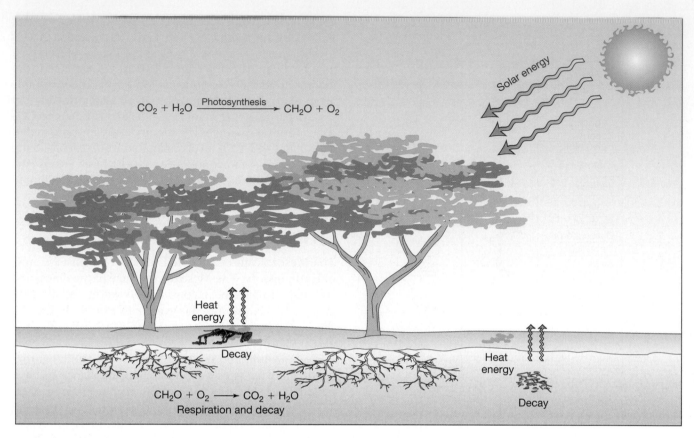

FIGURE 12-12

Land plants enhance the partial pressure of CO_2 in soils through root respiration and decay. (From T. McKnight. *Physical Geography: A Landscape Appreciation*, 6/e, 1999. Reprinted by permission of Prentice Hall, Upper Saddle River, N.J.)

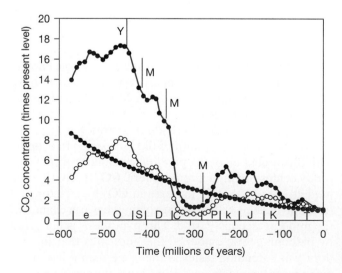

FIGURE 12-13

Phanerozoic atmospheric CO_2 levels predicted by the geochemical cycle model of R. A. Berner. Solid bars represent constraints on CO_2 from paleosol data. (From R. A. Berner. *Science* 261:68–70, 1993.)

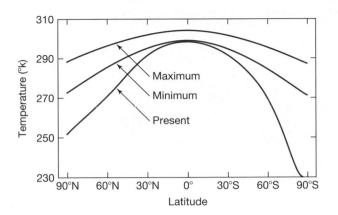

FIGURE 12-14

Estimated limits on longitudinally averaged surface temperatures during the mid-Cretaceous period, 100 m.y. ago, as compared with today. (After E. J. Barron and W. M. Washington. The Carbon Cycle and Atmospheric CO$_2$. *Geophysical Monograph* 32, AGU, Washington, D.C., 1985.)

Carbon Isotopic Evidence of High Mesozoic CO$_2$ Levels

Our discussion of paleo-CO$_2$ levels and climate has thus far been based mostly on theory. It has recently become possible, however, to test the theory with data for at least a few time periods in Earth history. Several different methods have been employed. One example, discussed earlier in the chapter, is based on the mineralogy of paleosols. That particular method is not applicable to more recent times because siderite (FeCO$_3$) is not stable under an oxygenated atmosphere. However, a similar type of analysis can be performed on some paleosols using trace carbonates found in the mineral goethite, Fe(OH)$_3$. The paleosol data shown in Figure 12-13 were obtained by this technique.

Another technique for inferring past CO$_2$ concentrations is based on carbon isotopes. As discussed in the previous chapter, photosynthetic organisms tend to take up ^{12}C faster than they take up ^{13}C. But they do this to a greater extent if CO$_2$ is relatively abundant in the organism's environment. So, photosynthetically produced organic matter created under high-CO$_2$ conditions tends to have a low ^{13}C/^{12}C ratio, or negative δ^{13}C. If CO$_2$ is scarce, organisms use whatever isotope is available and the relative isotopic abundances do not change as dramatically.

Carbon isotope abundances have now been measured in sediments of various ages by a number of different research groups. Figure 12-15 shows the difference in δ^{13}C values between carbonates and organic carbon. The story that has emerged is consistent with that told above: organic matter from sediments of Mesozoic age contains less ^{13}C than organic matter from sediments deposited during the last 20 m.y., indicating that atmospheric CO$_2$ levels were probably higher at that time. Although there is still

can be estimated by looking at magnetic patterns in the sea floor. Faster spreading rates would have led to faster rates of subduction of carbonate sediments and this, in turn, should have led to increased rates of CO$_2$ production from carbonate metamorphism. More CO$_2$ may also have been released by outgassing at the mid-ocean ridges themselves. And higher sea level at that time, itself caused by faster sea-floor spreading as well as the absence of polar ice, would have meant that there was less land area available on which to weather silicate rocks. These factors have been included in Berner's models and account for the high atmospheric CO$_2$ levels predicted for this time period (Figure 12-12).

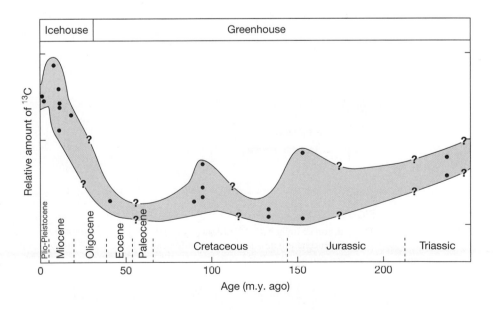

FIGURE 12-15

Carbon isotope data for the past 150 m.y. The vertical scale is a measure of the difference in ^{13}C content between carbonates and organic carbon. Large differences in ^{13}C correspond to lower atmospheric CO$_2$ values. The data are consistent with a decrease in CO$_2$ since the mid-Cretaceous Period. (After B. Popp et al. *American Journal of Science* 289:436, 1989.)

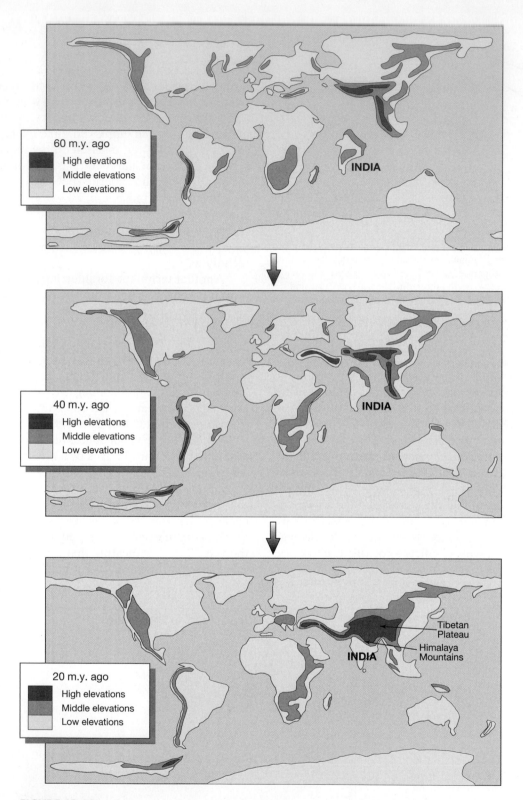

FIGURE 12-16

The collision of India with Asia, shown in a series of paleogeographic maps for the Paleocene (60 m.y. ago), the Eocene (40 m.y. ago), and the Miocene (20 m.y. ago) Eras. (After E. J. Barron. *Paleogeography, Paleoclimatology, Paleoecology* 50:45, 1985.)

considerable disagreement concerning how much CO_2 has decreased, the idea that CO_2 is a main driver of climate on these time scales has received additional support.

Other Possible Influences on Mesozoic Climate

Although higher atmospheric CO_2 levels can account for the overall warmth of the Mesozoic climate, this mechanism cannot by itself explain the extremely small latitudinal temperature gradient at that time. The equator-to-pole temperature contrast during the mid-Cretaceous Period was only 20–30°C, as compared to 50–60° today. Part of this difference can be explained by the absence of polar ice at that time. Recall that ice cover interacts with climate by way of a strong, positive feedback loop. Removal of the ice caps would cause a large decrease in the albedo of the polar regions that, in turn, should cause them to warm significantly.

Calculations with climate models suggest, however, that ice albedo feedback alone is not sufficient to explain the extreme warmth of the mid-Cretaceous poles. It seems likely that the atmosphere–ocean system was for some reason more effective in transporting heat from the equator to the poles than it is today. One possibility is that the thermohaline circulation of the oceans ran backward at that time: warm, but highly saline, deep water formed at low latitudes welled up near the poles, where it then warmed the climate through evaporation. Unfortunately, no one has yet demonstrated that this mechanism could work. A second possibility is that the tropical Hadley circulation extended further poleward than it does today. Because Hadley cells are very efficient at transporting heat, this mechanism could explain the low latitudinal temperature gradient. Moreover, it could explain the apparent absence of subfreezing temperatures in continental interiors (Siberia, for example) during the long polar night. Again, however, no one has demonstrated that this mechanism is dynamically feasible. Hence, the details of mid-Cretaceous climate remain largely unexplained.

Cooling During the Cenozoic Era

Starting about 80 m.y. ago, Earth's climate began to cool (except for a short-lived warming during the early Eocene Period). The initial decrease may simply have been caused by a decrease in mid-ocean ridge spreading rates, leading to a reduction in atmospheric CO_2. However, the cooling trend accelerated around 30 m.y. ago during the Oligocene Epoch in a way that does not correlate with the spreading-rate data. Thus, paleoclimatologists have searched for other explanations for the observed cooling. One intriguing theory, suggested by Maureen Raymo, then at the Massachutsetts Institute of Technology, and William Ruddiman of the University of Virginia, is that the carbonate–silicate cycle was perturbed by plate tectonics, but by a mechanism that differs from those discussed previously.

During the Mesozoic and early Cenozoic eras, India was a separate continent drifting slowly toward Asia. The two continents collided around 40 m.y. ago, a process that is still continuing today. The collision created a gigantic chain of mountains (the Himalayas) and a huge area of uplifted terrain called the Tibetan Plateau (Figure 12-16). The Himalayan Mountains provided fresh, readily erodable surfaces on which silicate weathering could proceed rapidly. At the same time, the uplift of the Tibetan Plateau created seasonal rainfall (the southeast Asian monsoon), which provided the water needed for weathering to occur on the face of the Himalayan range. The combination of these factors may have accelerated silicate weathering rates over a substantial portion of Earth's surface, thereby helping to bring atmospheric CO_2 concentrations down to the relatively low levels that prevail today.

The point to be drawn from this discussion is that plate tectonics probably does influence climate over long time scales, but not necessarily in the way that geologists have traditionally imagined. Continental-scale glaciers can indeed grow when land masses drift close to the poles but only if climatic conditions are ripe for such a development—specifically, if atmospheric CO_2 concentrations are relatively low. The main influence of plate tectonics on climate appears to be indirect: by changing the way in which the carbonate–silicate cycle operates, plate tectonics helps to modulate atmospheric CO_2 levels. This, in turn, affects climate by way of the greenhouse effect. Such changes, in combination with the long-term increase in solar luminosity, can account for the main features of the long-term climate record.

Chapter Summary

1. During the early parts of Earth's history, the faintness of the young Sun must have been offset by higher concentrations of greenhouse gases in the atmosphere.
   a. CO_2 may have dominated at first, but CH_4 was probably an important greenhouse gas as well, once it was being produced by methanogenic bacteria.
   b. The combination of CO_2 and CH_4 kept Earth's climate relatively warm until the early Paleoproterozoic Era.
   c. The rise of atmospheric O_2 at ~2.3 b.y. ago eliminated most of the CH_4 and thereby triggered Earth's first major glaciation.
2. Since that time, climate has been largely determined by the balance between increasing solar luminosity and decreasing atmospheric CO_2.

a. During most of this time, climate has been kept within moderate bounds by the negative feedback associated with the carbonate–silicate cycle.

b. This stabilization mechanism has broken down temporarily, however, at least three times, resulting in global glaciations near the beginning and the end of the Proterozoic. Life survived these Snowball Earth episodes by mechanisms that are currently being debated.

3. During the last 500 m.y. of Earth's history, climate has alternated between periods of warmth (the mid-Cretaceous) and periods of cold (the Late Ordovician, Permo-Carboniferous and Pleistocene glaciations). Earth is currently in a cold state—a brief interglacial period within the Pleistocene glacial epoch. Variations in atmospheric CO_2 levels caused by changes in plate tectonics and by biological innovations can explain the broad features of Phanerozoic climate history.

Key Terms

anti-greenhouse effect
cap carbonates
corona
dropstones
geothermal heat
glacial striations
ice-rafting
methanogenesis

Mie scattering
moraines
neutrinos
paleoclimate
polymerize
Rayleigh scattering
reduced gas

scattering
solar wind
sulfate-reducing bacteria
thermophiles
tillite
T-Tauri wind
vascular plants

Review Questions

1. Why does the Sun get brighter with time?
2. How might the carbonate–silicate cycle have helped to solve the faint young Sun problem?
3. Why is methane thought to have been an important greenhouse gas during the Archean Era?
4. What triggered the Huronian glaciation at 2.3 b.y. ago?
5. What types of geologic evidence are used to infer past glaciations?
6. How many separate episodes of glaciation have occurred during Earth's history?

7. What types of geologic evidence support the Snowball Earth model for the Late Precambrian glaciations?
8. How are carbon isotopes used to infer past atmospheric CO_2 concentrations?
9. How are atmospheric CO_2 levels affected by the presence of land plants?
10. What mechanisms might explain the warm climate of the Mesozoic Era? How might the equator-to-pole temperature gradient have been reduced?
11. Why did climate cool during the past 40 million years?

Critical-Thinking Problems

1. Evidence for low-latitude glaciation is found at both 0.6 Ga and 2.3 Ga. ("Ga" means "giga-annum," or billions of years ago.) These are two of the three possible Snowball Earth events mentioned in the text. (We will neglect the event at 0.75 Ga because it is similar to the first one.) Your job is to estimate how thick the ice was at those times.

a. The variation in solar luminosity with time can be approximated by the following formula (derived by fitting the results of a computer model of the Sun's evolution)

$$S = \frac{S_0}{1 + 0.4(t/4.6)}$$

where S = solar flux at time t

$S_o = 1370 \ W/m^2$ = present solar flux
t = time in Ga (billions of years before present)

Calculate the solar flux at 0.6 Ga and 2.3 Ga both in W/m^2 and as a percentage of its current clilaue.

b. As we have learned previously, the effective radiating temperature of the Earth can be found from the formula

$$\sigma T_e^4 = \frac{S}{4}(1 - A)$$

where A is the planetary albedo and $\sigma(= 5.67 \times 10^{-8} \ W/m^2/K^4)$ is the Stefan–Boltzmann constant. Calculate T_e at 0.6 and 2.3 Ga, assuming an albedo of 0.65 (the value for clean ice and snow). Then, calculate the global average surface temperature, T_s, assuming that the atmospheric greenhouse effect, ΔT_g, was the same as today (33 K). Recall that $T_s = T_e + \Delta T_g$.

c. The conductive heat flow through ice is given by

$$F = \frac{\lambda \Delta T}{\Delta z}$$

where $\lambda (= 2 \ W/m/K)$ is the thermal conductivity of ice, ΔT is the temperature difference between the top and

tions. We can sa
natural selectic
creased over the
Whether this ir
represented by
damental questi

Diversity, e
of species, has in
genetic mutatic
body plan, in bc
ished as other s
of all individual
thought of as a

rate of change
of species c

Researchers esti
year, and 10 to
the average rate
of natural select
Darwin. Howev
hibit much grea
tinction. In fac
origination and
tremely rapid. T
natural phenom
be the result of
ferent time scal

There is a cl
and the equatio
relationships ge
rapidly (expone
viduals is small.
the birth rate de
to shortages of
lation ultimatel
the *carrying cap*
death rate. Such
13-1). We might
logistic in shape
rate of extinctio

Biologists t
of the ability of
obviously cann
such as bacteria
leontologists ha
shape) to group
years of paleon
an effort to de
changed over E

The result
which presents
rine *invertebrate*

bottom of the ice layer, and Δz is the thickness of the layer. We know that the current geothermal heat flux, F, is about 0.06 W/m^2. Assume that F had this same value at 0.6 Ga, but was three times higher at 2.3 Ga. Assume also that the top of the ice is at temperature T_s and that the water below the ice has a temperature of $-2°C$ (the freezing point of water). What is the thickness of the ice at 0.6 Ga and at 2.3 Ga?

2. Suppose now that the ice is transparent enough so that some sunlight makes it through. Let's see how that would change the ice thickness.

a. Calculate the globally averaged solar flux incident on Earth's surface during the Neoproterozoic glaciation (0.6 Ga).
b. The solar flux at the equator is about 20% higher than the global average value. Suppose that 10% of this incident sunlight makes it through the ice. How thin must the ice layer be in order to conduct this heat back out? (Use the formula from Question 1-c.)
c. Is the ice thickness calculated in part (b) consistent with the assumption that it would transmit 10% of the sunlight through it? Determine this by consulting Box Figure 12-2.

Further Reading

General

Lovelock, J. E. 1988. *The Ages of Gaia: A Biography of Our Living Earth*. New York: W. W. Norton.
Gribbin, J. 1990. *Hothouse Earth: The Greenhouse Effect and Gaia*. New York: Grove Weidenfeld.

Advanced

Crowley, T. J., and North, G. R. 1991. *Paleoclimatology*. New York: Oxford University Press.

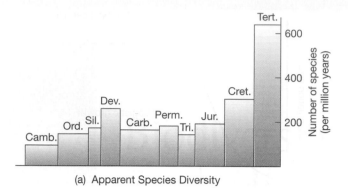

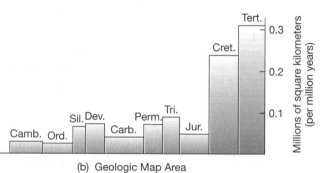

(a) Apparent Species Diversity

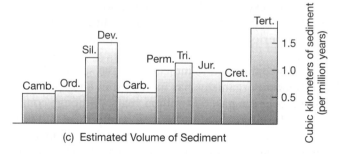

(b) Geologic Map Area

(c) Estimated Volume of Sediment

FIGURE 13-2

(a) The fossil record of species diversity of marine organisms, compared with the (b) outcrop area and (c) the volume of sedimentary rocks of the same age. All values are normalized to a million-year interval of Earth history. (After P.W. Signor. *Ann. Rev. Ecol. Syst.* 21:509–539, 1990.)

This chapter di geologic time. cient life) use r Shelly marine fossil record) a years ago. Thei speaking), reac dred million ye Earth experien Perhaps 95% o tinct. A subseq years ago, led t body of eviden cent mass extin of a 10-km-dia ronmental cons

ly, order, class, phylum (for animals and some protists) or *division* (for bacteria, fungi, plants, and some protists), *kingdom,* and *domain.* No other species in any of these other taxonomic levels need be found. Thus, it is unlikely that a higher taxonomic level will be unrepresented in the fossil record of a time when it did indeed exist. Paleontologists generally agree that the diversity record at the family level and above, for marine organisms with well-preserved skeletons and shells, is reasonably robust.

The most outstanding feature of the fossil record at higher taxonomic levels is the explosion in diversity of fossilizable organisms during the first few million to tens of millions of years of the Phanerozoic Eon (beginning about 545 million years ago), in the Cambrian and Ordovician Periods (see the geologic time scale in Chap-

ter 1). During this short interval, every phylum of fossil-producing marine invertebrates, and many of the classes and orders, appeared in the fossil record for the first time (Figure 13-3). Within 100 million years, the number of orders of marine organisms appears to have reached a rather constant value (Figure 13-4a) that has persisted for the subsequent 500 million years. Scientists debate the cause of this apparently sudden increase in diversity. The event followed upon the dramatic end of the "Snowball Earth" episodes of the Neoproterozoic, and may have taken place as the result of a significant increase in the amount of oxygen in the atmosphere and ocean. Higher oxygen levels would have enabled higher levels of aerobic metabolic activities and larger body plans. Other environmental factors may have promoted the evolution of hard parts. Recent studies show, however, that while the explosive diversification of hard parts and new body plans were early Cambrian phenomena, much of the evolutionary groundwork in terms of genetic diversification was occuring in the Neoproterozoic.

The shape of the time history of order-level diversity (Figure 13-4a) is reminiscent of the logistic growth curve described earlier in this chapter (Figure 13-1). By analogy, we can imagine that early in the Cambrian Period, as organisms evolved fossilizable hard parts, new species evolved rapidly as they developed distinctive ways to utilize their environment (see Figure 13-3). Common morphologies (related to form or function) arose that allowed them to be grouped into these higher taxonomic levels. As these ecological *niches* were filled and the opportunities for new lifestyles diminished, the rate of appearance of novel body shapes for marine organisms (as reflected in the number of new orders) diminished. The rest of the Phanerozoic had slower rates of origination of new orders, and this slowdown was very nearly balanced by the loss of orders (steady state was achieved).

The record of family-level diversity is somewhat richer (Figure 13-4b), with an early history comparable to that for orders, but with substantial perturbations later in the Phanerozoic. Of particular interest are the significant losses of families at the ends of the Ordovician (440 million years ago), Devonian (360 million years ago), Permian (251 million years ago), Triassic (206 million years ago), and Cretaceous (65 million years ago) Periods and the persistent increases in family diversity in the Cambrian–Ordovician boundary and the Mesozoic–Cenozoic transition. Extraordinary extinction events in which more than 25% of all extant families are lost are called **mass extinctions**. In Chapter 1 we learned that the end of the Cretaceous Period (the end of the Mesozoic Era) was marked by the extinction of the dinosaurs, along with many other species of organisms. This mass extinction pales, however, in comparison with that at the end of the Permian Period (the end of the Paleozoic Era), when approximately 95% of all species on Earth went extinct. The

USEFUL CONCEPTS
Taxonomy

Given the millions of species that exist on Earth today and the thousands of distinct fossils recognized in the geologic record, a system of organization is clearly needed. Hence the field of taxonomy (from the Greek *taxis*, meaning arrangement or order) was born. Two approaches are possible: one based on evolutionary, genetic relationships, and the other based on discernible physical similarities. The former is theoretically more satisfying, but the latter is usually more practical, especially for extinct organisms observable only as fossils. Species of extant (currently living) organisms are differentiated on the basis of their inability to interbreed. Higher levels of groupings (Box Figure 13-1) into genera (the plural of genus), families, orders, classes, phyla (the plural of phylum) or divisions, kingdoms, and domains are based on the degree of dissimilarity in all detectable characteristics between taxa (plural of taxon, or taxonomic group). For example, all families within an order should share some similarity that is distinct from all other orders.

Biochemists and evolutionary biologists are currently working feverishly to reveal the genetic sequences of a number of organisms, including both humans and bacteria. These investigators are finding that the established taxonomy based on morphology is in need of revision. Organisms that once seemed distant relatives are now surprisingly closely related on the "tree of life" (see Chapter 10). Moreover, there has been considerable transfer of genetic material across lineages, making the distinctions among groups a bit fuzzy. Finally, we have long known that the genetic make-up of the mitochondria of eukaryotes (including us) is distinct from the rest of the body's genetic sequences, probably indicating that eukaryotes evolved from an early symbiotic relationship among different bacteria (the "endosymbiosis" hypothesis of the origin of eukaryotes).

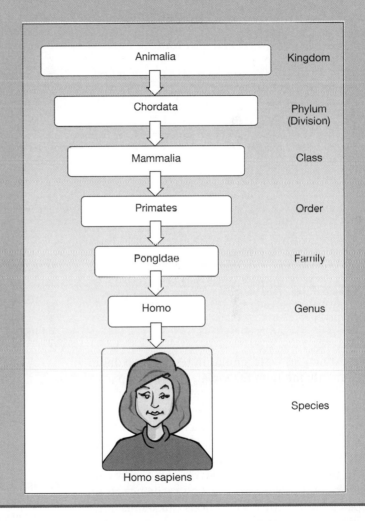

BOX FIGURE 13-1

A taxonomic tree for our species, *Homo sapiens.*

resiliency of higher taxa is demonstrated by their smaller losses: Only about 15% of the orders (Figure 13-4a) disappeared, as opposed to 50% of families (Figure 13-4b) and up to 85% of marine genera (Figure 13-4c). These mass extinctions seem to have been sudden, especially when compared with the recoveries, which required several million to tens of millions of years.

The general trends over Phanerozoic time indicate that although marine diversity at the family and order levels achieved steady state early on, the Permian mass extinction initiated an increase in family-level and genus-level diversity that has continued to the present. The persistent increase in biodiversity through the Mesozoic and Cenozoic Eras suggests that marine organisms have continued to find new ways of exploiting the environment. Some have developed new abilities as predators or as scavengers. Others have adapted to previously unfilled niches. In all cases they have evolved new biological characteristics that distinguish them from other groups of organisms, allowing their classification into new taxonomic groups. In this sense, not all extinctions are equivalent. For example, the Late Ordovician and Late

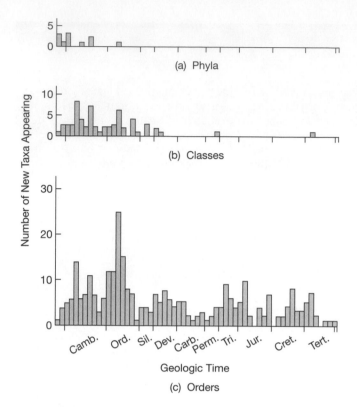

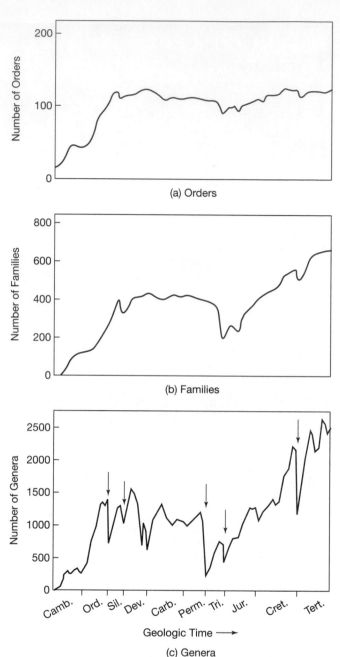

FIGURE 13-3

The number of new (a) phyla, (b) classes, and (c) orders of marine invertebrates that first appear in sedimentary rocks of a given age. The Cambrian and Ordovician were clearly a remarkable time of origination of a variety of new body shapes (which allow paleontologists to recognize the appearance of the various taxonomic groups). (After D.H. Erwin, J.W. Valentine, and J.J. Sepkoski, Jr. *Evolution* 41:1177–1186, 1987.)

FIGURE 13-4

The apparent diversity of fossilized marine invertebrates through geological time, grouped at the (a) order, (b) family, and (c) genus levels. Arrows show major mass extinctions. (After P.W. Signor. *Ann. Rev. Ecol. Syst.* 21:509–539, 1990.)

Devonian extinctions were similar in terms of taxonomic losses, but only the Late Devonian resulted in permanent restructuring of marine ecosystems.

The fossil record is remarkable in terms of the number of large losses of diversity the biota have suffered over the past 600 million years. The lack of fluctuation at high taxonomic levels and the rapid recovery from mass extinctions in the lower taxonomic levels exhibited by the fossil record point toward an amazing resiliency of the biota to perturbation. Mass extinctions cause substantial losses in diversity, but they also tend to stimulate origination rates as new organisms evolve and fill the ecological "job vacancies" resulting from the extinctions.

Although the response is geologically rapid, it takes up to tens of millions of years. The fossil record shows that if modern species diversity continues to decrease, it will be a long time before the system recovers.

The five major disruptions to the global biological diversity over the past 500 million years—the mass extinctions shown in Figure 13-4—share many similarities. All appear to be rather sudden, at least within the resolution of the fossil record. In all cases, at least 40% of genera

previously in existence became extinct. The effect at the species level was much larger. In the cases that have been well studied, there is an amazing lack of selectivity in the types of organisms that went extinct. In particular, older species (i.e., species that had been in existence for millions of years) seemed to suffer the same likelihood of extinction as did "young" taxa. In each case, the extinction event was followed by an interval of recovery in diversity.

Although the largest of all known mass extinctions occurred at the end of the Permian Period, 251 million years ago, the details of that extinction are just now becoming clear. In contrast, the Cretaceous–Tertiary (K–T) mass extinction (65 million years ago) has been the focus of a tremendous amount of research in the past few decades. It has also captured the attention of the general public, because it is the event at which the dinosaurs went extinct. For these reasons, our discussion centers on the K–T event.

The Cretaceous–Tertiary Mass Extinction

It has been estimated that 75% of all species went extinct in a very short interval of geologic time at the end of the Cretaceous Period. The species affected included both marine and land-based organisms. The most well known, of course, were the dinosaurs.

Two important questions arise concerning the K–T mass extinction: What caused it? and How did the Earth system respond to and recover from it? The first of these has been one of the most thoroughly studied questions in Earth science and has perhaps engendered more hypotheses than the origin of life itself. The second question is of particular importance to us because it returns us to our discussion of the resiliency of the Earth system to perturbations.

Possible Causes of the K–T Mass Extinction

As many as 20 hypotheses on the cause of the mass extinction at the end of the Cretaceous Period have appeared in the scientific literature. Add to that a similar number of wild speculations appearing in the tabloids (including invasions by extraterrestrial life forms), and you have the sort of scientific enigma that is certain to stimulate discussions in all forums for some time to come. Of these, only four survived the scrutiny of scientists working on the problem in the late 20th century: (1) sudden sea-level changes, (2) sharp temperature fluctuations, (3) volcanic eruptions, and (4) meteorite impacts. (We use the term *meteorite* liberally to mean a comet or asteroid that impacts Earth. Formal usage, however, would require us to refer to such objects as *meteoroids* until they strike Earth, at which point the materials that remain are referred to as meteorites.)

A central tenet of the scientific method is that hypotheses cannot be proved, only disproved. A hypothesis that continues to be consistent with observations may be elevated to the status of theory. Eventually, however, even well-established theories may be disproved as new observations are made that are inconsistent with theory. The debate about the causes of mass extinctions, especially that at the K–T boundary, provides a good example of how theories rise and fall from general acceptance within the scientific community.

Sea-Level Change and Climate Change. Until quite recently, changes in sea level or climate were the generally accepted theories for explaining mass extinctions including the K–T. Major drops in sea level are known to have contributed to extinctions of marine life by exposing vast regions of the continental shelves to the atmosphere, leading to the loss of habitat for shallow-marine organisms (which tend to dominate the fossil record). Climate excursions, especially glaciations or periods of extreme warmth, have also severely stressed marine communities, especially those in the tropics, which are especially temperature-sensitive, causing substantial losses of biodiversity.

However, some of the largest known sea-level drops were not associated with mass extinctions. It is also unclear how sea-level changes could have exterminated land-dwelling organisms, such as the dinosaurs. As for climate change, the best studied of the glaciations, those of the Pleistocene Epoch (see Chapter 14), were associated with only a modest level of extinction, despite rapid climate swings and sea-level fluctuations. Glacial sea-level fall seems to be the best explanation for the Late Ordovician mass extinction, though.

Meteorite Impact and Volcanic Eruptions. Astronomical explanations for mass extinctions have long existed but have suffered from lack of substantiation. One such hypothesis called for a nearby supernova, which would have destroyed Earth's ozone layer, leading to high levels of exposure to ultraviolet radiation. On similar, purely theoretical footing was the idea that a large meteorite impact caused the K–T extinction.

Anomalously high concentrations of **iridium**—an element that is rare at Earth's surface but is concentrated in Earth's interior and in extraterrestrial materials—were found in 1979 in unusual, fine-grained sediments deposited at the end of the Cretaceous (the K–T boundary clays of Gubbio, Italy). This iridium anomaly was the piece of evidence needed to establish meteorite impact as the currently accepted theory for the cause of the K–T extinction (see Chapter 1). However, a minority of scientists were still convinced that widespread volcanism was to blame. These scientists argued that volcanism would lead to many of the same environmental changes as a meteorite impact would have. We will examine this controversy as an example of how theories compete for acceptance by the scientific community.

Proponents of the meteorite-impact theory made a number of predictions, any of which, if invalidated, would disprove the theory. Luis Alvarez, one of the discoverers of the iridium enrichment layer, has enumerated predictions to test the theory and the ways in which previously made and new observations are consistent with the theory:

1. *An iridium enrichment should be found in K–T boundary sediments worldwide.* The K–T iridium anomaly has been documented in more than 75 localities around the globe in both marine- and land-based sediments. A 10-km-diameter meteorite is required to account for the amount of iridium deposited in all these regions.

2. *This enrichment should always be found within the same interval of geologic time.* All iridium layers found near the K–T occur in an interval defined by magnetic polarity reversals (see Chapter 7) known as *polarity chron C29R,* which has been precisely dated to span the interval from 65.6 to 64.9 million years ago. The K–T boundary is placed at 65.0 million years ago by these researchers.

3. *Large meteorites strike Earth sufficiently frequently to explain the extinction record.* The late Eugene Shoemaker of the U.S. Geological Survey (a codiscoverer of the Shoemaker–Levy comet that collided with Jupiter in 1994) demonstrated that a strong relationship exists between the frequency and size of impacts (Figure 13-5). The data he used span time and size scales ranging from observations of the rate of Space Shuttle pitting (one dust-sized impact every 30 μsec on average) to the frequency of large cratering events (every 10,000 years for 100-m-wide impactors). Recent satellite observations confirm the left side of the figure (objects smaller than 10 m in diameter), and near-Earth asteroid tracking studies confirm the diagram up to ~ 200 m diameter objects. According to this relationship, a meteorite 10 km in diameter should strike Earth on average every 100 million years. This is a rare event, but many such events have occurred since the origin of life. Five or six such events should have occurred since the beginning of the Phanerozoic. Impacts of the sort that caused the widespread devastation of Tunguska, Siberia, in 1908 occur about once every thousand years. In this event the bolide exploded in the atmosphere, flattening trees over hundreds of square kilometers (the area of a large city).

4. *On shorter time scales, such events should be rare.* In polarity chron C29R, there are no other iridium enrichments. In a study of a 34 million-year interval surrounding the K–T event, it is the only known anomaly. In fact, no iridium anomalies of comparable size have been detected anywhere else in the geologic record. According to prediction 3, several

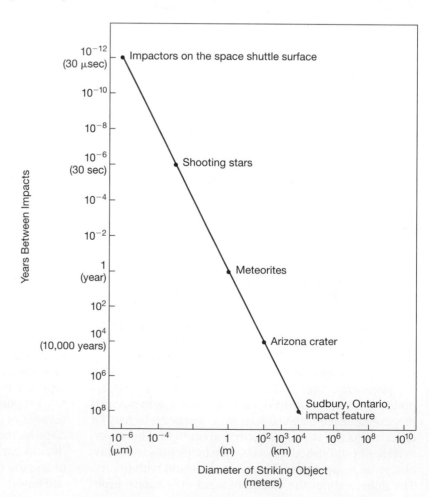

FIGURE 13-5

The Shoemaker curve: the frequency of collisions of extraterrestrial material with Earth, as a function of the size of the material. (After Luis W. Alvarez. *Phys. Today,* July 1987, 24–33.)

iridium layers should be lurking in rocks that have not been analyzed, unless most meteorites do not have large iridium abundances. (A comet, for example, would leave much less iridium for the same magnitude impact event because it is roughly half water ice—the iridium would be contained only in the rocky part—and because a typical comet impact would occur at much higher velocity than an asteroid impact. Comets originate from the outer solar system or beyond and often have orbits that are highly inclined to Earth's orbit, sometimes even in the opposite direction. A typical comet that hit Earth would do so at ~ 60 km/s, compared to ~ 20 km/s for a typical Earth-crossing asteroid. The kinetic energy of the impact is proportional to the square of the velocity: K.E. = ½ mv^2. Thus, for impacts of the same total kinetic energy, a comet would deliver only about 1/20th as much iridium as would an asteroid.)

5. *Plants as well as animals should have suffered as a result of the meteorite impact.* There is now clear evidence of significant turnover in the types of vegetation inhabiting the land surface, as recorded particularly well by spores and pollen (Figure 13-6).

6. *The gross chemical composition of the boundary clays should be identical worldwide, given that they all originated from the same excavated material.* The K–T boundary clays from Denmark and from the central Pacific are reported to be so similar in composition as to preclude diverse, local origins. The boundary clay is thought to contain debris from the impactor itself, along with a larger amount of material from Earth's crust and upper mantle that was excavated by the impact.

7. *The boundary clays should differ in composition from more typical clays deposited above and below the boundary clays at individual sites.* Compositional differences between the iridium-rich boundary clays and the surrounding sediments have been confirmed at a number of sites.

8. *Any chemical or isotopic signature in the boundary clay will have a significant extraterrestrial component.* The iridium anomaly is the best studied of these anomalies, although other geochemical and isotopic signatures are arguably extraterrestrial.

9. *The boundary clays should bear some evidence of the high temperatures generated during impact.* Small *spherules* of silicate minerals (Figure 13-7) have been found in the boundary clays. These spherules are thought to have formed when molten rock droplets, ejected into the atmosphere during the impact, cooled and solidified into glassy spheres before being redeposited at the Earth's surface.

10. *The boundary clays should bear some evidence of the high pressures generated during impact.* Small, fractured grains of *shocked quartz* are commonly found in boundary clays (see Figure 13-7). These grains form only when quartz is subjected to very high pressures—the sort we would expect from a meteorite impact. Under these conditions, the crystal structure of quartz is deformed in a characteristic fashion, so geologists can easily differentiate shocked quartz from quartz that is gradually subjected to the high pressures deep

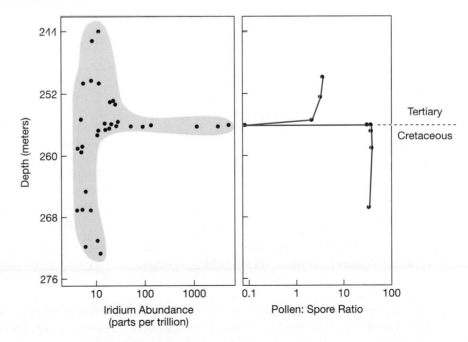

FIGURE 13-6

The Cretaceous–Tertiary iridium anomaly is coincident with a major change in the types of plants inhabiting the Raton Basin of New Mexico, as indicated by the fossil pollen-to-fern spore ratio. A diverse community of plants collapsed to one dominated by a few species of fern. Samples were drawn from a drill core at the depths indicated on the graph. (After Luis W. Alvarez. *Phys. Today*, July 1987, 24–33.)

(a)

(b)

FIGURE 13-7

[See color section] Shocked quartz and (b) microspherules from K–T boundary clays. (From Dr. David Kring/Science Photo Library, Photo Researchers, Inc.)

within Earth. Shocked quartz has also been found near known meteorite craters and in rocks surrounding the sites of underground nuclear explosions.

11. *The K–T event should have generated wildfires that might have left a sedimentary record of charred material.* Charcoal is indeed commonly found in K–T boundary clays. Calculations by Jay Melosh of the University of Arizona and colleagues provide a natural explanation for the apparent worldwide occurrence of wildfires during the K–T event. The impact excavated a huge amount of material in forming the crater, and this material was blasted in all directions away from the point of impact. Some of it was injected high into the atmosphere. The heating associ-

ated with reentry of ejected material through the atmosphere, estimated to be 50–100 times the solar flux, created unbearably hot conditions in cloud-free regions all over the world. Thus, a mechanism for rapid drying and ignition of vegetation on all continents would have existed. In cloudy areas, or beneath the sea or lakes, this energy would instead have gone into evaporating liquid water in cloud droplets.

12. *The iridium-rich layer should be just above the last dinosaur fossil.* At Gubbio, Italy, the iridium layer is less than a millimeter above the last occurrence of Cretaceous *foraminifera* (microscopic plankton; see Chapter 8). However, foraminifera are not dinosaurs. Dinosaur fossils are rare, and we are unlikely to find the last dinosaur fossil directly below the iridium layer (although, theoretically, if we keep looking we might find a charred dinosaur fossil buried in iridium-rich clay). Paleontologists have found articulated dinosaur fossils (fossils with joints intact) no closer than 2 m below the K–T boundary, as defined by the iridium layer. The low probability of preservation of fossils near boundaries can make an abrupt mass extinction appear to be gradual; this is called the *Signor-Lipps effect* and is well-known to paleontologists. Conversely, disarticulated dinosaur fossils have been found in sedimentary rocks well above the K–T boundary. This has generated some confusion and controversy, but these fossils likely have been eroded from older rocks and redeposited in younger sediments. No complete dinosaur skeleton has ever been found above the boundary.

13. *The pattern of extinction should show no evidence of preferential survival of species that were well adapted to the Cretaceous environment.* As mentioned before, mass extinctions in general, and the K–T one in particular, have been notably nonspecific. However, it does appear that large land animals were particularly hard hit by the K–T mass extinction; virtually all species with average weights greater than 25 kg (55 lb) went extinct. In contrast, mammals, which at the time were lightweights, fared much better. Perhaps smaller animals fared better simply because their population sizes were larger: Statistically speaking, some would survive the event. Or perhaps they were better able to avoid the immediate effects of the meteorite impact. After the extinction, mammals evolved, filling many of the niches left vacant by the demise of the dinosaurs.

The confirmation of the Alvarez predictions supports the meteorite-impact theory for the K–T extinction, but it does not prove it. The volcanic theory of mass extinction shares many features of the meteorite-impact theory. Large increases in volcanic activity could, in the short term, increase the amount of sulfuric acid aerosol in the stratosphere, cooling the climate. Indeed, several such volcanic cooling events have been identified in the climate record of the past few centuries (see Chapter 2). However, only explosive volcanoes like Mt. Pinatubo or El Chichon inject aerosols into the stratosphere. Large volcanic eruptions did occur during the Late Cretaceous; they formed great layers of basalt in India that are known as the Deccan Traps. If these flows resembled modern flood basalts, though, they were gentle outpourings of basalt that did not produce much aerosol. The settling ash from volcanic eruptions could create a worldwide clay layer similar to the K–T boundary clay and might also produce spherules and shocked quartz. In the long term, carbon dioxide released by volcanism would lead to global warming. These climatic fluctuations would present environmental stresses to which many groups of organisms could not adapt. However, according to Alvarez, the volcanic theory is inconsistent with three observations:

1. Sand-sized spherules, even if ejected by volcanoes, would not reach ballistic orbits (as do impact ejecta) or be distributed globally, as the K–T spherules are observed to be.

2. Shocked quartz of the sort found at the K–T boundary has never been found in deposits of volcanic origin but is common in deposits associated with known impact craters.

3. Volcanic ejecta tend to have very low iridium concentrations.

Who is right? The arguments are complicated, but the evidence strongly favors the impact hypothesis. Indeed, since the Alvarez team performed their study, additional evidence has emerged that provides further support for the impact hypothesis. One such new piece of evidence is the discovery of fullerenes or "buckeyballs" in K–T boundary sediments. *Fullerenes* are large cage-like molecules containing 60 or more carbon atoms arranged in a sphere. (They derive their name from architect Buckminster Fuller, who designed the first geodesic dome. Fullerenes resemble a tiny version of his visionary creation.) Fullerenes are formed in all sorts of environments where carbon is burned, including ordinary candle flames. However, the fullerenes found at the K-T boundary contain helium with isotopic ratios that are clearly extraterrestrial. (Recall that most of Earth's helium, ^4He, is produced from decay of uranium and thorium. By contrast, the K-T fullerenes contain mostly ^3He. This suggests that they were formed in the expanding envelopes of dying, carbon-rich stars.)

This argument, like most of the original arguments presented by the Alvarez team, is somewhat technical. It is convincing to most experts in the field, but is hard to explain to a nonscientist. As we will see, however, the evidence for the impact hypothesis has become quite down-to-Earth.

The Smoking Gun: The Chicxulub Crater

Another prediction from the meteorite-impact theory is that somewhere on Earth a very large crater should exist, representing the site of impact of the 10-km meteorite. Until very recently, scientists had explained their inability to locate such a crater by pointing out that it was probably excavated in oceanic crust, given that three-quarters of Earth's surface is ocean-covered, and thus might have already been subducted into the mantle by plate tectonics (see Chapter 7). Even if the impact had occurred in continental crust, they argued, the crater is likely to have been eroded. Nevertheless, the search continued. Several putative K–T craters were discovered but found to be of a different age or of insufficient size to explain the observed iridium anomaly.

Then, in the early 1990s, geologists discovered convincing evidence that a structure buried 1 km below the surface, located near the town of Chicxulub (sheek'-soo-loob), Mexico (Figure 13-8), was indeed of K–T age. Previous investigators had suspected, on the basis of remotely sensed geophysical anomalies, that this structure was an impact crater. The only expression at the surface of this subsurface feature was a ring of sinkholes, known in Mexico as cenotes. Then, exploratory wells previously drilled into the Chicxulub structure by Pemex, the Mexican oil company, provided samples containing shocked quartz and glass microspherules that closely resembled samples from known K–T intervals elsewhere. Further analysis of the drilled core materials revealed very strong enrichments of iridium. Isotopic age dating and paleomagnetic evidence constrained the age of the materials to polarity chron C29R, the interval containing all K–T iridium anomalies. Current estimates of at least a 200-km diameter for the crater indicate that it is the largest on Earth and one of the largest in the solar system.

Environmental Consequences of the K–T Meteorite Impact

What was Earth like during the late Cretaceous Period when the meteorite hit? How did the global environment change after the impact? If we could transport ourselves back 65 million years to the day of the impact, we would find ourselves in a world quite different from that of today. Equator-to-pole gradients in temperature were less, and the poles were essentially ice-free, most likely the result of high atmospheric CO_2 levels. Dinosaurs inhabited tropical to near-polar latitudes—not in the abundances of their peak some tens of millions of years before,

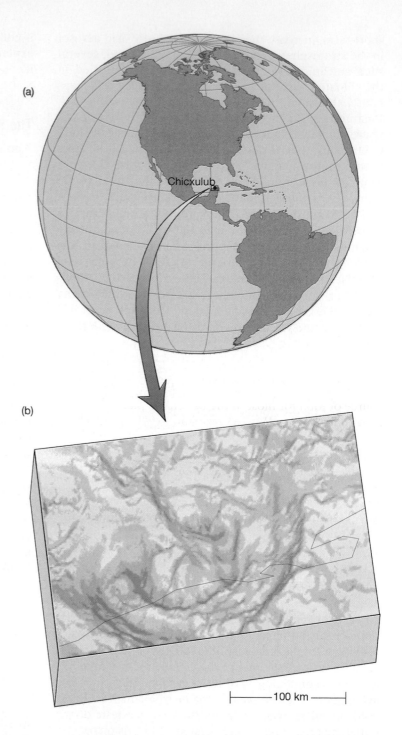

FIGURE 13-8

(a) The location of the Chicxulub impact crater. (b) The shape of the crater as inferred from gravity anomalies. (Courtesy of V.L. Sharpton, Lunar and Planetary Institute.)

but still in sufficient numbers to dominate higher levels of the food chain.

The impact would have been unexpected and sudden. The passage of the meteorite through the atmosphere would have converted nitrogen gas (N_2) to nitric oxide (NO), as does lightning on a smaller scale today (see Chapter 11). Produced in large quantities, the NO would have destroyed the stratospheric ozone layer by catalyzing reactions that consume ozone, allowing ultraviolet radiation to reach Earth's surface unimpeded for

as long as several years. Organisms unfortunate enough to be on the Yucatan Peninsula would have instantly been annihilated as an explosion of unimaginable intensity sent shock waves across the landscape. A huge crater would have been carved out immediately. The excavated material would have been sent on ballistic trajectories, enveloping Earth in a blanket of dust and debris. As the impact ejecta passed back through the atmosphere, the material would have been heated to temperatures sufficient to ignite wildfires in the affected areas, and those organisms not protected beneath the water surface or underneath clouds would have been subjected to air temperatures likened to putting your head in the oven on "broil." Soot would have been introduced into the atmosphere, further reducing the amount of sunlight reaching Earth. Because the bedrock of the Chicxulub area (mostly limestone) contained a thick layer of the calcium sulfate mineral *anhydrite,* a huge quantity of sulfuric acid aerosol would have been injected into the stratosphere. The scattering of sunlight from this aerosol layer would have reduced the amount of incoming solar radiation even further. Presuming a coastal impact, huge *tsunamis* (tidal waves) would have been initiated, wreaking havoc on Caribbean coastal environments. Evidence of tsunamis has been found on the island of Haiti and elsewhere in the Caribbean area.

In the next weeks, an unrelenting series of environmental insults would have ensued that few organisms could survive. The amount of solar radiation reaching Earth's surface would have been drastically reduced because of the presence of dense debris clouds and aerosol layers. The resultant dramatic cooling at all latitudes and cessation of photosynthesis would have removed the source of food for most of the life on the planet, both on land and in the sea. Where precipitation fell, it would have been in the form of rain and snow of sulfuric and nitric acids, produced from the rainout of sulfate aerosols and oxidation of NO. The acid rain may have been the least of the dinosaurs' problems, though.

Over the next several months, these tremendously detrimental effects would have lessened. The dust, soot, and much of the sulfate aerosol would have rained out of the atmosphere, and the acidity of the precipitation would have been reduced. However, very little life would have remained. The organisms that had survived the direct effects of the impact would have been subjected to a scarcity of food and suitable habitat. Cool climates would have continued for more than a year because of the approximately 6-month residence time of sulfate aerosol in the stratosphere.

Environmental recovery would have continued over the next few years, as the ozone layer was restored and the climate warmed. However, the warming would have continued well beyond the initial state as the result of thousands of gigatons of carbon dioxide released from the vaporization of limestone at the impact site. Atmospheric pCO_2 may have increased to a few thousand ppm, and global average temperatures could have risen 10–15°C. These warm conditions could have persisted for thousands to tens of thousands of years. Given that many organisms are intolerant of rapid temperature excursions, this warming (following the initial cooling) might have assisted in the mass extinction and been at least one of the factors delaying the reestablishment of an active

oceanic *biological pump* (see Chapter 9 and the Box "The K–T Strangelove Ocean"), for hundreds of thousands of years. Other factors that influence the recovery interval include persistently high levels of toxic metals brought to the ocean with the impactor and the inherent time it takes to reestablish ecosystems and biodiversity.

Extraterrestrial Influences and Extinction

There are two varieties of impacts: those caused by comets and those caused by asteroids. **Comets** are essentially large, dirty snowballs. They are composed of dust (metallic and rocky material) and solidified compounds that exist on Earth as gases: water, ammonia, methane, and carbon dioxide. Comets exist in stable orbits around the Sun, well beyond Pluto, in a region known as the **Oort cloud** (Figure 13-9), and closer in, outside of Neptune in the *Kuiper Belt.* The Oort cloud is a spherical reservoir of comets that extends more than one light year from the Sun. Comets in the Oort cloud can be perturbed by passing stars into orbits that pass through the solar system. Because they are arranged in a sphere, they can come in from any direction. These comets include many that have been observed over historic time, such as Halley's comet and comet Hale–Bopp. Kuiper Belt comets, by contrast, orbit the Sun in the same direction as do the planets. They are essentially pieces of material that never accreted to form a planet because they formed too far from the Sun and they did not collide frequently enough. Paradoxically, the Oort cloud comets are actually thought to have formed closer to the Sun, mostly in the Uranus/Neptune region, from which they were ejected early in solar system history by near-collisions with the giant planets.

Asteroids, by contrast, are composed of minerals and metallic elements characteristic of Earth and the other inner solar system planets (Mercury, Venus, and Mars). Most asteroids are in orbit around the Sun in a region known as the **asteroid belt,** between Mars and Jupiter (see Figure 13-9). The asteroid belt represents the remains of an inner planet that failed to form, probably because of the gravitational influence of nearby Jupiter. The largest asteroid, Ceres, is 1000 km in diameter, and two others, Vesta and Pallas, are both ~500 km in diameter. Were any of these three bodies to hit the Earth, they would likely vaporize the entire ocean and might sterilize Earth completely. Fortunately, most asteroids are much smaller than this. Many thousands of kilometer-sized bodies exist, and collisions between these bodies create millions of small, dust-sized fragments.

As mentioned earlier, iridium is an indicator of an asteroid impact because it is associated with the mineral

A CLOSER LOOK
The K–T Strangelove Ocean

Scientists studying the isotopic composition of carbon in skeletons of foraminifera deposited in sediments that span the K–T impact have found an intriguing perturbation of the marine ecosystem that persisted for hundreds of thousands of years after the K–T event. Normally, the operation of the oceanic biological pump (see Chapter 8) causes an enrichment in the carbon isotope ^{13}C in surface waters and an enrichment in ^{12}C in deep waters. The ^{12}C is preferentially incorporated into algal tissue during photosynthesis. It is then added to deep waters as the organic carbon in that tissue settles to the deep ocean and is decomposed by aerobic processes to dissolved CO_2. As a result, the $^{13}C:^{12}C$ ratio is enhanced in surface waters and diminished in deep waters. Foraminifera and other calcareous organisms (those producing $CaCO_3$ skeletons) record the ratio of ^{13}C to ^{12}C in their skeletons. Planktonic foraminifera, which live near the surface, record surface-water ratios, whereas benthic foraminifera, which live on the sea floor, record deep-sea ratios. In the modern ocean and in sediments recovered from layers deposited prior to the K–T event, the measured difference in the $^{13}C:^{12}C$ ratio between planktonic and benthic foraminifera is about 2 per mil (parts per thousand). However, as shown in Box Figure 13-2, the K–T event is marked by a collapse of this gradient. The implication is that the biological pump ceased to exist and that the ocean was essentially lifeless. Scientists have dubbed this the Strangelove Ocean, an allusion to the classic 1964 movie about nuclear war, *Dr. Strangelove.*

Perhaps even more intriguing than the collapse of the biological pump itself is its persistence. According to the isotopic record, the export of organic carbon from the surface ocean persisted at very low levels for hundreds of thousands of years, despite rapid rates of origination of new species and reestablishment of ecosystems in other settings. Although it is conceivable that slow rates of evolution retarded the reestablishment of the biological pump, other explanations seem to be required. One possibility is that toxic levels of trace metals such as copper, cadmium, and zinc resulted from the dissolution of the impact ejecta. Only after hundreds of thousands of years were these metals reduced to sufficiently low concentrations that metal-intolerant organisms could become reestablished.

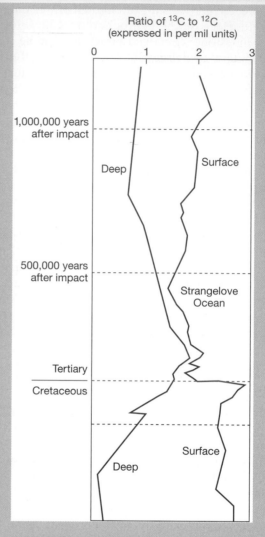

BOX FIGURE 13-2

Changes in the ratio of carbon isotopic composition between the surface ocean and deep ocean as a result of the Cretaceous–Tertiary mass extinction. The ocean's isotopic value is recorded in the skeletons of planktonic foraminifera, which represent surface waters, and benthic foraminifera, which represent deep waters. (After Zachos, J. C., Arthur, M. A., and Dean, W. E., *Nature* 337:61–64, 1989.)

and metallic material in extraterrestrial materials. Earth has lots of iridium as well; however, nearly all of it is in the core because iridium is a **siderophile** element that dissolves readily in molten iron. (The term "siderophile" means "iron-loving.") As we have seen, an asteroid impact should produce a significantly higher concentration of iridium than would a comet impact because of the differences in their composition and in their expected impact velocities. The large iridium spike in the K–T

boundary clay thus indicates that the impactor was most likely an asteroid. Some of the other Phanerozoic mass extinctions for which no Ir layer exists—which, if you recall, is all the rest of them—could conceivably have been caused by comets.

Most asteroids in the asteroid belt and comets in the Oort Cloud or Kuiper Belt pose no immediate threat of impact. Their orbits can be disturbed, however, sending them on paths through the inner solar system. For exam-

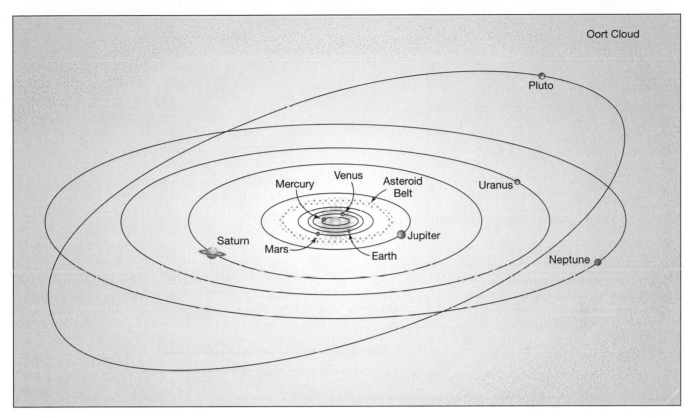

FIGURE 13-9

The solar system, showing the location of the Oort cloud (not drawn to scale) and the asteroid belt. (From T. McKnight. *Physical Geography: A Landscape Appreciation*, 6/e, 1999. Reprinted by permission of Prentice Hall, Upper Saddle River, N.J.)

ple, collisions within the asteroid belt occasionally cause asteroids to be deflected into a 3:1 *resonance* with Jupiter. At this distance the orbital period of the asteroid is precisely one-third that of Jupiter (3.95 Earth years, compared to 11.86 Earth years for Jupiter). This resonance occurs at an orbital distance of 2.50 astronomical units from the Sun, that is, 2.50 times the mean Earth–Sun distance. (In the next chapter, we show how to derive this distance using Kepler's third law.) Asteroids that find themselves within the 3:1 resonance do funny things. Because they always pass by Jupiter on the same side of the Sun, their orbits become elongated by the tug of Jupiter's gravity. Eventually, the orbits become **chaotic**. To a mathematician or celestial mechanician, this means that small changes in their positions at some initial time can lead to large changes in their positions at some later time. In practice, some of these asteroids get shunted into orbits that pass through the inner solar system and directly across the path of planet Earth. Some of these bodies, like the K–T impactor, end up hitting the Earth and causing mass extinctions.

Comets get deflected from their normally stable orbits by other means. Oort cloud comets can be perturbed by passing stars or by shifting tidal forces caused by the rotation of the galaxy. Comets on the inner edge of the

Kuiper Belt get "nibbled" away by Neptune's gravity, causing some of them to begin orbiting within the normal boundaries of the solar system. Such comets eventually pass near to one of the giant planets. This provides a gravitational "slingshot" effect that ejects most of them right out of the solar system. However, a few get kicked into Earth-crossing orbits, and a few of those wind up eventually hitting Earth. Sometimes, of course, the giant planets get hit as well, as happened in July 1994 when the comet Shoemaker–Levy collided with Jupiter.

Periodicity of Impacts and Extinctions?

Examination of the Phanerozoic history of global biodiversity reveals that mass extinctions have occurred about once every 100 million years (see Figure 13-4). In fact, a more detailed analysis of the fossil record of the past 250 million years has revealed an apparent 26 million-year **periodicity**—a time interval of regular recurrence—in the global rate of extinction (Figure 13-10). This regularity suggests an extraterrestrial cause for mass extinctions but requires another theory to explain why large asteroid or cometary impacts should occur every 26 million years.

At least three hypotheses have been proposed to explain the periodicity in the fossil data of extinction rates

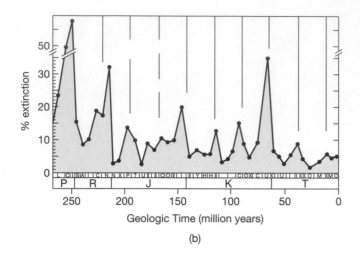

(b)

FIGURE 13-10

The fossil record of extinction rate, shown as the percentage of existing genera that went extinct in a particular interval of geologic time. The vertical line indicates the 26 million-year periodicity of extinction. (After J.J. Sepkoski, Jr. Special Paper of the Geological Society of America, 247:33–44.)

(Figure 13-11). Each hypothesis calls on gravitational disturbances to send comets or asteroids on Earth-crossing orbits, but the three differ in the causative factors. (1) According to the *galactic plane hypothesis* (Figure 13-11a), the oscillation of our solar system back and forth through the dense central plane of the spiral arm of the Milky Way galaxy leads to a periodically high likelihood of interactions with interstellar clouds of gas and dust, which could disturb the orbits of asteroids and comets in the Oort cloud, sending some on Earth-crossing orbits. (2) The *companion-star hypothesis* (Figure 13-11b) proposes that the Sun has a much smaller companion star, dubbed *Nemesis* by scientists who proposed the idea. In order not to have been seen already, Nemesis would have to be a *brown dwarf* (< 0.08 solar masses), too small to ignite nuclear fusion in its core. Nemesis is presumed to be on a highly elliptical orbit that passes through the Oort cloud every 26 million years. It is now thought to be 2 light years from Earth. (3) The *Planet X hypothesis* (Figure 13-11c) calls for a disturbance of the *Kuiper Belt* by a hypothetical tenth planet that orbits beyond Pluto and is known as Planet X.

None of these hypotheses has been disproved, but the evidence in support of any of them is scant. The existence of companion stars or a tenth planet must first be proven before we can argue that their presence is likely to lead to meteorite impacts. These objects are small, and thus faint; they could easily have escaped detection so far, or Nemesis could have been confused with more distant objects assumed to be brighter. Astronomers can distinguish close dim stars from distant bright stars using measurements of motion and parallax, but only a fraction of stars have been studied systematically in that way. Hypotheses concerning the cause of periodicity in the fossil record of extinction are likely to remain speculative for some time.

If none of these extraterrestrial mechanisms for producing periodic extinctions is found plausible, this implies that the observed periodicity in extinctions is itself an artifact of the geologic record. One possible reason for a 26 million-year periodicity: this time scale is approximately four times the length of an average geologic *stage*. (Recall that a stage is a subunit of a geologic period.) It may be that paleontologists created an apparent periodicity in extinctions simply by defining stages that spanned roughly the same amounts of time.

Future Impacts

What are the odds that a large impactor will hit Earth in our lifetime? They are higher than you might think. The direct correlation between impactor size and diameter (see Fig. 13-5) allows us to make this determination with some confidence. Let us assume that humans live about 100 years (admittedly, a bit optimistic). Because 100-m-diameter impactors (the size of the one that created Meteor Crater in Arizona) strike Earth about every 10,000 years, the likelihood of one striking Earth during your lifetime is about (100 years)/(10,000 years), or 1 in 100. The environmental consequences of such an impact are not tremendous, however, except close to the impact site. The impact frequency decreases by a factor of 100 for every factor of 10 increase in impactor diameter. Thus, the probability of a 1-km meteorite impact during your lifetime is about 1 in 10,000, and the probability of a K–T-size impact during that time is 1 in 1 million. By comparison, your chances of being struck by lightning are about 1 in 3000. Your chances of being killed by a lightning strike are quite high compared with your chances of being killed by the direct effects of a 1-km meteorite. But the likelihood that civilization will be destroyed by a meteorite impact is much larger than that of destruction by lightning strikes.

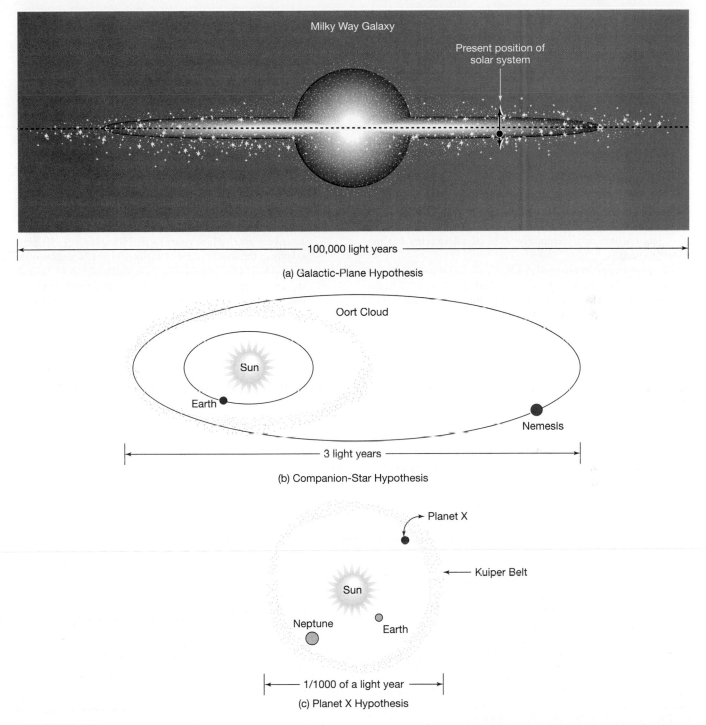

FIGURE 13-11

Three astronomical hypotheses explaining the 26 million-year periodicity in the fossil record of extinction. Figures not drawn to scale. (After D.M. Raup. 1986. See Further Reading.)

Chapter Summary

1. The diversity of life on Earth has oscillated in response to imbalances between the origination of new species and the extinction of existing species.
2. The fossil record of species-level diversity is biased by preservational artifacts associated with the likelihood of discovering fossils from a given interval of time.
3. The fossil record of family-level diversity is less biased by these artifacts and can thus be interpreted in terms of true trends in biodiversity through Earth history.
   a. Biodiversity of marine invertebrates increased dramatically about 544 million years ago, reaching a high level that persisted for 300 million years.
   b. The largest extinction known occurred 251 million years ago when approximately 95% of all species became extinct.
   c. Biodiversity recovered gradually after this event, and the general trend continues to the present. A significant disruption to this trend occurred 65 million years ago, at the end of the Cretaceous Period.
4. The Cretaceous–Tertiary mass extinction is the best studied of all major mass extinctions. Current scientific opinion favors the impact of a 10-km-diameter meteorite as the primary cause of this extinction.
   a. All available geological data are consistent with this hypothesis.
   b. A large crater dated as 65 million years old has been found in the subsurface at Chicxulub, Mexico.
   c. The meteorite impact would have destroyed the ozone layer, blocked out sunlight for days to weeks, ignited wildfires worldwide, baked organisms exposed to red-hot reentering ejecta, bathed the land surface in acid rain, and caused substantial global warming from elevated atmospheric CO_2.
   d. Large-meteorite impacts are predicted to occur on average every 100 million years or so. However, regular periodicity in the fossil record of extinction at a 26 million-year period might be caused by extraterrestrial forces, including close approaches of an undiscovered tenth planet, a companion star to the Oort cloud of comets, or crossings of our solar system through the galactic plane.

Key Terms

adaptation
asteroids
asteroid belt
biodiversity
chaotic orbit
comets

evolution
extinction
iridium
logistic growth
mass extinction
meteorite

natural selection
Oort cloud
periodicity
siderophile
taxon
taxonomy

Review Questions

1. What two processes cause the diversity of life on Earth to change through time?
2. How do paleontologists deal with the incomplete nature of the fossil record to establish a geologic history of biodiversity changes?
3. Why do diversity trends differ among taxonomic levels?
4. What explanations have been proposed for the mass extinction at the K–T boundary? Why is a meteorite impact the favored theory today?
5. What are the environmental consequences of the impact of a 10-km-diameter meteorite with Earth?
6. What are some hypotheses for the apparent 26 million-year regularity of the fossil record of mass extinctions?

Critical-Thinking Problems

1. Here you will calculate the change in a population of organisms (N) as a result of births and deaths. A mathematical expression can be used to calculate next year's population (N_{t+1}) on the basis of this year's population (N_t), where t = years

$$N_{t+1} = N_t + rN_t\left(1 - \frac{N_t}{K}\right)$$

Let us dissect this equation. The potential growth rate—that is, the birth rate minus the death rate—is r, and the carrying capacity—the population size that can be supported (given constraints of food or other resource availability, competition, and so on)—is K. If the population is small relative to K, then the term in parentheses is essentially 1, and the potential growth rate is achieved. In other words, next year's population would simply be some multiple of this year's population (exponential growth). As *N* approaches *K*, the population will tend to slow its growth, finally reaching the carrying capacity. This behavior is called *logistic growth*. You will also witness something bizarre: behavior that has been labeled chaos. PART 1 (Logistic growth): Fill in the following table, and then graph N_t versus t; assume that $r = 1.0$ and $K = 1000$.

TABLE 13-1

Time (years)	N_t	$1 - \dfrac{N_t}{K}$	$rN_t \times \left(1 - \dfrac{N_t}{K}\right)$	N_{t+1} (use as N_t next time)
1	20.0	0.98	19.6	39.6
2	39.6	0.96	38.0	77.6
3	77.6	0.92	71.6	149.2
4				
5				
6				
7				
8				
9				
10				

Describe the growth curve, and explain why it has the logistic growth shape (on the basis of the numbers you calculate).

PART 2 (Chaos): Now repeat your calculations (do 15 years' worth) for the following values of *r*:

$$r = 2.0$$
$$r = 2.8$$

Graph your results, either on separate graphs or using different symbols or colors on the same graph (be sure the graph(s) is(are) legible). If the population goes negative, call it quits on that series of calculations; the population has gone extinct.

Describe how the behavior changes as the growth rate increases from 1.0 to 2.0 to 2.8. When $r = 2.8$, the system is described as being chaotic. A scientist who observed this population might conclude that purely random factors are controlling the size of this population. What is wrong with this conclusion?

2. Perform the calculation that Luis Alvarez used to establish the size of the K–T impactor. Use the following information:
   a. Assume that the iridium layer was uniformly distributed around Earth by the impact.
   b. On average, the layer had a concentration of iridium of 10 parts per billion (10 ppb) by weight.
   c. On average, the layer was 4 cm thick.
   d. The density of the layer was 2.5 g/cm³.
   e. Assume the meteor was spherical, with a density of 6.0 g/cm³, and an iridium content of 0.5 parts per million by weight (0.5 ppm).
   f. The radius of Earth is 6378 km.
   What is the diameter of the meteorite?

3. Determine the probability that an asteroid of the following diameters will hit Earth during your (optimistic) 100-year lifetime: 1 m, 100 m, 10 km.

Further Reading

General

Erwin, D. H. 1993. *The Great Paleozoic Crisis, Life and Death in the Permian*. New York: Columbia University Press, 327 p.

Glen, W. 1994. *The Mass-Extinction Debates: How Science Works in a Crisis*. Stanford, Calif.: Stanford University Press, 370 p.

Raup, D. M. 1986. *The Nemesis Affair, a Story of the Death of Dinosaurs and the Ways of Science*. New York: Norton, 220 p.

Advanced

Alvarez, L. W. 1987. Mass Extinctions Caused by Large Bolide Impacts. *Physics Today*, July 1987, pp. 24–33.

Courtillot, V. 1999. *Evolutionary Catastrophes: The Science of Mass Extinction*. New York: Cambridge University Press, 173 p.

Hallam, A. and Wignall, P. B. 1997. *Mass Extinctions and their Aftermath*. Oxford: Oxford University Press, 320 p.

Ryder, G., Fastovsky, D., and Gartner, S. 1996. *The Cretaceous–Tertiary Event and Other Catastrophes in Earth History*. Boulder, Colo.: Geological Society of America, 569 p.

14

Pleistocene Glaciations

Key Questions

- What has caused Earth to oscillate in and out of glacial states over the past 2 million years?
- What geologic evidence substantiates theories about the causes of glaciation?
- How are glacial-to-interglacial cycles related to changes in Earth's orbit?
- Have components of the Earth system amplified the glacial climate response?

Chapter Overview

When viewed on multimillion-year time scales, Earth is presently in a glacial interval. Thick continental ice sheets cover Antarctica and Greenland, and only 20,000 years ago vast portions of northern North America and Scandinavia and other parts of northern Europe and Asia were covered with accumulations of ice that were many kilometers thick. Although polar ice has existed on Antarctica for tens of millions of years, only during the past 2.5 million years have ice sheets extended from the Arctic into the northern mid-latitudes. Thus, there is something peculiar about the operation of Earth's climate system during the past 2.5 million years that distinguishes it from the rest of the past 300 million years of Earth history.

In addition, there is now convincing evidence that this glacial interval, known as the **Pleistocene Epoch** (1.8 million years ago to 10,000 years ago), was characterized by regular cycles of growth and decay of Northern Hemisphere continental ice sheets. The oscillations are likely to continue. The warmer interval we are enjoying now, known as the **Holocene Epoch,** would under natural conditions end several thousand years from now as the Northern Hemisphere ice sheets return. Continued burning of fossil fuels may, however, forestall or even prevent this transition to the glacial state. What causes this cyclicity of glaciation?

As we will see, changes both in the distribution of sunlight across Earth's surface (insolation) and in atmospheric CO_2 seem to be involved. These changes in insolation are a predictable feature of Earth's orbit about the Sun—during warm as well as cold intervals of Earth history. They were not discussed in Chapter 12 because the magnitude of the forcing is small compared with that of the changes brought about by solar evolution on 100 million-year time scales. Yet as we narrow our focus to shorter time scales of climate change, these more subtle forcings increase in importance. During the Mesozoic, orbital changes caused cyclical variations in sedimentation that likely reflect climate change. But the Pleistocene climate system seems to have been especially attuned to these forcings, as indicated by the large oscillations of continental ice sheets.

Geologic Evidence of Pleistocene Glaciation

The scenic beauty of Canada and the northern portions of the United States, Asia, and Europe is to a great extent the result of the action of tremendously large sheets of ice that covered these regions during the Pleistocene Epoch (Figure 14-1). Buried under more than a kilometer of ice, the land surface was plucked, ground up, and excavated. With the melting of the ice, the depressions formed in this way were filled with water and became the thousands of lakes that pepper the North country. In this section we will see that the modification of the landscape by glaciers, together with their effect on the isotopic composition of seawater, provide convincing evidence that continental ice sheets expanded and decayed several times during the Pleistocene with almost clockwork regularity.

Glacial Deposits Document Major Glaciations

As we saw in Chapter 12, several types of geological features are characteristic of glaciation. As glaciers advance, rocks frozen to their base gouge the bedrock below. These gauges, called **glacial striations** (Figure 14-2a), indicate the direction of past glacial movement. The advance of continental ice sheets modifies the landscape in other ways as well. Ridges of sediment, known as **moraines** (Figure 14-2b), are deposited at the front and sides of the ice sheets. When the ice melts, these ridges are left behind, marking the farthest advance of the ice sheet. The indiscriminate nature of glacial erosion and deposition is reflected in the sediment contained in moraines and other glacial deposits. This sediment, called **till** (Figure 14-2c), contains a mixture of material of various sizes, from mud to boulder-sized rocks, and various composition, reflecting rocks eroded from a variety of places and transported to the point of deposition by the advancing ice sheet. Some of the material produced by glacial abrasion is silt-sized (about hundredths of a millimeter in diameter). When deposited in the typically arid region that surrounds continental ice sheets, this material is picked up by the wind and carried great distances. Such windblown deposits, known as **loess** (Figure 14-2d), form the rich soils of the midwestern U.S. grain belt.

Geologists of the 19th century used glacial deposits, including the geographical distribution of moraines, to define four main intervals of glaciation in Europe, from oldest to youngest: the Gunz, the Mindel, the Riss, and the Würm. In the early part of the 20th century, four glaciations were also identified in North America, from oldest to youngest: the Nebraskan, the Kansan, the Illinoian, and the Wisconsin. These glaciations on separate continents are now known to have been coincident with each other, and they represent the largest of the glacial

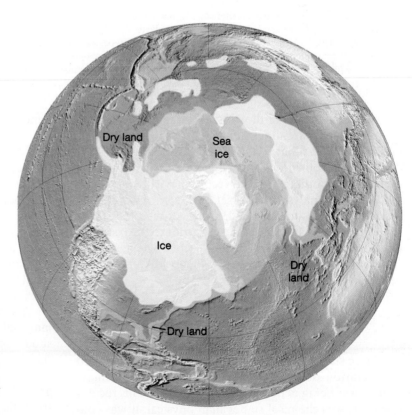

FIGURE 14-1

The Pleistocene ice sheet at maximum extent. (From W.K. Hamblin and E.H. Christiansen. *Earth's Dynamic Systems*, 8/e, 1998. Reprinted by permission of Prentice Hall, Upper Saddle River, N.J.)

FIGURE 14-2

Various geological features characteristic of glaciation. (a) Glacial striations are formed by the gouging out of bedrock by pebbles frozen onto the base of an advancing glacier. The lineations indicate the direction of glacial movement. (b) Moraines form by the bulldozing action of advancing glaciers. (c) Till is the mix of sediments of various types and sizes resulting from the indiscriminate transportation by glaciers. (d) Loess is fine-grained sediment produced by glacial abrasion and transported to the site of deposition by wind. (Part b from T. McKnight. *Physical Geography: A Landscape Appreciation*, 6/e, 1999. Reprinted by permission of Prentice Hall, Upper Saddle River, N.J.) (Part a, c, and d from T. McKnight. *Physical Geography: A Landscape Appreciation*, 5/e, 1996. Reprinted by permission of Prentice Hall, Upper Saddle River, N.J.)

episodes. Ice sheets in the Northern Hemisphere were advancing and retreating in unison. But the number four greatly underestimates how often this process has occurred. The problem is that each successive advance obliterates much of the geological record of earlier advances.

Detailed analysis of layers upon layers of windblown loess, preserved throughout Europe, proved that there had been at least 17 glacial intervals prior to the last. And evidence from marine sediments, described next, showed that such cycles have occurred with clocklike regularity for more than 2.5 million years.

The Oxygen Isotope Record of Glacial–Interglacial Oscillations

While ice sheets were destroying the sedimentary record of glaciation on land, a continuous record of climate change was being deposited on the sea floor. This record, however, was chemical rather than physical; it was con-

tained in the isotopic composition of the skeletons of marine organisms. The ratio of the stable isotopes ^{18}O and ^{16}O in the calcium carbonate skeletons of marine plankton depends on the temperature of that water. The colder the water, the greater the tendency for minerals to incorporate ^{18}O, and thus the larger the ratio of ^{18}O to ^{16}O in $CaCO_3$. Oxygen isotope ratios are typically reported in "delta" notation, where $\delta^{18}O$ is the ratio of ^{18}O to ^{16}O normalized to a standard (either seawater or a standard limestone) and multiplied by 1000 to accentuate the small differences observed. (See Chapter 5 for further discussion of isotopes.)

Theoretically, then, the isotopic analysis of skeletal material recovered from cores of deep-sea sediments should reveal the history of temperature fluctuations in the overlying surface waters. However, an additional isotopic effect is equally important. As continental ice sheets grow, significant quantities of water are removed from the oceans (Figure 14-3). Aside from causing consider-

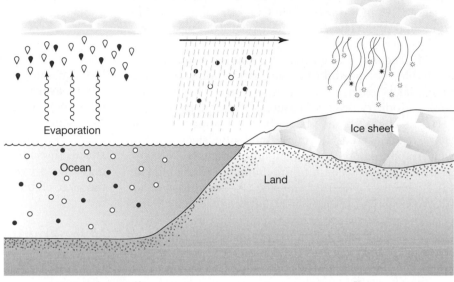

FIGURE 14-3

Changes in the oxygen isotopic composition of seawater during the growth of continental ice sheets.

○ H₂O containing ^{16}O

● H₂O containing ^{18}O

able drops in sea level (the last glaciation dropped sea level by about 130 m), significant changes in the oxygen isotopic composition of seawater occur as well. Evaporation transfers both $H_2^{16}O$ and $H_2^{18}O$ from the ocean to the atmosphere, but there is a preferential release of $H_2^{16}O$ to the vapor phase. Moreover, water-vapor molecules containing ^{18}O tend to condense more readily than do those containing ^{16}O, the lighter isotope. Thus, rain falling from the atmosphere preferentially removes ^{18}O, leaving the residual water vapor further enriched in ^{16}O. Snow falling onto the ice caps has traveled considerable distances and has had its ^{16}O content considerably enriched. The net preferential removal of ^{16}O from the oceans to ice sheets increases the $\delta^{18}O$ of the oceans. This effect adds to the direct temperature effect: Calcium carbonate precipitated from a glacial-age ocean has a larger $\delta^{18}O$, both because the water is cold and because the seawater is enriched in the heavier isotope.

In the 1950s the first deep-sea sediment cores were recovered, and isotopic analyses were performed on them. Instead of the four glaciations originally indicated by the continental record, the marine record indicated that dozens of climate swings have occurred over the course of the Pleistocene (Figure 14-4). The major intervals of Northern Hemisphere glaciation—**glacials**—of the past 700,000 years appear to have occurred every 100,000 years or so. During glacials, the globally averaged surface temperature was about 9–10°C (about 5–6°C cooler than today) and atmospheric CO_2 concentrations were about 200 ppm. These cold intervals are separated by warmer, shorter intervals known as **interglacials.** During interglacials, continental glaciation is limited to Greenland and

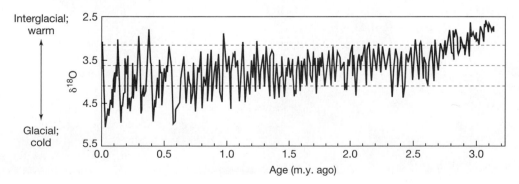

FIGURE 14-4

Deep-sea record of the $\delta^{18}O$ of seawater during the Pleistocene Epoch. The analyses were performed on two genera of bottom-dwelling foraminifera deposited in the sediments of the mid-latitude North Atlantic. Interglacials appear as peaks, with smaller values of $\delta^{18}O$; glaciations appear as valleys. Note that time proceeds forward to the left. (After M.E. Raymo, *Ann. Rev. Earth Plan. Sci.*, 22:353–383, 1994.)

Antarctica; globally averaged surface temperatures are about 15° C and atmospheric CO_2 concentrations are about 280 ppm. The Holocene Epoch (the past 10,000 years) represents one such interglacial.

The regularity of glacial–interglacial variation is remarkable. The **period** of a cyclical phenomenon is the time it takes to complete one cycle. Curiously, the dominant periodicity (cyclical nature) of glacial–interglacial variation seems to have been different prior to 700,000 years ago. Before that time, glacial–interglacial swings were smaller and occurred on approximately a 40,000-year time scale. Something fundamental to the climate system changed 700,000 years ago, and scientists are actively working to resolve what that fundamental change was. The pre-Pleistocene cooling and the onset of significant continental glaciation in the Northern Hemisphere (from 3.0 million years to 2.5 million years ago) are also apparent in the oxygen isotope record (see Figure 14-4).

Why has Earth's climate system been oscillating between these two states—glacial and interglacial—with apparent periods of 100,000 and 40,000 years? The answer seems to involve small changes in the way Earth orbits the Sun, changes that repeat in a predictable fashion over tens of thousands of years. Three sorts of changes are involved (Figure 14-5):

1. changes in the degree to which Earth's orbit around the Sun is elliptical (eccentricity);

2. changes in the tilt of Earth's spin axis with respect to the plane of its orbit around the Sun (obliquity); and

3. changes in the orientation of the spin axis with respect to Earth's orbit (precession).

The pacemaker that determines the variability of climate requires amplification, especially for the 100,000-year climate cycles, because only small changes in annual-average insolation result from eccentricity variations. We will first explore these orbital variations and will compare their predictions to the isotopic record from deep-sea sediment cores. We then discuss feedback mechanisms (including those affecting atmospheric CO_2) that may have provided the amplification necessary to create the large climate swings of the Pleistocene.

Milankovitch Cycles

What causes these remarkably regular shifts in Earth's climate? Long before the oxygen isotope evidence was obtained, scientists of the 19th century had suspected that the Pleistocene glacial–interglacial cycles were caused by variations in Earth's orbit around the Sun. In the early part of the 20th century, this hypothesis was put on a quantitative footing by the Serbian mathematician Milutin Milankovitch. He not only elaborated the mathematical theory of how orbital variations affect climate, but also calculated the changes in orbital parameters over the past several thousand years and demonstrated the connection between this theory and the rather scant geological record that existed during his time. The regular variations in Earth's orbit are often referred to as

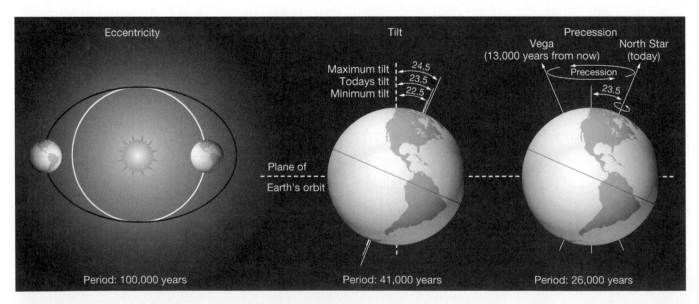

FIGURE 14-5

Aspects of Earth's orbit around the Sun that have implications for climate change. (a) The elliptical nature of the orbit (eccentricity) changes on 100,000- and 400,000-year time scales. (b) The tilt of the spin axis with respect to the plane of Earth's orbit around the Sun (obliquity) changes on a 41,000-year time scale. (c) The orientation of the spin axis in space wobbles (precesses) with a 26,000 period. (From J.P. Davidson, W.E. Reed, and P.M. Davis. *Exploring Earth: An Introduction to Physical Geology*, 1997. Reprinted by permission of Prentice Hall, Upper Saddle River, N.J.)

Milankovitch cycles in honor of this achievement. Milankovitch suggested that the critical factor for Northern Hemisphere continental glaciation was the amount of summertime insolation at high northern latitudes. High insolation leads to warmer summers, and the winter snowpack melts (as we see today). However, under low insolation the snowpack would survive over the summer, allowing snow and ice to accumulate and an ice sheet to form. Subsequent summers would allow further growth of the ice sheet toward lower latitudes.

Despite the elegance of Milankovitch's "astronomical theory of the Ice Age," it was strongly criticized by the scientific community in the 1920s and 1930s, in part because the available geologic record did not support the hypothesis of many Pleistocene glaciations. His response to these criticisms is perhaps best reflected in this excerpt from his 1941 book: "I do not consider it my duty to give an elementary education to the ignorant, and I have also never tried to force others to apply my theory, with which no one could find fault." Given the rather limited amount of tact with which Milankovitch presented his theory of the Ice Ages, it is not surprising that his brilliance was not widely acknowledged until well after his death.

Orbital Theory

Orbital theory itself predates Milankovitch; the fundamentals of this theory were developed in the 17th century by Johannes Kepler and Isaac Newton. The results are summarized by three rules that are known as *Kepler's laws* (see the box). The most important of these laws for our purposes is the first: The planets travel around the Sun in *elliptical* orbits with the Sun at one *focus*.

An ellipse is defined mathematically as the collection of points whose combined distance to two fixed points (the *foci*) is equal to a constant. The constant is equal to the length of the long, or *major,* axis of the ellipse (see Box Figure 14-1). This may be easily verified by adding up line segments along the major axis, remembering that the figure is symmetric about both the major and minor (vertical) axes. The degree to which the orbit of a planet (or other rotating object) is elliptical is called its **eccentricity.** For Earth, the distance from the center of the elliptical orbit to either focus is only 1.7% of the distance from the center to the edge of the ellipse along the major axis. In other words, the foci and the center are nearly indistinguishable, and Earth's orbit is very nearly circular. The eccentricity (often designated as e) is expressed numerically as this percentage in decimal form ($e = 0.017$).

THINKING QUANTITATIVELY
Kepler's Laws

First law: The orbit of each planet is an ellipse with the Sun at one focus.

Half of the major axis of an ellipse is called the *semimajor axis*. This is also the average planet–Sun distance. The semimajor axis is usually represented by the letter a. The distance from the center of the ellipse to one focus is equal to the length of the semimajor axis (a) multiplied by the eccentricity (e) (see Box Figure 14-1). In other words, the eccentricity of the ellipse can be determined by dividing ae by a. An ellipse with zero eccentricity is a circle (the two foci coincide with the center of the circle). Earth's current eccentricity is 0.017; Earth's orbit is nearly circular, but not quite.

Second law: A line joining a planet to the Sun sweeps out equal areas in equal times.

Box Figure 14-1 shows that to sweep out an equal area (each shaded region) in the same amount of time (Δt), Earth must travel farther around the perimeter of the ellipse when it is close to the Sun than when it is far from it. From a practical standpoint, then, this law means that the planet moves faster when it is closer to the Sun and slower when it is farther away.

Third law: The square of a planet's orbital period is proportional to the cube of its semimajor axis.

A planet's orbital *period* is the time that it takes the planet to go around the Sun. If we express the period P in Earth years and the planet's semimajor axis a in astronomical units (AU, the average Earth–Sun distance), we can replace the word *proportional* with *equal* and write Kepler's third law as: $P^2 = a^3$.

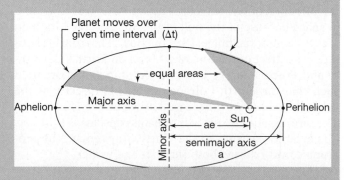

BOX FIGURE 14-1

Earth's elliptical orbit, showing the Sun at one focus. (From T. McKnight. *Physical Geography. A Landscape Appreciation,* 6/e, 1999. Reprinted by permission of Prentice Hall, Upper Saddle River, N.J.)

Because planetary orbits are eccentric, Earth is closer to the Sun at some times of the year than at others. The point of closest approach is called **perihelion**, and the point of maximum Earth–Sun distance is called **aphelion** (see Box Figure 14-1). The amount of sunlight hitting Earth is slightly greater at perihelion than at aphelion (as we will see in "Critical-Thinking," Problem 2). Perihelion occurs on January 3, 13 days after the Northern Hemisphere winter solstice, so Northern Hemisphere winters are somewhat milder than Southern Hemisphere winters. They are also somewhat shorter, as the planet moves faster at perihelion than at aphelion, according to Kepler's Second law. Conversely, Northern Hemisphere summers tend to be longer and milder than Southern Hemisphere summers.

Another factor that affects Earth's climate is the planet's **obliquity**, the fact that its spin axis is tilted 23.5 degrees from the perpendicular to the plane of its orbit. Obliquity creates the contrast between the seasons (Chapter 4); with no obliquity, the annual variation in the amount of solar insolation (resulting from the eccentricity of the orbit) would be very small. Winter and summer would basically not exist. Earth's relatively high obliquity means that there is a large **seasonal temperature contrast** between summer and winter. The eccentricity of Earth's orbit causes this seasonal temperature contrast to be slightly greater in the Southern Hemisphere than in the Northern Hemisphere.

Changes in Earth's Orbit through Time

Milankovitch's theory predicts that the gravitational influences of the Moon and the other planets, combined with Earth's slightly nonspherical shape, induce small but important variations in Earth's orbital parameters. These variations affect the amount of summertime insolation at high northern latitudes, triggering the onset and end of glacial intervals.

Precession of the Spin Axis

The most noticeable change in Earth's orbit has to do with the direction of its spin axis. The spin axis moves around in space because of the pull of the Sun and the Moon on Earth's equatorial bulge. (See the box "Thinking Quantitatively: Effect of the Sun and Moon on Earth's Obliquity and Precession.") Currently, the spin axis is oriented such that the North Pole points almost directly at the bright star *Polaris,* otherwise known as the North Star. The direction of the spin axis remains constant as Earth orbits around the Sun, so the North Star remains at geographic north during both summer and winter.

The spin axis has not always pointed in that direction, however. Egyptian pyramids built in 3000 B.C. were designed to observe the north star of the time, *Alpha Draconis,* not *Polaris.* Thirteen thousand years ago, the bright star *Vega* was approximately at geographic north. Over

time, the North Pole describes a circle in space as the spin axis points to different parts of the sky. The period of **precession** (i.e., the time it takes for the spin axis to *precess* one complete circle) is 25,700 years. However, the direction of the major axis of Earth's elliptical orbit is also precessing, but in the opposite direction, a phenomenon referred to as the *precession of perihelion.* Because perihelion is precessing in the opposite direction from the spin axis precession, the amount of time required to go through a complete precessional Milankovitch cycle is shorter than 25,700 years. The orbital precession is affected most strongly by two other planets, Venus (because it is close) and Jupiter (because it is big). Thus, two main periods result, at 23,000 and 19,000 years.

Precession modifies the relationship between the seasons and the distance from the Sun shown in Box Figure 14-1. Every half precession cycle, the hemisphere with the greatest degree of seasonal contrast switches between the north and the south. When the Southern Hemisphere has mild summers and winters, the Northern Hemisphere has hot summers and cold winters, and vice versa. Northern Hemisphere glaciation is promoted by a precessional state, as today, with northern summer at aphelion and thus low seasonal contrast. The maximum interglacial condition was achieved 9,000 years ago, with hot summers in the north.

Obliquity Variations

The same phenomenon that causes Earth's spin axis to precess also causes the obliquity to vary from 22 to 24.5 degrees, with a dominant cycle length of about 41,000 years. If you observe a spinning top carefully, you will notice that it, too, undergoes periodic changes in its tilt as its spin axis precesses. This effect becomes more pronounced as the spin rate of the top decreases: The top begins to wobble. Like precession of the spin axis, a change in obliquity does not alter the total amount of sunlight striking Earth. Rather, it determines the extent of seasonal contrasts: The warmth of summers and the coldness of winters is increased by higher obliquities (Figure 14-6). (See also the Box, "Thinking Quantitatively: Effect of the Sun and Moon on Earth's Obliquity and Precession.")

Eccentricity Variations

Earth's eccentricity also undergoes oscillations that can affect climate. The combined gravitational effect of all the planets causes Earth's eccentricity to vary periodically between 0 and 0.06. (Recall that the current value is 0.017.) As was true of the precessional cycle, two main periods are predicted. They are much longer in this case, however: about 100,000 years and about 400,000 years.

The eccentricity variations differ from the precessional and obliquity variations in one other significant respect: Eccentricity variations cause changes in the annually averaged amount of sunlight hitting Earth, where-

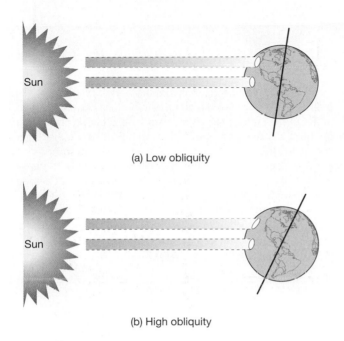

(a) Low obliquity

(b) High obliquity

FIGURE 14-6

(a) At low obliquity, Earth has less contrast in insolation between the seasons. (b) At high obliquity, the seasonal contrast is greater.

as precessional and obliquity variations do not. (The direction and steepness of tilt of a planet's spin axis have no direct effect on the total amount of sunlight the planet receives.) One can show mathematically that Earth receives about 0.2% more sunlight at maximum eccentricity than at minimum eccentricity. This difference is thought to be too small to cause major climate shifts by itself, but it might have some effect if it is amplified by a feedback mechanism (discussed later in this chapter).

Probably more important is the fact that eccentricity influences the climatic effect of the precession cycle. When Earth's eccentricity is nearly zero, there is no difference between the perihelion distance and the aphelion distance from the Sun, so it does not matter when summer or winter occurs. When the eccentricity is large, Northern Hemisphere glaciation is especially favored when precession causes Northern Hemisphere summer to occur at aphelion. Of course, within a half precession cycle the situation reverses, with Northern Hemisphere summer at perihelion. Nevertheless, analysis of past glaciations indicates that ice sheets survive this effect of high eccentricity. At present we are at low eccentricity, and according to the calculations of Belgian astrophysicist André Berger, the eccentricity will be decreasing to a minimum near zero in about 30,000 years from now. With eccentricity so low, the unusually cold winters needed to initiate Northern Hemisphere ice-sheet growth don't occur. Thus, climatologists predict that the present interglacial will be long-lived (at least 1.5–2.5 precession cycles). The

buildup of carbon dioxide in the atmosphere from fossil-fuel burning only serves to strengthen that prediction.

Comparing Orbital Forcing and Climatic Response by Means of Oxygen Isotopes

The combination of these various orbital forcings causes Earth's climate system to oscillate between two states. The situation can be displayed in a diagram similar to that developed for Daisyworld (Chapter 2). The glacial and interglacial states are represented as valleys separated by a ridge (Figure 14-7). Presumably the glacial state is situated in a deeper valley than the interglacial state, because a greater fraction of Pleistocene time was spent in glaciation. Orbital forcings continually rock the system back and forth. Since 700,000 years ago, the amplitude of this rocking has exhibited a strong 100,000-year periodicity. When these variations exceed a threshold, the system moves over the ridge into the other state. High eccentricity increases the amplitude of the variations on precessional cycles and thus is more likely to be associated with transitions from interglacial to glacial states (or vice versa).

We can now piece together the various parts of the astronomical theory of the Ice Ages described above and test the theory against the oxygen isotopic record of temperature and sea level changes during the Pleistocene. The precession, obliquity, and eccentricity variations can all be described mathematically, and the resulting equations can be solved for the amount of insolation received on a monthly or annual basis for any particular latitude. This is nothing new, of course; Milankovitch made these calculations many decades ago. However, some improvements have been made on the original calculations by Milankovitch. The particular result shown at the top of Figure 14-8 is the average monthly insolation (Q) for June, at 65° N latitude.

Shown at the bottom of Figure 14-8 is the past 400,000 years of the oxygen isotope record from Figure 14-4. Is the observed climate response ($\delta^{18}O$) the expected response to the climate forcing (Q)? Such a comparison is difficult; indeed, we might conclude from a visual inspection that the two curves are unrelated. However, the wiggles of these curves can be considered as the combination of a number of waves of different frequency (the mathematical inverse of period) and amplitudes, much as a musical chord is the combination of a number of notes, each of a different frequency (pitch). These curves can then be separated into their component periodic waves (or *bands*); the most important ones are shown in Figure 14-8. (The technical name for this procedure is *Fourier analysis*.) For Q these are the precession, obliquity, and eccentricity bands. As predicted by Milankovitch's astronomical theory, the dominant periodicities of the $\delta^{18}O$ record occur at these same bands (19,000 years and 23,000

THINKING QUANTITATIVELY
Effect of the Sun and Moon on Earth's Obliquity and Precession

The fundamental reason why Earth's spin axis precesses, and one reason why its obliquity varies, is because Earth is not perfectly round. Because it is spinning rapidly, it bulges slightly at the equator. The diameter through the equator is 12,756 km, while the polar diameter is about 43 km less. The Sun and the Moon pull on this bulge gravitationally, thereby exerting a torque on the Earth. (See Box Figure 14-2a.) A *torque* is a force applied perpendicular to the direction of the axis of rotation of a spinning body. This torque causes Earth's spin axis to precess in a circle and also to bob up and down slightly, giving rise to periodic variations in Earth's obliquity.

The same phenomena can be observed in a simple home experiment with a spinning top (Box Figure 14-2b). Earth's gravity is pulling the top toward the floor. If the top is standing straight up, the gravitational pull is along the top's spin axis, and the top spins smoothly, without precessing. If the top is tilted sideways, however, then the gravitational pull is partly perpendicular to the spin axis, and the top precesses around in a circle. As its rotation rate slows down (due to friction with the surface), the top will also begin to bob up and down, or *nutate*. This nutation is analogous to changes in Earth's obliquity.

What would happen to this system if one took away the Moon? (This is not just a hypothetical question; there may be Earth-like planets around other stars that lack large moons.) The Moon accounts for roughly two-thirds of the gravitational torque acting on Earth's equatorial bulge. The Sun accounts for the other one-third. If the Moon were not present, the net torque would be smaller, and Earth's spin axis would precess more slowly. Jacques Laskar and his colleagues at the University of Paris have shown that under these circumstances, Earth's obliquity would vary chaotically from 0° to as much as 60° on a time scale of tens of mil-

lions of years. (The reason is that the period of the spin axis precession would now match up with periods observed in the orbits of the other planets, such as the precession of their perihelia.) This would wreak havoc with Earth's climate. Continents located at high latitudes, like much of North America and Europe, would be subject to extreme seasonal variations. This has led some astronomers to suggest that a large moon may be necessary in order for a planet to have a stable obliquity and climate. In reality, the situation is more complicated than this because a more rapidly spinning Earth would *not* experience this problem, but it is still true that the Moon exerts a major influence on Earth's climate.

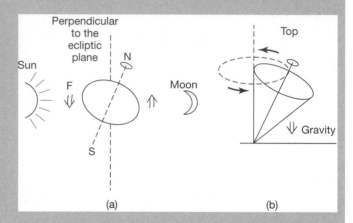

BOX FIGURE 14-2

The effects of the Sun and Moon on Earth's obliquity. a) Both the Sun and Moon exert a torque on Earth, causing it to precess and bob up and down. b) Analogous motion of a top spinning on its side.

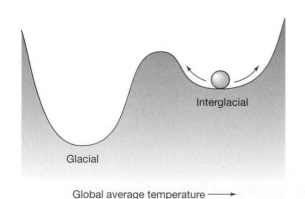

FIGURE 14-7

Stability of the glacial and interglacial states, relative to the changes in high-latitude Northern Hemisphere insolation that rock the state of the system back and forth.

years, 41,000 years, and 100,000 years, respectively). Thus, there is compelling evidence that the pacemaker for the ice ages is the "Milankovitch" signal.

Moreover, the amplitude of the response to the precession and obliquity changes seems to be roughly proportional to the amplitude of the forcing. These observations suggest that a simple link exists between fluctuations in the amount of radiation received at high latitudes and the extent of glaciation.

A closer inspection of Figure 14-8, however, reveals an important departure from a straightforward link between climate forcing and climate response: The direct forcing, that is, the average annual insolation change, in the eccentricity band is very small (some 10% of that in the other bands), yet the climate response is the largest of the three bands. The importance of eccentricity is evidently more indirect. Eccentricity modulates the insolation changes as-

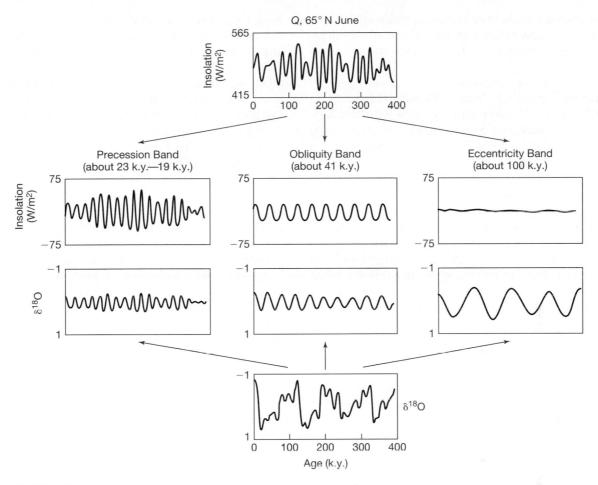

FIGURE 14-8

Northern Hemisphere June insolation (Q, the climate forcing) and marine oxygen isotopic composition ($\delta^{18}O$, the climate response) and their dominant periodic components. (After Imbrie et al., *Paleoceanography* 7:701–738, 1992.)

sociated with the precessional band, as can be seen in the envelope of variation for precession in Figure 14-8. Nevertheless, climate amplification of the 100,000-year forcing is considered necessary to create the climatic response from eccentricity forcing. Do positive feedback loops in the climate system amplify the weak eccentricity forcing into the major climatic response to orbital fluctuations? The remainder of this chapter explores this possibility.

Glacial Climate Feedbacks

The timekeepers for the glacial–interglacial climate fluctuations during the Pleistocene were subtle, periodic changes in Earth's orbital parameters. However, these changes have been small. Moreover, the dominant periodicity of glacial–interglacial fluctuations has been 100,000 years; if this phenomenon is the result of eccentricity changes, an amplifier is needed. The important climate variables that we need to consider are albedo and

the greenhouse effect. Clearly, the growth of continental ice sheets influences the albedo of the planet, so this effect must be incorporated into any model that attempts to explain Pleistocene climates. Clouds exert a major control on planetary albedo, and we may wonder whether the cloud albedo varies in concert with the Milankovitch cycles. Finally, we have seen how changes in the greenhouse effect of atmospheric CO_2 have affected climate on long time scales; perhaps the Pleistocene climate system has responded to more rapid fluctuations in atmospheric CO_2.

Ice–Albedo Feedbacks

Any change in the seasonal distribution of solar luminosity that affects the growth of ice during the winter or the meltback during the summer has the potential to affect the planetary albedo (see Chapter 3). As ice sheets begin to grow, they convert a region that previously had reduced albedo during the summer, as snow melted, to one that maintains high albedo throughout the year. The average annual albedo thus increases, which will lead to both a regional and a global cooling. This cooling will accelerate the growth of the continental ice and will allow it to spread to lower latitudes.

The positive ice–albedo feedback, involving global temperature, ice-sheet growth, and albedo (Figure 14-9), was introduced in Chapter 3. In Figure 14-9 the forcing is also indicated; note that a small change in the intensity of summer insolation at high northern latitudes could potentially lead to large changes in ice-sheet coverage and global temperature. Researchers have shown that the growth and destruction of the Northern Hemisphere ice sheet has a characteristic response time of about 100,000 years. Thus, the dynamics of glaciation are especially tuned to a frequency of one cycle per 100,000 years and should respond quite sensitively to eccentricity-induced changes. Numerical models show that instabilities develop as an ice sheet becomes very large, such that fairly subtle changes in high-latitude insolation can lead to its catastrophic destruction.

The ice–albedo feedback has significant effects on Northern Hemisphere climates, but can't explain why Southern Hemisphere climate changes are both large and in step with those in the Northern Hemisphere. Studies of polar ice cores indicate that carbon dioxide levels have also varied substantially. The greenhouse effect associated with these changes is not negligible, and may be the explanation for the link between northern and southern climate change. Ice cores also reveal that the number of cloud condensation nuclei in the atmosphere has changed with the Milankovitch cyclicity. We now explore some of the proposed mechanisms for large and rapid changes in atmospheric CO_2 levels and cloud condensation nuclei on glacial timescales.

Evidence from the Vostok Ice Core

During the Southern Hemisphere summer of 1982–1983, French and Soviet scientists met in Vostok, on the high plateau of Antarctica, to take samples from the longest, most continuous ice core ever recovered from the Antarctic Ice Sheet. With each length of ice removed from this 2-km-deep hole, our knowledge of Earth's climate history was extended millennia into the past. By the time the last, deepest section was recovered and sampled, 200,000 years of ice accumulation was revealed. The samples were taken to France, where the oxygen isotopic composition of the ice was analyzed. This analysis revealed a history that matched well what we would expect from the marine record of isotopic changes during the Pleistocene. Since then, the coring has been extended to 3.6-km depth to ice deposited some 420,000 years ago.

However, one of the most important discoveries made by the scientists studying the Vostok ice core was the realization that the ice contained air bubbles frozen into the glacier as it grew. Using extreme care, the scientists were able to measure the concentration of carbon dioxide in the bubbles. We saw the results of this effort in Chapter 1, and we also learned that air bubbles in ice cores have also been used to estimate atmospheric CO_2 concentrations on much shorter time scales (the past 1,000 years). The Vostok core gives us long time scales because the rate of snow accumulation at Vostok is very small, a few centimeters per year. By contrast, snowfall at Siple Station, where the data from Chapter 1 were collected, is many meters per year.

The results of the Vostok scientists' analyses provide a firm link between global climate change and variations in the quantity of greenhouse gases in the atmosphere. What they found is that the CO_2 concentration falls and rises in concert with variations in local temperatures (recorded in the hydrogen isotopic composition of the ice; see Figure 1-9), and both records are well correlated with global changes in temperature and ice-sheet size determined from $\delta^{18}O$ variations.

The rapidity of some of the changes is truly remarkable. The increase from glacial-stage levels (about 190 ppm) to nearly contemporary CO_2 levels (240 ppm) occurred over only 4,000 years, between 16,000 years and 12,000 years ago. Analyses of bubbles from ice created during the next-to-last deglaciation, approximately 145,000 years ago, display a similarly rapid rise in CO_2. The drop in CO_2 levels from the previous interglacial (about 130,000 years ago) to the height of the last glacial

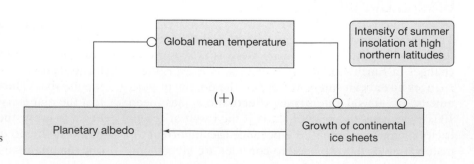

FIGURE 14-9

Feedback diagram showing the effect of changes in glacial growth on global temperature.

(about 20,000 years ago) was more subdued—about 1 ppm per millennium. The sawtooth nature of these changes is nearly identical to the oxygen isotope record, suggesting a close link between CO_2 changes, ice volume, and global temperature.

Feedbacks Affecting Atmospheric CO_2 on Glacial Time Scales

In Chapters 9 and 12 we found that on long time scales (millions of years) the carbonate–silicate geochemical cycle, together with the weathering and deposition of organic carbon-rich sedimentary rocks, determines the steady-state atmospheric CO_2 level. When studying glacial–interglacial fluctuations, however, other processes must be included because the assumption of steady state with respect to these processes may not be valid on shorter time scales (thousands to hundreds of thousands of years). The partitioning of carbon between the atmosphere and terrestrial biomass, and between the atmosphere and ocean, as affected by the oceanic biological pump (Chapter 8), are potentially of great importance to the CO_2 balance during glacial cycles. So, too, are the processes of limestone weathering and limestone deposition. On glacial–interglacial time scales, limestone weathering need not be balanced by limestone deposition, and any imbalance will affect atmospheric and oceanic concentrations of CO_2. Let us explore these feedbacks in greater detail.

Role of the Biological Pump. The photosynthetic conversion of dissolved carbon dioxide to organic matter in the surface ocean, the settling of this material through the water column, and its decomposition at depth (that is, the *biological pump,* described in Chapter 8) dominates the distribution of carbon throughout the world's ocean. Because the atmosphere equilibrates with the surface ocean, that photosynthetic conversion dominates the atmospheric CO_2 content as well. An atmospheric CO_2 pressure of 280 ppm (the preindustrial level) represents a biological pump that operates at intermediate efficiency, because regions of the ocean exist today (and presumably in preindustrial times as well) where nutrient concentrations are not completely depleted by biological uptake (Chapter 8). If nutrients were completely utilized—that is, if the biological pump were 100% efficient in removing nutrients and CO_2 from surface waters—the atmospheric CO_2 pressure would be reduced to about 165 ppm. At the other extreme, if the biological pump ceased completely, the atmospheric CO_2 level would rise to about 720 ppm as the CO_2-charged deep waters mixed with the surface waters and homogenized the chemical composition of the ocean. Thus, the low CO_2 concentrations of glacial intervals might be the result of a more efficient biological pump.

Why might the glacial ocean support greater biological productivity? Most of the answers proposed in the scientific literature involve increased nutrient supply through upwelling or riverine delivery. The hypotheses described next are intended to explain why nutrients might have been more available to the oceanic biota during glacial times.

Shelf Nutrient Hypothesis. The CO_2 concentration at the height of the last glaciation (20,000 years ago) was about 190 ppm. Might the biological pump have been more effective during the glaciation than it is today, perhaps as a result of a higher concentration of nutrients in the ocean as a whole? The nutrient concentration of the ocean represents a balance between supply by rivers (and, for nitrogen, by bacterial processes that convert nitrogen gas from the atmosphere to nutrient nitrate) and removal, primarily by sedimentation of organic matter (Figure 14-10). Thus, if the biological pump was intensified by higher nutrient concentrations during glacial intervals, either riverine fluxes were greater or sedimentation rates were lower.

As the glaciers grew, sea level fell, exposing the vast, low-relief margins of the continents (the **continental shelves**). Sediments of the continental shelves are rich in organic matter and nutrients, as the result of the highly productive nature of the overlying waters. When these sediments became exposed, weathering reactions released the nutrients (especially phosphate) to the rivers draining the shelves (Figure 14-11). This nutrient release enhanced the global delivery of phosphate to the oceans, causing an increase in oceanic phosphate concentrations, marine productivity, and carbon export from the surface ocean and a drop in atmospheric CO_2. The resulting effect on global temperature and ice volume created a positive feedback loop, as shown in Figure 14-12. An attractive aspect of this hypothesis is that the residence (response) time of phosphate in the ocean is about 40,000 to 100,000 years, which essentially matches the major periodicity of glacial–interglacial cycles. The oceanic phosphate cycle is thus "tuned" to a major Milankovitch frequency and should be able to provide at least part of the required amplification of the Milankovitch forcing.

However, there are problems with this *shelf nutrient hypothesis.* The distribution of the trace element cadmium throughout the ocean follows that of phosphate very closely. Cadmium is incorporated into $CaCO_3$ skeletons of bottom-dwelling organisms in proportion to its oceanic concentration, whereas phosphate is not. Thus, paleoceanographers use the cadmium content of the fossils of bottom-dwelling foraminifera as a proxy indicator of changes in the phosphate content of the oceans in the past. What these researchers find is that the cadmium content of these fossils does not indicate higher phosphate

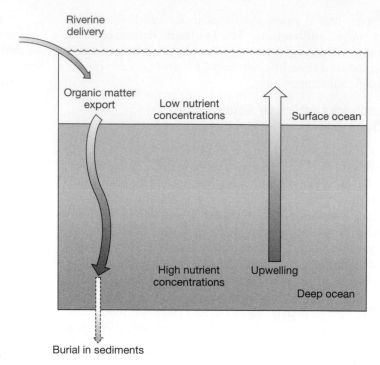

FIGURE 14-10

Simplified view of the nutrient throughput of the oceans.

concentrations during glacial intervals. This inconsistency has led scientists to shift the focus of their search for productivity changes to other nutrients, particularly iron.

The Iron Fertilization Hypothesis. Iron plays an important role in limiting primary productivity in certain regions of the ocean (Figure 14-13). In these regions the major nutrients are not depleted as they are in the rest of the surface ocean (Chapter 8), and productivity appears to be limited by trace nutrients such as iron. Moreover, nitrogen-fixing cyanobacteria have large demands for iron, which is an essential metal for the synthesis of the enzyme that catalyzes nitrogen fixation (see Chapter 11). Much

of the iron supplied to the oceans today comes from windblown dust particles, which typically have a coating of iron that dissolves in seawater. The Saharan and Gobi deserts today provide considerable quantities of dust to the Atlantic and Pacific Oceans, respectively. Aridity appears to have increased during glacial times. Furthermore, the east–west winds should have intensified in response to the greater equator-to-pole temperature gradient (Chapter 4). Both of these factors would have increased the flux of dust to the oceans during glacial times. The record of windblown dust accumulation in marine sediments supports this *iron fertilization hypothesis;* rates increase several-fold during glacial intervals.

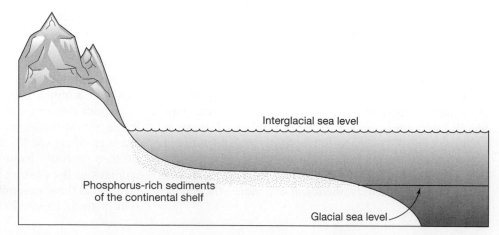

FIGURE 14-11

The exposure of nutrient-rich shelf sediments as a result of a drop in sea level, part of the shelf nutrient hypothesis for the cause of changes in biological productivity on glacial time scales.

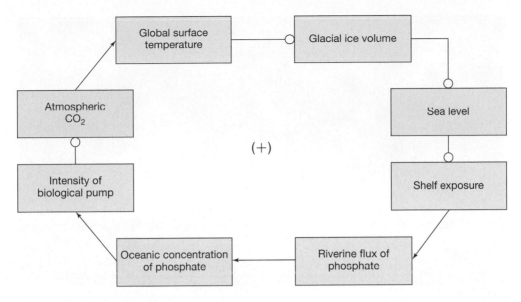

FIGURE 14-12

Systems diagram of the shelf nutrient hypothesis for the reduction of atmospheric CO_2 during glaciation.

Recognition that primary productivity in significant regions of the world's oceans is iron-limited today has led to the suggestion that fossil-fuel emission of carbon dioxide might be countered by iron-induced stimulation of biological uptake of CO_2 and its transfer to the deep sea (via the biological pump; see Chapter 8). Small-scale ocean experiments have indicated that iron fertilization is feasible, but also reveals that there may be detrimental and unexpected consequences of this environmental manipulation, including depletion of dissolved oxygen in the deep ocean.

The Coral Reef Hypothesis. The continental shelves between 30° N and 30° S latitude provide a habitat that is ideal for the growth of corals and other calcium carbonate-secreting organisms (Figure 14-14). As these organisms grow, they add incrementally to the rock framework of the reef, building laterally as well as vertically, until the water surface is reached.

As we saw in Chapter 8, the production of $CaCO_3$ can be written as follows:

$$Ca^{2+} + 2HCO_3^- \rightarrow CaCO_3 + CO_2 + H_2O.$$

Thus, the growth of coral reefs serves as an additional source of carbon dioxide to the atmosphere. (The effect is only temporary, however; after tens of thousands of years, this excess carbon dioxide becomes converted to bicarbonate, as the result of mineral weathering, and is redeposited as $CaCO_3$.) In contrast, when ancient reefs are exposed by a drop in sea level, chemical weathering leads to their dissolution. This process is the reverse of reef growth: Atmospheric CO_2 is converted into bicarbonate, which is carried by rivers to the ocean. Thus, the growth and destruction of coral reefs can affect atmospheric CO_2 on glacial–interglacial time scales.

The link to glacial–interglacial CO_2 fluctuations is once again through sea-level changes. As the glacial interval ends and the ice sheets begin to melt, sea level rises, flooding the continental shelves. In the tropics, reef growth resumes and CO_2 is released to the atmosphere. The increase in atmospheric CO_2 causes an increased greenhouse effect, thereby amplifying the original climate

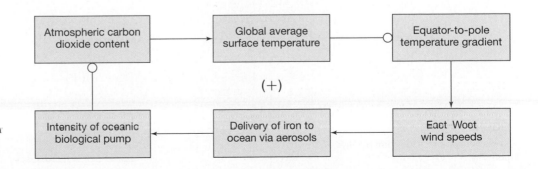

FIGURE 14-13

The iron fertilization hypothesis for the intensification of the biological pump during glaciations.

FIGURE 14-14

[See color section] Coral reefs may be responsible for the changes in atmospheric carbon dioxide concentrations that occur between glacials and interglacials. (From T. McKnight. *Physical Geography: A Landscape Appreciation*, 5/e, 1996. Reprinted by permission of Prentice Hall, Upper Saddle River, N.J.)

warming (Figure 14-15). Conversely, as sea level begins to fall at the end of the interglacial, reefs become exposed to the atmosphere, and rain, soil, and groundwaters begin the process of reef dissolution. Again, the feedback loop is positive.

The importance of this feedback depends on the rates of these processes and on how responsive the rates are to sea-level changes. Studies of rates of calcium carbonate formation by reef-building organisms indicate that the reef ecosystem can easily keep pace with sea-level rise. The rate of limestone dissolution, however, is relatively slow. Thus, it is possible that, in this *coral reef hypothesis,* there is an unbalanced response to sea-level rise and fall in terms of reef growth and dissolution.

Coral reefs are suffering from a host of diseases and other influences today. These are probably the result of multiple stressors, including warmer sea temperatures, in-

creased human perturbation (ship groundings, pollution, destructive collection of reef rock and reef organisms), and perhaps even an increase in the flux of dust from distant lands carrying pathogens and iron, which fertilizes the algal competitors of corals. Perhaps the most pervasive anthropogenic stress is the direct response of the ocean's carbon chemistry to increasing atmospheric CO_2. Recall that as atmospheric CO_2 increases, the pH of the surface ocean decreases and the carbonate ion concentration falls. Carbonate, together with calcium ion, is essential for corals to precipitate their skeletons. The historical rise of atmospheric CO_2 over the last century has reduced the carbonate ion concentration of ocean surface waters by a measurable amount and may already be causing significant stress to corals. Projections for the future are not encouraging: corals may lose their ability to precipitate skeletons by early in the next (22nd) century. Scuba divers,

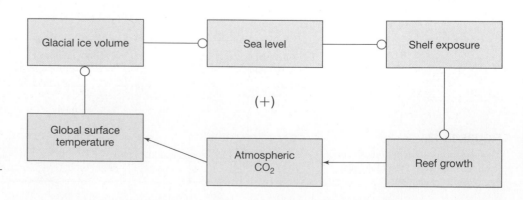

FIGURE 14-15

Systems diagram of the coral reef hypothesis.

or rather those who hope that their descendants will enjoy scuba diving, along with those who recognize the intrinsic and societal value of coral reefs thus have an additional reason to be concerned about future CO_2 increases.

Changes in Terrestrial Biomass: A Negative Feedback

In living tissue, the terrestrial biomass today contains about the same amount of carbon (600 Gton (C)) as does the atmosphere; about twice that much is contained in dead and decaying organic material in soils. On the basis of studies of plant fossils and other climatic indicators, it appears that the amount of forest coverage, and thus terrestrial biomass, was drastically reduced during the last glacial interval (Figure 14-16). Much of the northern forests were covered with ice, and tropical regions experienced greater aridity and thus the replacement of tropical rainforests with grasslands. Estimates have been made of the total amount of carbon that was transferred from the oceans to the terrestrial biomass at the end of the last glaciation. These numbers have large uncertainties, but

it is clear that the change (around 700 Gton (C)) was many times larger than the net change in the amount of CO_2 in the atmosphere (around 160 Gton (C)).

The growth of the terrestrial biomass during deglaciation and its destruction during the initiation of a glacial interval represent the only negative feedbacks that we have been able to identify to changes in atmospheric CO_2 on glacial timescales (Figure 14-17). That the CO_2 level rose and fell in concert with global temperature during the past 220,000 years indicates that positive feedback mechanisms have predominated. Note that this is just the opposite of what would be expected if organisms were modulating the climate system in such a way as to increase ecosystem stability. Gaia, if she exists, is destabilizing on glacial-to-interglacial time scales.

Cloud–Albedo Feedbacks

Recall from Chapter 4 that the process of cloud formation is critically dependent on the presence of small droplets (aerosols) known as *cloud condensation nuclei*. Besides dust (and pollution) over land and tiny sea-salt droplets

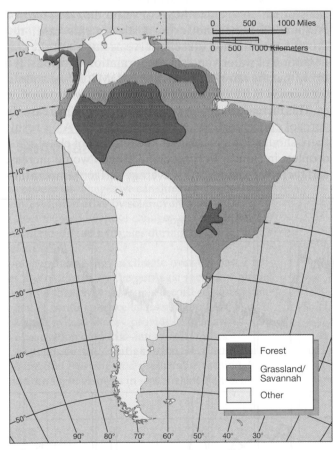

(a) Reconstructed vegetation cover, 18 k.y. ago

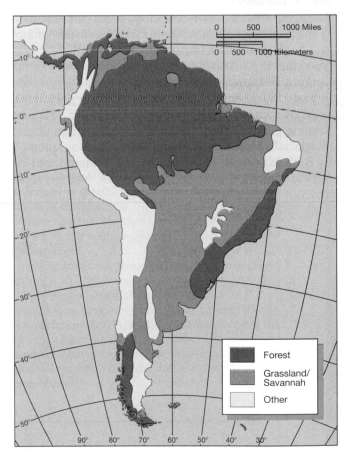

(b) Present-day "potential" vegetation cover.

FIGURE 14-16

The difference in South American vegetation between (a) the last glacial maximum and (b) today. Note the large increase in forest cover at the expense of grassland and savannah. The present-day map shows the "potential" vegetation cover; deforestation and other human activities have reduced the forest cover from its potential coverage shown here. (Courtesy J. Adams, Oak Ridge National Laboratory.)

Review Questions

1. What types of geologic evidence are diagnostic of glaciation?
2. What causes changes in the oxygen isotopic composition of seawater?
3. Which three characteristics of Earth's orbit around the Sun vary on the timescale of Pleistocene glaciations? How does each of these affect the amount of energy received from the Sun?
4. What orbital configuration favors glaciation? Why?
5. How is the oxygen isotopic record of marine limestones used to test Milankovitch's theory of the ice ages?
6. What role do biologically produced sulfur gases play in glacial climate fluctuations?
7. What factors might have caused atmospheric CO_2 variations that kept pace with glacial climate fluctuations?

Critical-Thinking Problems

1. Return to our Daisyworld analogy from Chapter 2. Construct a Daisyworld-like model of the MSA-climate feedback loop shown in Figure 14.19. First sketch a graph of how DMS production by algae would affect global temperature. Then sketch another graph of your view of how changes in global temperature might affect algal DMS production. Defend both graphs in writing. Then combine these graphs, and discuss the stability of the equilibrium states indicated.

2. An ellipse is defined as the locus of all points such that the sum of the distances to two fixed points, called the foci, is a constant. We can easily show that this constant is equal to $2a$, where a is the semimajor axis of the ellipse. (See Box Figure 14-1 and accompanying discussion.) The eccentricity e of the ellipse is defined such that the distance from one focus to the midpoint of the figure is ae. An ellipse with $e = 0$ is a circle.

   a. Kepler's first law states that the planets move around the Sun in elliptical orbits with the Sun at one focus. Earth's present orbit has a semimajor axis of 1 AU and an eccentricity of 0.017. The point of closest approach to the Sun is the perihelion; the point farthest away is the aphelion. How much closer is Earth to the Sun at perihelion than at aphelion? Express your answer in astronomical units.

   b. The Milankovitch theory of the ice ages holds that the most important forcing factor is the difference in solar heating at high latitudes when Northern Hemisphere summer occurs at perihelion as opposed to aphelion. Using the inverse square law (Chapter 3), find the solar flux at perihelion and at aphelion. Recall that the solar flux at 1 AU is 1370 Watts/m². How much higher is the solar flux at perihelion than at aphelion today? Express your answer as a percentage. How much warmer is the effective radiating temperature of Earth (Chapter 3)?

   c. The eccentricity of Earth's orbit varies with time as a consequence of gravitational perturbations caused by the other planets. Repeat Question 2b for e at its maximum value of 0.06.

   d. Kepler's third law states that the square of a planet's period P is proportional to the cube of its semimajor axis a. When P is expressed in Earth years and a is in AU, the relationship is simply $P^2 = a^3$. Venus and Mars have semimajor axes of 0.72 and 1.52 AU, respectively. How many Earth years does it take for them to go around the Sun?

Further Reading

General

Alley, R. B. 2000. *The Two-Mile Time Machine: Ice Cores, Abrupt Climate Change, and Our Future,* Princeton, NJ.: Princeton University Press, 229 p.

Broecker, W. S. 1995. Chaotic Climate, *Scientific American,* November 1995, pp. 62–68.

Broecker, W. S., and Denton, G. H. 1990. What Drives Glacial Cycles? *Scientific American,* January 1990, pp. 48–56.

Advanced

Imbrie, J. and others. 1992, 1993. On the Structure and Origin of Major Glaciation Cycles, Parts 1 and 2, *Paleoceanography,* 7, pp. 701–738, and 8, pp. 699–735.

Berger, A. 1995. Modeling the Response of the Climate System to Astronomical Forcing, In *Future Climates of the World,* A. Henderson-Sellers (ed.), World Survey of Climatology, Elsevier Science Publishers B.V., Amsterdam.

Liu, H.-S. 1995. A New View on the Driving Mechanism of Milankovitch Glaciation Cycles, *Earth and Planetary Science Letters,* 131, pp. 17–26.

Kump, L., and Lovelock, J. 1995. The Geophysiology of Climate, In *Future Climates of the World,* A. Henderson-Sellers (ed.), World Survey of Climatology, Elsevier Science Publishers B.V., Amsterdam.

CHAPTER 15

Short-Term Climate Variability

Key Questions

- How can we document climate changes that occur over shorter time periods than those we have discussed so far?
- What causes climate change on short time scales?
- Where in the climate system do we look for processes that might cause climate variability on the time scale of years to decades?

Chapter Overview

We have seen in previous chapters how the Earth system has undergone long-term changes in climate. In this chapter, we turn to shorter time scales and show how climate has fluctuated during the past 12,000 years. In this case, we are looking at variations that last for several hundred to a thousand years. For the most part, we are considering changes caused by processes that are internal to the Earth system rather than external forces such as the long-term variability of the Sun or changes in Earth's orbit about the Sun. We then move to even shorter time scales and discuss the year-to-year variability that we find in the global climate. Here we focus on two processes, both of which involve the oceans: one in the tropics (El Niño–Southern Oscillation events) and one at high latitudes involving sea ice-ocean–atmosphere feedbacks.

By the end of this chapter, we will observe that there have been large changes in climate (from a human perspective) over the past 12,000 years, but none of them were as large as the changes that are likely to happen in the future as a result of anthropogenic greenhouse warming.

Introduction

Time Scales of Change

In the past few chapters we have taken a rapid tour through 4.6 billion years of Earth's history, focusing on the processes that have shaped that history and examining the degree to which the physical environment has changed since the early days of Earth's formation. The underlying message of each chapter is the same: Earth acts as a single integrated system over any time scale we choose. There is increasing evidence that Earth has been entirely ice covered at several times in the past (a condition referred to as "Snowball Earth"). These time periods were infrequent and (geologically speaking) relatively short. With these exceptions, we can say that the Earth system has remained remarkably stable (staying within certain limits or boundaries) throughout its history, but within those limits Earth has experienced a considerable degree of change and variability. How we define these limits depends on the specific components of the system under discussion (global temperature, gas composition of the atmosphere, and so on). In general terms, we can say that for the past 3.5 billion years Earth has maintained an environment capable of supporting life—but life can exist under a wide range of conditions.

Indeed, when we look at Earth's history we find intervals that were much warmer than the present and intervals that were cooler (when huge masses of ice covered large areas of the land surface); we find that the atmospheric composition has changed considerably through time; and we find intervals when the distribution of land, ocean, and mountain ranges was very different from what we see today. From our earlier discussions we are also left with the impression that these large changes (such as changes in solar luminosity) take place over long time scales, but they occur very slowly—slowly enough that when we move to shorter time scales, these changes are not noticeable. On these shorter time scales, however, other processes are observed that produce smaller and more-frequent variations (such as the orbital changes that trigger ice ages). Finally, we also see from our earlier discussions that some of the forcing for these changes originates outside the Earth system itself. However, as we move to much shorter time scales, the situation is very different.

In focusing on short-term climate change and variability, we consider changes on a hundred- to a thousand-year time scale as well as variability on interannual to decadal scales. Our objective is twofold. On the one hand we present these processes as further illustrations of how system components interact; on the other hand, this discussion provides background material for our examination of global warming. We have seen the role long-term changes in the level of atmospheric greenhouse gases have played in Earth history, and we discuss the possible impacts anthropogenic greenhouse gases may have in the near future. Before we look at these possible future changes, however, let us set them in the context of the variability in the climate system that occurs naturally over these shorter time frames.

The last major continental glaciation reached its maximum extent about 21,000 years ago. The temperature difference between the last glacial maximum and today is not much greater than the temperature change we expect to see in response to greenhouse warming by the later part of the 21st century. The interval from the last glacial retreat (about 10,000 years ago) to the present is referred to as the **Holocene**. When we look at this epoch from the perspective of Chapter 14, the temperature record of the Holocene looks relatively uninteresting (Figure 15-1a). Once Earth warmed up from the last glacial maximum, nothing much else seems to happen. If we zoom in a little closer and look at the past 10,000 years (Figure 15-1b) or 1,000 years (Figure 15-1c), however, we see that a great deal of smaller-scale climate change has taken place.

It is not only the climate that has changed. Although our discussion of the Holocene focuses on climate change

and variability, there have been large ecological changes over this same interval. The species we find today existed prior to the last glaciation, but the characteristic groupings, or assemblages, of species that we find in specific environments has changed considerably. For example, the assemblage of species that we recognize as members of the boreal forest community in the high mid-latitudes came into being only during the Holocene, although the individual species existed prior to this time. What makes this interval even more interesting is that, for the first time, we also have to consider the impact that human societies have on the Earth system. Furthermore, many of the changes that appear to have been very significant in some regions during the Holocene actually represent smaller changes than those we project might occur in the future.

Humans have been adding greenhouse gases to the atmosphere in increasing quantities for the past 200 years. We might think that this should have had some sort of climatic impact by now, and one question that we asked in Chapter 1 was whether there is any evidence that warming is actually occurring. The 2001 Intergovernmental Panel on Climate Change (IPCC) report suggests that there is clear evidence of a global warming signal in temperature data in the last decade or so of the 20th century. Until then, it was difficult to distinguish the climate change signal from the natural variability of the system. Later in this chapter, we begin to see why this is so by moving to even shorter time scales and discussing processes that cause variability in the system on an annual to decadal basis.

Climate Change and Variability

Climate is an assemblage of weather conditions experienced at a location over some reference time frame. The World Meteorological Organization suggests that a 30-year averaging interval be used to describe climate. *Climate variability* refers to the fluctuations that take place within that interval. If there is a significant difference in the average conditions or in the pattern of variability between two time spans, we refer to this difference as a *climate change*. In evaluating climate change, we can compare one 30-year interval to another or to the average over the past 100 years or 1,000 years, or we can look at any month, year, decade, and so on within our 30-year period and compare that with the 30-year average. However, one thing that quickly becomes apparent when we carry out this sort of exercise is that there is no such thing as "normal" when it comes to climate. Climate varies continuously.

If we examine the climate of any 30-year (or longer) time span, we notice that conditions during that interval are seldom average: At any time the climate is almost always warmer or cooler, wetter or drier, than the average

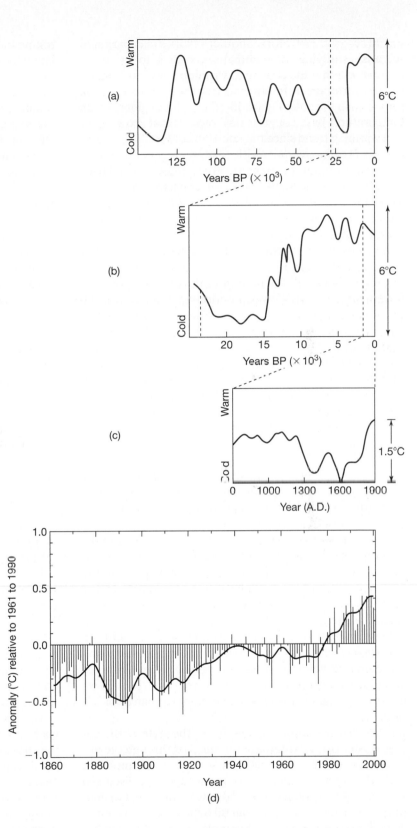

FIGURE 15-1

Mean global temperature change since the last glacial maximum. (a) Generalized oxygen isotope curve from deep-sea sediments. (b) General estimates from pollen data and alpine glaciers (emphasis on mid-latitudes from eastern North America and Europe). (c) General estimates from historical documents (emphasis on the North Atlantic region). (d) Annual anomalies of global average land-surface air temperature (°C), 1861 to 2000, relative to 1961 to 1990 values. The smoothed curve was created using a 21-point binomial filter giving near decadal averages. (*a-c* are adapted from U.S. Committee for GARP, 1975, Understanding Climate Change, National Academy of Sciences, Washington, D.C. *d* is from the *IPCC Third Assessment Report—Climate Change 2001: The Scientific Basis.*)

would suggest. Therefore, another favorite occupation of scientists studying these climate records is to try to determine whether the climate is warming or cooling, or getting wetter or drier. In other words, the researchers look for *trends* in the data. Figure 15-1d is a plot of global mean temperatures based on historical records that show a general warming trend since the late 19th century. This warming is in broad agreement with what we might expect from the increase in anthropogenic greenhouse gases. However, to extend the temperature series backward, we have to use **proxy data**—data that cannot be obtained by direct measurement but can be inferred from other evidence. When we do, we might interpret the warming trend in the early 20th century as part of the recovery from an earlier cool period (Figure 15-1c). Or, if we go further back, we can see it as a small fluctuation in the *cooling* trend following an even earlier warm period appoximately 6,000 years ago (Figure 15-1b). Hence, any analysis of trends is dependent on the time scale chosen: What looks like a trend at one time scale may simply be a short-term variation at another. Consequently, as we discuss short-term climate (climate over human time scales), we are much more concerned with the variations in climate than with the average conditions.

Proxy Climate Data

We have a vast quantity of data with which to describe and analyze the present-day climate system. An array of measuring stations covers much of Earth's land surface, and ships and floating buoys provide measurements for much of the world's oceans. Where there are gaps in the coverage, we can typically fill them in with satellite data. In fact, in terms of quantity, satellites now provide the bulk of the observational climate data set. Consistent and reliable satellite data, however, have been available only since the early 1970s. The data prior to the satellite era have huge gaps over the oceans, and over the land the climate record is highly variable—some regions have several hundred years of observations, others less than 30. With such a short record of observational data, how do we determine climate variability and climate change of the past?

We can make some estimates of the state of the climate system at different times in the past through the use of proxy data. We have already seen some examples in earlier chapters. In Chapter 14, for instance, we discussed the geological evidence of the advance and retreat of continental glaciers, such as glacial till deposits and moraines, as well as atmospheric and climate reconstructions derived from ice cores.

The chemical composition, physical structure, and sedimentation pattern of materials that make up sedimentary rocks provide some indication of the envi-

ronmental conditions that were in existence when the sediments were deposited. We can obtain these data both by examining sedimentary rocks on land and by recovering cores drilled in the sea floor that provide a record of the sedimentary history of the ocean basins. Fossils in these sediments can tell us much about the physical environment at the time the organisms lived. If we assume that fossil plants, animals, and protists lived in the same environment as those having similar physical characteristics that exist today, then by analogy we can make some assessment of what the environment was like at that location in the past. The principle that processes operating today also operated in the past, and that past geologic events can be explained in terms of processes active today, is known as **uniformitarianism**. This principle governs much of the work done in *paleoclimatology* (the study of past climates).

As we move to time scales covering the past 10,000 years, we make use of other types of evidence. One of the most useful data sources is ice cores, particularly for the earlier part of this interval. Numerous other techniques are also employed for climate reconstruction. Two of the most common are based on *palynology* and *dendrochronology*.

Palynology is the study of pollen and organic microfossils. Pollen grains are preserved in many different environments (e.g., lake sediments and peat bogs). If we drill a core into sediments from a lake or from a peat bog, divide the core into segments going back through time, and then extract the pollen from each layer, we can reconstruct the plant assemblages that lived in the area of that core at each time interval in the past. We then use the present-day distribution of those assemblages to place constraints on what the environment was like in the past. Pollen data from peat bogs have been used to reconstruct climate over the past 30,000 to 35,000 years in the British Isles. Radiocarbon dating (Chapter 5) can be used to date different levels of the core to associate environmental changes with particular time periods.

Dendrochronology is a method of dating trees by counting their annual growth rings. This method uses the growth characteristcs of certain tree species. The cross-section of a tree trunk consists of a series of rings, each ring representing growth over one year. By counting the annual rings, we know the age of the tree. The width of each ring indicates the amount of growth that occurred during the growing season. In certain circumstances, that amount can be related to the temperature during the growing season or to water availability, both of which can tell us something about the climate while the tree was alive. For example, a climate record that extends back almost 5,500 years has been reconstructed from tree rings of bristlecone pines in the White Mountains of California.

The Holocene

Assembling proxy climate evidence from around the world, we see that the Holocene displays a considerable range of climate change and variability. It is difficult to generalize, but there appears to be a dominance of temperature changes in the middle and high latitudes, whereas the tropics and subtropics appear to have experienced greater changes in moisture availability. These changes result partly from orbital effects that enhance seasonality and continentality (which directly affect the temperature regime) and partly from the resulting circulation changes (such as changes in the monsoon circulation) that affect precipitation patterns. Other factors may also have played a role in regional climate changes, as we will see later in this chapter.

Although there is evidence of widespread and frequent changes in climate throughout the Holocene, that evidence is not continuous in time or space. It is also difficult to assign accurate dates to some of the changes that we observe. Consequently, when we attempt to explain the climate record, it is not always easy to establish which changes are regional and which are tied to some larger-scale global forcing. Although we focus on only the broad patterns of Holocene climate change that are apparent in the record, it is still difficult to establish the degree to which these changes are global. Evidence of synchronous climate changes can be found in numerous locations around the world, but they are not found everywhere. Even where the evidence suggests that a change occurred in several different parts of the world at the same time, the evidence also indicates that the magnitude of the event was not the same at all locations. We will return to this fact in our discussion of global warming in Chapter 16. Although we tend to talk in terms of mean global changes, remember that there may be very large differences in the regional expression of these changes.

Another point to note is that relatively small changes in the mean global temperature are associated with relatively large changes in the physical environment. Mean global temperatures at the height of the last glaciation were probably only 5 to 6°C lower than the 20th century mean. Eight hundred years ago, when the Vikings were able to colonize parts of Greenland and grow sufficient crops to maintain a continuous settlement, mean global temperatures were only 0.5°C or so warmer than they are today. The significance of these numbers is highlighted by the fact that global temperature changes predicted for a doubling of atmospheric CO_2 are on the order of 1.5 to 4°C. In other words, the changes we predict for potential global warming in the 21st century are much greater than anything that has occurred in the past 10,000 years, and those changes are of comparable magnitude to the warming that took place between the last glacial maximum and the present day.

The Younger Dryas

In very broad terms, Earth began to warm between 10,000 and 15,000 years ago. The ice sheets that covered large areas of North America and northwest Europe began to retreat. Global sea levels rose as the glaciers melted, and the erosion from large volumes of meltwater reshaped the landscape around the edges of the former ice margins. Vegetation began to colonize the previously glaciated regions, and new vegetation patterns developed as soils formed and temperature and rainfall patterns changed. This spread of milder, more benign conditions came to an abrupt end 10,500 years ago in a climatic reversal known as the **Younger Dryas** event. The Dryas flower was widespread at that time—giving its name to the climatic event—but is currently found only in arctic and alpine tundra.

Some of the best evidence of climate and vegetation changes during the Younger Dryas comes from pollen and geologic analyses in northern Europe. As the climate warmed after the glacial retreat, there was a general increase in the density of vegetation, particularly grasses and sedges. This increase was followed by an increase in shrubs such as juniper and in willow; in some areas, the shrubs were later replaced by birch woodland. Pollen analysis shows a similar sequence of events in the British Isles, Ireland, and Scandinavia. Geologic evidence indicates that most of Scotland was probably deglaciated by 11,000 years ago. At this point a major climatic reversal occurred. By about 10,800 years ago, there was a new ice sheet several hundred meters thick over western Scotland, and there was renewed advance of valley glaciers in the upland regions of northern Europe. (*Valley glaciers* are individual glaciers that form at the head of a valley in mountainous regions and flow down the valley.) The pollen evidence shows a synchronous change in vegetation. The northern woodland diminished in area and was restricted to a few sites. The vegetation became more open, and the pollen data show a predominance of cold-tolerant vegetation types.

These changes represent a significant climate shift in northern Europe. However, the global impacts of this shift are more subtle. There is evidence of a similar climate shift in New England and along the east coast of Canada. There is little other evidence from North America except in the Gulf of Mexico and the Gulf of California, and there is only limited evidence from the Mediterranean. However, climatic reversals also appear to have occurred in the Andes and in Africa. The strongest evidence from Africa comes from lake levels, which increased after the northern deglaciation. During the early Holocene, what we know today as the Sahara Desert was primarily grassland (savannah), and the ecology of the region was very different from what we find today. However, lake levels declined again after 11,000 years ago; the evidence indicates that while the Younger Dryas event was taking place in northern Europe, much of tropical and subtropical Africa was experiencing increased aridity. Furthermore, data obtained from ocean cores taken in the western tropical Pacific Ocean and off the coast of Japan show some indication of a climate change at that time. Glaciers in the Southern Alps of New Zealand also readvanced during this interval.

The Younger Dryas thus appears to be centered primarily on the North Atlantic region, but nearly synchronous climate changes occurred in many other parts of the globe. What process might explain a shift in climate that has a strong regional, rather than global, focus yet is able to influence widely scattered regions across Earth's surface?

North Atlantic Deep-Water Formation.

A prime candidate for explaining the strong regional focus of the Younger Dryas climate change is the ocean circulation of the North Atlantic. The relatively high sea-surface temperatures in the northeast North Atlantic, which bring mild conditions to northern Europe today, result from the northward movement of the warm surface waters of the Gulf Stream and North Atlantic Drift. We saw in Chapter 5 that this movement was controlled in part by the atmospheric circulation and in part by the thermohaline circulation. Recall that the North Atlantic thermohaline circulation is driven by deep-water formation in the Norwegian and Greenland Seas: As the cold and highly saline water subsides and moves southward, it is replaced by warm, northward-moving water at the surface. Geochemist Wallace Broecker has suggested that some of the climate changes that accompanied deglaciation resulted from events that cut off or reduced this deep-water formation. Meltwater from the North American ice sheet, which normally flowed southward to the Gulf of Mexico, might have been blocked by a retreating lobe of ice as the ice cap melted. The meltwater then would have flowed eastward through the Gulf of St. Lawrence. The result would have been a large infusion of cold freshwater to the northern North Atlantic. Because freshwater is less dense than saltwater, this infusion would have produced a stable surface layer that would freeze very easily, pushing the sea ice margin southward and cutting off the formation of the North Atlantic deep water. Both the change in the thermohaline circulation and the southward expansion of the sea ice would have cut off the flow of warm surface water in the North Atlantic Drift, which would have resulted in a significant climate change in the region. Such a process could account for the climate reversal experienced during the Younger Dryas and would also explain the apparent focus on the North Atlantic region.

An On–Off Switch in the North Atlantic.

Although we have known about the Younger Dryas event for years and we have known that the event took place fairly rapidly (in geological terms), using ice core data obtained from the Greenland ice cap in the early 1990s, a team led by Richard Alley (a glaciologist at the Pennsylvania State University) revealed the startling information that these changes might have taken place in less than a decade. The snow accumulation record (Figure 15-2) shows increased accumulation in the warmer intervals; it also shows that the switch from cold to warm intervals occurred over a very short time span. Atmospheric dust deposited on the ice and recorded in the ice cores reveals similar rapid changes in deposition rates. More dust is deposited during glacials than during interglacials, because the increased north–south temperature gradient results in a stronger atmospheric circulation; that stronger circulation carries more dust. Both the snow accumulation and the dusty deposition, therefore, indicate a shift in the atmospheric circulation. The rapidity with which these changes occur suggests that the system switches almost instantly between two modes of circulation.

These data are of interest for two very different reasons. One obvious question concerns the processes or events that toggle the switch. The most likely cause is the one we have described here: a change in the oceanic circulation. Mechanisms that switch the thermohaline circulation on or off, or that cause sudden large movements in the latitude of the maximum sea-surface (and atmospheric) temperature gradient, could produce the sudden shifts in the atmospheric circulation that the ice core data suggest. So how did this result in glacial advance in New Zealand? At first glance it is difficult to see the connection. However, the ice core data show that the temperature record is almost perfectly matched by changes in atmospheric CO_2. Coming out of the last glacial maximum, as temperatures increased, atmospheric CO_2 increased. As temperatures dropped going into the Younger Dryas, CO_2 levels dropped—and both increased again at the end of the event as warming re-commenced. The changes in atmospheric CO_2 explain why the event was global—but what caused the CO_2 changes? This is open for debate, but the most likely explanation at present is that shutting down the thermohaline circulation in the North Atlantic triggered the event. The buildup of Northern Hemisphere glaciers intensified the north–south temperature gradient, which led to increased winds and increased erosion and the transport and deposition of aeolian dust in the oceans. Iron in the dust fertilized the oceans, resulting in increased phytoplankton growth that led to a draw-down of atmospheric CO_2—and a global climate change (refer back to the CO_2 feedback mechanisms discussed in Chapter 14). Observations show that the changes in the Southern Hemisphere do, in fact, lag slightly behind the start of the event in the North Atlantic.

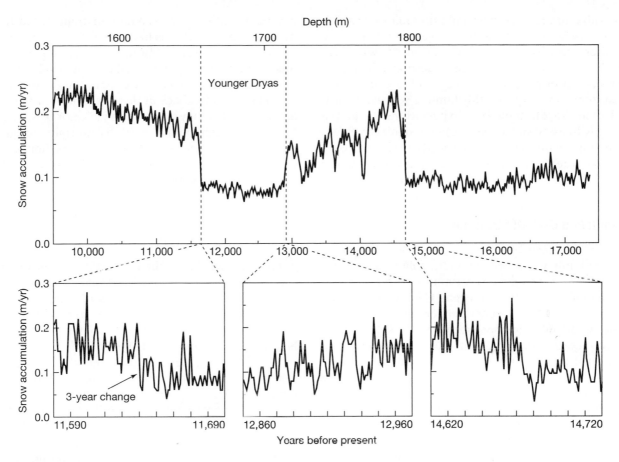

FIGURE 15-2

The snow accumulation record from a Greenland ice core. (After R. Alley et al. Abrupt Increase in Greenland Snow Accumulation at the end of the Younger Dryas Event. *Nature*, 362:527–529, 1993.)

The exact mechanism for explaining the shift in circulation is the subject of much speculation and discussion. One suggestion is that attributing the cause to any single process may be a mistake. An alternative approach is to view the system as chaotic. *Chaos theory* represents a rapidly emerging branch of science dealing with dynamic systems. Chaotic systems are iterative: The state of the system at one point in time is dependent on the state at the previous point. However, a characteristic of these systems is that very slight changes in the starting point are amplified through positive feedbacks, so the possible results diverge rapidly after only a short interval. Almost identical starting points can result in very different outcomes, and different starting points can produce outcomes that are very similar. The consequence is that, after a certain interval, the system becomes essentially unpredictable (refer back to "Critical Thinking," Problem 1, in Chapter 14).

In 1960, Edward Lorenz, a meteorologist at the Massachussetts Institute of Technology, was the first to recognize that the atmosphere is a chaotic system. Since then we have known why accurate long-range weather forecasting is impossible and why accurate daily weather forecasts can be effective only on time scales of days to a couple of weeks. One important attribute of chaotic systems, however, is that they exist in quasi-equilibrium states; in the case of the atmosphere, for example, we know in general what it will be like next year, even though we cannot predict exactly what it will be like on any given day. In the terminology of chaos theory, these quasi-equilibrium states are called *strange attractors:* The system is never precisely at that point, but it is always somewhere close to it. Another characteristic of chaotic systems is that they may switch rapidly between two or more of these quasi-equilibrium states. The point at which this switch occurs is called a *bifurcation point.*

We can view the climate system as having two (or more) stable steady states: glacial and interglacial, with the transition between the two representing the bifurcation point (see Figure 14-8). When the climate system is near the bifurcation point (e.g., at the end of the last glaciation), the system is unstable and any number of small perturbations could be amplified through positive feedbacks to push the system rapidly toward one stable state or the other. This possibility leads us to the second reason why the Greenland ice core data have generated considerable interest. If these data do indicate that climate can switch rapidly between two very different stable states after a relatively small perturbation, a new wrinkle is added to the greenhouse warming question: If increased greenhouse gases should lead to a rapid shift in the climate system (possibly to a third, much warmer, quasi-equilibrium state), then our expectations of global warming (discussed in Chapter 16) may significantly underestimate the strength and the speed of the climate response to greenhouse forcing (see "A Closer Look: Stochastic Resonance and Rapid Climate Change").

The Holocene Climatic Optimum

The short cold interval of the Younger Dryas event ended with a rapid shift to warmer conditions, followed by a constant climate or relatively slow warming over the next several thousand years—the **Holocene Climatic Optimum**. Evidence suggests that summer temperatures were slightly higher in the mid-Holocene (5,000 to 6,000 years ago) than are recorded in the recent (20th century) record. This interval was long thought of as a time of persistent mild conditions with very little climate change. This view originates from early studies in Europe, where the pollen record shows little evidence of major climatic shifts. Elsewhere in the world the picture is less straightforward.

Ancient lake levels in East Africa and the Sahara Desert show evidence of much wetter conditions than exist today. These high lake levels occurred when temperatures were higher—presumably, evaporation would have been greater then. This means that the amount of rainfall in the region must also have been greater than it is today. The evidence suggests a northward shift of the intertropical convergence zone in the Northern Hemisphere summer, bringing higher rainfall to the Sahara and East Africa and an enhanced monsoonal circulation (increased summer rain) in Arabia and northwest India. There is also evidence that the Mediterranean Sea experienced increased summer rainfall during this same interval. Turkey and western Iran, however, appear to have been more arid than at present.

That the mid-Holocene climate was very different from today's climate is also indicated by archeological evidence. The Tarim Basin, the site of the ancient silk route from China to Europe, is currently a desert, but between 5,000 and 6,000 years ago it was forested and populated with numerous settlements. Nomads grazed cattle in the central Sahara, and the Harappan culture flourished in the Indus River valley. This agricultural society, located in what is now the Rajasthan Desert, may have been the first to cultivate cotton. At first glance, changes in climate would appear to be a likely explanation for the decline of these earlier civilizations. Undoubtedly, the Sahara, for example, is now much drier than it was then. In several cases, however, it is possible that human land-use

practices, rather than a change in climate, led to land degradation. In fact, it is likely that the collapse of some of these cultures resulted from the interaction of both factors—where land degradation due to human land use was amplified by changing climate conditions. It would seem reasonable to look back at this, and other warm periods, and suggest that they might be an indicator of what our future climate may be like as global warming progresses. Unfortunately, analogs from the past are of limited use for predicting the future, as the forcing factors are different. The climate response, therefore, is also likely to be different.

The Medieval Warm Period

Temperatures fell after the Holocene Climatic Optimum, reaching a minimum about 3,000 years ago, but rose to a new maximum during the **European Medieval Warm Period**. Temperatures in Greenland appear to have increased after A.D. 600 to 650, with the temperatures in the North Atlantic (Greenland and Iceland) reaching a maximum around A.D. 1100. At this time the Vikings established a self-supporting colony on the southwest coast of Greenland. The colony lasted more than 400 years and, at its largest extent, had 280 farms and a population of 3,000. By the end of the 12th century, however, because of the decreasing temperatures, the sea ice east of Greenland grew more extensive. By the middle of the 14th century, ships had to take a more southerly route to avoid the ice. By A.D. 1410, communication with the Greenland colonies was lost completely.

Farther south, in northern, western, and central Europe, the Medieval Warm Period reached a maximum between A.D. 1150 and A.D. 1300. During this period, wheat was grown in Norway at about 64° N, oats and barley were grown in Iceland, vineyards were cultivated in England, and farm settlements spread to higher elevations in Norway, northern England, and Scotland—all evidence of milder climates than those areas have today. The average temperature of central England during this interval is estimated to have been 0.5–0.8°C above the mean for the first half of the 20th century.

From the beginning of the 14th century, the climate became more variable. Wetter (and probably colder) summers in Europe from A.D. 1313 to A.D. 1317 led to a succession of failed harvests and widespread famine, and the expansion of farms into the upland regions of northern Europe and Scandinavia came to an abrupt end. The interval from A.D. 1250 to A.D. 1350 was one of numerous large storms and floods. It has been suggested that the storminess was caused by a cooling at high latitudes that caused the sea ice to expand southward, resulting in an increased temperature gradient in the North Atlantic. Flooding along the North Sea coasts of Denmark and Germany was extensive, and 100,000 to 400,000 people

A CLOSER LOOK:
Stochastic Resonance and Rapid Climate Change

In the text we describe the rapid climate shift back into the Younger Dryas that occurred as the climate was warming from the last glacial maximum. We also explained how this could come about through shutting down the thermohaline circulation in the North Atlantic, followed by feedback processes that reduce atmospheric CO_2 concentrations. If we look in a little more detail at the climate record, however, we see that rapid climate changes such as these are actually very common (Box Figure 15-1). The temperature record from central Greenland (ice core data) reveals large swings in climate during the last glaciation. Warm episodes punctuate the generally cold conditions and, as we move out of the glaciation, we see the system gradually warming then suddenly dropping back into cold conditions, warming, and dropping back again. Some of these rapid cooling events are accompanied by the flooding of the North Atlantic with glacial meltwater as described in the text, but many of them are not. Some appear to be random, while others appear to be periodic. Nothing we have discussed in the text really accounts for the behavior that we see here. One explanation that is gaining in popularity is the concept of **stochastic resonance**. Looking at the temperature record up until the start of the Holocene we see the characteristics of what is referred to by physicists as an "excitable system"—a system that has a stable and an unstable mode (equilibrium state). It appears that there is a preferred "cold" mode and an unstable warm mode that the system cannot occupy for long.

Imagine a very weak but periodic forcing. This could, for example, be a very small change in solar output that occurs on a regular cycle, but is too weak to promote a significant change in climate. Superimposed on this are random variations that are inherent in the climate system. Taken together, these push the system across a threshold, and then what we need is a mechanism that can amplify the signal (Box Figure 15-2). The most likely mechanism that we can come up with at present is a change in the thermohaline circulation. It doesn't have to be anything as dramatic as shutting it down—maybe just a change in the location of the bottom water formation. What we then have is a situation with a stable cold mode (during the glaciation) with random variations (the "stochastic" part of stochastic resonance). At periodic intervals some other forcing gives a very weak push to the system, which "resonates" with the random variations and pushes the forcing beyond a critical threshold. At that point the change is amplified by the amplification mechanism (maybe ocean circulation) and pushes the climate into an unstable warm mode. The climate stays in this mode for a short interval before rapidly dropping back into the stable cold mode again. Analysis of the ice core data by Richard Alley and colleagues (see text) indicates that there is indeed a periodic forcing that occurs at an interval of about 1,500 years that accounts for much of the variation in the temperature record.

As we move out of the glacial episode, the reverse seems to happen. Now the system is warming up but we have sudden events like the meltwater release to the North Atlantic that temporarily pushes the climate back into a cold mode again. Where does this leave us? Climate scientists are looking at these ideas and asking, if these are correct, where does the periodic forcing come from? What is the amplification mechanism—is there more than one? Can the same thing happen during interglacial periods, when the planet seems to be in a more stable warm mode? Is there some other altogether different process waiting to be discovered that would explain what we see in the record? All of these are interesting questions, but why this is truly worth a closer look in the context of this text is the fact that the climate record clearly shows that rapid and large changes do occur that are essentially unpredictable. It also shows that as the system is transitioning between states (i.e., as Earth was warming coming out of the last glaciation) it is common for it to be pushed back in the other direction very rapidly. Neither of these possibilities is taken into account in any of the projections currently put forward for global warming. While Chapter 16 describes what we can project ahead in terms of global warming, at the back of your mind you should keep note of the fact that processes such as those described here have the potential to surprise us—and completely change the magnitude (and maybe even direction) of the changes we predict for the future.

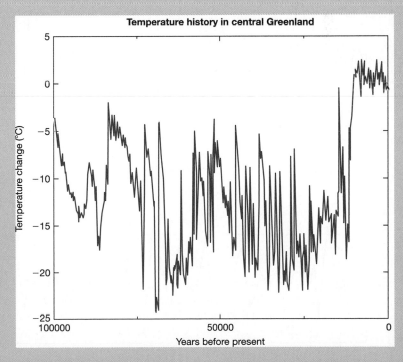

BOX FIGURE 15-1

Temperature reconstruction for Central Greenland based on ice core data. Very large fluctuations have been common except recently. (Courtesy of Richard Alley, The Pennsylvania State University.)

297

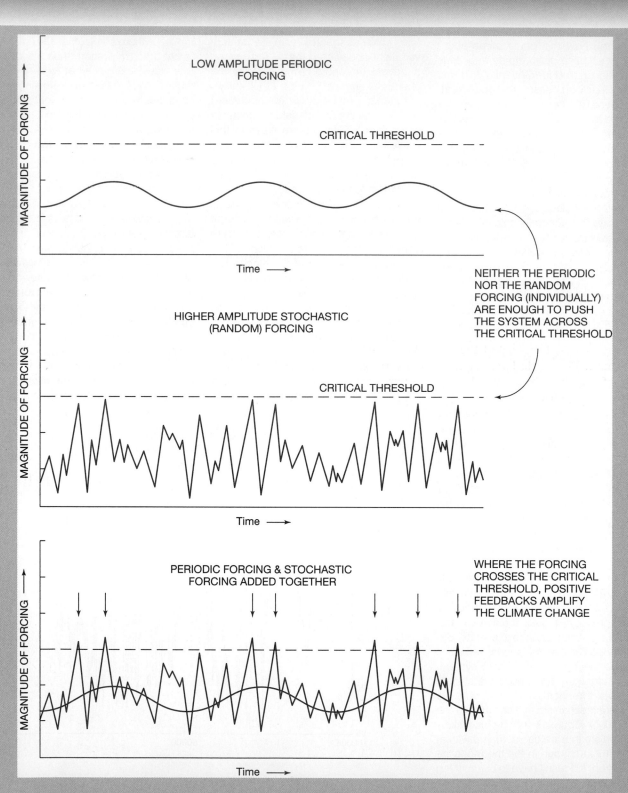

BOX FIGURE 15-2

Schematic representation of stochastic resonance.

were reported to have drowned in the floods that occurred at that time. The impact of these large environmental changes on human societies was then compounded by the arrival in Europe of the bubonic plague in 1346. The plague lasted until 1361, killing an estimated 25 million people (one-quarter of Europe's population).

The Little Ice Age

The extreme climatic fluctuations that marked the end of the Medieval Warm Period led to an interval of cooling until the early 1500s, after which the climate stayed relatively stable or even began to recover in some areas. The North Atlantic climate, however, entered a renewed episode of rapid cooling in the late 1500s, a period now known as the **Little Ice Age**. Although the cooling was originally thought of as a regional climate fluctuation centered on western Europe and the North Atlantic, a growing body of evidence from the European Alps, Asia and the Himalayas, South America, New Zealand, and Antarctica suggests that changes might have occurred over much of the globe. However, not all parts of the world show evidence of a climate change at this time. Where they do, it is not clear that the changes were synchronous or that they lasted the same length of time everywhere. The Little Ice Age continued through the middle of the 19th century, but the reduced temperatures were not continuous. The Little Ice Age is characterized by considerable variability, with episodic cold spells that varied in timing and duration from place to place.

The evidence of the Little Ice Age takes numerous forms, among them the readvance of mountain glaciers, the lowering of tree lines, increased erosion and flood-ing, sea-ice expansion, and the freezing of canals and rivers. The canals of Holland have long been used for transportation, and there are reliable records of freeze-up since 1633. The canals seldom freeze over in today's climate, but in the 15th and 17th centuries it was common for them to be frozen for 3 months at a time. There is also documented evidence of glaciers in the Swiss Alps advancing and covering houses on the outskirts of several villages (Figure 15-3). A climate change is also suggested by indicators of population decline: the abandonment of settlements and decreasing agricultural productivity. However, the societal evidence is not straightforward; population changes and declining productivity were almost certainly influenced by societal and political factors and by disease.

As with the Younger Dryas event described earlier, the Little Ice Age appears to have had a strong regional focus, yet evidence is mounting that many other parts of the world also experienced a similar cooling. In this case, however, there was no retreating continental ice sheet to help force large changes in the oceanic circulation. Are there any other likely explanations for this particular shift in the climate system?

Volcanoes and Climate. From earlier chapters we know that atmospheric composition can change and that these changes can have a significant impact on climate. Are there any mechanisms for changing the atmospheric composition on the time scales discussed in this chapter? Yes. Two primary mechanisms are worth discussing: (1) a change in gas composition due to anthropogenic activity and (2) volcanic activity. Anthropogenic activity, the sub-

(a)

(b)

FIGURE 15-3

The Argentiere glacier in the French Alps. (a) An etching made about 1850, showing the extent of the glacier during the waning phase of the Little Ice Age. (b) The same view, photographed in 1966. (Reprinted with permission from Understanding Climate Change, A Program for Action, National Academy of Sciences, Washington, D.C.)

ject of Chapter 16, cannot explain changes prior to the 20th century. Here we focus on the impacts of volcanoes on climate.

There has been an ongoing debate over the possible effects of volcanic eruptions on the global climate. We have already seen that volcanism played an important role in the origin and early evolution of Earth's atmosphere. It has also been suggested that volcanic eruptions have a direct impact on today's climate, and they might also account for the severity of the climate during the Little Ice Age. As early as 1784, Benjamin Franklin hypothesized that the abnormally cold winter of 1783–1784 might have been due to the 1783 eruption of Hekla, a volcano on Iceland. Actually, any climate anomaly in that year was more likely due to the 1783 eruption of Laki, another Icelandic volcano. Laki is, in fact, the largest effusive eruption in the historic record, producing about 12 km^3 of basalt lava flows.

Scientists once believed that the volcanic ash injected into the atmosphere during an eruption would result in an increase in global albedo, which would reduce the solar input to the surface and result in a surface cooling. The volcanic ash, however, aggregates very quickly and falls out of the atmosphere within tens to hundreds of kilometers of the eruption. More important to climate is the gas (primarily sulfur dioxide, SO_2) injected to high altitudes in the stratosphere. Sulfur dioxide gas oxidizes, forming sulfuric acid droplets, an atmospheric aerosol. Remember from Chapter 2 that this aerosol scatters and reflects solar radiation, reducing the solar radiation that reaches the surface (hence cooling the surface). The aerosol also absorbs upwelling infrared radiation, resulting in a warming of the stratosphere. These effects, although more important in a global sense than are those due to volcanic ash, are still relatively short-lived. The **residence time** is the length of time an aerosol is present in the stratosphere after a volcanic eruption. The aerosol concentration in the atmosphere decreases at a rate that is analogous to the rate of radioactive decay. You can think of the residence time as being similar to the half-life of the aerosol in the atmosphere. The residence time of a volcanic aerosol in the stratosphere is on the order of about a year. Consequently, any climatic effects from a single eruption are also very short-lived, lasting only 1–2 years after the event.

The degree to which a volcanic eruption affects climate depends in part on the location of the eruption and on the way the atmospheric circulation distributes the stratospheric aerosols globally. Mount Pinatubo in the Philippines experienced a major eruption in June 1991, when the equivalent of 3–5 km^3 of dense rock was ejected into the atmosphere (Figure 15-4). The ejecta also included approximately 20 Mton of SO_2 gas, which was subsequently converted to sulfuric acid aerosol particles, with the largest concentration occurring in the lower

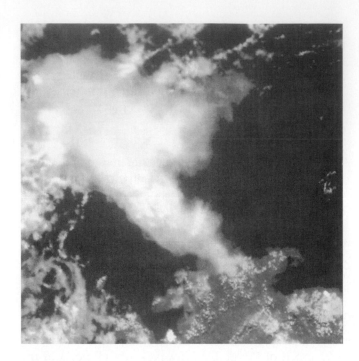

FIGURE 15-4

The eruption of Mount Pinatubo in the Philippines, June 1991. (From Robert M. Carey, NOAA/Mark Marten, Photo Researchers, Inc.)

stratosphere at an altitude between 15 and 20 km. The aerosol cloud was distributed very rapidly around the globe by the stratospheric circulation. After about 10 days, the cloud had been distributed fairly uniformly over an area of 15 million km^2, in a 10,000-km-wide band from Indonesia to central Africa (Figure 15-5; see color section). The cloud circled the globe in 22 days.

It is clear from the observations that a major volcanic eruption in the tropics can produce an aerosol cloud that is distributed rapidly around the globe and has an observable effect on the radiation budget of the stratosphere. But, can such an eruption also have a significant effect on the tropospheric climate?

One way to address this question is to measure or estimate the effect the aerosol cloud has on the radiation balance. A perturbation in the radiation budget caused by the presence of volcanic aerosols is referred to as **volcanic aerosol forcing**. To measure this forcing, we need long-term measurements of the global radiation budget. These long-term measurements then provide the background against which we can see the anomalies caused by the eruption. Such long-term measurements are now available from satellite observations, which, in this case, showed a dramatic increase in albedo over the tropics immediately after the Mount Pinatubo eruption. The satellites were able to measure strong anomalies in the radiation budget over much of the globe in the months that followed. The effects reached a maximum in late 1992 with a reduction in the mean global surface temperature

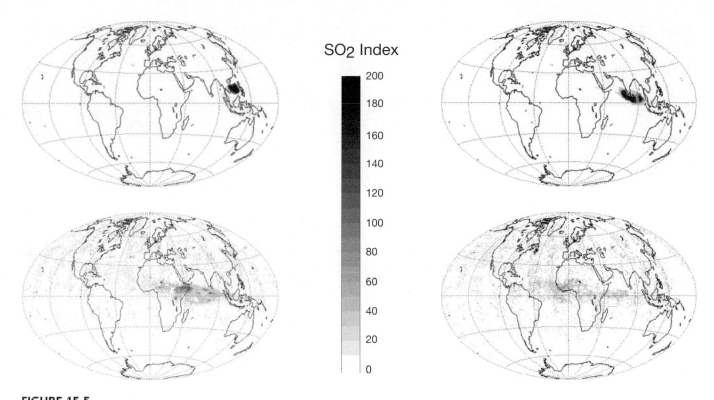

SO₂ Index

FIGURE 15-5

[See color section] Satellite observations of the Mount Pinatubo aerosol cloud, 1991. (a) June 16. (b) June 18. (c) June 23. (d) June 30. (After G.J.S. Bluth, et al. Global Tracking of the SO₂ Clouds from the June 1991 Mount Pinatubo Eruption. *Geophysical Research Letters,* 19:151–154, 1992.)

on the order of 0.5° C. This short-lived global cooling is evident in Figure 15-1d.

Interestingly, although temperatures tended to be lower in the year after the eruption, North America and parts of Europe and Asia had an extremely warm winter in 1991–1992. Such regional differences have been observed after several major eruptions during the past century. It has been suggested that the direct radiative impact of the aerosol produces cooling at the surface over the tropics but that the absorption of longwave radiation by the aerosol in the tropical stratosphere enhances the east–west circulation at mid-latitudes. This enhanced circulation draws more warm air over the continents, producing the warmer winters experienced by Europe and North America.

The historic record holds many accounts of volcanic eruptions. We can also obtain evidence of earlier eruptions from ice cores. Some of the acidic material injected into the atmosphere during a volcanic eruption may be deposited on a glacier's surface during snowfall events. When the snow is converted to ice, that ice layer has an anomalously high acidity. By counting the annual ice layers, we can obtain an accurate date for the eruption as well. For more recent eruptions, we can typically match the ice-core evidence with historic accounts. For earlier volcanic events, for which neither historic records nor ice-

core data are available, we can still obtain evidence of eruptions from the geologic record in the form of old lava flows and deposits of volcanic material, such as ash. These different data sources can be combined to present a record of the frequency and magnitude of past volcanic events. The climatic effects of these events are normally expressed in terms of temperature changes, with the temperature data being derived from a similar variety of sources (e.g., historic records, ice cores, and tree-ring reconstructions).

The relationships between volcanoes and temperature have been examined in numerous ways: by direct comparison of the two records, by making a composite of the temperature record from several eruptions (using several months or years before and after each eruption), and by using numerical models to predict the effects of any one eruption. When we consider several eruptions together, the results suggest a global cooling of 0.2–0.3°C for 1–3 years after an eruption (Figure 15-6). Individual eruptions producing a significant aerosol cloud can also result in a 1–3 year cooling of 0.3–0.7°C. Such eruptions include Tambora (Indonesia) in 1815 and Krakatau (Indonesia) in 1883. These effects are relatively small, but it is possible that the effects of larger or more frequent volcanic events in the past were much greater than those of more recent events. The eruption of Tambora in April

FIGURE 15-6

Cooling due to volcanic eruption: The global mean temperature changes for 5 years preceding and following a large volcanic eruption (at year zero). The temperatures are the average changes noted for five major eruptions: Krakatau, August 1883; Santa Maria, October 1902; Katmai, June 1919; Agung, March 1963; and El Chichón, April 1982. The effects of ENSO (discussed later in this chapter) on temperatures have been removed. (After A. Robock and J. Mao, 1995. The Volcanic Signal in Surface Temperature Observations. *Journal of Climate*, 8:1086–1103.)

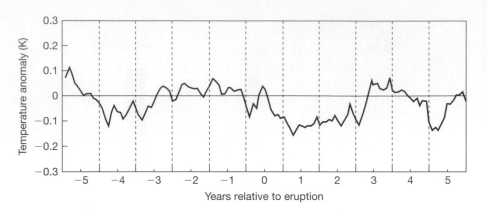

1815 was probably the largest in recent history, with about 50 km³ of erupted magma. The Toba (Indonesia) event of about 75,000 years ago, which was probably a series of eruptions, dispersed an estimated 2800 km³ of magma. The Laki (Iceland) eruption in 1783 produced 12 km³ of basalt, whereas the Roza flow (northwest United States) in the Columbia River Group (dated at 14 million years ago) released 700 km³ of basaltic lava in a single event.

The 1815 Tambora eruption is probably the most renowned eruption in terms of its effect on climate. Large acidity peaks for that time in ice cores from both Greenland and Antarctica show a global distribution of the aerosol cloud, which is estimated to have been about 10¹¹ kg (about five times larger than the cloud from the Pinatubo eruption). The following year, 1816, has been referred to as the "year without a summer" in Europe and northeast North America. Hudson Bay remained ice-covered that summer, and estimated average daily temperatures in this region were 5–6°C below the long-term average. Although these conditions were maintained only for 2 years, the cold weather from the winter of 1815–1816 through the summer of 1817 produced crop failures in China and bad harvests in India. The failure of the Indian harvest resulted in famine, which was followed by an outbreak of cholera that, over the next two decades, spread through Asia and Europe. However, the cold weather in the eastern Hudson Bay began with a series of anomalously cold years from the winter of 1811–1812, suggesting that not all the cold weather in 1816 was due to the eruption. Nevertheless, several other large eruptions occurred between 1811 and 1814, including the eruption of Vesuvius in 1813. The resulting cold temperatures could have been the cumulative result of the combined volcanic activity.

From the analysis of Greenland ice cores, we know that the interval from A.D. 1250 to 1500 and A.D. 1550 to 1700 were times of high volcanic activity, whereas A.D. 1100 to 1250 had far fewer volcanic events. The first two intervals coincide with the Little Ice Age; A.D. 1100 to 1250 lies in the Medieval Warm Period. Again, the evidence suggests a link between volcanic activity and climate, but that link is still not conclusive.

One problem with establishing the climatic impact of a volcanic eruption in historic times is that the perturbation due to the volcano is on the same order of magnitude as the natural (random) variability of the climate system. Despite the size of the Tambora eruption, it is still difficult to separate the volcanic signal from other sources of climate variability in the temperature record. As a further example, although the eruption of Mt. Agung (Indonesia) in 1963 was about two orders of magnitude less than that of Tambora, it resulted in a 6°C stratospheric warming that lasted 3 years and an estimated surface cooling of about 0.3°C in the Northern Hemisphere. Conversely, the eruption of El Chichón (Mexico) in 1982 produced about the same amount of magma and the same amount of aerosol as that of Agung (although there was a greater concentration of sulfuric acid), resulting in a 6°C stratospheric warming, but the El Chichón event had very little effect at the surface. Estimates ranging from 0.1–0.5°C were made for the expected surface cooling from El Chichón, but the signal, if present, was entirely swamped by a particularly large climatic event known as *El Niño–Southern Oscillation* (ENSO), which will be explained later in this chapter.

Other Causes of Holocene Climate Change

We can invoke volcanic eruptions and changes in ocean circulation to explain some of the more extreme changes in climate that followed the end of the last glacial period, but they do not explain all of the observed changes, particularly the changes to warmer conditions. What else do we need to account for these other changes? Although a large number of theories have been put forward, the actual causes of Holocene climate change are not very well understood. One suggestion is that the Milankovitch orbital forcing, which we discussed in Chapter 14, also plays a role here, with the precession of Earth's spin axis having the greatest influence.

Orbital Changes and Greenhouse Variations. The seasonal and latitudinal distributions of insolation 21,000 years ago were very similar to those of today (recall from Chapter 14 that the main precessional forcings are at 19,000 years and 23,000 years). Later in the Holocene, the precession changed in such a way that seasonality (the difference between summer insolation and winter insolation) increased in the Northern Hemisphere and decreased in the Southern Hemisphere. The atmospheric concentration of CO_2 was also at a minimum (200 ppm) 21,000 years ago, increasing to about 265 ppm by the mid-Holocene.

A group of scientists working on the Cooperative Holocene Mapping Project in the 1980s simulated the Holocene climate using a General Circulation Model. The model simulated many of the climate patterns that are indicated in the proxy record. At 21,000 years ago, the model produced colder temperatures over the continental ice sheet, together with steep temperature gradients at the ice margin, but little change in the tropics. At 6,000 years ago, temperatures in middle and high latitudes were generally warmer than present. There is no clear difference between the precipitation at 21,000 years ago and that today, except for a large reduction over Southeast Asia. At 6,000 years ago, however, summer rainfall was greater over the Sahara, the Middle East, and in the Southeast Asian monsoon region.

The broad pattern of change from the last glacial maximum through the Holocene Climatic Optimum to the present appears to be explained, therefore, by the same forcings that we outlined in Chapter 14. Explaining the Younger Dryas and the shorter-term changes within the Holocene (such as the Medieval Warm Period and the Little Ice Age), however, is not so simple. Volcanic activity and changes in the circulation of the North Atlantic provide part of the explanation, but there is also the possibility that some of the climate change was triggered by variations in our basic energy supply: the output of radiation from the sun.

Solar Variability. We have already seen that the output of the Sun has changed considerably over the long period of Earth's history. At the short time scales we are considering here, however, we tend to regard the solar output as constant. In fact, the output from the Sun probably varies slightly at *all* time scales—but is the solar variability sufficient to explain the observed climate changes?

The greatest changes in solar output at the decade-to-century time scale accompany variations in the number of sunspots. **Sunspots** are dark areas of lower-than-normal temperatures on the surface of the Sun. Imagine that magnetic field lines wind around the Sun like elastic bands. Where these bands are twisted together, the "knot" at the surface inhibits convection from below, resulting in

an area of lower temperatures (sunspots). Telescopes were first used by Galileo to observe sunspots in about 1610, but reliable sunspot data exist only for about the past 150 years. The historic record shows an 11-year cycle in sunspot activity (11 years between the peaks in sunspot abundance) superimposed on a 22-year cycle of magnetic reversals. Just like the magnetic field on Earth, the Sun's magnetic field also experiences periodic reversals, but the Sun's magnetic reversals are more frequent and regular.

The pattern of sunspot activity is illustrated by the *butterfly diagram* in Figure 15-7, which plots the location of sunspots on the Sun's surface through time. In late 1933, there were very few sunspots, but the number increased into 1937. Sunspots appeared first at higher solar latitudes, but, as they increased in abundance, the concentration shifted toward the solar equator. Sunspots then decreased in number until the next minimum in 1944, when, again, the activity switched to the higher latitudes. Plotted in this fashion, the cloud of points in Figure 15-7 resembles a butterfly wing. Note that the period of cycle is not always exactly 11 years. Even when the period is exactly 11 years from minimum to minimum, the curve is not symmetric: Less time passes between a minimum and a maximum than between the maximum and the next minimum.

An increase in the abundance of sunspots, which are darker (cooler) areas on the Sun's surface, corresponds with an *increase* in the amount of solar radiation Earth receives. The Stefan–Boltzmann law (Chapter 3) suggests that the opposite should occur: An increase in sunspot abundance should result in decreased luminosity and a drop in solar radiation. The reason for this behavior is that the sunspots are surrounded by bright areas of higher-than-normal temperature called **plages**, and the area of plages is larger than the area of spots. The net result is that the Sun is actually slightly brighter during periods when sunspots increase; hence, the amount of solar radiation emitted also increases. In reality, the dark areas of the sunspots and the bright areas of the plages almost cancel each other out.

How does this variation in sunspot activity relate to Earth's climate? Studies have discovered an 11-year or a 22-year cycle in numerous climate records (records of regional floods, droughts, temperatures, etc.). However, the change in solar output over the 11-year cycle is very small. Satellite measurements for the 1980s show that the incoming solar radiation decreased by only about 0.1% from the maximum to the minimum in sunspot activity. The direct climatic consequences of such a change are so small that they would be undetectable.

However, if we plot the mean global temperature changes over the past 100 years and superimpose on that the *length* of the sunspot cycle, we find a near-perfect

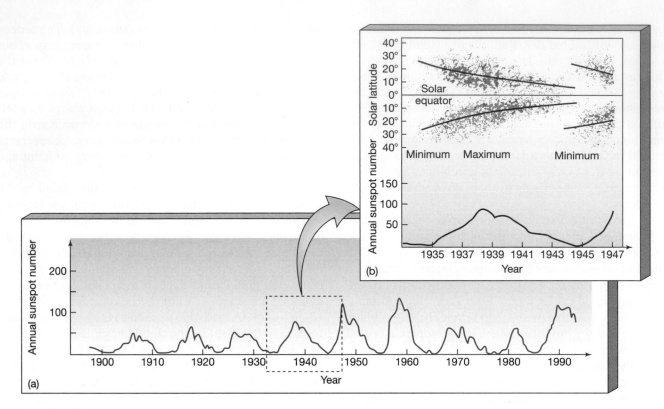

FIGURE 15-7

The butterfly diagram, showing the distribution of sunspots over the Sun's surface. (From E. Chaisson and S. McMillan. *Astronomy: A Beginner's Guide to the Universe*, 2/e, 1998. Reprinted by permission of Prentice Hall, Upper Saddle River, N.J.)

match (Figure 15-8). This correlation suggests a direct relationship between the sunspot cycle and fluctuations in Earth's temperature. But is the correlation real or only circumstantial? It has been suggested that the length of the sunspot cycle is related in some fashion to the efficiency of convection in the solar interior. More energy is transported upward when the cycle length shortens, thus raising the surface radiating temperature of the Sun and increasing solar luminosity. However, there is no evidence of

this transport, and no physical theory or model predicts that the sun should act in this way. Note also that the choice of scale on the vertical axes in Figure 15-8 is arbitrary, making the apparent match more obvious. However, this match may be purely coincidental, and there may be no physical relationship between these two parameters.

Systematic records of sunspot activity were initiated in 1848 by Rudolf Wolf, director of the Bern Observatory, after an amateur astronomer, Heinrich Schwabe, pub-

FIGURE 15-8

Variations of the sunspot cycle length (solid) and Northern Hemisphere temperature anomalies from 1861 to 1989. (After E. Friis-Christensen, and K. Lassen, 1991. Length of the Solar Cycle: An Indicator of Solar Activity Closely Associated with Climate. *Science*, 254:698–700.)

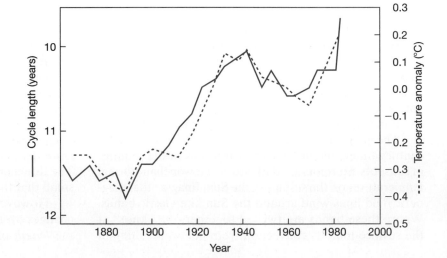

lished sunspot observations from 1826 to 1843. Wolf also compiled all of the pre-1826 data that were available, extending the record of sunspot abundance back to 1700. He also published the years of sunspot minima and maxima from 1700 back to the invention of the telescope by Galileo in 1610. The pre-1826 data are far from reliable, but they do indicate an interval between 1645 and 1715 when very few sunspots were recorded. The low sunspot abundance during this interval was noted by astronomers at the time and was discussed in detail by two solar astronomers in the late 19th century, Gustav Spörer and E. W. Maunder. This period of low sunspot activity is now referred to as the **Maunder Minimum**.

Large groups of sunspots, which occur during periods of maximum solar activity, can be seen by the naked eye (note that looking at the sun directly can damage your eyes; don't attempt to do this without the correct viewing equipment). Most of the earliest recorded sunspot observations, dating back to 28 B.C., were made in Japan, Korea, and China. More reliable data come from another form of proxy data: the ^{14}C content of tree rings. The variations in solar output over the sunspot cycle cause changes in the magnetic properties of the solar wind (the output of energetic particles from the Sun), which is thought to affect the production of ^{14}C in Earth's upper atmosphere (see "A Closer Look: Carbon-14—A Radioactive Clock"). High sunspot activity increases the magnetic field strength of the solar wind, causing a greater deflection of galactic cosmic rays away from Earth. The reduction in the number of cosmic rays entering Earth's atmosphere results in low ^{14}C production. Conversely, low sunspot numbers result in high ^{14}C production. The ^{14}C content recorded in annual tree rings is thought to be related to the amount of ^{14}C produced in the atmosphere.

The naked-eye observations and the proxy data indicate two other intervals when few sunspots occurred: The **Spörer Minimum** (1450–1534) and the **Wolf Minimum** (1282–1342). The fact that the Maunder and Spörer minima coincide with one of the major climate changes in the Holocene, the Little Ice Age, has led some researchers to speculate that the two may be related. Furthermore, the 12th and 13th centuries (the Medieval Warm Period) was the interval with the greatest sunspot activity in the record. Detailed studies of ^{14}C tree-ring records have also shown that mid-latitude glaciers advanced and temperatures fell when sunspot numbers were low and that temperatures increased and glaciers retreated when sunspot numbers were high. Unfortunately, not all of the changes in the ^{14}C record match suspected global climate changes.

Might the sunspot cycle have had a greater effect on solar luminosity in the past than it does today? At first glance, the answer is yes. Observations show that the Sun is significantly less variable in total light output than are other stars of similar magnetic activity, suggesting that our Sun could have been more variable in the past than it is now. However, the level of magnetic activity necessary to achieve this greater variability has not been observed in our Sun since the beginning of reliable sunspot records, so there is little chance that solar variability can explain climate changes over the past 150 years. In addition, the proxy data suggest that the present-day level of solar variability is as high as any achieved in the past several millennia. The changes in solar output estimated for the Holocene are still less than the amount estimated to be necessary for the observed climate changes.

In summary, we have considerable evidence that solar activity varies over a wide range of time scales, with the most regular variation being an approximate 11-year cycle in sunspot activity. Recent observations indicate that the total solar output over this cycle varies only slightly, and this difference is not sufficient to explain observed variations in climate. Ultraviolet radiation varies by a factor of three over the sunspot cycle and exospheric temperatures vary from about 1,000 K at solar minimum to 2,500 K at solar maximum. This may have an indirect effect on climate, but a reasonable explanation of how this may happen has yet to be advanced. If we look at variations in the length of the sunspot cycle, we see a close match with global temperature changes over the past century. There is also clear evidence that the major climate changes that have taken place over the past 1,000 years coincide with extremes in solar activity (as indicated by the number of sunspots observed). Again, we have no direct evidence that these changes in solar activity result in variations in solar luminosity large enough to explain the climate changes, yet the possibility still exists. Scientists are now investigating how sunspot cycles may modulate the cosmic x-ray flux to Earth, which may influence cloud albedo and thus climate.

We can thus present a picture of climate change over the past 20,000 years in which the broad trends from the last glacial maximum (at 21,000 years ago) to the present may be explained by Milankovitch orbital variations, possibly modified on the decade-to-century time scale by changes in solar activity. On shorter time scales, further variablity may be introduced by changes in ocean circulation and volcanic activity. The nature of the climate changes that occurred at any point in time may be a result of some combination of various factors. The Little Ice Age, for example, was a time of low solar activity, with extreme cold spells possibly caused by periods of enhanced volcanic activity. The combined effects of volcanic eruptions, orbital changes, and increasing atmospheric CO_2 concentrations are well illustrated in recent energy balance modeling studies by Tom Crowley at Texas A&M University. Crowley compared a 1,000-year record of global temperatures with the results of the climate model incorporating solar and volcanic forcing (Figure 15-9). The volcanic and solar forcing explains much of the cli-

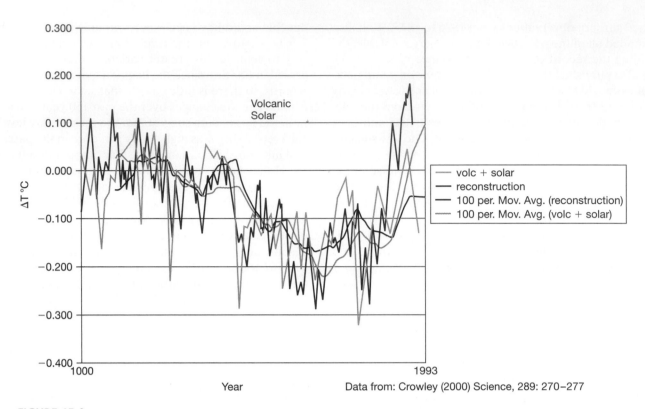

FIGURE 15-9

A thousand-year reconstruction of Northern Hemisphere temperatures and the results of an energy balance model forced with orbital variations and volcanic eruptions. The colored lines are 100-year moving averages, which smooth out the year to year variability. (From data described in T.J. Crowley, 2000. Causes of Climate Change Over the Past 1000 Years. *Science,* 289: 270–277.)

mate variability before the industrial revolution; greenhouse gas forcing recreates much of the more recent record. When all three are combined (Figure 15-10) we find a good match to the record of the past 1,000 years. These model results lend support to the view that much of the climate history over this time period can be explained by orbital variations and volcanic eruptions, until we approach the present when we need to include the effects of increasing atmospheric CO_2 concentrations.

Present-Day Climate Variability

Figure 15-1d presents a compilation of mean global temperatures since 1860. The record shows a steady rise in temperature through the first part of the 20th century, peaking in the 1940s (encompassing the Dust Bowl period in the United States). Temperatures then dropped in the 1960s and 1970s before rising again in the 1980s. In Figure 15-8, we showed that the temperature record closely matches the length of the sunspot cycle. Yet, the temperature trend is also consistent with the suggestion of CO_2-induced global warming. We can explain the 1940–1970 temperature decrease by increases in sulfate

aerosols as a result of the burning of coal, and it is now generally accepted that the high temperatures at the end of the 20th century were influenced to some degree by anthropogenic greenhouse gases. However, there are shorter-term variations in the climate record not explained by any of these mechanisms and, for the remainder of this chapter, we are going to look at this year-to-year variability in the record.

Figure 15-11 is a schematic illustration of the major components of the climate system. We have seen that as we change time scales, different parts of this system take on different levels of importance. Over very long time scales, the input of energy to the system is variable, and the climate system is regulated by silicate weathering, plate tectonics, and changes in atmospheric composition. The distribution of oceans and continents changes on time scales of tens of millions of years, and the discussions in Chapters 4 and 5 indicate that this distribution would have had some effect on climate. We saw in Chapter 14 that on time scales of thousands to hundreds of thousands of years, climate changes are associated with atmospheric composition, Earth's orbit about the Sun, and changes in the extent of large ice sheets. In the first part of Chapter 15 we considered several processes that affect climate on

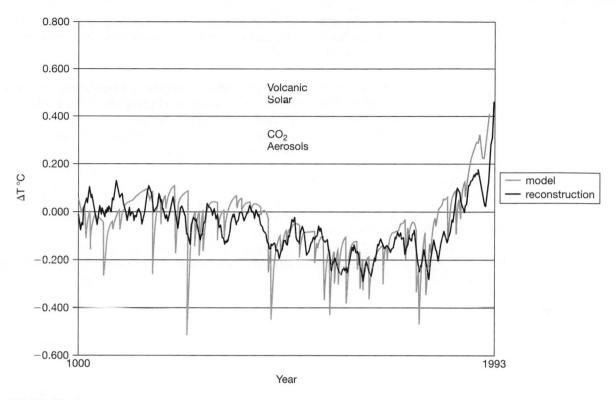

FIGURE 15-10

A thousand year reconstruction of Northern Hemisphere temperatures and the results of an energy balance model forced with orbital variations, volcanic eruptions, and greenhouse gas concentrations. (From data described in T.J. Crowley, 2000. Causes of Climate Change Over the Past 1000 Years. *Science*, 289:270–277.)

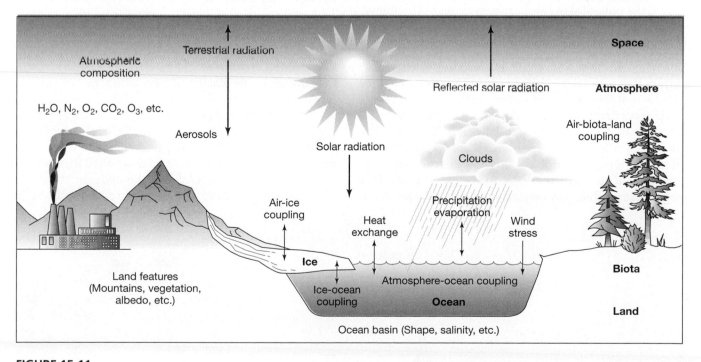

FIGURE 15-11

Schematic representation of the climate system. (After S. Schneider, and R. Londer. 1984. *The Coevolution of Climate and Life.* San Francisco: Sierra Club Books.)

the order of decades to centuries, or thousands of years. When we turn to time scales of years to tens of years, however, we see that most of the features that are significantly variable at long time scales are not significant at short time scales. Similarly, although the land surface can change (through cultivation, deforestation, and so on), these changes are not likely to have a major global impact on a year-to-year basis. Given this, all we have left to consider are the atmospheric state itself (temperature, cloud cover, etc.), and the condition of the oceans.

In practice, the state of the atmosphere is not particularly important in this context. Recall from Chapter 4 that the atmosphere transports heat very rapidly, and any anomalies are quickly dissipated. The atmosphere has very little "memory" of change. In contrast, the oceans absorb and store large amounts of heat, and they release this heat very slowly. Consequently, the ocean's memory of any change is much longer than the atmosphere's. Transient anomalies that develop in sea-surface temperatures can be expected to have a lingering impact on climate for some time afterward. The oceans, therefore, are the most likely place to look for processes that might cause climate anomalies on the interannual or decadal time scales.

So, where do we look in the oceans? One obvious place to start is in the tropics, where we find the large convective towers of the Hadley cells that drive the atmospheric circulation over a large portion of the planet. We will also look at the opposite extreme, and examine the role of the polar oceans in short-term climate variability.

El Niño–Southern Oscillation (ENSO) Events

The name **El Niño** was originally given to a warm ocean current that appeared off the coasts of Peru and Ecuador. The current flowed for only a few weeks, and, because it usually occurred near Christmastime, local fishermen named it El Niño after the Christ child. The name has taken on a different meaning more recently; it is now used by researchers to describe a major shift in the oceanic circulation that occurs in this region every 2 to 10 years. This broad oceanic shift is associated with large changes in the circulation of the tropical atmosphere, which give rise to significant climate anomalies over much of the tropics and mid-latitudes.

The Equatorial Atmospheric Circulation. Superimposed on the north–south Hadley circulation (described in Chapter 4) is a significant east–west circulation in the troposphere that is most prevalent over the equatorial Pacific. The western equatorial Pacific has the highest sea-surface temperatures on the globe. This region, which encompasses Australia and Indonesia, is a site of intense atmospheric convection. As is true of Hadley cells, the rising air diverges at high altitudes, but, in this case, we are concerned with a component of the flow that moves eastward and westward along the equator rather than northward and southward (Figure 15-12). The eastward-moving air crosses the Pacific, where it subsides off the west coast of South America. The circulation is completed by an easterly flow at the surface. This circulation is linked to other, smaller cells driven by convection over South America and Africa. Figure 15-12 shows the normal pattern of the equatorial east–west circulation. The figure also indicates the normal pattern of precipitation, with heavy precipitation in the convective regions and drier conditions in the areas of subsidence. The circulation cells produce an oscillation in the sea-level pressure distribution between the western and the central/eastern portions of the tropical Pacific Ocean. When pressures are low in the west, they tend to be higher in the east, and vice versa. This oscillation in sea-level pressures is referred to as the **Southern Oscillation** (SO).

The Ocean Circulation. The persistent easterly wind at the surface in the Pacific Ocean produces a westward-flowing ocean current, which results in the water piling up in the western part of the ocean. This causes very warm water to accumulate in the western Pacific (Figure 15-13). Figure 15-13 shows that water piles up in the west, which causes the ocean surface to slope downward from west to east. The slope is exaggerated in the diagram; the difference in surface elevation from west to east is only on the

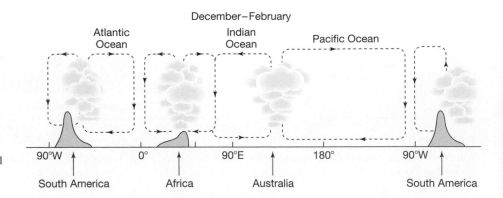

FIGURE 15-12

The east–west circulation in the equatorial troposphere. The shaded areas represent heavy precipitation.

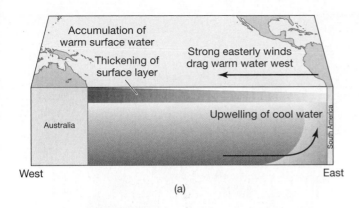

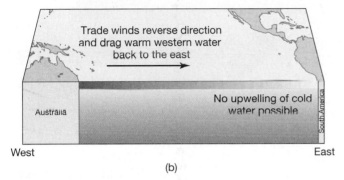

FIGURE 15-13

The ocean surface layer in the tropical Pacific Ocean.

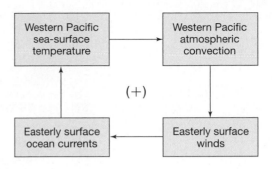

FIGURE 15-14

Schematic diagram of the interaction between the atmosphere and ocean in the tropical Pacific Ocean.

order of a couple of meters. This east-to-west movement of water thickens the warm surface layer in the west and thins it in the east. The thinner surface layer in the east allows the upwelling of colder, nutrient-rich water from below, which promotes high levels of biological productivity and large fish populations.

We can regard this pattern of atmospheric and oceanic circulation as the norm, but in any year we see substantial differences. In some years this pattern intensifies: The sea-surface temperatures in the central and eastern Pacific are colder than normal, and convection over Indonesia is enhanced. These conditions are referred to as **La Niña** conditions—the pattern is similar to the "normal" conditions, but the circulation is enhanced. More drastic changes occur when the pattern breaks down in what is referred to as an **El Niño—Southern Oscillation** (ENSO) event. The Southern Oscillation may also be referred to as being in a "cold" phase during La Niña (or *anti-ENSO*) events and in a "warm" phase during ENSO events.

Before we describe what happens when an ENSO event occurs, let us look at this circulation again. The atmospheric convection in the western Pacific occurs as a result of the high sea-surface temperatures, but the high sea-surface temperatures are a result of the atmospheric circulation, which is driven by the convection. In other words, it is a classic chicken-and-egg situation. It is not a case of the ocean forcing the atmosphere or vice versa. Instead, it is a single integrated system with a positive

feedback loop (Figure 15-14). If we perturb the system at any point, we should expect to see changes throughout all of its components.

ENSO Events. Scientists are still debating what causes ENSO events to occur. For the purposes of this discussion we will simply break into the cycle in Figure 15-14 and ask what happens if, for some reason, there is a decline in the strength of the easterly winds.

If these winds weaken or reverse direction, which happens in some ENSO events, there is nothing to restrain the pile-up of warm water that has accumulated in the western Pacific. This water then comes sloshing back across the ocean in what is known as a *Kelvin wave*. It takes about 60 days for this wave to travel back across the Pacific. When it does, it has two major consequences. First, it shifts the pool of high sea-surface temperatures from the western to the central Pacific (Figure 15-15), which then completely changes the atmospheric circulation. Second, it shuts off the upwelling in the eastern Pacific, which has drastic consequences for biological productivity. The loss of the nutrient-rich water leads to a massive die-back of marine organisms and the bird life that feeds on them.

The changes in the atmospheric circulation are shown in Figure 15-16. The greatest area of convective activity during ENSO events lies over the central Pacific. The rising air diverges to the east and west, meeting and subsiding over Africa, although there is also localized uplift on the western side of the Andes. In a non-ENSO year, there is low pressure (rising air) at the surface over Australia and Indonesia and high pressure (subsiding air) at the surface in the central and eastern Pacific (Figure 15-12). In an ENSO year, this pattern reverses: Pressure increases over Australia and decreases in the central Pacific. We can calculate the pressure difference between these two locations (Figure 15-17a) and plot this difference through time to produce the **Southern Oscillation Index** (SOI) (Figure 15-17b). This index is a measure of the pressure difference between the western and eastern parts of the

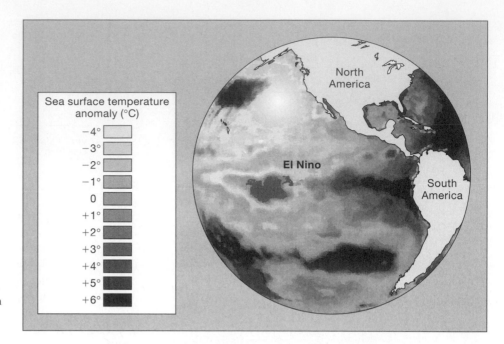

FIGURE 15-15

[See color section] Sea-surface temperature anomalies in the central and eastern Pacific during the 1997–1998 ENSO event.

tropical Pacific Ocean. Strong positive values indicate La Niña (non-ENSO) conditions; strong negative SOI values indicate ENSO conditions, which have occurred throughout the interval for which recorded data are available. We also find evidence from ice cores that ENSO events of various magnitudes have been a feature of the climate system for at least the past 500 years.

Climatic Impacts of ENSO. The most dramatic impacts of ENSO events are seen in their effects on rainfall patterns. In non-ENSO years, summertime convection and rainfall occur over Australia and Indonesia. There is also convective activity and rainfall over equatorial Africa and the Amazon Basin. In contrast, there is subsiding air (dry conditions) west of the Andes. However, in ENSO years the dominant convective region shifts toward the central Pacific, and convection and rainfall over Australia and Indonesia diminish. Figure 15-16 shows that there is also subsidence over Africa and some localized convection over the western Andes. The result is drought in central

America, Brazil, Australia, Indonesia, and southeast Africa and anomalously high rainfall amounts in the central Pacific and on the western slopes of the Andes in Ecuador and Peru. These high rainfall amounts typically result in floods and landslides, with their accompanying high levels of soil erosion. ENSO events also appear to have some effect on the monsoon circulation over India, resulting in increased rain over southern India and reduced rainfall over northern India and the Himalayas.

The impact of ENSO events is not confined to the tropics. The changes in atmospheric circulation can also influence the mid-latitudes. We saw in Chapter 4 that the location and strength of mid-latitude weather systems are controlled in part by the subtropical highs and in part by the sea-surface temperature gradients. Because both of these factors change during ENSO events, it is not surprising that they should have some effect on mid-latitude climate. The general pattern of temperature and rainfall anomalies associated with an ENSO event are shown in Figure 15-18 for the Northern Hemisphere winter. In

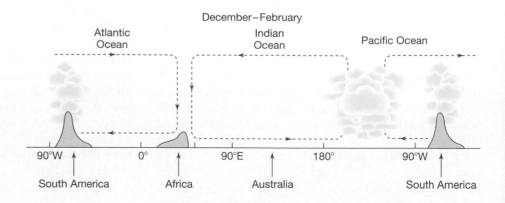

FIGURE 15-16

Atmospheric circulation during an ENSO event.

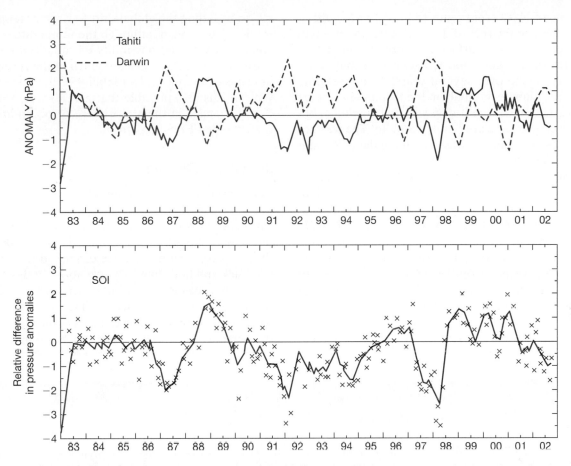

FIGURE 15-17

(a) The sea-level pressures at Tahiti and Darwin, Australia, and (b) the Southern Oscillation Index (SOI). The SOI is computed from the sea-level pressure differences. Negative values of the SOI indicate warm (El Niño) events. Note the strengths of the 1982–1983 and 1997–1998 events. (From the NOAA Climate Prediction Center. http://www.cpc.nccp.noaa.gov/data/indices/)

December–February

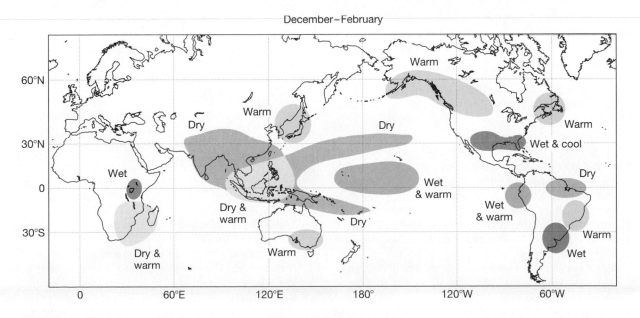

FIGURE 15-18

Rainfall and temperature anomaly patterns associated with an ENSO event. (From the NOAA Climate Prediction Center. http://www.cpc.ncep.noaa.gov/products/analysis_monitoring/impacts/warm.gif)

practice, each ENSO event tends to be somewhat different from others in the record. For example, the strength and location of the sea-surface temperature anomalies are not always the same, and ENSO events may last 1 year or may extend over several years. As a result, the pattern of the climate anomaly in the tropics tends to be fairly consistent, although it may vary in magnitude, but the mid-latitude effects are highly variable. The western United States, for example, was very dry in the 1976 event but very wet in 1982.

It is interesting to note that the Southern Oscillation was in a warm phase (low SOI values) almost continuously from 1976 through to the end of the 20th century, and that there were five ENSO events between 1976–1977 and 1997–1998. There was only one La Niña event during this interval. Furthermore, the system was in an almost constant El Niño mode from 1990 through the winter of 1997–1998 (except for a short break in 1995–1996). Climate model simulations that use the observed sea-surface temperature distribution in the tropical Pacific recreated much of the temperature variability of the 1970s and 1980s, including the global shift to warmer temperatures in the 1980s. Again, although the temperature record is consistent with the possibility of enhanced greenhouse warming, we can still explain much of the recent pattern in terms of the natural variability of the system. Nevertheless, it is also possible that greenhouse-warming-induced changes in the atmosphere or ocean are causing the extended El Niño pattern.

Sea Ice and Climate

In the last part of this chapter we turn to the polar oceans to examine climate variability induced by sea-ice modifications to atmosphere–ocean interactions at high latitudes. In earlier chapters we hinted at the importance of the ice distribution when we discussed the ice–albedo feedback and bottom water formation. In this section we discuss the seasonal distribution of sea ice and some of the factors that control this distribution, and we examine some of the ways in which changes in the ice cover affect the polar ocean climate. These effects are particularly in-

A CLOSER LOOK
The 1982–1983 and 1997–1998 ENSO Events

The 1982–1983 ENSO event was one of the most severe of the 20th century. It developed unexpectedly. The normal early warning sign of changing tropical sea-surface temperatures was missed because of the preceding eruption of a Mexican volcano (El Chichón). The volcanic aerosols reduced the outgoing infrared radiation, resulting in lower satellite-derived sea-surface temperature measurements. The resulting pattern of climate anomalies was also slightly different from a "normal" ENSO: They were much more intense than earlier ENSO events. The impact of this particular event was considerable, with major droughts and floods occurring throughout the tropics. It is estimated that climate-related catastrophes resulting from the 1982–1983 ENSO event left more than 1,000 people dead and caused almost $9 billion worth of damage:

- Ecuador and northern Peru experienced floods and landslides that left 600 dead and resulted in crop and property losses totaling approximately $400 million. Guayaquil, Ecuador, had 20 times its normal rainfall in May 1983.
- In Indonesia, there were crop failures and starvation.
- In Botswana, the ENSO-induced drought followed two previous years of drought and eventually led to the loss of thousands of livestock.
- Eastern Australia suffered the worst drought of the century. Animal feed supplies were so diminished that thousands of sheep, dying of starvation, had to be shot. The dry conditions resulted in huge dust storms, one of which deposited 11,000 tons of topsoil on the city of Melbourne.

- Tahiti and French Polynesia had last experienced a typhoon at the beginning of the 20th century. The warm-water pool that formed in the central Pacific during the 1982–1983 ENSO event generated several large storms, and the islands were hit by six typhoons in 5 months.
- In the United States, increased rain in the Midwest resulted in the flooding of the Mississippi River. An increase in storms on the West Coast resulted in severe flooding and landslides in California, where 10,000 houses were lost or damaged and farm losses totaled half a billion dollars. There was a record snowfall in the Rockies, which, when it melted, resulted in flooding in Salt Lake City and along the lower Colorado River.

The 1997–1998 event was equally severe. Anomalous weather patterns occurred in many parts of the world, including extensive impacts in the equatorial Pacific, North and South America, and East Africa:

- Flooding, mud slides, and disease killed more than 80 people in Peru, and flooding in Ecuador resulted in 90 deaths and the evacuation of 22,000 people.
- The central Pacific had eight tropical cyclones, but only two the year before.
- Drought conditions produced hunger in Indonesia and Papua New Guinea. The extreme drought in Indonesia resulted in forest fires that burned over 1 million acres of forest.
- In the United States, this event produced heavy rains, flooding, and mudslides along the California coast but a very mild winter in the northeastern region.

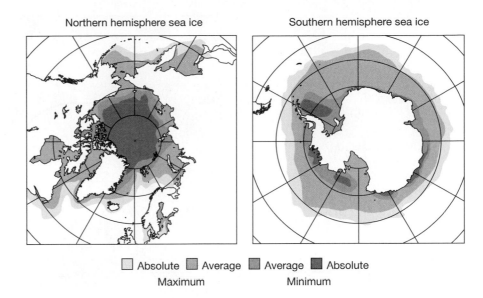

Northern hemisphere sea ice   Southern hemisphere sea ice

FIGURE 15-19

The seasonal distribution of sea ice in the
Northern and Southern Hemispheres.

☐ Absolute ▨ Average ▨ Average ■ Absolute
Maximum   Minimum

teresting because, as we will see in Chapter 16, they tend
to dominate the high-latitude response to increasing atmospheric CO_2 levels in global climate models.

Seasonal Distribution of Sea Ice. The seasonal distribution of the sea ice of polar oceans is shown in Figure
15-19. The seasonal range of the ice cover is extreme: The
Northern Hemisphere ice cover almost doubles in size
from approximately 8.5×10^6 to 15×10^6 km² between
summer and winter; the Southern Ocean ice cover grows
from 4×10^6 to 20×10^6 km². Sea ice forms when the temperature of the ocean surface drops below the freezing
point ($-1.8°C$ for typical ocean salinities). Sea ice grows
in thickness by new ice forming on the bottom of the ice
pack, typically reaching thicknesses of about 1 m during
a single winter. The ice cover melts and freezes seasonally at the equatorward margins of the pack but lasts
throughout the year at higher latitudes. This permanent
pack ice in the Arctic Ocean reaches equilibrium thicknesses of about 5 m; the Southern Ocean ice is much thinner, because very little survives through the summer melt.

Several features of this ice distribution are important
to our discussion. The large seasonal range is a result of
thermodynamic controls on the ice cover (i.e., the ice
cover is responding to the seasonal changes in temperature). The asymmetric nature of the ice margin in the
Arctic Ocean, however, reflects the influence of the continental configurations and the ocean circulation. We find
the ice margin extending well to the south in both the
western North Atlantic and western North Pacific. At the
eastern margins of both oceans, open water extends much
farther north off Norway and Alaska. Note that the shape
of the ice margin reflects the pattern of ocean currents
you saw in Chapter 5. The warm water of the North Atlantic Drift and the Kuroshio Current prevent the ice margins from extending farther south in the eastern oceans,

whereas the East Greenland Current, the Labrador Current, and the Oyoshio Current bring colder water and
carry increased ice cover southward in the west. The ice
distribution helps illustrate one of the important facts
about the sea-ice cover: Sea ice moves. The ice does not
simply form and melt in place; it is in a constant state of
motion. The typical pattern of sea-ice drift in the Arctic
Ocean is shown in Figure 15-20. The map reveals two primary features of the ice circulation: a clockwise circulation, or gyre (the Beaufort Sea Gyre), in the west and the
Transpolar Drift Stream, which flows across the pole from
the Siberian coast to the East Greenland Sea. This ice
flows into the East Greenland Current, where it is transported southward out of the Arctic Ocean. On average it
takes about 5 years for ice that forms off the Siberian
coast to be transported across the Arctic Basin. Ice that
forms in the Beaufort Gyre, however, may last much
longer, possibly circling the western Arctic several times
before melting in the summer or becoming caught in the
Transpolar Drift.

Moved around by winds and ocean currents, the ice
is broken into individual pieces *(ice floes)* that continuously join and break up. Where floes collide, large
mounds of ice are thrust up into pressure ridges; where
floes move apart, they produce areas of open water called
leads or *polynyii* (singular, *polynya*). Leads are linear
open-water features, and polynyii are irregular areas of
open water. In winter, these open-water areas freeze very
rapidly, but because the ice pack is so dynamic, there is
always a small amount of open water present, even in mid-
winter. This is important for two reasons: One, the presence of open water allows for the constant production of
new ice, thus releasing salt to the upper ocean and increasing the density of the surface layer. Two, the open
water has a considerable impact on the Arctic energy budget: The ocean loses heat to the atmosphere about 100

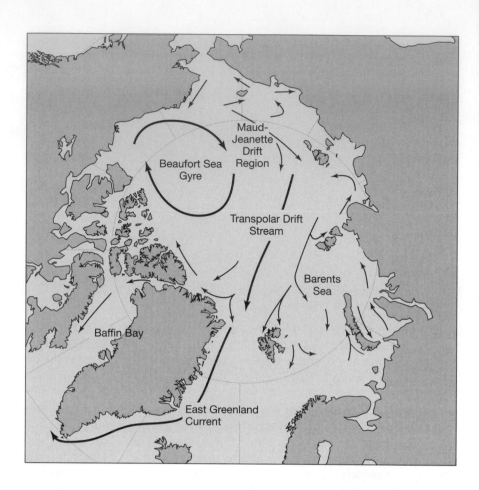

FIGURE 15-20

The typical pattern of sea-ice drift in the Arctic Ocean.

times faster from the open water than it does through the insulating ice cover. Polynii may have provided a refuge for metazoan life during the Snowball Earth (see Chapter 12).

Ice–Climate Interactions. We have seen in Chapters 4 and 5 that the sea-ice cover forms as a response to ocean temperatures and that the distribution is then modified by winds and ocean currents. However, the ice cover is not simply a response to the climate: The ice cover modifies the atmospheric and oceanic circulations. We have seen that ice production in the North Atlantic contributes to the formation of North Atlantic Deep Water, which is a major factor in driving the oceanic thermohaline circulation. We have also discussed the ice–albedo feedback, whereby a change in temperature causes a change in the ice cover and the surface albedo, which further modifies the temperature. Recall that this is a positive feedback: Increasing temperature → decreasing ice cover → decreasing albedo → increasing temperature. This feedback can operate year-round near the sunlit ice margins but can play a role only in summer at higher latitudes. (Recall that there is no solar radiation at high latitudes in winter.) Another sea-ice feedback, however, is important year-round. The ocean surface is a source of thermal energy and latent heat for the polar atmosphere, and the

most rapid heat transfer from the ocean to the atmosphere occurs through the areas of thin ice or open water. Thus, we can picture a second positive feedback (Figure 15-21): Increasing temperature leads to decreasing ice cover, which leads to increasing ocean heat flux to the atmosphere, which leads to increasing temperature. Both the ice–albedo feedback and the ocean heat-flux feedback are related to the ice extent and concentration (i.e., the area of ice-covered ocean). The heat-flux feedback is also controlled to some degree by ice thickness: The heat loss from the ocean to the atmosphere increases as ice thickness decreases. The extent of the ice cover, the ice concentration, and the ice thickness are all determined

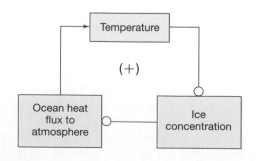

FIGURE 15-21

The sea ice–ocean heat flux feedback loop.

by the energy budget and the ice dynamics. If we want to generate numerical models of these processes, we need to be able to account for both thermodynamic and dynamic interactions among the ice, ocean, and atmosphere.

Several models with this degree of sophistication exist, but, few have been incorporated in atmospheric or coupled atmosphere–ocean global climate models.

Chapter Summary

1. Earth's climate changes over all time scales from decades and centuries to millions of years.
   a. Since the last glacial maximum, Earth has undergone several changes in climate that appear to be global in extent, even though the magnitude of the changes varies from region to region.
   b. Although the evidence for these changes is conclusive, our explanations for why they occurred are ambiguous.
2. As Earth warmed up from the effects of the last glaciation, a sudden and very rapid climate reversal occurred (the Younger Dryas event), resulting in the readvance of ice sheets and glaciers over much of northern Europe and other parts of the globe.
   a. Several hypotheses have been proposed to explain the Younger Dryas, the most likely of which involves the switching on and off of the thermohaline circulation in the North Atlantic Ocean.
   b. Regardless of the exact cause of this shift in climate, the Younger Dryas is of particular interest because recent ice-core data show that the change occurred very rapidly, over years to decades; the apparent ability of the climate system to shift rapidly between two very different states poses some interesting questions about how the climate may respond to future greenhouse warming.
3. Two major warm periods occurred during the Holocene: one in the mid-Holocene and the other, the European Medieval Warm Period. The forcing for these events is not clear, but they can probably be attributed to the Milankovitch orbital forcing discussed in Chapter 14, with the precession of the poles having the greatest influence.
4. A return to colder conditions, the Little Ice Age, took place from the late 1500s through the mid-19th century. One of the contributing causes of the colder temperatures during the Little Ice Age may be the effect of a higher frequency of volcanic eruptions.
   a. The sulfur dioxide injected into the stratosphere by an eruption hydrolyzes, forming sulfuric acid droplets, which both reflect solar radiation and absorb some of the long-wave radiation emitted from the troposphere.
   b. The combined effect of an increased albedo and increased long-wave absorption is to heat the stratosphere and cool the troposphere.
   c. The result is lower-than-normal surface air temperatures for 1 to 2 years after a major eruption.
5. Further factors contributing to a change in climate during the Little Ice Age may include changes in the abundance of sunspots and the length of the sunspot cycle, which may affect the solar output and Earth's climate.
   a. Although there is no direct evidence of a climate forcing by sunspots, two of the longest time periods for which no, or few, sunspots were observed occurred during the Little Ice Age.
6. On century-long time scales, Earth's climate experiences large variations over intervals of years to decades. Most of the forcing for these variations originates in the oceans.
   a. ENSO events and sea ice–atmosphere–ocean interactions are two examples of how the ocean may have an impact on climate on these time scales.
   b. ENSO events are large-scale changes that occur in the atmospheric and oceanic circulations of the tropical Pacific Ocean and have a significant impact on tropical precipitation patterns. They also seem to produce anomalous climate conditions in different parts of the mid-latitudes.
   c. At higher latitudes, the interactions taking place between the ocean and the atmosphere are modified by the sea-ice cover through several positive feedback processes.
   d. The global consequences of these feedbacks are not as great as those of the ENSO events, but they still have a significant impact on polar and mid-latitude climates.

Key Terms

dendrochronology
El Niño
El Niño–Southern Oscillation
European Medieval Warm Period
Holocene
Holocene Climatic Optimum
La Niña
Little Ice Age

Maunder Minimum
palynology
plages
proxy data
residence time
Southern Oscillation
Southern Oscillation Index
Spörer Minimum

stochastic resonance
sunspots
uniformitarianism
volcanic aerosol forcing
Wolf Minimum
Younger Dryas

Review Questions

1. What is the Holocene?
2. Why are climatologists usually not very interested in "average" climate conditions?
3. What are proxy data? Describe several examples of proxy climate data.
4. Briefly describe the Younger Dryas, the Holocene Climatic Optimum, the European Medieval Warm Period, and the Litte Ice Age.
5. Explain why the formation of North Atlantic Deep Water might have played a role in causing the Younger Dryas event.
6. How do volcanoes affect climate?
7. What are sunspots? Why are they thought to have a possible effect on climate?
8. Use a diagram to help describe the "normal" east–west atmospheric circulation in the tropical Pacific Ocean.
9. Explain what happens to the atmospheric and oceanic circulations in the tropical Pacific during an ENSO event.
10. How do ENSO events affect precipitation patterns in the tropics? Explain.
11. Describe the major feedback processes between sea ice and climate.

Critical-Thinking Problems

1. In this chapter we suggested that the most likely place to look for processes that would force short-term climate variability is the ocean, because the ocean has a longer "memory" of its previous states than does the atmosphere. As an example we pointed out that temperature anomalies persisted for much longer in the ocean. Atmospheric temperature anomalies are dispersed very rapidly, so they affect the climate for only a very short time. As a further example of the rapid transfers that take place in the atmosphere, use the information from the hydrologic cycle from Figure 4-21 to calculate the residence time of water in the atmosphere. How many times a year does the water in the atmosphere go through the cycle of evaporation and precipitation?

2. In the section "Sea Ice and Climate," we described a positive feedback among ice concentration, ocean heat flux, and temperature. However, we can also assume that there will be considerable evaporation from the open water surface, which, because of the cold air temperatures over polar oceans, should result in condensation and the formation of low-level clouds. Clouds have a high albedo, which tends to lower temperatures, but they are also very efficient at trapping infrared radiation (i.e., they have a strong greenhouse effect). Given this, modify Figure 15-21 so that it includes the effect of clouds. You may need to construct different diagrams for summer and winter.

Further Reading

General

Wigley, T. M. L., M. J. Ingram, and G. Farmer. 1981. *Climate and History.* Cambridge, England: Cambridge University Press.
Knox, P. N. 1992. A Current Catastrophe: El Niño. *Earth,* 1(5):31–37.

Advanced

Rind, D., and J. Overpeck. 1993. Hypothesized Causes of Decade-to-Century-Scale Climate Variability: Climate Model Results. *Quaternary Science Review,* 12:357–374.

Global Warming

Key Questions

- How does the amount of CO_2 produced by the burning of fossil fuels compare with natural CO_2 sources?

- By which processes will anthropogenic CO_2 eventually be removed from the atmosphere?

- How much are atmospheric CO_2 and other greenhouse gases expected to rise over the next few decades to centuries, and how will this rise affect Earth's climate?

- How is sea level projected to change over the next century?

- How will forests and other ecosystems respond to atmospheric CO_2 increases?

- What policies might be adopted to reduce future CO_2 emissions?

Chapter Overview

Atmospheric CO_2 concentrations are increasing as a consequence of the combustion of fossil fuels and the cutting down of tropical forests. If no actions are taken to reduce emissions, CO_2 levels are expected to more than double over the next century and could increase by a factor of 7 to 10 over the next few centuries. Other greenhouse gases, including CH_4, N_2O, and various CFCs, are also increasing in concentration in the atmosphere. Temperature increases this century are predicted to be only a few degrees Celsius, but long-term climate changes could be substantially larger than this. Even a few degrees of warming could have substantial effects on water availability, on speciation within terrestrial ecosystems, and on sea level. Whether to take action to limit greenhouse gas emissions is an issue that is currently being hotly debated.

Introduction

We have seen in earlier chapters that Earth's climate has varied on a number of different time scales. Over the past few billion years, the Sun has brightened considerably and Earth's climate has gone through a series of warm and cold periods. The Earth has even experienced brief periods where the surface seems to have been completely frozen. But the system has recovered from these catastrophes and has remained within a range conducive to the continued presence of life. Over the past few million years, the climate has oscillated between glacial and interglacial intervals triggered by variations in Earth's orbit and amplified by internal feedbacks within the climate system. Over the past hundred years, the climate has warmed by about 0.7°C, but this change has been superimposed on other, shorter-term warming and cooling trends that make interpretation of the recent climate record difficult.

In this chapter, we use the information and techniques described earlier to try to predict how climate may change over the next few decades to millennia. The possibility of future global warming is, after all, why both scientists and policymakers are so interested in climate and why we have devoted so much attention to the subject. In trying to make such a prediction, we show why an interdisciplinary approach is essential and how a knowledge of climate history can help compensate for the deficiencies in our physical models of the climate system.

Carbon Reservoirs and Fluxes

We saw in Chapter 1 (Figure 1-2) that atmospheric CO_2 concentrations have increased by about 25% since the beginning of the 19th century. This change is larger than any natural fluctuation that has occurred since the retreat of the glaciers 11,000 years ago and is almost certainly attributable to human-induced causes, principally the burning of fossil fuels and deforestation. To understand why the human influence is so noticeable, it is useful to compare the sizes of the various carbon reservoirs and the

rates at which carbon cycles in both the natural and perturbed systems.

Natural Reservoirs and Fluxes

The carbon cycle is diagrammed in Figure 16-1. This diagram combines the essential features of the organic and inorganic carbon cycles described in Chapter 8. It also includes a box labeled "fossil fuels," which includes several types of compounds—most importantly coal, oil, and natural gas. Fossil fuels are, in many respects, a natural part of the carbon cycle. (It is the rate at which they are being oxidized that is unnatural.) They were created over many millions of years by the accumulation of dead organic matter in soils and sediments. Table 16-1 lists the relative amounts of different types of fossil fuel. Most of it is coal, which is stored in vast quantities in many parts of the world.

The total amount of fossil fuels is uncertain, because new reservoirs are continually being discovered and be-

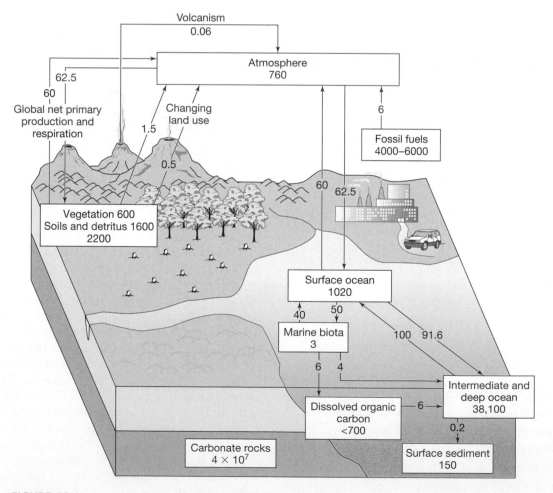

FIGURE 16-1

The major reservoirs and fluxes in the global carbon cycle. Units are Gton (C) for reservoir sizes and Gton(C)/yr for the fluxes. (After *Climate Change, 1994: Radiative Forcing of Climate Change.* Cambridge: Cambridge University Press, p. 21.)

TABLE 16-1

Fossil Fuel Reservoir Sizes and Burning Rates		
Reservoir	*Size,* * *Gton(C)*	*Burning rate,* ** *Gton(C)yr*
Coal	3500	2.4
Oil	670	2.7
Natural gas	500	1.3
Total	4670	6.4

*Reservoir sizes from H.-H. Rogner. Paper presented to the International Agency Workshop on Climate Change Damages and Benefits of Mitigation, Feb. 26-28, 1997 (International Institute of Applied System Analysis). Quoted on Greenpeace webpage: *http://www.greenpeace.org/~climate/science/reports/carbon/clfull-3.html#Heading22*
**2000 burning rates from U. S. Department of Energy website: *http://www.eia.doe.gov/emeu/international/environm.html#IntlCarbon*

cause whether a given reservoir is counted depends on the price of fuel. Estimates of the amount of fossil fuel that is economically recoverable at 2002 market prices range from 4000 to 6000 Gton(C), 7 to 10 times the amount of carbon that was present in the preindustrial atmosphere as CO_2, 590 Gton(C). The value quoted in Table 16-1, 4670 Gton(C), is in the middle of this range. This gives us our first indication of why the burning of fossil fuel is a potential problem: Simply, a lot of it is available. In "Critical-Thinking" Problem 1, we shall compute what would happen to atmospheric CO_2 concentrations if all the fossil fuel were to be burned instantaneously.

Two additional points should be noted about the fossil fuel reservoirs listed in Table 16-1. First, the value given for oil reserves, 670 Gton(C), includes both conventional and unconventional reserves. Conventional oil reserves, that is, oil that can be pumped directly out of the ground, make up only 240 Gton(C) of this amount. The rest of the oil is in the form of oil shales, tar sands, and heavy crude oil, which is more difficult to access and process. Second, the natural gas values also include both conventional (easy to access) and unconventional reserves, but they do not include methane hydrates. Over the past decade, oceanographers have discovered vast quantities of methane-rich ice on the sea floor just off the continental shelves. Technically, this material is termed **methane clathrate hydrate.** It is a solid in which CH_4 molecules are encased in a lattice of 5–6 H_2O molecules. Such methane hydrates form when organic matter decomposes anaerobically at great depths. The overlying water must be at least 300 m deep and must be cold as well in order for methane hydrates to be stable. The total amount of methane hydrate on the sea floor is uncertain but some estimates are as high as 18,000 Gton(C), or more than three times higher than the rest of the fossil fuel reserves! We do not list this in Table 16-1 because it is not recoverable

at current market prices, but if someone discovers how to extract it economically the problem of global warming will become even more difficult.

Of equal importance to the carbon reservoirs themselves are the fluxes that connect them. These fluxes are in many cases wholly disproportionate to the size of the reservoirs involved. For example, the huge carbonate rock reservoir is connected to the rest of the system by only a tiny flux, about 0.03 Gton(C)/yr. Shown in Figure 16-1, this volcanic CO_2 flux must, on average, equal the rate at which silicate rocks are converted into carbonates. Despite its large size, this reservoir exerts little influence on the rest of the carbon cycle on any time scale shorter than hundreds of thousands of years. So, processes such as silicate weathering and volcanism, which were extremely important on geologic time scales, should have little effect on atmospheric CO_2 levels over the next few centuries.

The largest fluxes in the natural carbon cycle are the exchange of carbon between the atmosphere and the terrestrial biosphere, about 60 Gton(C)/yr, and between the atmosphere and surface ocean, about 60 Gton(C)/yr. The terrestrial biosphere, largely composed of terrestrial vegetation and soils and detritus, exchanges CO_2 with the atmosphere by way of the short-term organic carbon cycle. In that cycle, the photosynthetic uptake of CO_2 by forests is balanced by the respiration and decay of plant material. This cycle is responsible for the seasonal fluctuations in atmospheric CO_2 (the Keeling curve) discussed in Chapters 1 and 8. The ocean exchanges CO_2 with the atmosphere by diffusion through the surface interface. Once in the surface ocean, CO_2 is taken up by marine photosynthesizers. The net rate of photosynthesis in the oceans, about 50 Gton(C)/yr, is comparable to that on land, but it does not affect the atmosphere in the same way because the amount of living biomass is very small: The marine biota contain only about 3 Gton of carbon, as compared with approximately 600 Gton of carbon in forests. Free-floating marine organisms and seaweed do not need the massive amounts of structural carbon required by trees. As a result, the marine organic carbon cycle is more closely balanced than the terrestrial cycle and does not contribute appreciably to the seasonal fluctuations of CO_2 in the atmosphere.

Rates of Fossil-Fuel Burning and Deforestation

The currently observed increase in atmospheric CO_2 concentrations is attributable, at least in part, to the combustion of fossil fuels. For natural gas (which is mostly methane), the combustion reaction can be written as

$$CH_4 + 2\,O_2 \rightarrow CO_2 + 2\,H_2O$$

The combustion reactions of coal and oil are more complicated, as these fuels consist of mixtures of more complex hydrocarbons, but the main reaction products are the same, CO_2 and H_2O. The rate of fossil-fuel consumption is known fairly accurately (to within 10% or better), because records are kept by companies that produce fossil fuels and by the countries in which these companies operate. The 1999 value was about 6.1 Gton(C)/yr (Table 16-1). World population is currently about 6 billion, so this consumption rate amounts to 1 metric ton of carbon per person per year. The rate of fossil-fuel burning is about 10 times less than the rate of CO_2 exchange with the terrestrial biosphere but 100 times larger than the rate of CO_2 release by volcanos. Its effect on the atmosphere is disproportionately large, however, because this part of the global carbon cycle is not in balance.

All three of the major fossil fuels are being consumed at appreciable rates, but oil leads the way at 2.7 Gton(C)/yr, followed by coal at 2.1 Gton(C)/yr and natural gas at 1.3 Gton(C)/yr (Table 16-1 and Figure 16-2). This implies that the different forms of fossil fuel have vastly different projected lifetimes. Recall from Chapter 8 that the residence time of a reservoir is just the reservoir size divided by the outgoing flux. As fossil fuels are not forming at appreciable rates, their residence times indicate how long they will last at current burning rates. From Table 16-1, we see that oil has the shortest projected lifetime, ~ 250 years, while coal has the longest, ~ 1700 years. However, these values can be misleading. Conventional oil will be exhausted in less than 90 years at current burning rates and may be gone much sooner than that if world oil consumption increases. Global warming is not the only potential problem on the horizon. Depletion of oil reserves is also something that must be considered.

Fossil fuel consumption is distributed unequally among the various nations of the world. As shown in

TABLE 16-2

Fossil Fuel Consumption by Geographic Region	
Region	*Consumption rate,* * *Gtons(C)/yr*
North America	1.832
Central and South America	0.269
Western Europe	1.000
Eastern Europe (incl. Russia)	0.844
Middle East	0.288
Africa	0.240
Far East (incl. China and Japan)	1.970

*2000 burning rates from U. S. Department of Energy website: *http://www.eia.doe.gov/emeu/international/environm.html#IntlCarbon*

Table 16-2, most of the fossil fuel is burned in the Northern Hemisphere, with North America and the Far East essentially tied for the lead at nearly 1.8 Gtons(C)/yr apiece. The United States, with only about 5% of the world's population, accounts for some 1.52 Gtons(C)/yr, or 25% of global CO_2 emissions. Thus, our *per capita* emissions (emissions per person) are about five times the world average. The large amount of CO_2 emitted reflects the high standard of living and energy-intensive lifestyle. Other countries are catching up, however. China, in particular, has a rapidly growing economy and an increasing appetite for fossil fuels to go with it. China recently surpassed Russia to become the second biggest coal-consuming nation in the world, after the United States. With

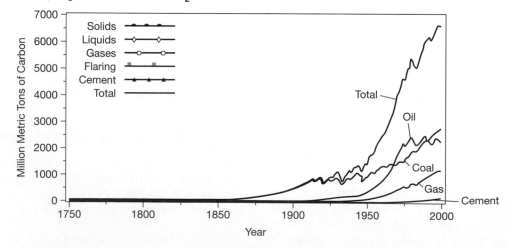

FIGURE 16-2

Coal, oil, and natural gas consumption rates, 1750–present. (From the Carbon Dioxide Information Analysis Center, Oak Ridge National Labs, available online at *http://cdiac.esd.ornl.gov/trends/emis/em_cont.htm*)

its large population and extensive coal reserves, China could become the world's largest CO$_2$ emitter within the next two decades.

Of the 6.1 Gton(C)/yr produced by the combustion of fossil fuels, about 3 Gton(C)/yr is currently accumulating in the atmosphere. This number can be calculated directly from the observed 1.5 ppm/yr rate of atmospheric CO$_2$ increase. (We shall do so in "Critical-Thinking" Problem 2.) About 2.5 Gton(C)/yr is apparently being absorbed by the oceans, as the imbalance in fluxes between the atmosphere and the surface ocean (Figure 16-1) indicates. The remainder is thought to be accumulating in forests and soils.

To further complicate matters, fossil-fuel burning is not the only anthropogenic source of CO$_2$. Significant amounts of CO$_2$, equivalent to 1.5 ± 1.0 Gton(C)/yr, are also being produced by deforestation, most of which is currently occurring in the tropics. The clearing of forests and the utilization of land for agricultural purposes generally results in a substantial release of carbon into the atmosphere, both from the trees themselves and from the soil beneath them. If the actual emissions are at the high end of the estimated range, tropical deforestation may be contributing as much to atmospheric CO$_2$ increase as is the burning of either coal or oil. Surprisingly, however, the biota as a whole are currently acting as a net sink for carbon. The reasons are discussed below.

CO$_2$ Removal Processes and Time Scales

How fast the CO$_2$ released from fossil-fuel burning will disappear and where it will ultimately go are perhaps the most confusing aspects of the whole global warming issue. The confusion arises because the carbon cycle is complex and involves processes that operate on a variety of time scales. Scientists have developed computer models of the carbon cycle that are capable of simulating many of these processes and that can be used to make projections of future atmospheric CO$_2$ concentrations. Some of these model predictions will be described in the following section. To understand these predictions, however, and to gain some idea of how reliable they might be, we must consider the physical processes involved. Next, we describe the major CO$_2$-uptake processes in decreasing order of the speed at which they are expected to occur.

Northern Hemisphere Reforestation

The fastest uptake process for anthropogenic CO$_2$ is photosynthesis by the terrestrial biota. As discussed in Chapter 9, we can represent photosynthesis by the simplified reaction

$$CO_2 + H_2O \rightarrow CH_2O + O_2$$

where CH$_2$O represents more complex forms of organic matter. This organic matter can accumulate either in living biomass or in soils. When it does so, carbon is removed from the atmosphere, at least temporarily. This carbon is eventually returned to the atmosphere when trees are cut down and burned or when soil organic matter decays.

Whether photosynthesis will act as a sink for anthropogenic CO$_2$ over the next few centuries depends largely on whether forests expand or shrink in size. Deforestation causes them to shrink, so the terrestrial biota act as a CO$_2$ source. Reforestation causes them to expand, so the biota act as a CO$_2$ sink. Studies suggest that a significant amount of reforestation, about 0.5 Gton(C)/yr, is occurring in temperate parts of the Northern Hemisphere. Recall from Chapter 1 that deforestation of North America during the 19th century, the *pioneer effect*, was responsible for most of the rise in atmospheric CO$_2$ between 1800 and 1850. Much of the deforested land was converted to farms and remains farmland today, so that land cannot be where the carbon is going. But the mountain ridges in Pennsylvania were stripped of trees to provide fuel for making steel and for powering trains. The demand for wood is now lower, and many areas are protected from logging by the state and federal governments. Consequently, the forests are regrowing and are probably contributing to the Northern Hemisphere CO$_2$ sink. Most of these forests will not reach maturity for another century or more, so this sink should remain active for some time into the future.

Northern Hemisphere forest regrowth, however, cannot be the only place where CO$_2$ is disappearing. The total anthropogenic CO$_2$ source is about 7.5 Gton(C)/yr, and we have accounted for only about 5.5 Gton(C)/yr of carbon sinks: 3 Gton(C)/yr in the atmosphere, 2 Gton(C)/yr in the oceans, and 0.5 Gton(C)/yr in Northern Hemisphere forests. An additional sink of approximately 2.0 Gton(C)/yr is required to balance the global carbon budget. What could this "missing sink" be?

CO$_2$ and Nitrogen Fertilization

A substantial part of the missing carbon may be going into existing forests as a consequence of higher atmospheric CO$_2$ concentrations. Most plants raised under greenhouse conditions (where plenty of water and other nutrients are available) grow faster when exposed to higher CO$_2$ levels (Chapter 9). This process is termed **CO$_2$ fertilization**. The increased growth rate is caused, in part, by the fact that CO$_2$ is a limiting nutrient under these conditions. But there is an indirect stimulation effect as well. Plants tend to use water more efficiently under elevated CO$_2$ conditions. Plants have small openings, called *stomata,* on the undersides of leaves that allow air to pass in and out of them; these stomata need not open as wide as normal when CO_2 levels are high. As CO_2 from the atmosphere enters the leaf, water vapor from inside can escape. Not opening their stomata as wide allows plants to survive under drier conditions, allowing the plants to grow faster under high CO_2 levels. The number of stomata per leaf also tends to decrease in plants grown under elevated CO_2 concentrations.

Whether forests and other natural ecosystems should respond similarly to CO_2 fertilization is a question that has aroused considerable debate. Ecologists have noted that many natural ecosystems are limited by other factors, such as nutrient availability and competition for sunlight. In such cases, higher atmospheric CO_2 concentrations should have little effect on plant growth. Even if CO_2 fertilization is occurring, other factors might come into play in the future as the climate warms. One concern is that soil carbon might decay more rapidly under such conditions. Tropical soils, for example, are deficient in carbon as a result of rapid rates of decomposition. Because most of the carbon in temperate forests resides in the soil rather than in the trees themselves, the total amount of carbon stored in forest ecosystems could actually decrease in the future even if the trees themselves grow faster.

A related phenomenon that might be encouraging CO_2 uptake by the terrestrial biota is **nitrogen fertilization**. Nitrogen is an essential nutrient for all organisms, and it is often in short supply because it is difficult to convert atmospheric N_2 into fixed nitrogen that organisms can use. Humans have been helping out in this regard by adding nitrogen fertilizer to agricultural fields and by emitting large amounts of nitrogen oxides from combustion. Agricultural activity does not normally lead to net uptake of CO_2, because the crops that are grown are harvested and eaten and the remaining organic matter is burned or decays. But the nitrogen oxides released into the atmosphere are oxidized to nitric acid, HNO_3, and are removed by precipitation. If the concentration of nitric acid in rainwater becomes too high, the resulting acid rain is harmful to plants and to other organisms, especially fish (although nitric acid generally contributes less to acid rain than does the sulfuric acid generated from SO_2 released by coal burning). At more modest concentrations, the nitric acid becomes a source of fixed nitrogen, which can stimulate plant growth. So, some of the increased forest growth currently taking place may be a response to anthropogenic nitrogen emissions.

Dissolution in the Oceans

The next-fastest mechanism for removing anthropogenic CO_2 from the atmosphere is dissolution in the oceans. The chemical process by which the oceans dissolve CO_2 is described in the Box "A Closer Look: The Chemistry of CO_2 Uptake." According to Figure 16-1, the flux of CO_2 between the atmosphere and oceans is on the order of 90 Gton(C)/yr, so the residence time for atmospheric CO_2 with respect to this process, t_{OA}, is approximately 8 years.

This time scale, however, is somewhat misleading. Recall that the ocean can be thought of as consisting of two layers: a well-mixed surface layer approximately 75 m thick and a poorly mixed deep-ocean layer nearly 4 km thick. These layers can be represented schematically by separate boxes, as shown in Box Figure 16-1. Only the shallow surface layer is capable of rapid CO_2 exchange with the atmosphere. Because of its small volume, its CO_2 uptake capacity is relatively low. The uptake capacity is determined largely by the abundance of carbonate ion. The deep ocean has a much larger volume and, hence, a much larger CO_2 uptake capacity, but its turnover time, t_{SD}, is on the order of 1,000 years. Therefore, CO_2 exchange between the atmosphere and the oceans occurs on a variety of time scales, ranging from 8 years to more than 1,000 years. Because the uptake of CO_2 by the oceans is a chemical process, the lifetime of anthropogenic CO_2 depends on how much of it we produce. The first puffs of CO_2 released at the dawn of the industrial age were taken up almost immediately by the surface ocean. But the more CO_2 we release, the deeper it must penetrate into the ocean in order to be buffered by reaction with carbonate ion. The lifetime of CO_2 released today is estimated to be only about 60 years, but the lifetime of CO_2 released in the future is predicted to be much longer. Computer models that take this chemistry into account are required to calculate how fast the ocean will actually take up anthropogenic CO_2.

Dissolution of Sea-Floor Carbonates

Carbon dioxide can also be taken up by the dissolution of carbonate sediments on the sea floor. This may at first seem surprising, as we learned previously that the precipitation of carbonate sediments, in conjunction with silicate weathering, removes CO_2 from the atmosphere—ocean system. Carbonate sediments are the long-term sink for CO_2. On shorter time scales, however, just the opposite occurs: Atmospheric CO_2 is taken up when carbonate sediments dissolve, because both compounds are converted to bicarbonate.

Carbonate sediments dissolve when CO_2-enriched seawater comes in contact with the sediments. For this to occur, the CO_2-rich water needs to be carried down into the deep ocean. As discussed above, this is a slow process, requiring many hundreds of years. Furthermore, the sediments on the sea floor need to be stirred to expose fresh surface area for the reaction. This stirring, or **bioturbation,** is accomplished by burrowing organisms, such as

A CLOSER LOOK
The Chemistry of CO₂ Uptake

The rate at which CO_2 can be taken up by the oceans depends on ocean chemistry as well as ocean mixing. The reason is that CO_2 does not simply dissolve in seawater as would a gas, such as N_2 or O_2. As we discussed in Chapter 8, when CO_2 dissolves in water, it forms carbonic acid, H_2CO_3, which then dissociates into bicarbonate ions, HCO_3^-, and carbonate ions, $CO_3^=$. Long before humans began perturbing the Earth system, the ocean contained substantial quantities of carbonate and bicarbonate ions as part of the natural inorganic carbon cycle. The presence of these ions in solution makes it possible for seawater to absorb more anthropogenic CO_2 than would otherwise be possible. The reason is that these ions, carbonate ions in particular, moderate the change in the ocean's acidity as CO_2 is added. A chemist would say that they serve as a pH buffer, a dissolved substance that helps maintain a stable pH. At pH values that are typical of the surface ocean (pH of about 8), the chemical reaction that occurs can be written as

$$CO_2 + CO_3^{2-} + H_2O \rightarrow 2\,HCO_3^-$$

Each anthropogenic CO_2 molecule that enters the ocean combines with one carbonate ion and one water molecule, yielding two bicarbonate ions. A similar reaction converts borate ion ($H_2BO_3^-$) into boric acid (H_3BO_3). The presence of borate increases the ocean's buffering capacity by an additional 25%. The fact that such chemical reactions occur implies that the ocean's capacity to absorb CO_2 is limited. If the amount of anthropogenic CO_2 added to the ocean exceeds the amounts of carbonate ion and borate ion that were initially present, the ocean's buffering capacity will be exhausted and its ability to absorb CO_2 will be greatly diminished. We can estimate the CO_2 uptake capacity of the ocean by measuring the dissolved carbonate ion concentration in the surface and deep ocean and multiplying by the volumes of the respective reservoirs. (We shall do so in "Critical-Thinking" Problem 3). The results are shown in Box Figure 16-1. The surface ocean has only a small buffering capacity compared with the amount of CO_2 that could be produced from the burning of fossil fuels. The deep ocean contains enough carbonate and borate ion to react with approximately 30% of the fossil-fuel reservoir.

The oceans can also absorb anthropogenic CO_2 by dissolving carbonate sediments on the sea floor. The chemical reaction involved is

$$CO_2 + CaCO_3 + H_2O \rightarrow \\ Ca^{2+} + 2\,HCO_3^-.$$

This reaction is similar to that by which seawater itself absorbs CO_2, except that the required carbonate ion is initially attached to a calcium ion. Neither of these two processes is a permanent sink for CO_2 because, even after the reactions have occurred, the CO_2 is still present in the oceans as bicarbonate. This bicarbonate will eventually be removed when enough calcium ions have been provided by silicate weathering to reprecipitate it as carbonate sediments. Only then will the anthropogenic CO_2 truly be gone.

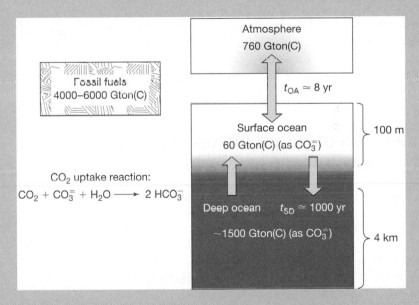

BOX FIGURE 16-1

Two-box ocean model illustrating the capacity of the ocean for CO₂ uptake. The numbers in the ocean boxes represent the amount of carbonate ion that can react with CO₂.

worms, that make their homes in the sediments. Bioturbation occurs primarily in the uppermost 10 cm of sediments, but repeated episodes of carbonate dissolution followed by renewed burrowing can eventually cause the uppermost 40–50 cm of marine sediments to dissolve. As Wallace Broecker of Lamont-Doherty Earth Observatory has observed, "Even the worms will do their part in [taking up fossil fuel CO₂]!" Thus, sea-floor carbonate dissolution may eventually play an important role in CO₂ removal, but it is not likely to prevent atmospheric CO₂ from rising over the next few decades to centuries.

Weathering of Continental Rocks

The slowest, but most permanent, sink for anthropogenic CO₂ involves the weathering of silicate rocks on the continents, followed by the precipitation of carbonate sediments on the sea floor. As we discussed in Chapter 8, the combination of these two processes can be represented by the chemical reaction

$$CaSiO_3 + CO_2 \rightarrow CaCO_3 + SiO_2$$

where $CaSiO_3$ represents a variety of more complicated silicate minerals. This process is much slower than the carbonate dissolution process discussed above, because it requires that Ca^{2+} ions produced by weathering accumulate in the ocean. The time required to precipitate anthropogenic CO_2 as carbonate can be estimated by dividing the total amount of carbon in the combined atmosphere—ocean system, about 38,000 Gton, by the rate at which CO_2 is consumed by silicate weathering, 0.06 Gton(C)/yr. This time scale, which is in excess of half a million years, is the characteristic response time of the carbonate—silicate cycle. Our human-induced perturbation to the natural carbon cycle is likely to last at least that long.

Carbonate rocks on the continents can also be weathered and dissolved. This process is analogous to sea-floor carbonate dissolution and provides another "temporary" sink for CO_2. The CO_2 is not really gone because it is stored in the oceans as bicarbonate. Carbonate rocks dissolve more rapidly than do silicate rocks, so this loss process could become important on time scales of only a few thousands of years.

Projections of Future Atmospheric CO₂ Concentrations and Climate

Once we have understood the present-day sources and sinks for CO_2, the really difficult task begins: we need to project this information into the future to try to estimate future atmospheric CO_2 levels. Doing so involves making various assumptions about how much fossil fuel people will consume and how much deforestation/reforestation will take place. If we input this information into a computer model of the global carbon cycle that includes the various CO_2 removal processes just described, we can attempt to predict how atmospheric CO_2 concentrations will change. Then, this information can be used in global climate models to predict how future climate may be affected. However, we must consider other greenhouse gases as well because CO_2 is not the only greenhouse gas that is increasing.

Atmospheric CO₂ Levels for Different Emission Scenarios

Rather than try to make our own estimates for how much fossil fuel will be burned over the next few decades, we will rely on projections that other researchers have made. A group of scientists called the Intergovernmental Panel on Climate Change (IPCC) has been actively involved in making such projections and in determining their consequences. The IPCC is an international organization whose purpose is to produce a "consensus" scientific view on global warming. Although there are scientists who disagree with some of the IPCC's conclusions, as well as differences of opinion within the IPCC itself, our impression is that they provide a balanced, and useful, view of this topic. A new IPCC report was issued in 2001, and we will follow their approach.

Four possible CO_2 emission curves from the IPCC report are shown in Figure 16-3a. Curve *a* is the most optimistic scenario. It assumes that total CO_2 emissions (fossil fuels plus deforestation) will rise from their value of 7

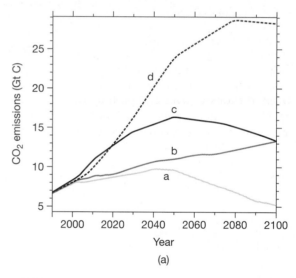

(a)

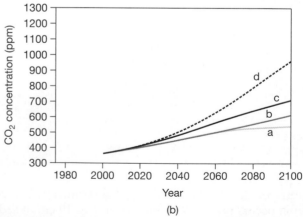

(b)

FIGURE 16-3

Estimated CO_2 emissions (panel *a*) and atmospheric CO_2 concentrations (panel *b*) for the next century for different assumptions about population and economic growth. (From *IPCC Report: Climate Change 2001: The Scientific Basis*. Available online at *http://www.ipcc.ch/*)

Gtons(C)/yr in 1990 to just under 10 Gtons(C)/yr in 2050 before declining to 5 Gtons(C)/yr by the end of the next century. This level of emissions is possible if society takes strong steps to curb fossil-fuel consumption and encourage reforestation. Curve *b* is slightly more pessimistic. It assumes that CO$_2$ emissions will increase at about 0.7%/yr, reaching just over 13 Gtons(C)/yr by the turn of the century. Scenarios *a* and *b* both assume modest growth in world population and economic output. By contrast, curves *c* and *d* assume more rapid population growth and faster economic development. The case *c* scenario assumes that the energy needed to fuel this economy comes from a variety of different sources. In this case, the CO$_2$ emission rate peaks at 17 Gtons(C)/yr in 2050 and then slowly declines. Case *d* assumes that most of our future energy will be derived from fossil fuels, as it is today. In this scenario, CO$_2$ emissions reach 28 Gtons(C)/yr (four times the present value) by 2080. The cumulative CO$_2$ emissions over the next century in case *d* are a little over 2000 Gtons(C), or almost half the recoverable fossil fuel reserves.

Figure 16-3b shows the predicted atmospheric CO$_2$ concentrations corresponding to these four different emission scenarios. Even in the optimistic case *a*, CO$_2$ reaches concentrations of over 500 ppm by the year 2100. In the most pessimistic case *d*, the predicted CO$_2$ concentration is nearly twice this high, close to 1000 ppm. This is almost two doublings compared to the preindustrial CO$_2$ level of 280 ppm. The first doubling of CO$_2$ would occur by about the year 2050 in this scenario.

By how much would we expect climate to change in these various cases? We can derive a preliminary answer from the one-dimensional, radiative—convective climate models discussed in Chapter 3. Doubling the atmospheric CO$_2$ concentration in such a model produces about a 2.5°C increase in global mean surface temperature when the water vapor feedback is included. This temperature increase could occur by the year 2050 in case *d* or by 2100 in case *a*. Doubling CO$_2$ a second time would produce approximately the same amount of warming. Thus, global temperatures could increase by as much as 5°C (9°F) over the next century in the worst-case scenario. In the most favorable case, the warming would be about 2°C. Both of these numbers should be compared with the warming of 0.7°C that has taken place over the past 100 years. Evidently, climate change during the next century is likely to be considerably more rapid than it was in the last century. A prediction of this nature, however, is not considered good enough for making policy decisions because it overlooks several factors that should affect Earth's climate over the next few decades. These factors include radiative forcing by trace gases other than CO$_2$, the oversimplification of one-dimensional models, and the thermal properties of the ocean.

Radiative Forcing by Other Trace Gases

Carbon dioxide is not the only atmospheric greenhouse gas that is currently increasing in concentration. Methane and nitrous oxide have also been increasing in concentration over the past 200 years (Figure 16-4). Methane has strong anthropogenic sources, mostly cattle raising and rice cultivation, that account for 60–80% of its total emissions. About 30% of nitrous oxide emissions come from bacterial denitrification in fertilized soils; the remaining 70% is natural. Emissions of both methane and nitrous oxide are expected to increase over the next century as human population increases and agricultural output increases accordingly. Chlorofluorocarbon compounds (CFCs) have also been increasing over the past several decades, but here the future projections are quite different. Most conventional CFCs, such as freon-11 and

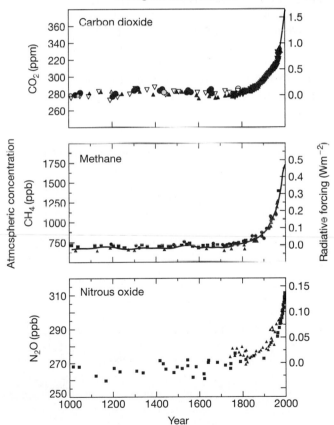

Indicators of the human influence on the atmosphere during the Industrial Era

(a) Global atmospheric concentrations of three well mixed greenhouse gases

FIGURE 16-4

Atmospheric concentrations of CO$_2$, CH$_4$, and N$_2$O since 1000 A.D. The earlier data come from ice cores; more recent data are from direct atmospheric measurements. The additional radiative forcing relative to the preindustrial level is shown by the scale on the right. (From *IPCC Report: Climate Change 2001: The Scientific Basis.* Available online at *http://www.ipcc.ch/*)

freon-12, have now been banned in order to protect the ozone layer. (See Chapter 17.) Thus, they are not expected to contribute to future global warming. Freon replacement gases are beginning to accumulate, however, and these gases could eventually contribute to global warming. We return to this issue in Chapter 17.

The contributions of each of these different gases to the atmospheric greenhouse effect can be measured in terms of a quantity termed **radiative forcing**. Radiative forcing refers to the change in the outgoing infrared flux caused by a change in the concentration of a particular greenhouse gas. A doubling of atmospheric CO_2 levels produces a radiative forcing of about 4.4 W/m². We have already seen that such a forcing produces a surface temperature increase of about 2.5°C in radiative–convective climate models, or 1.5–4.5°C in GCMs (see Chapter 6).

The right-hand scale in Figure 16-4 shows the radiative forcing produced by the increases in different greenhouse gases over the past 1,000 years. As one can see, the increases in radiative forcing are about 1.5 W/m² for CO_2, 0.5 W/m² for CH_4, and 0.15 W/m² for N_2O. Other factors contribute to radiative forcing as well (Figure 16-5). Some of these forcings are positive, like those from CO_2, CH_4, and N_2O. Others, however, are negative. Increases in SO_2 emissions, for example, lead to increased concentrations of sulfate aerosols. Such particles cool the surface by in-creasing Earth's albedo. Sulfate aerosols may also cool the surface by acting as **cloud condensation nuclei** (CCNs), thereby increasing the reflectivity of clouds (see bar labeled "aerosol indirect effect"). The different radiative forcing factors are listed in Figure 16-5 in order of how well we understand them. Those on the left-hand side of the diagram (which includes the greenhouse gases) are relatively well understood. Those on the right-hand side are understood only poorly.

AOGCM Predictions of Global Warming

The IPCC has made estimates for how much atmospheric CH_4 and N_2O will change over the next century and for how future production of SO_2 will affect sulfate aerosol concentrations. These estimates (which are not shown here) are generally in accord with the assumptions made regarding emissions of CO_2. In the most pessimistic case (case *d* in Figure 16-3b), CH_4 increases by a factor of about 2 over the next century (from 1.7 ppm to ~3.4 ppm), while N_2O increases by about 50% (310 ppb to 450 ppb). Increases in CH_4 by more than a factor of 2 seem unlikely. Methane emissions are closely linked to agriculture, which in turn is linked to human population. Human population may double over the next century, but it is unlikely to go much higher than that because of competition for food and other resources. Nitrous oxide

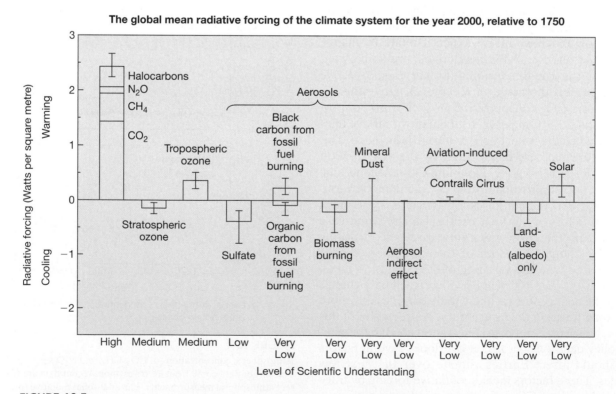

The global mean radiative forcing of the climate system for the year 2000, relative to 1750

FIGURE 16-5

Net radiative forcing by different atmospheric trace gases and other factors. The confidence in these estimates decreases from left to right. (From *IPCC Report: Climate Change 2001: The Scientific Basis.* Available online at *http://www.ipcc.ch/*)

should increase by a smaller amount because a significant fraction of its present sources are nonanthropogenic. These factors are taken into account in the IPCC estimates. Various assumptions are made regarding SO$_2$ emissions as well. In the low-growth model (case *a*), SO$_2$ emissions decline gradually with time over the next century, whereas in the other three models, SO$_2$ emissions increase at first and then slowly decline. In cases *c* and *d* in particular, the growth in SO$_2$ emissions significantly reduces the climate warming expected by the middle of the next century.

To convert these estimates for future trace gas emissions into predicted temperature changes, it is necessary to use numerical climate models. However, as mentioned earlier, the most reliable climate model predictions do not come from simple, one-dimensional climate models, but rather from three-dimensional general circulation models, or GCMs. We discussed these models in some detail in Chapter 6 because of their importance to this topic. GCMs can predict the geographical distribution of future climate change, along with changes in other important variables such as precipitation. Furthermore, GCMs that include the ocean as well as the atmosphere, so-called AOGCMs (atmosphere–ocean general circulation models), can simulate another important effect as well. As the global climate warms, the ocean is expected to heat up more slowly than the atmosphere. It then acts as a brake on how fast the atmosphere itself can warm. Technically, it exerts a **thermal drag** on the system. Thus, the *transient* (time-dependent) response of the atmosphere to greenhouse gas increases is smaller than the *equilibrium* response.

The thermal drag of the oceans is sufficiently important in simulations of global warming that it has given rise to a new standard, or metric, for comparing climate models. In equilibrium GCM calculations, the accepted metric is the response to a doubling of atmospheric CO$_2$. We have already seen that the range of responses for various GCMs is 1.5–4.5°C, with most of the variation being accounted for by differences in how clouds are treated. For AOGCMs that are being run in a time-dependent manner, the metric is taken to be the increase in global average surface temperature at the time of CO$_2$ doubling *for an assumed CO$_2$ increase rate of 1%/yr.* CO$_2$ is actually increasing at only about half this rate presently, but the new metric accounts for the fact that other greenhouse gases are increasing as well and that the radiative forcing from the combination of these gases is about equal to that of CO$_2$. The range of *transient climate response,* as this metric is termed, for various AOGCMs is 1.1–3.1°C. It is significantly smaller than the equilibrium climate response to doubled CO$_2$ because of the thermal drag of the oceans.

The various climate-modeling groups within the IPCC have taken the greenhouse gas scenarios described above and used them as input for time-dependent climate models. Some of these models are true AOGCMs. AOGCMs are very time-consuming to run, however, so not all of the scenarios have been modeled by multiple AOGCMs. Some climate predictions have been generated with simpler climate models that are "tuned" so as to reproduce the behavior of a true AOGCM. All such models include the thermal drag of the ocean, which reduces the rate at which the global climate can warm. The results of these simulations for the four cases described earlier are shown in Figure 16-6. Somewhat surprisingly, the results for the global average temperature are remarkably similar to those given earlier, which were based on simple one-dimensional climate model calculations. For the optimistic case *a,* the climate warms by about 2°C over the next century. For the pessimistic case *d,* the predicted warming is about 4.5°C. The various curves represent the average surface temperatures calculated by a number of different climate models. Not all of the models agree with each other. The range of uncertainty for the amount of warming produced by the year 2100 is shown by the error bars to the right of the figure. The total range of uncertainty for all climate models and all greenhouse gas scenarios is shown by the shaded region of the figure. When all the uncertainties are taken into account, the predicted range of climate warming by the year 2100 is 1.4–5.8°C.

Long-Term Climate Warming

The climate calculations discussed so far extend only to the year 2100. But if you look at the Box "A Closer Look: Long-Term CO$_2$ Projections" you will see that atmospheric CO$_2$ could continue to rise for several centuries beyond that time and that its concentration might eventually reach 2100 ppm—almost eight times its preindustrial value. What effect would this extended rise have on Earth's climate?

The greenhouse effect of CO$_2$ is roughly logarithmic, which means that each factor-of-2 increase in CO$_2$ produces roughly the same amount of warming. An eightfold increase in CO$_2$ should therefore cause about three times as much warming as would a twofold increase (because $8 = 2^3$). Thus, a GCM that predicted 3°C of warming for doubled-CO$_2$ conditions should produce about 9°C of warming for an eightfold CO$_2$ increase.

The actual range of GCM responses to doubled atmospheric CO$_2$ is between 1.5 and 4.5°C of warming. The corresponding range for an eightfold CO$_2$ increase is 4.5 to 13.5°C of warming. It is instructive to compare these numbers to estimated surface temperature changes during the past 100 million years of Earth history. The warmest part of the Mesozoic is thought to have been about 6–10°C warmer than today on a global average. The coldest part of the Pleistocene was probably about 5–6°C cooler than today. Thus, regardless of whether we use the low model estimates or the high ones, the warm-

A CLOSER LOOK
Long-Term CO₂ Projections

What will happen to atmospheric CO_2 levels in the distant future if we continue to burn fossil fuels? This question can be studied by using specially designed computer models that are able to take large time steps. For illustrative purposes, let us assume that most of the fossil fuel reserve listed in Table 16-1, 4200 Gton(C), is consumed during the next 400 years. Let us further assume that current deforestation trends continue until only 30% of the world's forests remain.

This particular scenario has been investigated with a computer model that includes a six-box ocean, along with separate boxes to represent the atmosphere, forests, and carbonate sediments on the sea floor. The results of the simulation are shown in Box Figure 16-2. Box Figure 16-2a shows calculated atmospheric CO_2 concentrations over the next 3,000 years. The computer model predicts that atmospheric CO_2 levels will increase to a peak of about 2100 ppm by the year A.D. 2300. At that time, the fossil fuel reserves will be exhausted and atmospheric CO_2 will begin to decline.

If we extend this calculation further into the future, the atmospheric CO_2 concentration must eventually return to its preindustrial value of 280 ppm, because the model assumes that the carbon cycle was balanced at that time. The amount of time required to return to that steady-state level is shown in Box Figure 16-2b. There, the date is displayed on a logarithmic scale extending millions of years into the future. The processes that are responsible for CO_2 uptake on different time scales are indicated on the figure. Most of the CO_2 is removed over the next few thousand years by dissolution in the deep ocean and by the dissolution of carbonate sediments and rocks. The last vestiges of the CO_2 pulse are removed by silicate weathering over a period of more than 1 million years.

Although the calculations shown here are speculative, they suggest an intriguing possibility. If the climate record of the past few million years were extended into the future, we would expect that Earth should experience at least 10 major glacial–interglacial cycles over the next 1 million years. But atmospheric CO_2 levels were relatively low, 200 to 280 ppm, during the previous glacial cycle, whereas they are predicted to exceed 350 ppm during most of the next 1 million years. Could the addition of this much anthropogenic CO_2 break the glacial–interglacial cycle? If it did, the results would be at least partly beneficial; after all, no one looks forward to the beginning of the next Ice Age. Yet, the accompanying increase in sea level over this time could cause shorelines to move substantially inland and force massive relocations of people. The possible long-term effects of the burning of fossil fuels on the Earth system are evidently quite large and will eventually need to be considered.

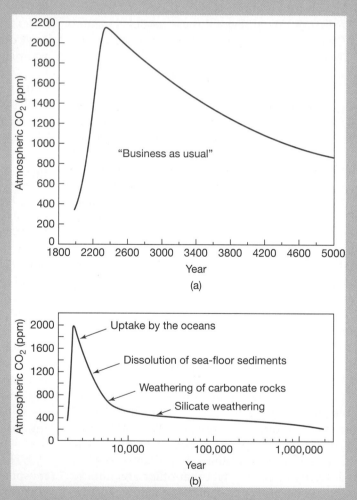

BOX FIGURE 16-2

Long-term projections for atmospheric CO_2 for (a) the next 3,000 years and (b) the next several million years. The total amount of fossil fuel consumed is equivalent to 4200 Gton (C). (After Walker and Kasting. *Global and Planetary Change* 97:151, 1992).

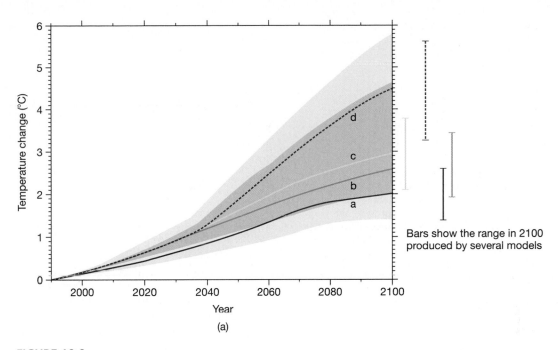

Bars show the range in 2100
produced by several models

FIGURE 16-6

Predicted trends in global surface temperature over the next century for the four cases shown in Figure 16-3. The shaded area shows the total range of temperatures predicted by different climate models for the various scenarios. (From *IPCC Report: Climate Change 2001: The Scientific Basis.* Available online at *http://www.ipcc.ch/*)

ing from an eightfold CO_2 increase could make Earth warmer than it has been for tens of millions of years.

Changes in Sea Level and Ocean Circulation

How concerned should we be about the prospect of future global warming? The answer depends not only on the magnitude and rate of the climate change itself, but also on the effects of that change on other parts of the Earth system.

Sea-Level Change During the 20th Century

One factor that we must consider in any discussion of long-term global warming is sea level. Researchers have estimated that sea level has increased by approximately 12 cm since 1880 (Figure 16-7). The data come from tide gauges on various shorelines around the world. Interpreting such measurements is complicated because the solid Earth itself is moving up or down in some locations. Parts of Canada and Europe, for instance, are moving upward at a rate of a few centimeters per century because they are still rebounding from the weight of the great ice sheets that covered these regions as recently as 11,000 years ago. In contrast, the Nile River delta is subsiding because silt from the Nile has increased the loading in this region. These localized trends in land surface elevation

must be subtracted from the tide gauge data before we can draw any inferences about sea-level change.

The increase in global sea level during the 20th century parallels the rise in global mean surface temperature (see Figure 1-4). Indeed, approximately half of the rise in sea level can be attributed to simple **thermal expansion** of surface ocean water. Water, like most other materials, normally expands as it warms. (More correctly, pure water expands when it warms except when its temperature is between 0 and 4°C, when warming causes it to contract. Seawater, however, does not exhibit this unusual behavior.) Because the lateral boundaries of the ocean are largely fixed, any expansion of ocean volume must result in a rise in sea level. The 0.7°C atmospheric temperature rise during the 20th century is expected to have warmed the surface ocean by this same amount, which should have increased sea level by about 7 cm. An increase of 0.7°C in the temperature of the deep ocean would cause a much larger increase in sea level, but such a change would take many centuries to occur because the thermohaline circulation is very slow (Chapter 5). Changes in deep-ocean temperatures caused by Milankovitch cycles may be responsible for periodic sea-level fluctuations of about 5 m that are recorded in carbonate platforms formed during the Mesozoic and Paleozoic Eras.

Most of the remaining increase in sea level during the 20th century is thought to have been caused by melting of **mountain glaciers**, ice fields formed on the cold, upper reaches of mountains (Figure 16-8). Several glaciers in

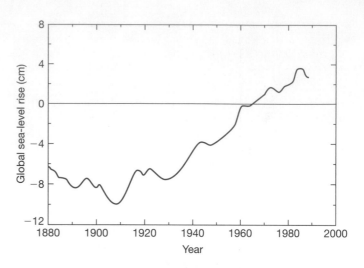

FIGURE 16-7

Global sea-level rise since 1880 (relative to the interval 1951–1970). (From T. P. Barnett. Global Sea Level Change. In *NCPO, Climate Variations Over the Past Century and the Greenhouse Effect.* A report based on the First Climate Trends Workshop, Sept. 1988, Washington, D.C. Reproduced in J. T. Houghton et al. *Climate Change: The IPCC Scientific Assessment.* Cambridge: Cambridge University Press, 1990.)

the Alps are known to have retreated during the past two centuries as Earth emerged from the Little Ice Age (Chapter 15). Glaciers in the high Andes Mountains in Peru are also known to be receding rapidly at present. If this trend were to continue, sea level might eventually rise by another 40 cm from this source alone.

Sea-Level Rise in the Future

A serious concern for the future is that the polar ice caps will begin to melt. Only the ice that is now on land is important in this respect. The melting of sea ice does not increase sea level, because floating ice displaces an amount of seawater that is precisely equal to its mass. (You can prove this by filling a glass with water and ice cubes and allowing the ice to melt. If the ice cubes were initially floating, the water level in the glass will not change.)

Today, large continental ice sheets are found on Greenland and Antarctica. The Greenland ice sheet contains enough water to raise sea level by approximately

7 m, were it to melt entirely. The Antarctic ice sheet contains much more water—some 60–70 m of equivalent sea-level rise. However, these two ice sheets are expected to behave quite differently as the climate warms. The island of Greenland extends to lower latitudes than does the continent of Antarctica, so the climate in southern Greenland is considerably warmer than the Antarctic climate. The Greenland ice sheet is therefore much more likely to experience increased melting as the climate warms than is the Antarctic ice sheet. Over most of Antarctica increased snowfall (resulting from warmer ocean temperatures and increased evaporation rates) is expected to cause the ice sheet to thicken over the next 50 to 100 years. This phenomenon could, paradoxically, cause global sea level to decrease as atmospheric CO_2 levels increase.

The West Antarctic Ice Sheet

The actual situation in Antarctica is even more complicated than we have indicated. The Antarctic ice sheet can be divided geographically into an eastern and a western part. The East Antarctic ice sheet contains most of the water and is the part that might thicken as the climate warms. The West Antarctic ice sheet contains less water (about 5–6 m of equivalent sea level), but its response to greenhouse warming could be quite different. This ice sheet flows primarily into the Ross and Weddell Seas, where it forms **ice shelves,** large expanses of floating sea ice formed at the margins of continents. That in the Ross Sea is called the Ross Ice Shelf, and that in the Weddell Sea is called the Filchner-Ronne. Both ice shelves are grounded at several points on offshore islands.

Glaciologists have speculated that an increase in water temperature of just a few degrees in the Ross and Weddell Seas could melt enough ice off the bottom of these ice shelves to cause them to become completely free-floating. As a result of this melting, those parts of the West Antarctic ice sheet that feed these shelves might

FIGURE 16-8

A mountain glacier. (From Gilbert S. Grant, Photo Researchers, Inc.)

flow much more rapidly, because the contact with the off-shore islands currently inhibits the glaciers' flow. Such a sudden, rapid increase in a glacier's flow rate is called a **glacial surge.** Glacial surges are occasionally observed in mountain glaciers, and there is indirect evidence (from ice-rafted debris in North Atlantic sediments) that they occur in continental-scale glaciers as well. Once started, a glacial surge tends to perpetuate itself, because the increased flow rate causes frictional heating at the base of the glacier. This heating in turn produces a thin layer of water that allows the glacier to slide more smoothly over the surface. If such a positive feedback process were to be triggered by global warming, the West Antarctic ice sheet might thin relatively rapidly and could contribute significantly to sea-level rise over the next few centuries.

Projections of Future Sea-Level Rise

Given all these possible effects, what can we say about sea-level change in the near future? Will the observed upward trend of the 20th century continue?

Projections of possible sea-level increase during the 21st century are shown in Figure 16-9. The four curves shown correspond to the four scenarios illustrated in Figures 16-3 and 16-6. These calculations take into account thermal expansion of the oceans and melting of

mountain glaciers, but do not include estimates of polar ice sheet melting (because those estimates are so uncertain). The projected increases in sea level range from 30 cm for case *a* to 50 cm for case *d*. These rates of increase are 2–3 times faster than the rate of increase that occurred in the 20th century. Sea-level increases of this magnitude could pose problems for low-lying areas such as the Gulf Coast of North America, Bangladesh, and numerous islands in the South Pacific and Indian Oceans.

A potentially more serious problem is the change that might occur in the more distant future. The amount of water now tied up in polar ice is so large, and the projected time scale for global warming is so long, that sea level could ultimately increase by many meters. As one concrete example, the IPCC extended several of their global warming calculations out to the year 3000 and attempted to estimate what would happen to the Greenland ice sheet. Southern Greenland is not that far below freezing even now, so increases in surface temperature could cause significant amounts of melting. The resulting effect on sea level is shown in Figure 16-10. Three curves are shown, corresponding to low, moderate, and high rates of climate change. The temperatures shown on the diagram represent the average temperature change over Greenland by the year 3000. These changes are larger than the global average temperature changes by factors of

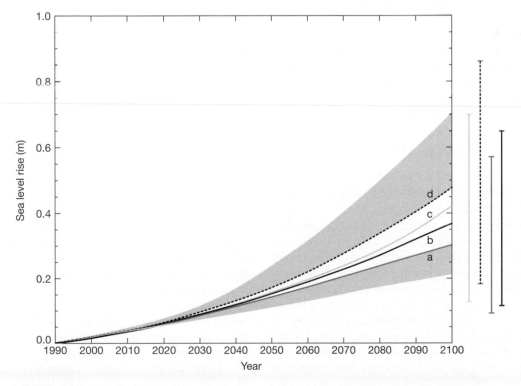

FIGURE 16-9

Predicted changes in sea level over the next century for the four different cases shown in Figures 16-3 and 16-6. (From *IPCC Report: Climate Change 2001: The Scientific Basis.* Available online at *http://www.ipcc.ch/*)

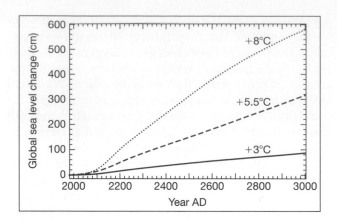

FIGURE 16-10

Long-term changes in sea level resulting from melting of the Greenland ice sheet. The numbers on the curves show the predicted temperature changes over Greenland by the year 3000 A.D. for three different scenarios of global warming. (From *IPCC Report: Climate Change 2001: The Scientific Basis.* Available online at *http://www.ipcc.ch/*)

up to 3. In the worst-case scenario, sea level increases by 600 cm (6 m) by the year 3000, indicating that most of the Greenland ice sheet has disappeared.

How serious would such a change be? A sea-level increase of this magnitude would wreak havoc with Earth's present geography. A 6-m rise in sea level, like the one shown here, would submerge the southernmost one-third of Florida. In the very long term (thousands of years in the future), it is conceivable that the polar caps could melt entirely if we were to consume all of the available fossil fuels. This melting would increase sea level by 70–80 m and would submerge roughly 20% of the present continents. So, our decisions about fossil-fuel usage over the next few centuries could have major repercussions for our descendants.

Changes in the Thermohaline Circulation

Sea level is not the only aspect of the oceans that might change as a consequence of global warming. It is possible that ocean circulation might change as well. One particular aspect of ocean circulation that has been studied extensively with regard to global warming is the Atlantic Conveyor Belt. This is the thermohaline circulation pattern described in Chapter 5 in which deep water forms in the North Atlantic, spreads out globally at depth, then is upwelled and returns to the North Atlantic as a warm surface current. Recall that this circulation pattern is responsible for keeping Western Europe warm in the wintertime. Without it, winters there would be much colder. In the previous chapter, we saw that this circulation pattern probably did cease for almost a thousand years during the Younger Dryas event at the end of the last Ice Age. The cause of that shutdown is thought to have been a pulse of freshwater from the melting of the Laurentide

ice sheet that flowed down the St. Lawrence River and into the North Atlantic. This created a *freshwater cap* that was too buoyant to sink easily. Recall that Europe grew much colder again during this time, leading to the reappearance of alpine flowers like the dryas.

In at least some AOGCMs, a similar phenomenon is predicted to occur in the future, although for a somewhat different reason. Melting of the Greenland ice cap could, of course, supply freshwater that might duplicate the effect of the meltwater from the Laurentide ice sheet at the end of the last Ice Age (Chapter 15). That effect, if it occurred, would likely be several centuries in the future. In some AOGCMs, however, another similar phenomenon takes place much earlier, within the next 100–200 years. Global warming causes increased evaporation in the low- to mid-latitude Atlantic ocean, putting more moisture into the air. This moisture is transported northward by winds and falls out as rain in the North Atlantic. The increased rainfall freshens the surface layer, reducing its density, and making it less likely to sink. In some models, the thermohaline circulation slows, then stops entirely within the next century and a half. This is not a robust prediction at this time, as this type of couple ocean–atmosphere calculation is still fraught with uncertainties but it is at least a possibility. If it did, something really paradoxical might happen: Europe might cool substantially as the rest of the world warmed! It is unlikely that Europe would experience a new Ice Age, as the global climate by this time would be significantly warmer than it was 11,000 years ago, but the predicted temperature drops could still be severe. This is yet one more example of why we must consider the entire Earth as a system. It takes a tightly coupled systems model, an AOGCM in this case, to predict this type of counterintuitive behavior.

Effects on Ecosystems

Also of great interest is the effect that atmospheric CO_2 increases and global warming might have on terrestrial and marine ecosystems. We have already mentioned several such effects. Higher atmospheric CO_2 concentrations are expected to cause increased rates of plant growth. Plants are also expected to use water more efficiently at high CO_2 levels, because they do not need to open their stomata as wide to obtain CO_2 for photosynthesis and because stomatal density also decreases in plants grown under elevated CO_2. This increased efficiency might be offset in some regions by decreased soil moisture during the summertime growing season.

If we examine ecosystems in more detail, the possible changes induced by higher CO_2 levels and higher temperatures become more and more complex. This should not come as too much of a surprise, as biological systems are incredibly complicated by comparison with most physical systems. Here, we mention only a few of the many changes that are expected to take place.

C₃ and C₄ Plants

As we saw in Chapter 9, different types of plants have different mechanisms of **carbon fixation**, the biochemical process that occurs during photosynthesis by which atmospheric CO_2 is converted to organic carbon. Most photosynthetic organisms alive today (about 95% of all terrestrial plants) fix carbon by a biochemical pathway called the **Calvin cycle.** The first step of this cycle involves the production of a stable, intermediate compound (3-phosphoglyceric acid) that contains three carbon atoms. Hence, this process is called C₃ *photosynthesis,* and plants that metabolize in this way are termed **C₃ plants.**

Some plants, however, including corn, sugarcane, and many tropical grasses, begin the photosynthetic process by producing a four-carbon compound. (Actually, corn and sugarcane are themselves species of grasses, although we do not usually think of them as such.) Plants of this type are termed **C₄ plants.** As a consequence of biochemical modifications, C₄ plants are able to photosynthesize at much lower CO_2 concentrations than are C₃ plants. Indeed, C₄ plants became widespread only about 7 or 8 million years ago, possibly in response to decreased atmospheric CO_2 levels caused by the uplift and weathering of the Himalayan Mountains (see Chapter 12). Also, C₄ plants are much less responsive to CO_2 increases than are C₃ plants. Thus, whereas CO_2 fertilization might have an important effect on the growth rate of C₃ plants, it is not expected to have a large effect on C₄ plants. Thus, certain agricultural crops, such as corn, could be at a disadvantage compared with C₃ weeds in a high-CO_2 world.

Changes in Speciation within Forests

Even within the C₃ world, different types of plants are expected to exhibit different responses to changes in temperature and moisture availability that might accompany CO_2 increases. Figure 16-11 shows the predicted response of various tree species in Minnesota to estimated climate

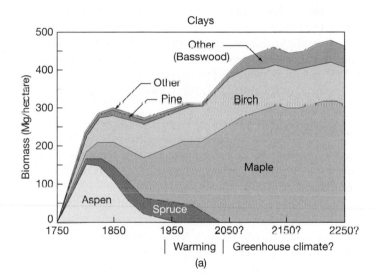

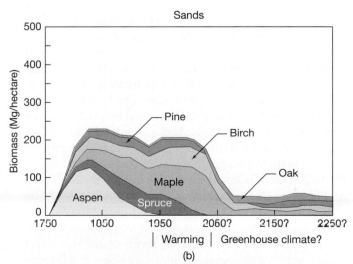

FIGURE 16-11

Predictions of species composition, in terms of (aboveground) biomass in Minnesota forests under doubled-CO_2 conditions, (a) for a clay soil with a high water-holding capacity, and (b) for a sandy soil with a low water-holding capacity. (After J. T. Houghton et al. *Climate Change: The IPCC Scientific Assessment.* Cambridge: Cambridge University Press, 1990.)

changes since 1750 and projected to A.D. 2250. The results depend strongly on whether the local soil has a high water-holding capacity (Figure 16-11a) or a low one (Figure 16-11b). In the first case, some species (pine and birch) thrive at high CO_2 levels, whereas others (aspen and spruce) decline when the temperature becomes too warm. In the second, all species do poorly after A.D. 2050, because the soil becomes too dry to support them. A general conclusion that we can reach, irrespective of the details of the calculation, is that species distributions within ecosystems are likely to change as the global climate warms.

A particular concern related to this change in species distribution is whether mid-latitude forests will be able to keep pace with the anticipated rate of climate change. Climatic warmings have occurred many times in the past, most recently at the end of the last glaciation around 10,000 years ago. However, the modern world is different from the postglacial world in several respects. Perhaps most importantly, forests in regions such as North America and Europe have been dissected by farms and highways, which tend to inhibit the migration of species. Thus, the poleward spread of species adapted to warmer climates might not occur fast enough to keep pace with the rate at which the climate warms. The extinction of many species of trees and animals could result unless humans intervene actively to facilitate the migration process.

Other Concerns

Accompanying the poleward migration of temperate and subtropical vegetation would be a migration of animal and insect species that live in warm climates. Insects, in particular, pose a significant problem for agriculture. If mid-latitude winters become less severe, insect pests that are confined to the tropics today could expand their ranges to mid-latitudes. For example, in the United States the potato leafhopper (Figure 16-12), a serious pest in soybean and other crops, is currently confined to a narrow band bordering the Gulf of Mexico. Warmer winters might allow this pest to double or triple its range. The expansion of insect ranges might well be one of the most economically damaging consequences of climate change.

Various human diseases, such as malaria, that are currently confined to the tropics might become a problem at mid-latitudes as well. Malaria is spread by the *Anopheles* mosquito, whose life cycle does not allow it to survive cold winters. If wintertime temperatures remained above freezing in the continental United States, as might happen 100 to 200 years from now if atmospheric CO_2 levels continue to increase, malaria-bearing mosquitos might be able to live there, presenting a whole new set of potential health problems.

Increased atmospheric CO_2 levels could also affect marine ecosystems. Aquatic photosynthesizers, which in-

FIGURE 16-12

A potato leafhopper. This insect, along with other agricultural pests, could expand its range poleward if the climate warms. (From Donald Specker, Animals Animals/Earth Scenes.)

clude both phytoplankton and larger, multicellular organisms such as kelp, are not expected to have increased growth rates from CO_2 fertilization, because their growth rates are generally limited by other nutrients, especially phosphorus and nitrogen. Changes in water temperature could affect some marine species, such as corals, that are adapted to a relatively narrow range of temperature. But the most significant influence of global warming on marine ecosystems is the possible change in ocean currents and upwelling zones that might accompany climate change. Upwelling of nutrient-rich deep water is the key factor that accounts for regions of high productivity in the modern oceans, such as the Grand Banks off the coast of Labrador or the fisheries off the coast of Peru. If the intensity or location of upwelling were to change as the climate warms, it could have major effects on both the high-productivity ecosystems themselves and on the fishing industry that takes advantage of them.

Economic Consequences of Global Warming

Thus far, we have confined our discussion of the effects of global warming to physical and biological aspects of the Earth system, specifically, climate, sea level, and ecosystems. Much of the current policy debate, however, is centered on possible economic consequences of global warming. Each of the physical effects that we have talked about will have some effect on the future economy of the United States and of the world as a whole. Agricultural output, for example, is almost certain to be affected. Some of the changes, such as decreased soil moisture in continental interiors, will be detrimental to agricultural output while other changes, CO_2 fertilization of plant growth for example, are expected to be beneficial. By now, many economists have developed forecast models that attempt to calculate how the projected changes in CO_2 and climate will affect the global economy.

Cost–Benefit Analysis

The way that such calculations are typically done is by what is often termed **cost–benefit analysis**. At least some of the changes anticipated from global warming are likely to cost money to deal with. Relocating cities away from continental coastlines is one obvious example. On the other hand, reducing CO_2 emissions is likely to cost money, too. As discussed further below, alternative means of producing energy are available, but none of these is currently as convenient or inexpensive as fossil fuels. Thus, if society chooses to reap the economic benefits of limiting climate change, there will of necessity be certain costs that will be incurred. The goal of economic models of global warming is to determine how these projected costs and benefits balance out. From an economic standpoint, the optimum solution is one in which the benefits less the costs is at a maximum.

A detailed discussion of such economic models is beyond the scope of the present discussion. It may nonetheless be useful to make a few comments about them because of their importance to climate policy. The first comment is that all of the economic models are highly uncertain—more so, even, than the physical models of global warming. This should come as no surprise. Predicting the behavior of humans is inherently more difficult than predicting the behavior of physical systems. However, this does not mean that economic models are of no use. As with physical models that contain large inherent uncertainties (clouds in climate models, for example), we can still use these models to help guide our choices.

Economic models, however, often contain value judgments in addition to other uncertainties. As an example, consider the ongoing debate about how much money to put into (or pay out of) the Social Security system. Social Security is predicted to go bankrupt some time before the middle of this century unless Social Security taxes are raised or benefits are reduced. The government could fix this problem by taking one or both of these two actions now, or it could postpone dealing with the problem until 10 or 20 years from now. In the first case, the present generation of taxpayers and beneficiaries would be affected. In the second case, the future beneficiaries of the system would bear the costs. Which course should the government take? The answer involves a value judgment. One has to decide how much economic hardship is worth putting up with now in order to avoid economic hardship for a somewhat different group of people several decades in the future.

Economic Discounting

In economic models of global warming, the key question is: How much should we pay now in order to avoid damages that may be incurred in the distant future? Global warming is a slow process. As we have already seen earlier in this chapter, the biggest changes in climate, and hence the largest economic damages, are not likely to occur until more than 100 years from now. In a typical cost–benefit analysis, such as one to decide whether or not to build a dam or a power plant, future damages or benefits are *discounted* at a rate of as much as 10% a year. That is, a benefit of $100 that is realized one year from now is valued at only $90.91 [= $100/(1 + 0.1)]. The same benefit reaped two years from now is valued at $82.64 [= $100/(1 + 0.1)^2], and so forth. This **discount rate** takes into account two factors: (1) If one saves money now by not building the dam, one could invest this money somewhere else and make a profit. This is called *growth discounting;* (2) most people would rather have a dollar today than a dollar 10 years from now (after adjustment for inflation). This is called *time preference discounting.*

In models of global warming, growth discounting is generally accounted for in one way or another. This makes a certain amount of sense: If society is richer 100 years from now than it is today, then people at that time can afford to pay more than they can at present. (On the other hand, if the change one is considering is essentially irreversible, like sea level rise, then no amount of economic growth may compensate for it.) Economic models of global warming usually include time preference discounting as well. According to analyses of past economic behavior, societies as a whole exhibit a preference for having money right now, as opposed to receiving it some time in the future. Yale economist William Nordhaus has studied investment behavior in U.S. society over the past 40 years and has determined that the *pure rate of social time preference,* as it is formally termed, is about 3% per year. Thus, a benefit (or cost) that was worth $100 today would be valued at $97.08 one year from now, $94.26 two years from now, and so on.

A discount rate of 3% per year may not sound like much, and indeed significantly larger discount rates (7–10%/yr) are often used in short-term, cost–benefit analyses. Consider what happens over long time spans, however. In 50 years, the assigned value of a $100 benefit (or damage) would be $100/(1 + 0.03)^{50} = $22.81. In 100 years, the value drops to $5.20. In 200 years, it is $0.27. Thus, even if some truly catastrophic change were predicted to occur 200 years from now, its influence on a typical cost–benefit analysis would be minimal. The damages could be real and could be large in real economic terms, but time preference discounting ensures that they would be largely neglected.

A Coupled Climate–Economy Calculation with and without Discounting

As a concrete example of the importance of discounting, let us examine a calculation performed with a slightly modified version of Nordhaus's own Dynamic Integrated Climate–Economy (DICE) model. The DICE model is typically run over a time span that extends 400 years into the future. It attempts to predict global economic growth, taking into account such factors as increases in population, new developments in technology, and climate change. The output from the DICE model is a time-dependent factor labelled μ (the Greek letter "mu") that represents the optimal reduction in greenhouse gas emissions compared to the "business-as-usual" case. In other words, if $\mu = 0$ at some time t, then the model is telling us that we should make no attempt to reduce greenhouse gas emission. If μ is close to 1, however, then we should try to eliminate almost all of our emissions.

To illustrate the effects of time preference discounting, the DICE model was run twice: once with the standard discount rate of 3%/yr and once with a discount rate of zero. The results are shown in Figure 16-13. For the standard case, the DICE model predicts that the economically optimal solution is to reduce greenhouse gas emissions by 10–13% (Figure 16-13a—solid curve) compared to business-as-usual emission rates. The projected atmospheric CO_2 concentrations are shown by the solid curve in Figure 16-13b. In this calculation, atmospheric CO_2 rises to about 1500 ppm by the year 2220. The calculation then ceases because the recoverable fossil-fuel resources (5000 Gton(C) in this case) are exhausted. However, when the same model is run with a 0% discount rate, the results are completely different. The DICE model then suggests that greenhouse gas emissions should be cut by more than 50% immediately, increasing to nearly 97% by the year 2200. Projected atmospheric CO_2 con-

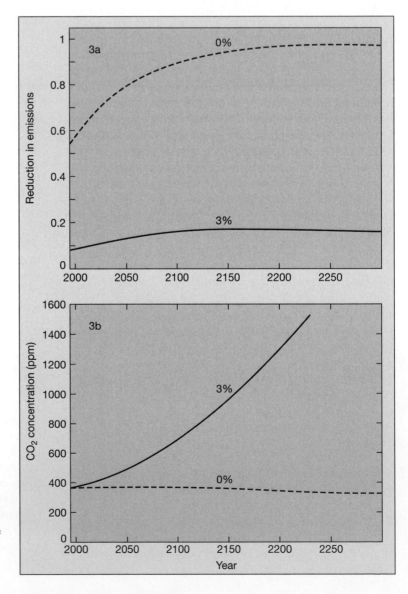

FIGURE 16-13

(a) Optimal reductions in greenhouse gas emissions over the next three centuries predicted by a modified version of Nordhaus's DICE model. The solid curve is for a 3%/yr time preference discount rate; the dashed curve is for a discount rate of zero. (b) Predicted atmospheric CO_2 concentrations for the two calculations shown in panel *a*.

centrations actually decline with time in this scenario because the emissions reductions are so drastic.

The recommended reductions in greenhouse gas emissions may be somewhat too severe in the 0% discount rate case because Nordhaus's DICE model was not designed with these types of drastic cuts in mind. However, the comparison still illustrates the importance of time preference discounting. Evidently, whether or not we choose to cut greenhouse gas emissions depends critically on how much we value the welfare of future generations compared to our own or on how much faith we have in future technological "fixes." Such issues of **intergenerational equity** are a lively topic at economic workshops focused on long-term climate change.

Policies to Slow Global Warming

Should we take action now to slow or halt global warming? Whether immediate action is warranted is one of the most hotly debated scientific issues of our time. The theoretical argument that warming will eventually occur is solid. Whether CO_2-induced global warming has already occurred is more debatable, although we argued in Chapter 1 that it probably has. Because of the high cost of addressing the problem, there is a great reluctance in most societies to act hastily. The attitude of many citizens and governments is to wait and see how bad the problem is before we commit any substantial resources to addressing it.

Although prudence is often justified, many climatologists feel that it will be a matter of only a decade or two before the evidence of global warming becomes impossible to refute. So, it is not too early to consider which measures might be undertaken if this indeed proves to be the case. One step on which virtually everyone involved in the debate agrees is to encourage energy conservation. Every kilowatt-hour of electricity or gallon of gasoline saved is one less that contributes to CO_2 buildup in the atmosphere. Indeed, many conservation measures have already been implemented. Homes built in the United States today are typically much better insulated than were homes built a generation ago. New cars get significantly better fuel mileage than did cars built in the 1960s, although the recent trend toward larger cars and SUVs has rolled back some of the gains in fuel efficiency that had occurred during the 1970s and early 1980s.

Energy-conservation measures are to be lauded, but those who have thought seriously about the global warming problem realize that conservation by itself is not likely to solve it. One way of demonstrating this is by performing "inverse" calculations with carbon cycle models. In an inverse calculation, one *assumes* a specified stabilization level for atmospheric CO_2 and then calculates what level of CO_2 emissions would be required to produce it.

For example, suppose we wanted to stabilize atmospheric CO_2 at 450 ppm. According to Figure 16-14, this stabilization would require a decrease in net CO_2 production from 6 Gton(C)/yr to just over 1 Gton(C)/yr by the year 2300. Cutting global fossil fuel usage by this amount would be a daunting task. The task might be made somewhat easier if the terrestrial biota were still absorbing CO_2 at that time, but such absorption may or may not occur. Thus, we can conclude that large reductions in fossil fuel consumption would be necessary to stabilize atmospheric CO_2 anywhere near its present value. With world population increasing, and with developing countries eager to raise their standard of living to levels comparable to that in the West, the demand for energy is likely to increase despite our best efforts to conserve. If we wish to fulfill this demand and still reduce CO_2 emissions, we will have to develop nonfossil energy sources.

Alternative Energy Sources

What alternative, nonfossil energy sources might we turn to? For the production of electricity, nuclear energy is one option that is currently available. Conventional nuclear power plants (Figure 16-15) produce energy from the **nuclear fission** of uranium atoms, the splitting of an atomic nucleus into two fragments, accompanied by the release of energy. This process, which is the basis for the atomic bomb, produces no CO_2 (although some CO_2 is produced in mining the uranium fuel by means of conventional, gasoline-powered machinery). Many environmentalists, however, feel that nuclear power poses its own threat to the global environment. A large part of their concern stems from the problem of disposal of long-lived radioactive wastes. Although several methods have been proposed for handling wastes, none of them is totally without risk. Nuclear waste disposal sites are vigorously opposed by citizens in areas where the potential sites are to be located. Local opposition has so far prevented the United States from opening its planned nuclear waste storage facility at Yucca Mountain in Nevada. Although many Nevadans see such a delay as positive, it is not without negative consequences. From a practical standpoint, most of the waste produced by existing nuclear reactors is still stored on-site in large underground "swimming pools" rather than in a long-term facility designed for that purpose.

Acceptance of nuclear power varies widely from one country to another. The United States produces about 20% of its energy from nuclear power, but no new nuclear plants have been ordered by utilities for more than 25 years, partly as a result of pressure from vocal antinuclear lobbying groups. Germany and Sweden have also almost abandoned this energy option. Conversely, both France and Japan have active nuclear programs, and both produce the majority of their electricity from nuclear

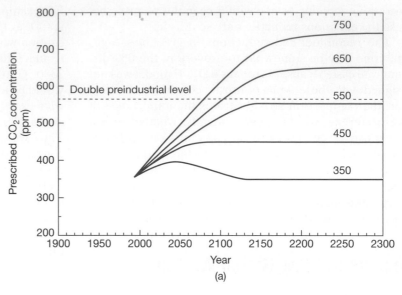

(a)

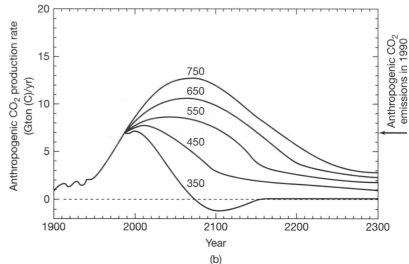

(b)

FIGURE 16-14
(a) Various prescribed levels of atmospheric CO_2. Stabilization at these levels requires (b) these net CO_2 emission rates. (After *Climate Change 1994: Radiative Forcing of Climate Change.* Cambridge: Cambridge University Press.)

power. One reason for their different outlook may lie in the fact that neither country has appreciable domestic reserves of coal or oil.

A second type of nuclear power that shows some hope for the future is **nuclear fusion,** the combining of lightweight atomic nuclei into a heavier nucleus, with an accompanying release of energy. The fusion of hydrogen atoms into helium is the energy source that powers the Sun. Prototype fusion reactors on Earth use the hydrogen isotopes deuterium (^{2}H) and tritium (^{3}H) as fuel, because these atoms can fuse at lower temperatures than can the normal (^{1}H) isotope. Deuterium is fairly abundant on Earth; 16 of every 100,000 atoms of hydrogen in seawater consist of this isotope. Unfortunately, designing a successful fusion reactor is a technologically daunting task. If ongoing efforts to do so succeed, this process could eventually be an important source of energy.

Other non-CO_2-producing energy sources include wind power, tidal power, geothermal power, and biomass-based fuels. **Wind power** generates electricity by means of windmills, which utilize Earth's solar-energy-driven winds. **Tidal power** produces electricity by using long, floating booms to harness the energy of ocean tides. **Geothermal power** utilizes temperature gradients within the solid Earth as an energy source for generating electricity. **Biomass-based fuels** are liquid fuels, such as methanol (CH_3OH) or ethanol (C_2H_5OH), that are produced from fast-growing plants. These fuels release CO_2 when they are burned, but the plants from which they are made absorb CO_2 while they are growing, so the net effect on the atmospheric CO_2 budget is zero. Although each of these clean alternative energy sources may be useful in certain locales, none appears to be capable of supplanting fossil fuels on a global basis.

Perhaps the most promising energy source in the long term is solar energy. Sunlight is a clean and virtually inexhaustible energy source that is on the verge of becoming economically competitive. In some favorably situated

FIGURE 16-15
A nuclear power plant. (From Jerry Irwin, Photo Researchers, Inc.)

areas, such as southern California, solar power is already competitive and contributes a substantial amount of electrical energy to local power grids. In sunny areas, **solar thermal power** is an efficient method for producing electricity (Figure 16-16). In this method, sunlight is used to heat a fluid that, in turn, drives a turbine. In more northern regions, the production of electricity from **photovoltaic cells,** specially designed panels that can directly convert sunlight into electricity, may be the most practical solution. Photovoltaic cells, similar to those flown on satel-

FIGURE 16-16
A solar thermal power plant. (From Hank Morgan/Science Source, Photo Researchers, Inc.)

lites, are currently more expensive than conventional energy sources, however. Ultimately, huge orbiting **solar-power satellites** may collect the Sun's energy in space and beam it to Earth's surface as laser or microwave radiation. Such satellites could be built by the more industrially developed nations to provide power across the entire globe. Much work would need to be done to determine whether solar-power satellites could be designed to be both economical and safe. They appear to be exorbitantly expensive today, but they might be more cost-effective in the future as space technology advances.

Specific Policies That Might Be Adopted

How could a shift to nonfossil energy sources be promoted if such action is deemed necessary? One mechanism would be to impose a tax on any energy source that produces CO_2. Such a **CO_2 tax** would make other forms of energy more economically competitive and, hence, more likely to be exploited. It would also encourage people to drive cars that are fuel-efficient and to commute shorter distances to work. The CO_2 tax could be augmented by a "gas-guzzler" tax on fuel-inefficient automobiles. Such taxes need not represent an additional burden on the average citizen, because all or part of the revenue that they generate could be returned to the public in the form of income tax breaks. It would be costly to some industries and localities, however, particularly those involved in mining coal or in producing oil. In the United States, the energy-rich western states, along with some eastern coal-producing states, including Pennsylvania and West Virginia, might lose jobs if such a policy were implemented. Thus, measures such as these are not likely to be accepted unless the population at large becomes alarmed about global warming.

Carbon dioxide emissions could also be reduced by the imposition of direct governmental regulations. An example that is already in effect in the United States is the CAFE (Combined Automobile Fleet Emissions) standard that governs cars sold. Each manufacturer's fleet must meet an average fuel economy rating specified by the federal government. Although this standard undoubtedly increases the number of small, fuel-efficient cars that are sold, it does little to encourage people to drive less or to live closer to their workplace. It also irritates manufacturers, because it limits their flexibility to design cars that will sell. Thus, taxes, as distasteful as they appear to most of us, might be a more effective way of decreasing CO_2 emissions.

Efforts have also been made at the international level to impose restrictions on CO_2 emissions from fossil fuel burning. The Kyoto Protocol, which was presented at a meeting in Kyoto, Japan, in December, 1997, called for all of the industrialized countries of the world to roll back their CO_2 and other greenhouse gas emissions to 7 percent below 1990 levels during a "trial" time period of 2008–2012. Given the growth in emissions that has occurred since 1990 and that is predicted to occur in the near future, this could imply as much as a 20–30 percent reduction compared to the "business as usual" scenario. Then-President Clinton signed the agreement when it first came out, as did the leaders of many other nations, but the Protocol was never submitted to the U.S. Senate for approval. The United States objected to several elements of the proposal, including the specific greenhouse gas emission targets, the fact that developing countries were exempted from emissions reductions, and the fact that trading of emission "permits" was not allowed. The difficulty in obtaining ratification for the Kyoto Protocol may be contrasted with the relative ease with which the Montreal Protocol and subsequent international agreements concerned with protecting the ozone layer have been accepted. The difference in attitudes stems partly from the imminence of the ozone depletion problem compared to global warming, and partly from the much higher cost of reducing CO_2 emissions compared to freon emissions. This underscores the difficulty in taking action to solve the global warming problem.

One measure that would arouse little or no public opposition would be to encourage the widespread planting of trees. As we discussed earlier in the chapter, existing forests and soils contain three to four times as much carbon as is currently in the atmosphere in the form of CO_2. Expanding the amount of forested area could, in principle, absorb a substantial amount of fossil fuel CO_2. We say "in principle" because, in practice, the amount of forested land is decreasing as tropical forests are cleared for agriculture. Halting tropical deforestation and encouraging reforestation around the globe would be beneficial for its own sake and might be at least a start toward reducing global warming.

Although all these measures might help reduce the rate at which CO_2 is accumulating in the atmosphere, none is likely to stop the buildup entirely. In the long run, stabilizing climate may well require changes in our lifestyles, including where we live and how we work. In the United States, urban growth over approximately the past century has relied on the availability of automobiles and the relatively low cost of gasoline. Many people prefer to live in distant suburbs, far from their workplaces. For example, in Los Angeles, the average commuter spends an hour and a quarter *each way* driving to and from work alone every day in a private car. Switching to highly fuel-efficient cars can do only so much to reduce CO_2 emissions in such a situation. Cutting back seriously on emissions would require either that the fuel used to power automobiles be derived from renewable (nonfossil) resources, such as biomass-based fuels, or that efficient mass-transit systems be developed (preferably ones powered by nonfossil energy sources). Mass-transit systems tend to be efficient, however, only if the population density is high, so the current practice of living in uncrowded suburbs might have to be abandoned. Alternatively, with better electronic communications capabilities becoming available over the Internet, more people might work at home instead of commuting to the office each day. Such lifestyle changes are likely to be adopted only if we become very worried indeed about the problem of global warming.

Chapter Summary

1. The observed rise in atmospheric CO_2 concentrations is caused chiefly by the burning of fossil fuels and, to a lesser extent, by tropical deforestation. The anthropogenic CO_2 source is much smaller than the rate at which CO_2 is released by respiration and decay but much larger than the rate at which CO_2 is emitted by volcanoes. It therefore represents a substantial perturbation to the global carbon cycle.

2. The CO_2 generated by human activities can be removed from the atmosphere by several mechanisms.
   a. The fastest of these mechanisms is photosynthesis, but this sink will be effective only if forests are replanted or if CO_2 fertilization of plant growth continues to cause additional carbon to be stored in forests and soils.
   b. Much of the fossil fuel CO_2 will dissolve in the oceans. The rate of CO_2 uptake is limited by the mixing rate of the deep ocean and by the chemical buffering capacity of seawater. Only about 30 to 40% of the available carbon in fossil fuels can be absorbed in this manner.
   c. Additional CO_2 can be removed by the dissolution of carbonate sediments on the sea floor and of carbonate rocks

on land, but these processes occur over hundreds to thousands of years.
   d. The fossil fuel CO_2 pulse would be completely removed by silicate weathering on a time scale of about 1 million years.

3. Computer models of the global carbon cycle predict that atmospheric CO_2 levels will double within the next 50 to 100 years and that CO_2 concentrations could exceed 2000 ppm within a few centuries if nothing is done to limit emissions. Radiative forcing of climate could be accelerated by increases in methane and other trace greenhouse gases. Earth's climate could warm by several degrees Celsius over the next century and by as much as 10–15°C in the long term. The warming will be unevenly distributed, with the polar regions warming the most and the tropics warming the least. Potentially damaging consequences of such warming include the drying out of continental interiors, the spread of insect pests and tropical diseases, and substantial increases in sea level.

4. Slowing future climate change will require drastic reductions in greenhouse gas emissions, particularly those of CO_2. Emissions reductions of 85% or more would be required to stabilize atmospheric CO_2 anywhere near its current level.

Such a reduction could be encouraged by taxation measures that promote energy conservation and the development of alternative energy sources. Whether such measures are warranted is the subject of ongoing debate.

Key Terms

biomass-based fuels
bioturbation
buffer
C_3 plants
C_4 plants
Calvin cycle
carbon fixation
cloud condensation nuclei
cost–benefit analysis
CO_2 fertilization

CO_2 tax
discount rate
equilibrium surface warming
geothermal power
glacial surge
ice shelves
intergenerational equity
methane clathrate hydrate
mountain glaciers
nitrogen fertilization

nuclear fission
nuclear fusion
photovoltaic cells
radiative forcing
solar-power satellites
solar thermal power
thermal drag
thermal expansion
tidal power
wind power

Review Questions

1. How does the amount of CO_2 produced by fossil fuel consumption compare to the natural flux of CO_2 in the carbon cycle?
2. What are the major processes that can remove CO_2 from the atmosphere? What are the approximate time scales for these processes to be effective?
3. What is the size of the fossil fuel reservoir compared with the atmospheric CO_2 reservoir?
4. Why does the ocean have a limited capacity for CO_2 uptake?
5. By how much is global temperature predicted to rise over the next century?

6. Why is predicting future sea-level change such a tricky task?
7. Why do some climate models predict that the thermohaline circulation might shut down?
8. How does the expected effect of elevated atmospheric CO_2 levels differ for C_3 and C_4 plants?
9. What is the meaning and significance of the economic discount rate?
10. By how much would we need to reduce CO_2 emissions to stabilize atmospheric CO_2 over the next century?
11. What measures might be taken to reduce future CO_2 emissions?

Critical-Thinking Problems

1. a. The present atmosphere contains approximately 700 Gton(C) in the form of CO_2. Earth's total recoverable fossil fuel reserves contain at least 4200 Gton(C), mostly in the form of coal. (We shall use the value 4200 Gton(C) to be specific.) At present, about half the CO_2 produced by the burning of fossil fuels stays in the atmosphere. The other half dissolves in the oceans or is taken up by the terrestrial biosphere. If this ratio remained constant and we burned up all of our fossil fuels instantaneously, by how much would atmospheric CO_2 concentrations rise? (Express your answer in terms of the new CO_2 level divided by the old one.)

   b. Climate models predict that each doubling of the atmospheric CO_2 concentration will cause the mean global temperature to increase by 1.5–4.5°C. (The range is due largely to uncertainties about how clouds will respond.) By how much would the mean temperature increase for the scenario described in part (a)? Express your answer as a temperature range in degrees Celsius and in degrees Fahrenheit.

   c. The actual problem of global warming could be more severe than we have just calculated. Forests and soils together contain an additional 2100 Gton of carbon that might go into the atmosphere if deforestation is not prevented. The ocean becomes more acidic as it absorbs CO_2, so it might not be able to continue taking up as much CO_2 as it has been until now. If we burned up all our fossil fuels and deforested one-third of the globe without losing any CO_2 to the ocean (or to CO_2 fertilization), by how much would atmospheric CO_2 and temperature increase?

2. The atmospheric CO_2 concentration is currently increasing by about 1.5 ppm/yr. How many gigatons of carbon are being added to the atmosphere each year? (Hint: The total mass of the atmosphere is 5×10^{18} kg, and its mean molecular weight is about 29. You will need to do the calculation in moles and then convert back to mass units.)

3. The surface ocean contains about 2.6×10^{19} liters of water with a carbonate ion content of about 2×10^{-4} mol/L. The deep ocean contains about 1.4×10^{21} L of water with a car-

bonate ion content of roughly 9×10^{-5} mol/L. If each mole of carbonate reacts with 1 mol of CO_2 according to the reaction

$$CO_2 + CO_3^= + H_2O \rightarrow 2\ HCO_3^-$$

what percentage of the fossil fuel reservoir, 4200 Gton(C), can be neutralized by the surface ocean? By the deep ocean?

Further Readings

General

Philander, S. G. 1998. *Is the Temperature Rising?: The Uncertain Science of Global Warming.* Princeton, N.J.: Princeton University Press.

Schneider, S. H. 1989. *Global Warming: Are We Entering the Greenhouse Century?* San Francisco: Sierra Club Books.

Advanced

IPCC Report: Climate Change 2001: The Scientific Basis (Technical Summary and Summary for Policymakers). Available online at *http://www.ipcc.ch/*

Kasting, J. F. 1998. Long-Term Effects of Fossil Fuel Burning. *Consequences,* 4; 15–27. Available online at *http://www.gcrio.org/ CONSEQUENCES/vol4no1/carbcycle.html*

Levi, B. G., Hafemeister, D., and Scribner, R. 1992. *Global Warming: Physics and Facts.* Washington, D.C.: American Institute of Physics.

17

Ozone Depletion

Key Questions

- How is ultraviolet radiation categorized, and what are its biological effects?
- How is the thickness of the ozone layer measured, and how does this layer vary from place to place?
- How do trace chemicals catalyze the destruction of ozone?
- What is the cause of the Antarctic ozone hole?
- What is being done to prevent ozone depletion in the future?

Chapter Overview

Solar ultraviolet radiation between 200 and 320 nm poses significant health hazards if not effectively blocked by stratospheric ozone. Ozone is formed by reactions initiated by the splitting of O_2 and can be destroyed by various catalytic cycles involving the elements nitrogen, chlorine, and bromine. The latter two elements have large anthropogenic sources and therefore have generated concern. Chlorine, in particular, has been directly implicated in the formation of the Antarctic ozone hole and may be responsible for a slow, long-term decrease in mid-latitude ozone levels. International agreements that are already in place are ex-

pected to halt the depletion of ozone and to restore the stratosphere to its natural state.

Introduction

We have discussed in detail the evolution of Earth's climate on different time scales. Our reason for dwelling on climate is twofold: First, the prospect of global warming over the next few decades to centuries is probably the most intractable environmental problem that we currently face. Second, the many different aspects of climate serve as an excellent case study for illustrating the interactions among the various elements of the Earth system, along with the system's capacity for self-regulation.

No book on global change, however, would be complete without a discussion of Earth's ozone layer and the possibility that it could be depleted. We learned in Chapter 11 that the development of a protective ozone screen to shield out harmful solar ultraviolet radiation was an important step in the evolution of advanced forms of life. The ozone layer arose naturally as a by-product of the evolution of photosynthesis and the rise of atmospheric oxygen. It is a relatively fragile feature, however, that is now threatened by chemicals released into the atmosphere by industrial activities. Unlike global warming, ozone depletion is widely recognized as a serious problem, and significant steps have already been taken to reduce its impact. It is important to understand why the ozone layer is so essential and why we must remain committed to protecting it.

Ultraviolet Radiation and Its Biological Effects

We learned in Chapter 11 (see Figure 11-15) that ozone absorbs ultraviolet (UV) radiation in the 200- to 400-nm region, where few other atmospheric gases absorb. Here, we look at this spectral region in greater detail.

UVA and UVC Radiation

Ultraviolet radiation between 200 and 400 nm is usually subdivided into three distinct spectral ranges, as shown in Table 17-1. The longer wavelengths are termed *UVA,* the middle wavelengths are termed *UVB,* and the shortest wavelengths are termed *UVC.* The solar flux increases with increasing wavelength throughout this spectral region, as shown in Figure 17-1a. Thus, more UVA photons are available at the top of the atmosphere than are UVB or UVC photons. The ozone *absorption coefficient* is also low in the UVA region. The absorption coefficient is a measure of how strongly a molecule absorbs electromagnetic radiation of a given wavelength. Because the ozone absorption coefficient is low at UVA wavelengths, most of the incident photons make it down to the ground.

Fortunately, UVA radiation appears to be relatively harmless to humans and other forms of life. Many tanning parlors use UVA radiation to tan their patrons "safely." However, whether UVA radiation is really safe for humans is not fully understood. Overexposure to UVA radiation may lead to premature aging of the skin, and there is some evidence that it can damage the immune system. We are certain, though, that UVA radiation is much less dangerous than shorter-wavelength UV radiation.

At shorter, UVC wavelengths, the solar flux is lower and the ozone absorption coefficient is high. Thus, relatively few UVC photons hit the top of the atmosphere, and even fewer make it to Earth's surface. This is a good thing for humans and other organisms, because UVC radiation is extremely dangerous to most forms of life. Indeed, the absorption peak for DNA, the molecule that contains the genetic information for all organisms, is right in the middle of the UVC region. Some single-celled, prokaryotic microorganisms have developed highly efficient DNA-repair mechanisms and can tolerate substantial doses of UVC radiation. However, more advanced, eukaryotic organisms, including all multicellular life forms, are extremely sensitive to radiation at these wavelengths.

UVB Radiation and Its Biological Effects

The wavelengths between about 290 and 320 nm, UVB radiation, are the ones of current concern to us. At these wavelengths, the solar flux is relatively high and the ozone absorption coefficient is relatively low. Thus, a substantial radiation flux reaches Earth's surface. The biological effect of this radiation is determined by the **dose rate,** which is the number of UV photons per unit time that lead to a specific biological response, such as sunburn or skin cancer. The dose rate is the product of the surface UV flux and the action spectrum for the particular response being studied. The **action spectrum** of a biological response is the relative efficiency with which UV photons at different wavlengths contribute to that response in a specific organism. For example, the **erythemal action spectrum** shown in Figure 17-1b describes the appearance of sunburn in humans. The dose rate (in relative units) leading to sunburn is indicated by the shaded area in the figure. Clearly, most of the UV rays that cause sunburn are in the UVB spectral region.

Although UVB radiation is not quite as harmful to organisms as is UVC radiation, it is still capable of causing substantial damage. In addition to sunburn, overexposure to UVB radiation can lead to skin cancer in humans. This radiation is also harmful to the eye, where it can cause cataracts and damage the retina. Many animals other than humans, such as hippopotami, are susceptible to sunburn, and at least half of the terrestrial plants that have been studied exhibit slower growth and smaller leaves when exposed to enhanced UVB fluxes. Increases in UVB radiation could be detrimental to aquatic life, including phytoplankton, zooplankton, larval crabs and shrimp, juvenile fish, and corals. Thus, almost all inhabitants of the Earth system have an interest in ensuring that the UVB flux does not increase above its present value.

Relationship between UVB Flux and Stratospheric Ozone

How do we know that decreasing stratospheric ozone increases the UVB flux at the ground? In general, this inference is derived theoretically. (We shall do so in "Critical-Thinking" Problem 1.) In a few cases, however, scientists have been able to measure both quantities

TABLE 17-1

Classification of UV Radiation Wavelength		
Range (nm)	*Name*	*Biological Effect*
320–400	UVA	Relatively harmless; causes tanning but not burning
290–320	UVB	Harmful; causes sunburn, skin cancer, and other disorders
200–290	UVC	Extremely harmful but almost completely absorbed by ozone

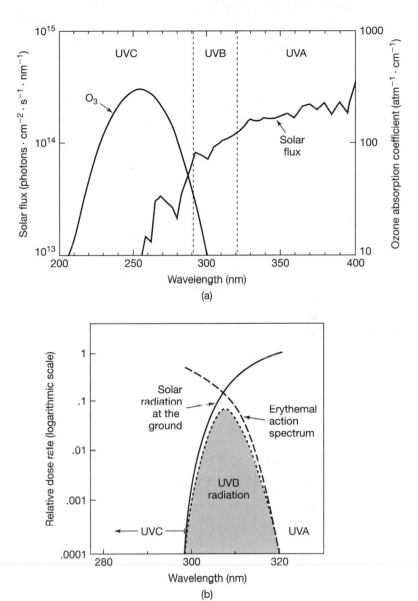

FIGURE 17-1

(a) Diagram showing the three different categories of solar UV radiation. The curve labeled O_3 is the ozone absorption coefficient. (b) Graph illustrating the significance of UVB radiation. The shaded area represents the dose rate of the UV radiation that causes sunburn. (After R. Turco, *Air Pollution*. New York: Oxford University Press, 1997.)

directly and to show that these quantities behave as expected. Figure 17-2 shows simultaneous measurements of atmospheric ozone content and ground-level UVB flux at Melbourne, Australia, both before and after the intrusion of ozone-poor air from Antarctica in mid-December 1987. (The units in which ozone is measured, Dobson units, were introduced in Chapter 1 and are explained further below.) By January 1, ozone levels had dropped by about 16% and the UVB flux had increased by about 25%. A similar inverse relationship between ozone and UVB radiation has been observed more recently in Antarctica. The strong negative correlation between the two quantities gives atmospheric scientists great confidence that UVB fluxes will go up if ozone levels go down.

Ozone Vertical Distribution and Column Depth

As we discussed in Chapter 3, most of Earth's ozone (O_3) is confined to the stratosphere. Indeed, the reason the stratosphere exists is because the absorption of solar UV radiation by ozone heats the air and causes the temperature to increase with altitude. The ozone layer is not as uniform, though, as the discussion in Chapter 3 may have implied. Rather, it varies with both latitude and time of year.

Measurements of Ozone Column Depth

Several methods exist for measuring the vertical distribution of ozone. Two of the most common methods are re-

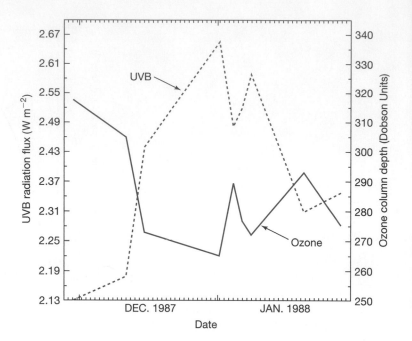

FIGURE 17-2

Comparison of solar UVB radiation flux at ground level and ozone column depth observed over Melbourne, Australia, from December 1987 to January 1988, during an intrusion of ozone-poor air. (After *Scientific Assessment of Ozone Depletion,* World Meteorological Organization, 1991.)

mote sensing from satellites and in-situ measurements from **ozonesondes,** balloon-borne instruments that measure the concentration of stratospheric ozone. A comparison of the results of these two methods is shown in Figure 17-3. Ozone concentrations are given in units of **number density,** or number of molecules per cubic centimeter. At its peak in the stratosphere, near 20 to 25 km altitude, the ozone number density is typically about 5×10^{12} molecules/cm³. In terms of relative concentration,

this amounts to a few parts per million. In the example shown here, the two methods for measuring ozone agree to within a few percent at all altitudes above about 16 km, which is where most of the ozone resides. The satellite does not produce accurate results at low altitudes, because it is looking down through the bulk of the ozone layer.

The flux of solar UV radiation that reaches Earth's surface depends on the **solar zenith angle**—that is, the angle of the Sun from the vertical (Figure 17-4)—and on the vertical column depth of ozone. The **column depth** is the total amount of ozone per unit area above a certain location at the surface. It is measured in several ways. The simplest unit is molecules of O_3 per square centimeter. In this unit, the sum of the ozone number densities at all altitudes multiplied by the height of the atmosphere is equal to the ozone column depth. A typical, mid-latitude ozone column depth is about 8×10^{18} O_3 molecules/cm², meaning that there are 8×10^{18} molecules over 1 cm² of Earth's surface.

A second way of measuring column depth is to express it as the thickness that a layer of pure ozone would have at 1 atm pressure. One *atmosphere-centimeter* (1 atm-cm) is equal to 2.687×10^{19} molecules/cm². Thus, a typical mid-latitude ozone profile would have a column depth of about 0.3 atm-cm. Physically, this means that the ozone in Earth's atmosphere is equivalent to a 0.3-cm-thick layer of pure ozone at the surface. The small magnitude of this number gives a preliminary indication of why the ozone layer is so fragile.

The unit atmospheric chemists use to measure ozone column depth is called the **Dobson unit** (DU). One Dobson unit is equivalent to a layer of pure ozone 0.001 cm thick at 1 atm pressure. Because 1 atm-cm is equal to

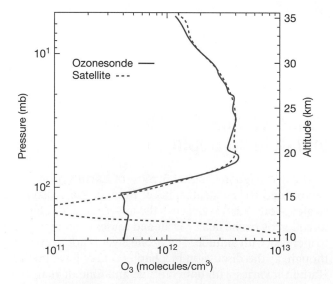

FIGURE 17-3

Ozone vertical profile measured in number density (molecules per cubic centimeter) by ozonesonde (solid curve; data from World Ozone Data Center) and by satellite instrument at sunset over Wallops Island, Virginia.

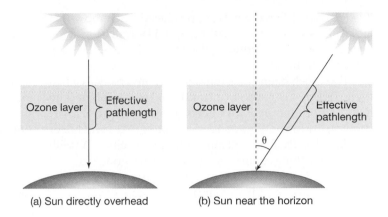

FIGURE 17-4

The pathlength through the ozone layer depends on the solar zenith angle, θ (theta).

1000 DU, a typical mid-latitude ozone column depth is about 300 DU. The Dobson unit is named after George Dobson, an English physicist who in 1931 became the first person to measure ozone column depth accurately. The ground-based device that he developed has become known as a *Dobson spectrophotometer.* The instrument compares the different amounts of ozone absorption at two wavelengths of sunlight. By measuring the solar UV flux at both wavelengths and at two or more solar zenith angles, it is possible to calculate the ozone column depth directly.

Spatial Distribution of Ozone

Satellite measurements represent a tremendous advance over ground-based measurements because satellites make it possible to determine ozone column depth at all points over the globe during all seasons of the year. The results from a composite of several years of such mapping by the TOMS instrument (see Chapter 1) aboard the *Nimbus-7* satellite are shown in Figure 17-5. Ozone column depths are highest during springtime at mid-to-high latitudes and lowest over the equator. These data were collected during the early years of TOMS, before the Antarctic ozone hole had become as pronounced as it is today; that is why the ozone hole is not visible in the figure.

The reason ozone is distributed in this manner is complex. As we shall soon see, the greatest production of ozone occurs in the tropics, where the solar UV flux is the highest. Both the mid-to-high latitude peak and the equatorial minimum in column depth are caused by stratospheric circulation: Ozone-rich air from the tropical upper stratosphere is transported poleward by north–south winds. The details of this stratospheric circulation are beyond the scope of our discussion. Figure 17-5 does explain, however, why it is easier to get sunburned in Florida (25–30° N) than in New York (about 43° N). Not only is the Sun higher in the sky at low latitudes than it is farther north, but the ozone column depth is also smaller than at higher latitudes. The combination of these two effects allows a much higher percentage of the incident UV photons to make it to the surface at low latitudes.

Tropospheric Ozone

Not all of Earth's ozone is confined to the stratosphere. Roughly 10% is in the troposphere, where it plays a key role in tropospheric chemistry. Its most important function is to provide a source of oxidizing radicals (highly reactive molecules) that help cleanse the lower atmosphere of pollutants such as carbon monoxide (CO) and sulfur

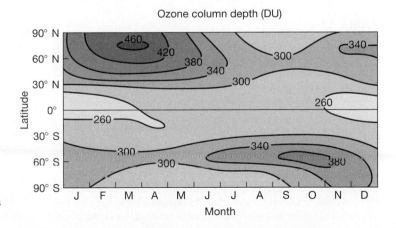

FIGURE 17-5

Latitudinal and seasonal distribution of ozone column depth as measured by satellite, composited over several years prior to the development of the "ozone hole" over Antarctica. Contours represent column depths in Dobson units.

dioxide (SO_2). The radicals are produced when the ozone molecule is split by short-wavelength UV radiation. So, a certain amount of ozone in the troposphere helps keep our air fit to breathe.

Too much ozone at ground level, conversely, is a bad thing. Ozone in high concentrations is toxic to plants and irritates both our eyes and lungs. Indeed, ozone is a key component of the photochemical smog that forms over Los Angeles, Denver, and many other large cities. As with many other things in life, ozone has both a good side and a bad side. Ozone that is found in the stratosphere, however, is always good from our human perspective.

The Chapman Mechanism

To understand why the ozone layer is susceptible to alteration by anthropogenic processes, we must examine the chemistry by which ozone is formed and destroyed. The first person to describe these processes was Sydney Chapman, an atmospheric chemist who published a ground-breaking paper on ozone photochemistry in 1930. The four chemical reactions that he proposed for a simplified (pure oxygen-nitrogen) atmosphere have become known as the **Chapman mechanism** (Table 17-2).

Production of Ozone

The first step in the Chapman mechanism is the splitting apart of molecular oxygen, O_2, by an ultraviolet photon to form two atomic oxygen (O) atoms. The splitting of a molecule by the absorption of light or by UV radiation is called **photolysis**, or **photodissociation**. UV radiation at wavelengths shorter than 240 nm is required to photolyze O_2. One UV photon in this wavelength range will split apart one O_2 molecule. Here is where the particulate behavior of light, mentioned in Chapter 3, becomes evident. Two visible-light photons with the same combined energy as one UV photon will not be able to split O_2, because photolysis reactions are discrete events.

TABLE 17-2

The Chapman Mechanism of Ozone Production and Destruction		
Reaction*		*Rate*
1) $O_2 + UV\ photon \rightarrow O + O$ ⎫ production		Slow
2) $O + O_2 + M \rightarrow O_3 + M$ ⎭		Fast
3) $O_3 + photon \rightarrow O_2 + O$ ⎫ destruction		Fast
4) $O + O_3 \rightarrow 2\ O_2$ ⎭		Slow

* The symbol *M* represents a third molecule necessary to carry off the excess energy of the collision between an O atom and an O_2 molecule.

Once it has formed in the atmosphere, an atomic oxygen atom reacts quickly with another O_2 molecule to form ozone (reaction 2 of the Chapman mechanism). This process cannot occur in isolation, however, because the colliding molecules have too much energy to stick together unless some third molecule, represented by *M,* is available to carry it away. Molecule *M* can be N_2, O_2, ^{40}Ar, or any of the other molecules in Earth's atmosphere.

The situation is analogous to a billiards shot (Figure 17-6). If you shoot at a single, isolated ball (Figure 17-6a), chances are that the cue ball and target ball will roll away in different directions. At best, if you hit the target ball straight on with the proper amount of backspin, the cue ball will remain motionless after contact, while the target ball will roll in the forward direction. If you place a third ball in contact with the target ball, however, and hit the pair dead on, the third ball will roll away and the cue ball and the target ball will remain together (Figure 17-6b). This, essentially, is what happens in the atmosphere when an ozone molecule is formed by the collision of an O atom with an O_2 molecule.

Destruction of Ozone

Once formed, an ozone molecule can be photolyzed by the absorption of another photon (reaction 3 of the Chapman mechanism). Photolysis reverses the process that occurred in reaction 2. Unlike O_2, O_3 can be split by radiation in the visible-light range. Because many more visible photons than UV photons are available, O_3 is photolyzed much faster than O_2. Furthermore, O_3 can be photolyzed all the way down to Earth's surface, whereas O_2 can be photolyzed only above about 20 km. All of the short-wavelength UV radiation required to split O_2 is absorbed above this height. That is why the ozone layer is located in the stratosphere and not near Earth's surface.

The fourth reaction in the Chapman mechanism is the reaction of an O atom and an O_3 molecule to yield two O_2 molecules. This is a slow reaction, as measured in the laboratory, but as we will see, it is the key to understanding ozone photochemistry. The reason is that when ozone reacts with atomic oxygen (reaction 4), the ozone is destroyed permanently, whereas when ozone is photolyzed (reaction 3), the resulting O atom is free to combine with another O_2 molecule, reforming O_3.

Odd Oxygen

We can express the difference in reactions 3 and 4 of the Chapman mechanism more precisely by defining a quantity called **odd oxygen,** or O_x. As its name implies, odd oxygen includes all pure, oxygen-containing atoms or molecules that have an odd number of oxygen atoms. Thus, the concentration of odd oxygen, denoted by $[O_x]$, is equal to the sum of the concentrations of atomic oxygen and ozone: $[O_x] = [O] + [O_3]$. Ordinary molecular oxygen,

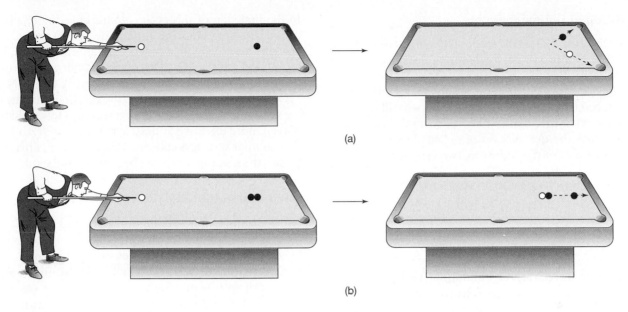

FIGURE 17-6

The ozone formation reaction (Chapman reaction 2) is analogous to hitting a billiards shot. (a) Hitting a target ball with the cue ball will result in the movement of both in different directions. (b) A third ball is required to allow the cue ball and the target ball to stick together.

O_2, contains an even number of oxygen atoms; it is not counted as O_x.

With the concept of odd oxygen, we can analyze the Chapman mechanism at a deeper level. Table 17-3 shows the change in odd oxygen for each of the four reactions. Reaction 1 produces two O atoms from one O_2 molecule, so the change in O_x is +2. Reactions 2 and 3 merely interconvert two forms of odd oxygen, O and O_3, so the change in O_x is 0 for each reaction. Reaction 4 destroys both an O atom and an O_3 molecule, so the change in O_x is −2.

The significance is the following: The two slow reactions, 1 and 4, are the most important, because they control the abundance of odd oxygen. The two fast reactions, 2 and 3, determine the ratio of the O abundance to the O_3 abundance, but they have no direct effect on the concentration of odd oxygen. As we will see in the next section, it is the destruction of odd oxygen that really matters in ozone photochemistry.

Catalytic Cycles of Nitrogen, Chlorine, and Bromine

The Chapman mechanism provides the basis for understanding ozone chemistry, but it does not by itself provide an accurate description of Earth's stratosphere. When the Chapman mechanism is included in a computer model of the stratosphere, the model predicts about 30% more ozone than is actually present. Other processes must be destroying ozone as well or, equivalently, destroying odd oxygen.

The shortcoming of the Chapman mechanism is that it ignores the effects of atmospheric trace constituents such as nitrous oxide (N_2O), water vapor, and freons. These trace gases can be photolyzed, producing highly reactive radicals that keep ozone abundances lower than they would otherwise be.

TABLE 17-3

The Chapman Mechanism and Odd Oxygen		
*Reaction**	*Rate*	*ΔOx*
1) O_2 + UV photon → O + O ⎫ production	Slow	+2
2) $O + O_2 + M → O_3 + M$ ⎭	Fast	0
3) O_3 + photon → O_2 + O ⎫ destruction	Fast	0
4) $O + O_3 → 2 O_2$ ⎭	Slow	−2

* The symbol M represents a third molecule necessary to carry off the excess energy of the collision between an O atom and an O_2 molecule.

The Nitrogen Catalytic Cycle

One such radical is nitric oxide, NO. In Chapter 11 we noted that NO produced by lightning discharges constitutes a natural source of fixed nitrogen in the oceans. Lightning does not occur in the stratosphere, but NO can be produced from nitrous oxide, N_2O (as we will see in the next section). Nitric oxide is one of several radicals that can facilitate the destruction of ozone. The destruction process consists of the following two reactions:

$$NO + O_3 \rightarrow NO_2 + O_2 \qquad \textbf{fast}$$
$$NO_2 + O \rightarrow NO + O_2 \qquad \textbf{fast}$$

Net: $\qquad O_3 + O \rightarrow 2\,O_2 \qquad\qquad \textbf{fast}$

In the first reaction, nitric oxide reacts with ozone, forming nitrogen dioxide (NO_2) and molecular oxygen. Nitrogen dioxide is a brownish gas that is a major component of photochemical smog. It is visible in the "brown cloud" that hangs over Denver and many other large cities on weekdays. In the second reaction, nitrogen dioxide reacts with atomic oxygen, reforming nitric oxide and producing a second O_2 molecule.

We can determine the net effect of the two reactions by adding the reactions together and cancelling out atoms or molecules that occur on both sides of the reaction arrows. When we do this, we find that the net reaction is exactly the same as step 4 in the Chapman mechanism. Thus, this reaction results in the destruction not just of ozone but of odd oxygen as well.

This destruction process is an example of a **catalytic cycle,** a set of chemical reactions facilitated by the presence of a catalyst. A **catalyst** is a substance that increases the rate of a chemical reaction but is itself unchanged by the reaction. In the nitrogen catalytic cycle, the catalyst is the NO molecule. The NO molecule is destroyed in the first step of the cycle, but it is reformed in the second step and, hence, is free to react again. In the lower stratosphere, one NO molecule can destroy hundreds or thousands of O_3 molecules. Furthermore, both steps in the cycle are fast reactions, as measured by laboratory experiments. As a catalyzed reaction should be, the net reaction is therefore also fast in comparison with the direct reaction between O_3 and O.

The Chlorine Catalytic Cycle

Similar ozone-destroying catalytic cycles can be created by other radicals. One very important cycle involves atomic chlorine, Cl:

$$Cl + O_3 \rightarrow ClO + O_2 \qquad \textbf{fast}$$
$$ClO + O \rightarrow Cl + O_2 \qquad \textbf{fast}$$

Net: $\qquad O_3 + O \rightarrow 2\,O_2 \qquad\qquad \textbf{fast}$

In the first step of this cycle, atomic chlorine reacts with ozone, forming chlorine monoxide (ClO) and O_2. The chlorine monoxide then reacts with atomic oxygen, forming Cl and another O_2. This cycle is analogous to the nitrogen catalytic cycle, with NO replaced by Cl and NO_2 by ClO. The individual reactions in the chlorine catalytic cycle are faster than those in the nitrogen cycle. Hence, this cycle is even more effective at destroying ozone. Or, to look at it in another way, less chlorine is needed than nitric oxide before the loss rate of ozone becomes significant.

Other Important Catalytic Cycles

Nitrogen and chlorine compounds are not the only compounds that affect stratospheric ozone. Ozone can also be destroyed by catalytic cycles involving *bromine (Br) radicals* and **hydroxyl (OH) radicals.** Bromine radicals have both natural and anthropogenic sources (as we will see next), but hydroxyl radicals are entirely natural. Thus, the photochemistry of even the unperturbed stratosphere is, in reality, quite complex. Elaborate computer models are needed to simulate all of the possible catalytic cycles that affect the ozone concentration.

Sources and Sinks of Ozone-Depleting Compounds

The Odd Nitrogen Cycle

The NO and NO_2 that participate in the nitrogen catalytic cycle are referred to as **odd nitrogen** (NO_x) compounds, which contain an odd number of nitrogen atoms to distinguish them from N_2. Odd nitrogen is similar in concept to the *fixed nitrogen* discussed in Chapter 11. The first term is used by atmospheric chemists, the second by biologists. In both cases, the important characteristic of the molecules is that the strong bond between the two nitrogen atoms in N_2 has been split. Hence, odd nitrogen molecules are much more reactive than is N_2.

Stratospheric odd nitrogen derives primarily from nitrous oxide, N_2O. The reaction that produces NO is

$$N_2O + O^* \rightarrow 2\,NO.$$

Here, O^* is an electronically excited atomic oxygen atom produced by the UV photolysis of ozone. It is much more reactive than normal, ground-state, atomic oxygen. (Normal O atoms are incapable of reacting with N_2O.) The N_2O itself comes from Earth's surface, where it is produced by microbial activity in soils and in the ocean (Figure 17-7.) This activity is enhanced in places by the addition of nitrate fertilizers to the soil. Much of the nitrate in such fertilizers is taken up by growing plants. Howev-

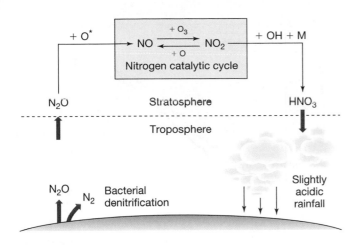

FIGURE 17-7

The atmospheric odd nitrogen cycle.

er, a substantial fraction undergoes bacterial denitrification (Chapter 9) and is subsequently released as either N_2 or N_2O.

The nitrous oxide makes its way to the stratosphere, where some of it reacts to form NO. The rest is photolyzed back to N_2 and O. The NO so produced participates in the ozone-destroying nitrogen catalytic cycle, forming NO_2 in the process. Every once in a while, however, the resulting NO_2 molecule, instead of reacting with atomic oxygen, encounters a hydroxyl radical instead. The hydroxyl radical combines with the NO_2 molecule, producing nitric acid, HNO_3:

$$NO_2 + OH + M \rightarrow HNO_3 + M.$$

The M in this reaction is a third molecule that carries off the excess energy of the collision. Nitric acid then diffuses down into the troposphere, where it dissolves in cloud droplets and is removed by precipitation. The nitric acid makes the rain slightly more acidic, but the amount of acid formed is so small that it poses no environmental problem. Acid rain itself is a problem in some areas, but it is caused by nitric and sulfuric acids formed within the troposphere.

Although N_2O is currently the largest source of stratospheric odd nitrogen, most of the concern about these compounds stems from potential increases in their abundance caused by high-flying, *supersonic transport* airplanes, or SSTs. Jet airplanes produce nitric oxide during the process of combustion. The reaction is similar to the production of NO by lightning. In each case, the high temperatures that are generated cause N_2 and O_2 to react with each other to form two NO molecules. Conventional jets fly in the upper troposphere and, hence, are not a threat to the ozone layer. The SSTs, however, inject nitric oxide

and other exhaust products directly into the stratosphere. The French and British had been flying an SST called the *Concorde* since 1977, mostly on trans-Atlantic flights. In the face of declining revenues and one fatal accident, the Concorde was taken out of service in late 2003. Other governments, however, including the United States, may consider developing this type of aircraft. Because of the possible impact on ozone, careful environmental impact studies should be performed before large fleets of SSTs are built and flown.

The Chlorine Cycle

The attention of most stratospheric chemists is currently on chlorine. Chlorine is introduced into the stratosphere by several different gases produced at Earth's surface. The chlorine-containing gases that occur naturally are methyl chloride (CH_3Cl), and hydrogen chloride (HCl). Methyl chloride is produced in large quantities by marine plankton. Most of the methyl chloride released at the surface reacts in the troposphere, however. The amount that makes its way up to the stratosphere is enough to produce a stratospheric chlorine concentration of only about 0.6 ppb. By comparison, the current stratospheric chlorine concentration is approximately 3.3 ppb. So, most of this chlorine must derive from other sources.

Hydrogen chloride has received a great deal of attention because it is released in large quantities during volcanic eruptions. Particularly violent eruptions inject gases directly into the stratosphere. This fact has led some skeptics to suggest that most stratospheric chlorine derives from volcanoes and that we need not worry about anthropogenic sources. However, detailed studies of the El Chichón eruption in Mexico in 1982 and of the Mt. Pinatubo eruption in the Philippines in 1991 have led to the conclusion that very little of the HCl released in such events reaches the stratosphere. Most of the emitted HCl dissolves in water droplets that condense out of the volcanic plume. It is removed when these droplets fall out as rain.

Hydrogen chloride is also emitted by less violent volcanic eruptions and by the evaporation of sea spray. Seawater contains chloride ion, Cl^-, which forms HCl when the water evaporates. However, the HCl from these sources typically does not reach the stratosphere, because it, too, is removed by precipitation before it can leave the troposphere. The same is true of the chlorine released from swimming pools. This chlorine comes off the water surface as molecular chlorine, Cl_2, but is quickly converted to HCl by photochemical reactions in the lower troposphere. Hence, despite what you may hear on "talk radio," nearly all of it is removed from the atmosphere before it can damage the ozone layer.

The largest sources of stratospheric chlorine today are *chlorofluorocarbons* (CFCs), or *freons,* which are an-

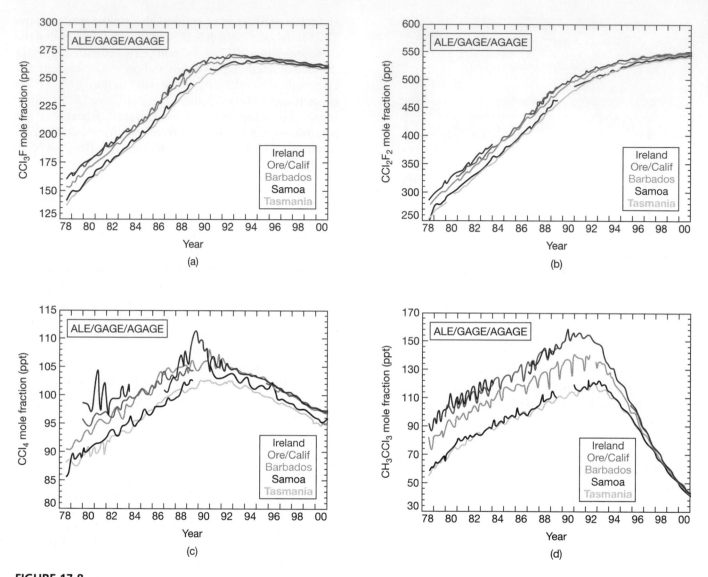

FIGURE 17-8

Atmospheric concentrations of (a) freon-11, (b) freon-12, (c) carbon tetrachloride, and (d) methyl chloroform since 1977. (Data from R.G. Prinn, R.F. Weiss, P.J. Fraser, P.G. Simmonds, F. N. Alyea, and D.M. Cunnold. The ALE/GAGE/AGAGE database, *DOE-CDIAC World Data Center* (E-mail to: *cpd@ornl.gov*), Dataset No. DB-1001, 2002.)

thropogenic compounds. The two most common of these are freon-11 (CCl_3F) and freon-12 (CCl_2F_2). These compounds are also known as F-11 and F-12, respectively. Their atmospheric concentrations have increased by about a factor of 2 since first measured in 1977 (Figure 17-8). We mentioned these gases in Chapter 3 and again in Chapter 16 because they contribute to Earth's greenhouse effect. Freon-11 has been used as a propellant in spray cans and as a blowing agent for producing foams (e.g., Styrofoam®). It is also used to clean semiconductor chips for computers and other electronic devices. Freon-12 is used as a refrigerant and until 1990 was the working fluid in most car air conditioners. However, the uses of freons are changing rapidly. Freon-11 has been banned from use in

spray cans in the United States since 1978. Freon-12 has already been replaced by more ozone-friendly compounds in all new cars. This decrease in freon usage has caused the atmospheric growth rates of these compounds to decrease substantially in the 1990s. Indeed, freon-11 concentrations peaked in 1993 and have begun a slow decline (see Figure 17-8a). Freon-12 concentrations were still increasing in 2000 but not nearly as rapidly as they had been earlier.

Also shown in Figure 17.8 (panels *c* and *d*) are the abundances of two chlorocarbons, carbon tetrachloride (CCl_4) and methyl chloroform (CH_3CCl_3). Carbon tetrachloride is used in dry cleaning and methyl chloroform is used as a solvent in various manufacturing processes. These gases are also regulated under the Montreal Protocol (discussed later in the chapter) and, as one can see, their concentrations have decreased dramatically since about 1991. These chlorocarbons have shorter atmospheric lifetimes than their cousins, the chlorofluorocarbons; consequently, they have responded more quickly to decreases in their emission rates.

When freons were first introduced in the 1930s, they were considered to be wonder chemicals. Not only did they have useful thermodynamic properties, but also they were inert and nontoxic to humans. In contrast, early refrigerants such as ammonia and sulfur dioxide could be quite dangerous if they leaked. It is this very property of being inert, however, that makes freons dangerous to the ozone layer (Figure 17-9). Freons do not react in either the troposphere or the lower stratosphere. Hence, freon gases released at the surface diffuse all the way to the upper stratosphere (above 40 km altitude). Once there, they are photolyzed by short-wavelength UV radiation, which breaks the molecules apart and releases their chlorine as atomic Cl. The Cl then proceeds to destroy ozone by way of the chlorine catalytic cycle.

Skeptics of the prospect that ozone depletion is anthropogenic occasionally suggest that freon gases, being heavier than air, should tend to stay in the lower atmosphere rather than diffusing upward to the stratosphere. (The molecular weight of freon-11, for example, is 136, whereas that of air is only about 29.) This idea, however, is based on a misconception about how the atmosphere mixes. In actuality, winds and eddies cause the atmosphere to be well mixed up to an altitude of about 100 km. Only above that height do heavier gases begin to separate from lighter ones. So, freons are able to deliver ozone-destroying chlorine atoms to the stratosphere despite their high molecular weights.

As in the case of the odd nitrogen cycle, the chlorine cycle in the stratosphere is eventually broken. In this case, Cl reacts with methane (or with H_2), forming hydrogen chloride:

$$Cl + CH_4 \rightarrow HCl + CH_3.$$

Hydrogen chloride, like nitric acid, is relatively unreactive and diffuses downward into the troposphere, where it is removed by precipitation. Dissolved in water, HCl is the very strong acid hydrochloric acid. However, the actual amount of hydrochloric acid formed is relatively small, so it has little effect on the acidity of rainwater.

The Bromine Cycle

Atmospheric bromine, like chlorine, has several different sources. Methyl bromide (CH_3Br) like methyl chloride, is produced naturally as a by-product of biological activity in the oceans. However, it is also used as a fumigating agent for soil pests, including termites. Some 15% of manufactured methyl bromide is used in California for agricultural fumigation. The other major source of atmospheric bromine involves artificial chemical compounds known as **halons,** which are used in certain types of fire extinguishers. The two most common of these are halon-1211 (CF_2ClBr) and halon-1301 (CF_3Br).

Some types of fire extinguishers use CO_2 instead of halons. Both gases are nonflammable and heavier than air, so they are relatively efficient at keeping oxygen away from a flame. Being heavier than air does make a difference in this case, because the gas released from a fire extinguisher does not have time to disperse before it reaches the flame.

The atmospheric bromine cycle is analogous to the chlorine cycle: Most of the methyl bromide reacts in the troposphere, whereas the halons diffuse up into the stratosphere. There, they are photolyzed by short-wavelength UV radiation, and bromine atoms are released. Bromine is removed by the formation and eventual rain-out of hydrogen bromide, HBr.

The Antarctic Ozone Hole

Science, like life, is full of surprises. Perhaps the biggest surprise for most atmospheric chemists was the discovery of the Antarctic ozone hole in 1985. Recall from Chapter 1 that the ozone hole was overlooked for 6 years in the *Nimbus-7* satellite measurements because the observed column depths were considered too low to be real. Computer models also failed to predict its occurrence until long after it was observed. By taking a closer look at what causes the ozone hole, we can see why it caught atmospheric scientists off guard.

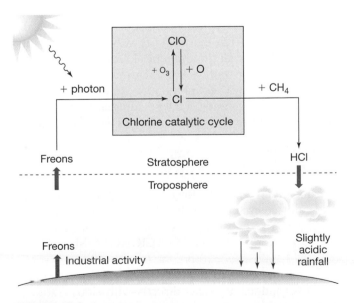

FIGURE 17-9

The atmospheric chlorine cycle.

A CLOSER LOOK
How the Link between Freons and Ozone Depletion was Discovered

The first step in establishing the link between freons and ozone depletion was made by James Lovelock, the same Lovelock who formulated the Gaia Hypothesis (see Chapter 1). Long before he became known as a theorist, Lovelock established his scientific reputation by inventing a device called the *electron capture detector*. When mounted on an instrument called a gas chromatograph, this device can be used to measure the concentrations of extremely dilute atmospheric gases. Lovelock used his detector to measure the concentration of freon-11 and other chlorine-containing gases. He published his measurements in 1970.

By comparing his measured concentrations with estimates of freon emission rates, Lovelock was able to show that most of the freon-11 that had been produced up until that time was still present in the atmosphere. This meant that freon-11 must have a very long atmospheric lifetime. The lifetimes of freon-11 and freon-12 have since been determined to be about 60 years and 130 years, respectively. Ironically, Lovelock was not the least bit concerned about this finding, because freon gases were at that time considered to be totally harmless. Indeed, he wrote that "The presence of these compounds [CFCs] in the atmosphere constitutes no conceivable hazard." This sentence was added to ward off environmentalists, who were at that time quick to warn about the health effects of newly discovered chemicals.

Lovelock, though, was wrong on this last count for reasons that were entirely different from what he had in mind. Chemists had known since the 1930s that Cl and ClO were capable of destroying ozone. It did not take long after Lovelock's discovery for atmospheric scientists to put two and two together. The critical paper was published in 1974 by Sherwood Rowland and Mario Molina, two atmospheric chemists at the University of California at Irvine. Rowland and Molina pointed out that the long lifetimes of freon gases would allow them to diffuse up into the stratosphere, where the chlorine released by their decomposition could wreak havoc on the ozone layer. This idea, obvious in retrospect but quite novel at the time, caught on quickly in the scientific community and earned them a share of the Nobel Prize in Chemistry in 1995. (The prize was shared with Paul Crutzen, who developed the theoretical foundation for understanding ozone-destroying catalytic cycles.) More than 10 years passed and many pitched battles were fought, however, before it was finally accepted by the chemical industry. These battles have been referred to as the "ozone wars." We owe a debt of gratitude to all the scientists who participated in those wars and helped bring us to the understanding that we have today.

To begin, recall that the ozone hole occurs only during the month of October; it apparently did not occur at all prior to about 1976; and it is found primarily over Antarctica. A corresponding Arctic ozone hole has been seen in March of some years, but it is not nearly as pronounced as the Antarctic hole and is not seen at all in other years. So, perhaps we should not be surprised that the explanation of the ozone hole turns out to be rather complicated.

Homogeneous and Heterogeneous Reactions

The chemistry that we have touched on in the last few sections is reasonably well understood. The chemical reactions that we have discussed are all **homogeneous reactions,** that is, reactions between molecules that are in the gas phase. We can study gas-phase reactions in the laboratory if we keep the reacting molecules away from the walls of their container. In contrast, the chemistry that causes the ozone hole involves **heterogeneous reactions,** which are reactions that occur on solid surfaces, such as particles. These reactions are more difficult to study experimentally; consequently, much less is known about them.

The particles involved in the formation of the Antarctic ozone hole are collections of droplets called **polar stratospheric clouds** (PSCs). These clouds were first discovered by high-flying spy planes. Over most of the globe, clouds form only in the troposphere; the stratosphere is too dry for condensation to occur. However, in winter the polar stratosphere is so cold ($-80°C$ or below) that certain trace atmospheric constituents can condense. The particles that form typically consist of a mixture of water and nitric acid. Although they are very tenuous by comparison with normal, tropospheric clouds, these PSCs alter the chemistry of the lower stratosphere in two fundamental ways: (1) by coupling between the odd nitrogen and chlorine cycles and (2) by providing surfaces on which heterogeneous reactions can occur, as we will see next.

Coupling between the Odd Nitrogen and Chlorine Cycles

Throughout most of the lower stratosphere, the nitrogen and chlorine cycles are coupled by way of the reaction

$$ClO + NO_2 + M \rightarrow ClONO_2 + M$$

The product formed in this reaction, chlorine nitrate ($ClONO_2$), does not react directly with either ozone or atomic oxygen and can be converted back to ClO only with difficulty. Thus, it serves as a relatively inert storage reservoir for chlorine, keeping it out of the more reactive forms, Cl and ClO. Because they can directly catalyze ozone destruction, Cl and ClO, along with compounds that can be easily converted into them by photolysis, are collectively termed **reactive chlorine.**

Under normal conditions in the polar stratosphere, NO_2 is always sufficiently abundant to tie up a significant fraction of the available chlorine in the form of chlorine nitrate. In the wintertime Antarctic stratosphere, however, NO_2 concentrations become very low, because most of the odd nitrogen has been converted into HNO_3 and subsequently incorporated into cloud droplets as PSCs. This removal allows reactive chlorine concentrations to increase, because less chlorine is bound up as chlorine nitrate.

The PSC particles also help convert unreactive forms of chlorine into reactive chlorine by providing surfaces on which heterogeneous reactions can occur. For example, one reaction that is thought to be important is

$$ClONO_2 + HCl \rightarrow Cl_2 + HNO_3$$

Molecular chlorine, Cl_2, does not react directly with ozone itself but is readily photolyzed to atomic chlorine:

$$Cl_2 + photon \rightarrow Cl + Cl$$

Once formed, atomic chlorine can execute its destructive effect on ozone.

The Polar Vortex

We now have the chemistry we need to understand how the ozone hole forms. Some knowledge of stratospheric circulation is required, however, to complete the story. We saw in Chapter 4 that the prevailing surface winds at polar latitudes blow from east to west. The opposite is true, however, in the wintertime stratosphere: the winds blow from west to east. When viewed from above the South Pole, this wintertime circulation pattern appears as a gigantic whirlpool called the **polar vortex** (Figure 17-10). The cold, dense air in the middle of the vortex is subsiding in much the same way as water going down a drain. The sinking air carries cloud particles along with it, permanently removing odd nitrogen from the stratosphere. The polar vortex also effectively isolates the Antarctic stratosphere from the rest of the globe. The air inside the vortex is rotating much faster than the air outside, making it difficult for air to pass through the vortex boundary. Some air does get through and replaces the air that is sinking in the center of the vortex, but the amount is small compared with normal rates of latitudinal mixing. Thus, the odd nitrogen removed from the polar stratosphere during winter is not replaced by odd nitrogen from lower latitudes. The same phenomenon also ensures that very little new ozone can be brought in once the ozone hole forms.

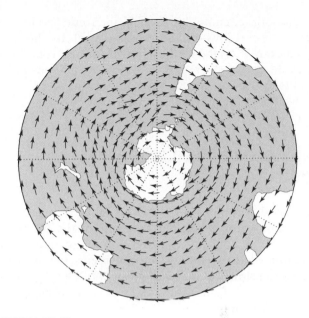

FIGURE 17-10

Schematic diagram of the Antarctic circumpolar vortex. (From Climate Prediction Center/NOAA.)

The full story of the ozone hole involves an intricate coupling of chemistry, atmospheric circulation, and the availability of sunlight. During the long Antarctic winter (May through September), the stratosphere becomes cold enough to allow PSCs to form. These PSCs deplete the polar stratosphere of odd nitrogen and help convert unreactive forms of chlorine ($ClONO_2$ and HCl) into more reactive forms, such as Cl_2. The reactive chlorine, however, remains bound to the surfaces of cloud particles until late September, when the Sun first peeks over the Antarctic horizon. The sunlight releases reactive chlorine from the particle surfaces, and the chlorine wreaks havoc with the ozone layer during October. Near the end of the month, the polar vortex breaks down, allowing fresh ozone and odd nitrogen to be brought in from lower latitudes; the ozone hole disappears until the following October.

Why does this process not occur to the same extent over the Arctic during Northern Hemisphere spring? The answer seems to lie in the stratospheric circulation. As a result of geography, the Arctic polar vortex is less fully developed than the Antarctic vortex and breaks up earlier in the spring. The Southern Hemisphere contains a roughly symmetric continent (Antarctica) surrounded by a large ocean. The Northern Hemisphere contains a mixture of land and ocean, and it also has large, north—south mountain ranges, such as the Rocky Mountains in North America. The mountains disrupt the east—west tropospheric circulation and create atmospheric waves that propagate up to the stratosphere and deposit energy there. This energy warms the stratosphere and disrupts the smooth, circumpolar flow. If an Arctic ozone hole were to occur, it would form during April. By this time of year, however, the Arctic polar vortex has typically already disappeared. Thus, the unique combination of conditions that occur in the Southern Hemisphere is not duplicated in the north.

Evidence of Mid-Latitude Ozone Depletion

The fact that we do not observe an ozone hole over the Arctic does not mean that residents of the Northern Hemisphere should be complacent about ozone depletion. Indeed, there is considerable evidence that ozone is decreasing slowly at mid-latitudes in both hemispheres. The data come from three satellite-borne ozone-monitoring instruments.

Sources of Natural Variability in Ozone

Detecting trends in ozone column depths at low- and mid-latitudes is difficult, because the rates of change are not nearly as large as those seen over Antarctica. Furthermore, before we can identify any long-term trends, we must first correct for natural variability. Figure 17-11, which shows satellite measurements of ozone column depth over Hohenpeissenberg, Germany, illustrates the problem. The first signal that we see in the data is a strong annual cycle of ±75 Dobson units. For reasons that are related to stratospheric circulation patterns, ozone is most abundant during the spring and least abundant during the fall. (Verify this by referring back to Figure 17-5.) But there are other, more subtle oscillations in the data as well. These oscillations include the solar cycle—the 11-year cycle in sunspot number—and the *quasi-biennial oscillation,* or QBO, which is a 27-month cycle in the direction of winds in the equatorial lower stratosphere.

The sunspot cycle affects ozone because the amount of ultraviolet radiation emitted by the Sun varies with sunspot activity. When sunspots are abundant (termed **solar maximum**), as they were in 2002, the Sun gives off more UV energy. The added UV radiation comes not from the spots themselves but from bright areas called *plages* surrounding the spots (see Chapter 15). Higher UV fluxes cause more O_2 to be photolyzed, resulting in increased production of ozone. At times of low sunspot activity (**solar minimum**), as in 1986 and 1997, the solar UV flux is low and ozone production decreases.

The effect of the solar cycle on stratospheric ozone is illustrated in Figure 17-12. The solid curve shows the flux of 10.7-cm radio emission from the Sun, which happens to be well correlated with the solar UV flux. The dashed curve shows the percentage change in average ozone column depth (40° N to 40° S) after the seasonal cycle, QBO, and longer-term trend have been removed. Ozone column depth is about 2–3% higher during solar maximum than during solar minimum. As this variation is of the same magnitude as the trend in the satellite data, it, too, must be removed before the trend can be determined.

Long-Term Trends in Ozone

By removing all the known sources of natural variability from the satellite data, it is possible to identify long-term trends in ozone column depths. The trend from 1979–1997 for the entire region between 60° N and 60° S is shown in Figure 17-13. The average ozone column depth apparently decreased by about 6% at high northern and southern latitudes while remaining virtually unchanged near the equator. The data, along with the uncertainties in the measurements, are displayed in terms of percent decrease per decade in Table 17-4. The trend at mid-latitudes is statistically significant and is an obvious cause for concern. Surface UV fluxes would increase substantially if this trend were to continue for even another decade or two into the future.

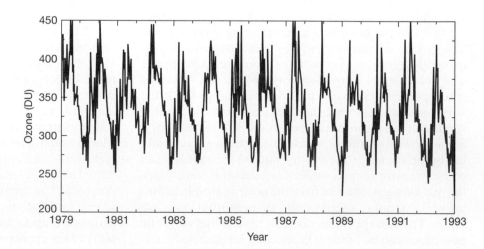

FIGURE 17-11

Weekly averaged satellite data on ozone column depth over Hohenpeissenberg, Germany. A strong annual cycle is observed in the data. (After World Meteorological Organization, 1994.)

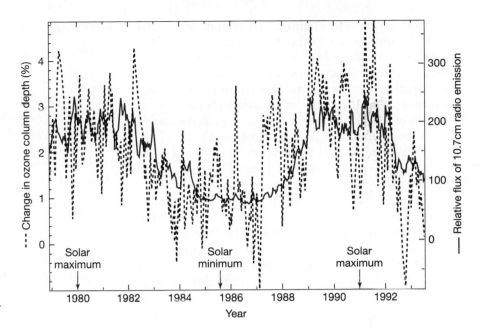

FIGURE 17-12

Response of average ozone column depth (40° N to 40° S) to the solar cycle after the seasonal cycle, quasi-biennial oscillation, and long-term trend have been removed. (After World Meteorological Organization, Scientific Assessment of Ozone Depletion: 1994.)

What could be causing the downward trend in ozone at mid-latitudes? The obvious candidates are stratospheric chlorine and bromine, which have been increasing throughout the past few decades as a consequence of freon and halon emissions. However, atmospheric chemists who have tried to simulate this effect with computer models have been able to account for only about one-third of the observed rate of ozone decrease. The models are fairly accurate during summer and fall, but they miss badly during winter and spring, which is when most of the observed depletion seems to be occurring. Either the chemistry of ozone is still not fully understood or the details of stratospheric circulation are even more complex than we have already discovered.

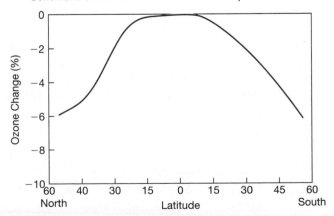

Schematic of the North-to-South Ozone Depletion: 1979–1997

FIGURE 17-13

Percentage deviation of monthly averaged ozone column depth (60° N to 60° S) from the 1979 value. (After World Meteorological Organization, Scientific Assessment of Ozone Depletion: 1998.)

Mechanisms for Halting Ozone Depletion

We began this chapter by pointing out that ozone depletion is now widely recognized as a global threat and that steps have already been taken to reduce or eliminate the problem. These steps fall into two categories. (1) the ratification by the international community of treaties designed to reduce freon and halon emissions and (2) the development of "ozone-friendly" substitutes for freons by the chemical industry.

International Agreements

The first and most important step in terms of diplomacy was the signing of the **Montreal Protocol** in 1987. This treaty, which has now been ratified by all of the major industrial nations of the world, placed strict limits on the amount of freons and halons that could be released into

TABLE 17-4

Percentage Change in Ozone Column Depth per Decade, 1979–1997*	
Latitude Belt	*Trend (%/decade)*
North 50–65°	−3.7 ± 1.6
Mid North 30–50°	−2.8 ± 1.7
Equatorial 20–20°	−0.5 ± 1.3
Mid South 30–50°	−1.9 ± 1.3
South 50–65°	−4.4 ± 1.8

Source: Scientific Assessment of Ozone Depletion: 1998, World Meteorological Organization

the atmosphere by any country. The original treaty was agreed on prior to the crucial measurement that linked the Antarctic ozone hole to anthropogenic chlorine (see Figure 1-6). That discovery, however, solidified international opinion on the issue and has led to several subsequent revisions of the treaty, which ultimately placed even stricter limits on freon emissions. The expected effect of these treaties up to A.D. 2075 is shown in Figure 17-14. The figure shows that in the absence of any kind of international agreement ("business as usual"), stratospheric chlorine levels would have been expected to have risen quite rapidly over the next few decades. Under such circumstances, computer models predict that ozone column depths could decrease precipitously. The Montreal Protocol alone would have slowed this process, but stratospheric chlorine concentrations would still have continued to rise. With the most recent international treaty (the Montreal 1997 Accord) in place, the outlook is much improved. As shown earlier (Figure 17-8), tropospheric concentrations of most anthropogenic chlorine-containing gases peaked in 1994 or earlier and are beginning a slow decline. Gas concentrations in the stratosphere lag those in the troposphere by 3 to 5 years, so stratospheric chlorine and bromine abundances will probably continue to increase until about the turn of the century. By A.D. 2060, stratospheric chlorine should be back down to about 2 ppb, its value in the late 1970s when the Antarctic ozone hole first appeared. Hopefully, the hole will disappear at that time. Within the following century, stratospheric chlorine should reach the background level of 0.6 ppb that results from naturally produced methyl chloride.

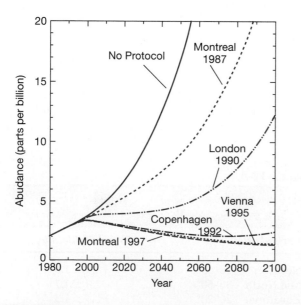

FIGURE 17-14

Projected atmospheric chlorine concentrations under the various international agreements. The horizontal dashed line shows the estimated preindustrial level of chlorine. (After Dr. M. McFarland at DuPont.)

Freon Substitutes

Can the world make do without conventional freon gases? The answer appears to be yes. Substitutes for freons are already being developed by major chemical companies. At least two different strategies are being employed. One of these is to replace one of the chlorine or fluorine atoms with a hydrogen atom to make compounds referred to as *HCFCs*. An example is the compound HCFC-22 ($CHFCl_2$), which can be used as a substitute for freon-12. Attaching a hydrogen atom to a CFC makes the CFC susceptible to chemical attack by the hydroxyl radical, OH (Figure 17-15). Recall from earlier in the chapter that radicals play an important role in removing pollutants from the atmosphere. Because of its increased reactivity, an HCFC-22 molecule released at the surface has only a 5% probability of reaching the stratosphere, compared with a virtually 100% probability for freon-11 or freon-12. The other 95% of the HCFC-22 is destroyed in the troposphere, where it can do no harm to ozone. So, this compound is 20 times more ozone-friendly than normal freon gases.

A potential downside of this strategy is that the more reactive HCFCs may pose direct health threats. Hence, thorough testing of the biological effects of these compounds is required. On the bright side, HCFC-22 has a short lifetime (about 15 years), which means that it contributes far less greenhouse-enhanced warming than does a long-lived gas such as freon-12. Thus, replacing CFCs with HCFCs can help slow global warming while it reduces the destruction rate of stratospheric ozone.

A second strategy is to eliminate chlorine from the CFC molecule, creating an HFC compound. An example of this type is the compound HFC-134a (CF_3CH_2F), which is already in use as a substitute for freon-12 in some new cars. Because HFCs contain no chlorine, they pose no significant threat to ozone. (The reason is somewhat paradoxical: Fluorine radicals are *more* reactive than Cl radicals. Indeed, they are so reactive that they combine directly with water vapor to form hydrofluoric acid, HF. Like HCl, HF is a tightly bound molecule that drifts down to the troposphere, where it is removed by rainout. So, fluorine radicals are removed from the stratosphere before they do much damage to ozone.) Also, HFC-134a is

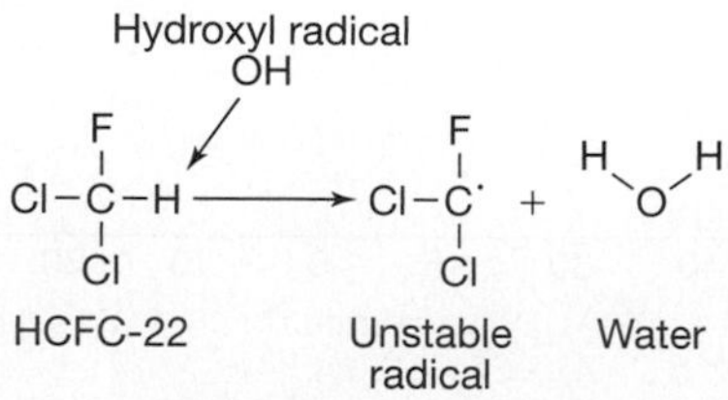

FIGURE 17-15

Hydroxyl (OH) radicals can react with hydrogen atoms in HCFCs to form unstable radicals plus water.

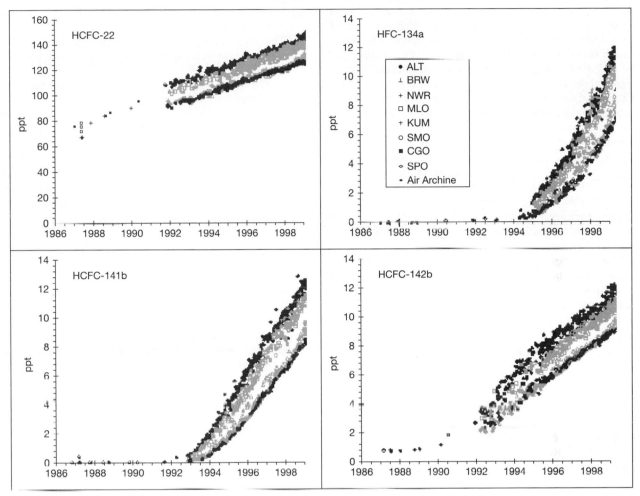

FIGURE 17-16

Abundances of various freon replacements: (a) HCFC-22, (b) HFC9134a, (c) HCFC-141b, (d) HCFC-142b. (Data courtesy of NOAA/CMDL, *http://www.cmdl.noaa.gov/info/ftpdata.html*)

relatively short-lived (about a 16-year lifetime), so it is a good choice in terms of climate as well. That said, the abundances of this gas and other freon substitutes are currently increasing at a rapid rate (Figure 17-16). Clearly, it would be wise to monitor the concentrations of these gases carefully to make sure that we do not create a new environmental problem while eliminating an old one.

Lessons Learned from the Ozone Experience

The manner in which the international community and the chemical industry have responded to the threat of ozone depletion shows that we *can* deal successfully with global environmental problems. This success, however, did not come without a fight. In the 1970s, when the danger posed by freons had just been discovered, the chemical industry vigorously opposed the idea that their

products might be harmful. A concerted effort by atmospheric scientists associated with universities and national laboratories caused the industry to look more closely at the problem. Eventually, companies such as DuPont hired their own atmospheric chemists and confirmed for themselves that action needed to be taken. Now, DuPont is leading the chemical industry in the design and marketing of freon substitutes.

It remains to be seen whether energy companies and consumers will follow a similar path with regard to global warming. That problem, unfortunately, is much more difficult to correct. Fossil fuels are a much bigger part of the global economy than are freons, and adequate substitutes might be harder to develop or be prohibitively expensive. Hopefully, though, the precedent that has been set on ozone depletion will eventually help guide us to a solution to the global warming problem.

Chapter Summary

1. Solar ultraviolet radiation between 200 nm and 320 nm poses significant dangers to humans and other organisms if not effectively blocked by stratospheric ozone. The wavelength region of greatest concern is the UVB region, between 290 nm and 320 nm, where the solar flux is relatively high and the ozone absorption coefficient is relatively low.

2. The flux of solar UV radiation that reaches Earth's surface depends on the ozone column depth, which is a measure of the total amount of ozone overhead. The column depth varies regionally; it is generally highest near the poles and lowest near the equator. This distribution, combined with the fact that the Sun is generally higher in the sky near the equator, produces much higher ground-level UV fluxes in the tropics than at high latitudes.

3. Ozone chemistry in a pure oxygen–nitrogen atmosphere can be described by a set of four reactions known as the Chapman mechanism.
   a. The first and last of the Chapman reactions affect the abundance of odd oxygen (O_3 and O); the middle two reactions cause the different forms of odd oxygen to interconvert.
   b. In the real atmosphere, catalytic cycles involving nitrogen, chlorine, and bromine provide alternative, faster ways of destroying odd oxygen. Both chlorine and bromine have large anthropogenic sources (freons and

halons, respectively) that are considered to be a threat to the ozone layer today.

4. The most striking evidence of ozone depletion comes from Antarctica, where an ozone hole forms each year during October.
   a. The ozone hole is apparently caused by a complex series of reactions that occur on the surfaces of polar stratospheric cloud particles. These same particles remove odd nitrogen from the polar stratosphere, breaking the coupling between the odd nitrogen and chlorine cycles and allowing free chlorine to wreak havoc on the ozone layer.
   b. The Arctic stratosphere behaves differently than the Antarctic stratosphere, because the wintertime polar vortex is not as well developed or long-lasting in the Arctic.
   c. Ozone has also been decreasing at a lesser rate at mid-latitudes in both hemispheres for reasons that are not fully understood.

5. International agreements, namely the Montreal Protocol and various follow-on accords, have placed strict limits on future freon and halon emissions. If these accords are enforced and obeyed, stratospheric chlorine concentrations should decline to their natural levels during the course of the next century, and the ozone hole should disappear. It is important that we continue to recognize the wisdom of these agreements and the reasons why they were established.

Key Terms

absorption coefficient
action spectrum
catalyst
catalytic cycle
Chapman mechanism
column depth
Dobson unit
dose rate
erythemal action spectrum

halons
heterogeneous reaction
homogeneous reaction
hydroxyl radical
Montreal Protocol
number density
odd nitrogen
odd oxygen
ozonesondes

photodissociation
photolysis
polar stratospheric clouds
polar vortex
reactive chlorine
solar maximum
solar minimum

Review Questions

1. What are the three categories of UV radiation? Which of these are considered to be biologically harmful?
2. What is ozone column depth? In what units is it measured?
3. What are the four chemical reactions that comprise the Chapman mechanism? Which ones affect odd oxygen?
4. What is a catalyst?
5. How do nitrogen and chlorine catalyze the destruction of ozone?

6. What role do polar stratospheric clouds play in the formation of the Antarctic ozone hole?
7. Why is a springtime ozone hole observed over the Antarctic but usually not over the Arctic?
8. What is the long-term trend in mid-latitude ozone column depths? Can it be explained by the observed increase in stratospheric chlorine and bromine?
9. What strategies have been adopted for reducing or eliminating the use of freons?

Critical-Thinking Problems

1. a. According to Figure 17-5, the maximum ozone column depth occurs at high northern latitudes during late winter. The column depth there is 460 DU. How many molecules of ozone per square centimeter does this correspond to?

b. The minimum ozone column depth, which occurs in the tropics, is 240 DU. How many ozone molecules are in a 1-cm² vertical column there?

c. What was the approximate ozone column depth (in Dobson units) over Wallops Island, Virginia, at the time of the measurements shown in Figure 17-3? (Hint: Ozone column depth is ozone concentration [in molecules/cubic centimeter] multiplied by the height of the column. You may want to use a different average ozone concentration for the troposphere and the stratosphere.)

2. The absorption of solar ultraviolet radiation of a given wavelength l by atmospheric ozone follows Beer's Law:

$$\frac{F}{F_0} = \exp\left[-\frac{kN}{\cos\theta}\right]$$

where F = the UV flux at altitude z; F_0 = the UV flux at the top of the atmosphere; k = the absorption coefficient at wavelength l; N = the ozone column depth above height z; and θ = the solar zenith angle.

The function $\exp(x)$ (or e^x on some calculators) is the exponential function, which also describes radioactive decay, population growth, and other interesting phenomena.

a. The wavelength region where changes in the solar UV flux have the most potential to do harm is around 290 nm, in the UVB range. The absorption coefficient of ozone at this wavelength is about 10 atm^{-1}-cm^{-1}. The average column depth of ozone from the ground up to the top of the atmosphere is about 0.3 atm-cm. By what factor is the incident solar UV flux at 290 nm attenuated today—that is, what is the current value of F/F_0 at ground level? Evaluate your answer for a solar zenith angle of 45°.

b. If the ozone column depth were to be reduced by 1% as a consequence of increasing concentrations of chlorofluorocarbons, what would be the resulting percent increase in the ground-level UV flux at 290 nm? What about for an ozone decrease of 10%? 50%? Assume a solar zenith angle of 45° in each case.

c. Suppose that we decided that we could tolerate a 10% increase in the UV flux at 290 nm for $\theta = 45°$, but no more. What would be the maximum percentage decrease in ozone that we could allow? (Hint: The inverse of the exponential function is called the natural logarithm, abbreviated as ln. If $y = \exp(x)$, then $x = \ln(y)$.)

3. New York City is at 43° N. Miami, Florida, is at 25° N. In March, when most colleges have their spring break and students go to Florida, the ozone column depth is about 280 DU over Miami and 350 DU over New York, according to Figure 17-5. Using the data from the previous problem, calculate the ground-level solar UV flux at 290 nm in Miami compared to that in New York at noontime on March 21. Note that March 21 is the vernal equinox, so the Sun is directly over the equator. (Hint: It may help to draw a picture to help you visualize the geometry.)

Further Reading

General

Dotto, L., and Schiff, H. 1978. *The Ozone War.* Garden City, NY: Doubleday.

Turco, R. P. 1997. *Earth Under Siege* (Ch. 13). New York: Oxford University Press.

Advanced

Albritton, D. L., Aucamp, P. J., Mégie, G., and Watson, R. T. Assessment Co-Chairs., Scientific Assessment of Ozone Depletion: 1998, World Meteorological Association. Available online at: http://www.al.noaa.gov/WWWHD/pubdocs/assessment98.html

Human Threats
to Biodiversity

Key Questions

- Where is species loss occurring, and why?
- Why should we care about species loss?
- Could the loss of biodiversity have important consequences for humans and for the health of the Earth system?

Chapter Overview

Humans are responsible for a rate of species extinction that rivals any of the mass extinctions Earth has experienced in the past. Most of this species loss is occurring as a result of human land-use practices that destroy natural habitats, and we may soon reach the point at which not only are species becoming extinct, but also destroying whole ecosystems. This species loss is particularly severe in tropical rainforests. Focusing on ecosystems rather than simply considering individual species is important. The loss of biodiversity could affect the health or stability of the planet and may well threaten world food supplies.

Introduction

In Chapters 16 and 17 we saw two ways in which human activity is having, or has the potential to have, a significant impact on the Earth system. In this chapter we address a

third issue, that of biodiversity and the loss of species. In Chapter 9 we introduced the concept of biodiversity and in Chapter 13 we discussed the mass extinctions that have occurred in the past. We also illustrated the dramatic changes that took place in the biota as a result of these extinction events. What you may not have realized is that the rate of species loss due to human activity in the present is equal to, or greater than, the rate of species loss that marked these past mass extinctions.

The implications of species loss can be harder to grasp than are those of other environmental threats we have discussed previously. It is relatively easy to see the impact of the loss of stratospheric ozone or of a change in climate. Most of us have experienced extreme summers or winters in the past and can imagine what it might be like if those extremes were to become the norm in the future. Few of us, however, have any direct experience of species loss or of the consequences that it might entail.

Recall from Chapter 9 that a definition of a species is that it consists of closely related organisms that can potentially interbreed. On evolutionary time scales, new species continuously develop and others go extinct; in fact, the average lifetime of most species is between 1 million and 10 million years. When we consider that life has been present on Earth for the past 3.5 billion years, it is apparent that many species have evolved and disappeared over that time period. Indeed, most of the species that have ever lived are now extinct. However, the fossil record shows that the rate of **speciation** (the origination of new species) is slightly greater on average than the rate of extinction. Consequently, there are more species today than there have ever been at any single time in the past.

- How many species are living today? We do not know for certain. Approximately 1.4 million species have been described, and new species are being found faster than they can be catalogued.

- How many more species have yet to be discovered? Estimates range from 10 million to 100 million. Beetles make up a large proportion of the known species, prompting J. B. S. Haldane's famous quote (referring to the work of Darwin), "From the fact that there are 400,000 species of beetles on this planet, but only 8,000 species of mammals, he [Haldane] concluded that the Creator, if He exists, has a special preference for beetles" (report of a lecture given by Haldane in 1951). However, despite the preponderance of beetles, there are certainly still many plant and invertebrate species yet to be discovered.

- How many species go extinct each year? Again, we do not know. The estimates are based on the proportion of species that are lost; hence they also depend on the estimates used for the existing number of species.

It is difficult to obtain reliable numbers on extinction rates. We do not know how many species there are and it is likely that many species go extinct without us ever knowing they existed. Based on what little we do know, some estimates suggest that at the present rate of extinction, one-quarter of all species on Earth may be lost within the next 50 years. If we accept 100 million as the number of existing species, then the extinction rate is half a million species per year. If 10 million is closer to reality, then the extinction rate is approximately 50,000 species each year. This is a somewhat circular argument! In addition, we are multiplying the number of species (which we don't know) by an extinction rate (that we also don't know) to arrive at a number of extinctions per year that obviously has little real quantifiable meaning. However, it is probably safe to say that a substantial number of species disappear every day as a direct consequence of human activity.

Why is this massive species loss occurring, and does it make a difference to us? These are the questions that we address in this chapter. Before we do, it is important to note that there is likely to be some uncertainty in many of the numbers presented here. While some information is factual—we have evidence of certain species going extinct in a particular location at a particular time—much of what we report here concerns estimates of species loss and habitat destruction. How these estimates are made varies by country and region and over time. Estimates of habitat destruction may be made by field surveys, satellite analyses, or interviews with landowners. Estimates of species loss depend on how well we know the existing species, which again varies by region, and what we know about the relationship between extinction and habitat loss

for a particular ecosystem in any given region. Frequently, these estimates are made on detailed studies of small regions, which are then extrapolated to the larger scale. Furthermore, any large-scale analysis is expensive and time consuming, so they tend to be infrequent. The result of all this is that there is large uncertainty in these numbers, and the numbers you see quoted may be fairly old, which means that you often find little agreement in the published estimates from different sources. This is not a major problem for our purposes here—the estimates of species loss and habitat destruction are sufficiently large that it is obvious that a problem exists, even though precise numbers may not be available.

Much of our discussion in this chapter revolves around ecosystems, and it is possible to analyze individual ecosystems by using the same systems approach that we have used in previous chapters. Note, however, that we know very little about ecosystems from an Earth systems perspective. Although some studies describe the cycling of a particular nutrient through a given ecosystem or the transfer of energy or water through that ecosystem, there are very few studies of how any particular ecosystem functions in its totality. Furthermore, we have little idea of how a particular assemblage or distribution of ecosystems interacts at the global scale. In earlier chapters we were often able to reduce the complexity of the Earth system to a few relatively simple concepts and systems diagrams, but that is not really possible here. Ecologists have been studying ecosystems for some time and have made extensive use of systems theory. We could fill this chapter with diagrams illustrating the different functions of numerous ecosystems, but none of them would answer the questions posed earlier: Why is the species loss occurring? Should we care about it?

In Chapter 9 we introduced the idea of a possible relationship among diversity, community stability, and environmental stability; however, very little work has actually been done along these lines. In the remainder of this chapter, we describe some of the ecosystems in which species are being lost and some of the reasons why this loss is occurring. We then talk about the value of species, revisiting the idea that biodiversity is vital for the long-term health of our planet.

The Importance of Ecosystems

Why is the concept of an ecosystem important in our discussion of biodiversity and species loss? When we deal with systems, a perturbation to one part of the system can have impacts throughout the system. Because there is an interdependency among species in an ecosystem, if one species is removed, the loss of other species may follow. How significant a loss will depend on the role that the removed species plays in the ecosystem. In Chapter 9 we suggested that biodiversity should not be measured simply in terms of the number of species but should also describe the interactions among those species. We can suggest further that to know the number of interactions might not be sufficient; it is quite likely that interactions among some species will be more important than others. Ecologists use the term **keystone species** to describe a species that plays a vital role in the operation of an ecosystem.

The El Niño–Southern Oscillation (ENSO) events described in Chapter 15 provide a graphic illustration of the importance of ecosystems. When the westward-flowing ocean current in the tropical Pacific Ocean slows, warm water flows back toward the South American coast, preventing the upwelling of cold, nutrient-rich water off the coasts of Peru and Ecuador. The loss of these nutrients prevents the growth of the phytoplankton that form the basic food source for the local fish population. The reduction of the fish population in turn results in the loss of the birds that feed on those fish. The end result is a massive drop-off in the marine and bird populations and a drastic change in the local ecosystem.

These losses are not permanent: After several years the populations gradually grow back to their previous levels. If there were a drop-off of species found only in this locality, however, such a disturbance could easily lead to their extinction. In fact, the warmer waters that appeared off the western coast of South America during the 1982 ENSO event also killed large areas of coral, and three coral species that lived only in this area went extinct.

The Modern Extinction

The Beginnings

The modern mass extinction episode began when the first humans evolved and spread to colonize larger and larger areas (Figure 18-1). Estimates suggest that approximately 70% of the large mammal genera of the late Pleistocene Epoch are extinct. A similar percentage of the large bird genera has also been lost. The archeological evidence

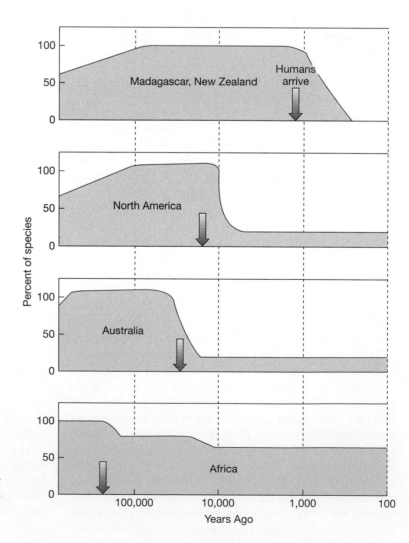

FIGURE 18-1

The extinction of large mammals and birds corresponds to the spread of human populations. Humans evolved first in Africa; in fact, they coevolved with other animals there over millions of years. As a result, the curve for Africa stands out from the others. (After E. O. Wilson. 1992. *The Diversity of Life*. New York: W. W. Norton.)

shows that the losses coincide with the spread of human populations:

- When the first humans arrived in Madagascar in about A.D. 500 there were 6 to 12 species of large flightless birds, 17 genera of lemurs, and numerous species of other large birds and mammals. Within a very short time the flightless birds were gone, as were seven of the lemur genera, plus one species of aardvark, a pygmy hippopotamus, and two species of large land tortoises.
- Humans arrived in New Zealand around A.D. 1000. Since then, 20 species of large land birds and 22 other species of flightless birds have been hunted to extinction.
- As they spread through the western Pacific, the Polynesians killed off more than half of the native species they found. Species that are indigenous, or native, to a particular location are called **endemic species.** The distribution of these species is limited to relatively small geographic spaces and are, therefore, most at risk of extinction if their habitat is lost or degraded.
- The native Hawaiians extinguished 35 to 55 species of land birds. Fifty species remained when Captain Cook arrived in 1778, and one-third of those are now extinct.

There is evidence of a similar loss of species in North America. The first humans who crossed from Siberia into North America found vast grasslands with herds of large mammals, including bison, antelope, and mammoth, together with numerous large birds and other mammals that are now extinct. However, there is an additional explanation for the North American extinction besides hunting. The Pleistocene was a period of large glacial advances and retreats, which accompanied significant swings in mid- and high-latitude climate. Fossil evidence suggests that anomalously high extinction rates coincided with large climate changes. Humans arrived in Australia 30,000 years ago, and there is fossil evidence of the existence of a large number of species that are now extinct. However, the extinction was more gradual than elsewhere, and Australia suffered a long period of drought between 26,000 and 15,000 years ago, when the most rapid species loss also occurred.

When attributing blame for the extinctions in North America and Australia, a good case can be made for both climate change and overhunting. However, in other parts of the world, such as New Zealand, the Pacific islands, and Madagascar, there is no indication of a significant climate change. In those areas, the evidence supporting overhunting appears overwhelming.

The Present Day

As we move forward in time, the predominant agent of destruction has changed from overhunting to habitat destruction, and the pace of species loss has increased dramatically. Across Europe, from the hedgerows of southern Britain to the vine-covered hills of Cyprus, are widely diverse landscapes and ecosystems; all of them, except for the high peaks of the Alps and the Pyrenees, have been created by human activity. It is becoming increasingly difficult, if not impossible, to find landscapes anywhere in the world that have not been modified in some way by human actions. Today this activity almost always involves changes in land use, resulting in habitat destruction and species loss. With few exceptions, human land use leads to a reduction in biological complexity and reduced biodiversity. One interesting exception is the hedgerows of southern England. The original forest was cleared and the land used for cultivation. Later, the large medieval fields were subdivided, and hedgerows were planted to separate the individual fields. These hedgerows actually support a more diverse ecosystem than did the original forest. However, changes in farming practices are leading to the removal of many of these hedgerows in the latest chapter of human modifications to the British landscape.

Tropical Deforestation. The greatest rate of species loss today is found in the tropical forests (Figure 18-2). The climates of these forest regions are characterized by high rainfall (in excess of 2 m/yr), high mean annual temperatures, and low seasonal contrasts. For the most part, the forests are located in the areas of trade-wind convergence associated with the Hadley circulation, as described in Chapter 4. These forests cover approximately 6% of the land surface, yet they are thought to contain over half of the planet's plant and animal species. Great Britain has 1,430 species of flowering plants and 35 native tree species, whereas the Malay Peninsula, with only half the area of Great Britain, has 7,900 flowering plants and 2,500 native tree species. Of the slightly more than 9,000 known bird species in the world, almost half live in the Amazon Basin or in Indonesia. Studies in Peru found 300 tree species in a 2.5-acre plot (approximately 10,000 m²), and 1,000 tree species were found in a combined census of ten 2.5-acre plots in Borneo. In contrast, there are only 700 native tree species in the whole of the United States and Canada. Quite obviously, the greatest potential for species loss is in the tropical forests. However, although this chapter focuses on loss of biodiversity, it is important to recognize that tropical deforestation has other regional and global consequences (see the Box "A Closer Look: Other Consequences of Tropical Deforestation").

IMPACT OF FOREST CLEARANCE. There are no hard and fast rules, but in general the number of species living in a

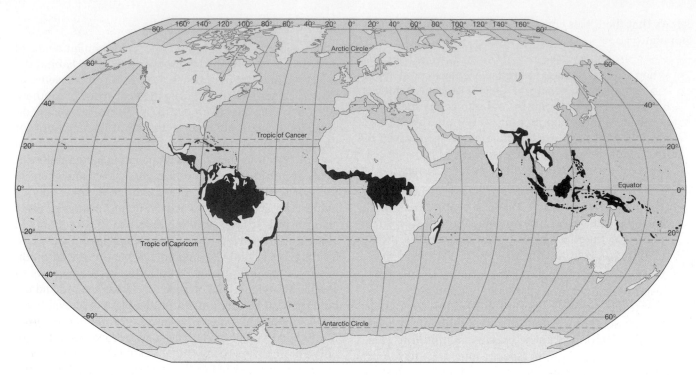

FIGURE 18-2

Distribution of tropical rainforest. (Adapted from the United Nations Food and Agricultural Organization, Global Forest Resources Assessment, 2000.)

tropical rainforest is related to the area of forest coverage: A tenfold increase in area results in double the number of species. Applying these numbers to the Atlantic coast of Brazil, where the forest has been reduced to less than 1% of its original cover, suggests that the forest biota may have already decreased by 75%. Similar losses are occurring throughout the tropics. Harvard University biologist Edward O. Wilson concludes that, even using the most optimistic and conservative estimates, 27,000 species are lost each year—that is, 74 species per day or 3 species every hour—simply as a consequence of tropical deforestation.

If an area of temperate forest is cleared and then abandoned, something very similar to the original forest will grow back within 100 years, and it will have similar levels of biodiversity as the original forest. There is little information available on the regeneration time of tropical rainforests, but the indications are that when large areas are cleared the forest takes hundreds of years to grow back to its original state. In some cases it is likely that the original rainforest will never return. Why are tropical rainforests so sensitive to change?

When a large area of temperate forest is cleared and then abandoned, grasses will spread, shrubs and a few trees will appear, and eventually the spreading vegetation cover will produce the temperature, moisture, and shade conditions that frequently promotes the return of the original forest species. This sequence of events is fa-

cilitated by the nature of the trees' seeds, which tend to be fairly resistant to stress. They are able to lie dormant for long periods, ready to germinate when the right environmental conditions return. Furthermore, a large fraction of the organic material in the forest is returned to the soil through the fall of litter and the decay of dead organic matter. Over time, this recycling produces a nutrient-rich soil layer, which, after the forest has been cleared, is still available to help promote new forest growth. Consequently, it is typically possible for temperate forests to return to something similar to the original forest cover.

The situation in tropical forests is very different. Figure 18-3 illustrates the nutrient flow through a forest–soil system. In simple terms, there are three primary nutrient reservoirs: the trees, the litter, and the soil. The transfer of nutrients to the litter is controlled by the death rate, the transfer to the soil is determined by the rate of decomposition, and the transfer back to the trees depends on the photosynthetic uptake. Tropical forests return little organic matter and few nutrients to the soil. When trees die and decompose, the nutrients in them are returned to the living biomass very quickly. This scenario implies a very efficient nutrient recycling system: higher temperatures and moisture conditions favor rapid decomposition, whereas extensive root systems (roots constitute 60% of the biomass) extract the nutrients from the soil very quickly. In tropical forests, therefore, high rates of decomposition and photosynthetic uptake cycle the nu-

A CLOSER LOOK
Other Consequences of Tropical Deforestation

Clearing tropical forests and converting the land to other purposes, such as agriculture and grazing, has other consequences besides contributing to an overall loss of biodiversity. At the local scale, deforestation tends to result in soil degradation and increased erosion. Without the continuous addition of fertilizer, the nutrient-poor soils rapidly lose their usefulness for agriculture. Where before much of the rainfall was intercepted by the tree canopy, now more of it reaches the surface directly, leading to erosion, which increases as agricultural productivity and surface cover decrease. The result is very rapid soil loss. Deforestation also has an effect on local rainfall. Estimates from the Amazon Basin indicate that approximately half its precipitation is derived from recycled water. In other words, rainfall occurs and the water is intercepted by the canopy; some is used by the trees and some is returned to the atmosphere through evapotranspiration, the combined effect of water loss by evaporation and by transpiration (the transfer of water, in vapor form, from plants to the atmosphere through pores in the leaves). The water that is returned to the atmosphere falls again as rain elsewhere in the basin. When large areas of forest are cleared, a large proportion of the rainfall is lost as runoff, evapotranspiration decreases, and rainfall over the basin as a whole decreases. Several groups of researchers have used atmospheric general circulation models to examine these processes. The models suggest that deforestation has its greatest impact on the local climate by changing the albedo, the surface hydrology, and the surface roughness. (Surface roughness is important because it affects the windflow, which, in turn, affects evaporation and the transfer of latent and sensible heat.)

One such study was carried out in 1989 by J. Lean and D. Warrilow at the British Meteorological Office. They simulated the replacement of all the tropical forests and savannahs in South America (north of 30°S) by tropical pastures and examined the effect that this simulation had on the climate model. Box Table 18-1 presents some of the results, comparing observational data with the model simulation before (the control run) and after the deforestation. We can see from the table that the control run produces a little too much rainfall and runoff and slightly underestimates evaporation, but it generally produces a climate very similar to the observed climate.

More interestingly, when we compare the deforestation experiment with the control run, we see some major differences and some interesting contradictions. Overall, the hydrologic cycle is weakened, which is what we would expect, except that runoff decreases instead of increases. Also, the net radiation (the total amount of available energy) decreases, but temperature goes up. The reasons for this lie in the multiple interactions that occur in the climate system. The pasture has a higher albedo than that of the forest or the savannah, so deforestation increases the albedo and reduces the incoming solar energy. These effects contributed to the decrease in net radiation. Temperatures go up instead of down, however, because, prior to deforestation, evaporation had a cooling effect on the forest; when the forest is removed and evaporation decreases, the cooling effect is reduced and temperatures increase despite the reduction in net radiation.

The surface runoff is determined by several factors, including the rate at

BOX TABLE 18-1

Impacts of Deforestation on Local Climate			
Surface Variable	*Observed*	*Control**	*Deforested**
Evaporation (mm/d)	3.34	3.12	2.27 (−27.2%)
Precipitation (mm/d)	5.26	6.60	5.26 (−20.3%)
Soil moisture (cm)		16.13	6.66 (−58.7%)
Runoff (mm/d)	2.76	3.40	3.00 (−11.9%)
Net radiation (W/m^2)		147.3	126.0 (−14.5%)
Temperature (°C)	24.0	23.6	26.0 (+2.4°C)

*Model simulations are averaged over 3 years.
Source: Adapted from J. Lean and D. Warrilow. Simulation of the Regional Climatic Impact of Amazon Deforestation. *Nature* 342:411–413, 1989.

trients rapidly through the soil and back to the trees, making the living biomass the major nutrient reservoir. In temperate forests the rate of decomposition is lower, as is the rate of photosynthetic uptake, and nutrients tend to accumulate in the organic litter and in the soil.

There is a further important difference between tropical and temperate forests. After rainforests have been cleared, heavy rains, characteristic of the tropics, wash away much of the organic material that does enter the soil, leaving behind soils that are very acidic and nutrient deficient. In contrast to the seeds of temperate forest plants, the seeds of the tropical forest plants tend to be less resistant to stress and typically germinate within a few weeks. For small clearings this is not a problem: Dead organic matter is rapidly broken down, releasing nutrients that are used for new growth. Where the clearing is extensive, however, the nutrients are removed very quickly, the cleared areas are usually very hot (because there is no longer a forest canopy to shade the surface), and the germinating seeds cannot survive. Eventually vegetation

which water infiltrates the soil (which is the only soil parameter changed in this model for the deforested case). The infiltration rate is reduced for the deforested soil, so runoff should increase. However, runoff is also affected by the intensity and frequency of the precipitation events. Infrequent, high-intensity events deliver rain to the surface faster than the soil can remove it, and there is high runoff. If the same amount of rain falls in more-frequent, less-intense events, water is delivered to the surface at a slower rate, more infiltrates, and runoff is reduced. In the deforestation case, runoff is reduced because, although less water infiltrates the soil and is held by the vegetation, there is also a decrease in frequency and intensity of rainfall events. The soil moisture shown in Box Table 18-1 is not the total amount stored in the soil but actually the amount that is available to the vegetation. The decrease that the model shows in available soil moisture is due to the rooting depth of the vegetation: The shallower roots of the pasture cannot reach down far enough to tap the water that exists at greater depth. These and additional interactions are illustrated in a systems diagram (Box Figure 18-1). Here we see that runoff and temperature have both positive and negative couplings from other system components, thus allowing for their apparent contradictory response to what we might, at first, have expect-

ed. Numerous other model experiments have been carried out since the late 1980s. Some show similar temperature changes while others show a smaller change, but all show a large impact on precipitation and evaporation—indicating that the presence or absence of the forest has a significant impact on the regional climate.

In addition to these local effects, deforestation has an impact on the global climate. The tropical forests represent a large store of organic carbon. When the forests are cut and burned to clear the land for agricultural use, this carbon is returned to the atmosphere as carbon dioxide, which contributes to the buildup of atmospheric greenhouse gases. Approximately half of the postindustrial increase in atmospheric CO_2 has come from tropical deforestation.

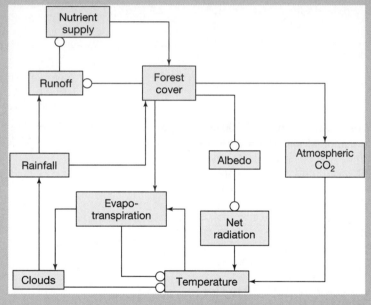

BOX FIGURE 18-1

Systems diagram linking tropical forest cover with global climate as well as local climate and hydrologic parameters.

and a forest cover return, but the forest composition is normally very different from what was there before the land was cleared.

In terms of species loss, there is yet another major difference between temperate and tropical forests. Whereas some temperate forest species are confined to relatively small geographic locations, many of the plant and animal species have a wide geographic distribution. If the forest is cleared in one area, many of the species will continue to thrive elsewhere and can eventually return to recolonize the cleared regions. Some tropical forest species also have extensive distributions, but a significant proportion of the plant and animal species live only in very small geographic areas. Over time, the overall forest structure might not change and the same fami-

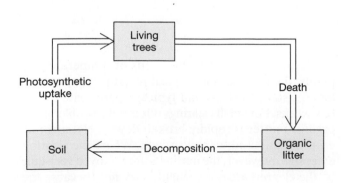

FIGURE 18-3

Nutrient flow through a forest–soil system.

lies of plant and animal life will be present, but the actual species composition can change very rapidly. Thus, when a large area of tropical forest is cleared, all the species that were found only in that area go extinct.

There were about 8 million km² of tropical rainforest in 1989—half of what existed in prehistoric times. In the late 1980s, this forest was being removed at a rate of 1.8% (140,000 km²) per year. The total global cover of tropical rainforest in 1989 was approximately equal to the area of the contiguous 48 states of the United States; at a clearance rate of 1.8%, an area of tropical forest the size of Florida was being removed each year. The Food and Agriculture Organization (FAO) of the United Nations estimates that the rate of forest clearance in the tropics was slightly lower in the 1990s, but that the difference is not statistically significant. What effect does this have on biodiversity and species loss? If a million years is the typical lifetime of a species, then the normal extinction rate is one in 1 million species per year. If we use 10 million as a conservative estimate of the number of species that live in the rainforests, then the background extinction rate should be 10 species per year. What is the actual extinction rate? No one knows for certain, but we presented Wilson's estimates earlier: 27,000 species per year due to forest clearance alone. Even with these conservative estimates, the extinction rate due to human activity is 3,000 times greater than the background rate of extinction.

REASONS FOR FOREST CLEARANCE. Why is so much forest being destroyed? The superficial response is to say that the forests are cleared for logging; to open new areas for agriculture and cattle ranching; and for large-scale development, which includes hydroelectric projects, exploitation of mineral resources, resettlement projects, and so on. Approximately 8 million hectares (80,000 km²) of forest are cleared for agriculture each year. These numbers (and the ones that follow for logging and ranching) are from the 1980s. The proportions may have changed slightly in the 1990s, but they still give a good estimate of how the bulk of the forest clearing is occurring. For agricultural production, the traditional procedure is to cut and burn the forest and then use the cleared area to grow crops, a practice known as *slash-and-burn* agriculture. The burning releases nutrients to the soil and provides fertilizer for the crops. After several seasons the nutrients are depleted, and the farmer moves on to slash and burn a new area. This is not necessarily as wasteful as it sounds: If the areas are small and the old fields are left fallow for a sufficient time (up to 30 years), enough vegetation grows back that nutrient levels are replenished and the area can be burned and cultivated again. Problems occur when large areas are cleared or when the vegetation is not given enough time to regenerate. In these cases the soil loses its fertility, crop yields decrease, and permanent soil degradation is commonly the end result.

This type of damage is normally caused by population increase; however, in some areas extensive deforestation is caused by the production of commercial crops (cash crops), which are typically used for export, by large agricultural plantations.

Apart from agriculture, approximately 50,000 km² of forest is cleared for timber, pulp, and other wood products each year. The most extensive logging is taking place in Asia and West Africa. Approximately half of what is cut is exported, with Japan taking the largest amount and the United States being the next largest importer. Wood is also cut for fuel use: For almost half the world's population, wood is the principal energy source for cooking and heating. Large areas are also being cleared for cattle ranching—approximately 20,000 km² per year in Latin America. Much of the beef once went to the United States, but most Latin American beef is now consumed in Latin America or in Europe. Cattle ranching is particularly destructive, because once large areas are cleared for pasture, soil degradation, erosion, and compaction from hooves and vehicles soon makes the land unusable, requiring more forest to be cleared.

The above description reflects the obvious transition from one land cover, the forest, to something else, as the human use of the land changes. What it hides is a multitude of social, economic, and political forces that are causing the land use to change. The particular combination of factors varies from place to place and through time, but examples might include the following:

- Growing population pressures that force an increasing number of people into the forested areas, requiring an expansion of food crops.

- Unequal division of land that results in a very small proportion of the population owning a very large fraction of the existing developed land. This arrangement forces people who do not own land to develop new areas, encroaching farther and farther into the forest.

- Deliberate resettlement programs in which governments encourage migration into the forests to reduce population pressures elsewhere. Such programs themselves may be driven by the hope of solving (or hiding) political and economic problems in other areas—poverty and unemployment in large cities, for example.

- A growing need in the industrializing nations to produce cash crops and to develop mineral resources to pay ever-increasing international debts.

Understanding the human factors that are causing forest clearance in a particular location is essential if efforts are to be made to stem the tide of tropical deforestation. Preaching conservation and offering plans for the sustainable use of forest resources do little good if the plans do not address the problems that led to the defor-estation in the first place.

Much of the world's attention has focused on the Amazon forest, the world's largest remaining area of undisturbed rainforest. In the 1980s, the Brazilian gov-ernment was very active in promoting the development of the forest. It granted 250 acres of land to people willing to migrate to the forest, provided subsidies for cattle ranch-ers, and promoted two huge development projects: the Grande Carajas project, a $62 billion industrial enterprise in eastern Brazil expected to involve an area the size of France and Great Britain; and Polonoreste, a develop-ment scheme built around a highway that has been ex-tended through the forest. The Brazilian government has since recognized some of the dangers involved with this type of development. It has halted the subsidies to cattle ranchers, and it has placed 7.5 million acres of rainforest into extractive reserves—land that is protected and where the two major rainforest products, rubber and brazil nuts, can be extracted in a sustainable fashion.

Hotspots. The tropical rainforests have the largest ex-tinction rates, but they are not the only places where species are at risk. Around the world, numerous areas have large numbers of endemic species threatened by the loss of habitat. In the late 1980s, Norman Myers of Oxford University identified 18 such areas. In each case the habi-tat had been reduced to less than 10% of its original cover or was expected to be reduced to that amount within the next few decades. This work was expanded and updated in the late 1990s. Myers and colleagues defined regions containing a distinct and identifiable assemblage of plant and animal species as "biogeographic units." Hotspots in this study were then defined as biogeographic units con-taining at least 0.5% of the world's 300,000 known vas-cular plants (i.e., at least 1,500 different plant species). In addition, to be considered a hotspot, the region must have also lost 70% or more of its primary vegetation. Once an area was defined as a hotspot, further data were also col-lected on the vertebrate species (mammals, birds, rep-tiles, and amphibians—fish were excluded because of generally poor data availability).

Twenty-five hotspots were identified—up from 18 in the earlier study. Their locations are shown in Figure 18-4 and some of their characteristics in Table 18-1. These 25 hotspots comprise only 1.4% of Earth's land surface, yet contain almost half of all the world's species of vas-cular plants and 35% of all vertebrate species (again ex-cluding fish). These regions are, at the same time, amongst the most biologically diverse terrestrial systems on the planet—and also the most threatened. The primary veg-etation in these regions originally covered approximate-ly 17.5 million km^2. This area has been reduced to a little over 2 million km^2—just 12% of the original cover. A con-siderable amount of the world's biodiversity (at least as counted by numbers of endemic plant and vertebrate spiecies) is, therefore, confined to just 2 million km^2 of

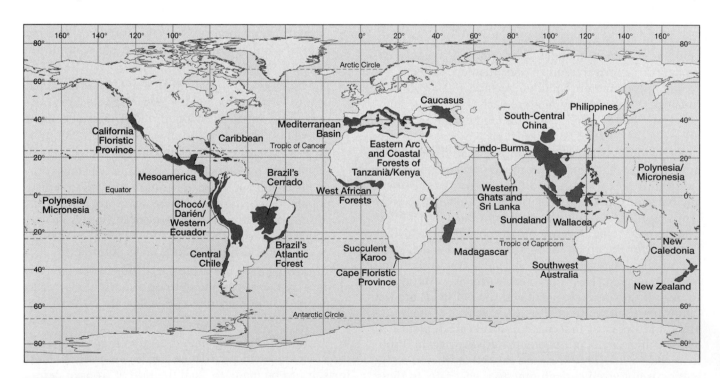

FIGURE 18-4

Hotspots of habitat loss—areas with many species that live nowhere else and that are in greatest danger of extinction as a result of human activi-ties. (From N. Myers et al. 2000. Biodiversity Hotspots for Conservation Priorities. *Nature*, 403: 853–858.)

TABLE 18-1

The 25 Hotspots identified by Norman Myers and Colleagues						
Hotspot	*Remaining primary vegetation (km²) (% of original extent)*		*Plant species*	*Endemic plants (% of the 300,000 global plants)*		*Endemic vertebrates (% of 27,298 global vertebrates)[1]*
Tropical Andes	314,500	(25.0)	45,000	20,000 (6.7)		1,567 (5.7)
Mesoamerica	231,000	(20.0)	24,000	5,000 (1.7)		1,159 (4.2)
Caribbean	29,840	(11.3)	12,000	7,000 (2.3)		779 (2.9)
Brazil's Atlantic Forest	91,930	(7.5)	20,000	8,000 (2.7)		567 (2.1)
Choc/Darien/Western Ecuador	63,000	(24.2)	9,000	2,250 (0.8)		418 (1.5)
Brazil's Cerrado	356,630	(20.0)	10,000	4,400 (1.5)		117 (0.4)
Central Chile	90,000	(30.0)	3,429	1,605 (0.5)		61 (0.2)
California Floristic Province	80,000	(24.7)	4,426	2,125 (0.7)		71 (0.3)
Madagascar[2]	59,038	(9.9)	12,000	9,704 (3.2)		771 (2.8)
Eastern Arc & Coastal Forests of Tanzania/Kenya	2,000	(6.7)	4,000	1,500 (0.5)		121 (0.4)
Western African Forests	126,500	(10.0)	9,000	2,250 (0.8)		270 (1.0)
Cape Floristic Province	18,000	(24.3)	8,200	5,682 (1.9)		53 (0.2)
Succulent Karoo	30,000	(26.8)	4,849	1,940 (0.6)		45 (0.2)
Mediterranean Basin	110,000	(4.7)	25,000	13,000 (4.3)		235 (0.9)
Caucasus	50,000	(10.0)	6,300	1,600 (0.5)		59 (0.2)
Sundaland	125,000	(7.8)	25,000	15,000 (5.0)		701 (2.6)
Wallacea	52,020	(15)	10,000	1,500 (0.5)		529 (1.9)
Philippines	9,023	(3.0)	7,620	5,832 (1.9)		518 (1.9)
Indo-Burma	100,000	(4.9)	13,500	7,000 (2.3)		528 (1.9)
South-Central China	64,000	(8.0)	12,000	3,500 (1.2)		178 (0.7)
Western Ghats/Sri Lanka	12,450	(6.8)	4,780	2,180 (0.7)		355 (1.3)
SW Australia	33,336	(10.8)	5,469	4,331 (1.4)		100 (0.4)
New Caledonia	5,200	(28.0)	3,332	2,551 (0.9)		84 (0.3)
New Zealand	59,400	(22.0)	2,300	1,865 (0.6)		136 (0.5)
Polynesia/Micronesia	10,024	(21.8)	6,557	3,334 (1.1)		223 (0.8)
Totals	2,122,891	(12.2)	*	133,149 (44)		9,645 (35.0)

[1]Excludes fish.

[2]Madagascar includes nearby islands of Mauritius, Reunion, Seychelles, and Comores.

*Totals cannot be calculated because of overlap between hotspots.

earth's surface. For comparison, this is only a little more than one-fifth the size of the United States.

Myers and colleagues analyzed the biodiversity data in a number of ways and identify the regions most threatened if we continue present rates of habitat destruction. Madagascar, the Philippines, and Sunderland rise to the top of the list. Also in the most threatened category are Brazil's Atlantic forest, the Caribbean, the tropical Andes, and the Mediterranean basin. Organizations such as Conservation International *(http://www.conservation.org/xp/CIWEB/home)* are using this information to try and target conservation funding. Myers' assessment is based on biodiversity, size of area, and degree of degradation. The analysis could be expanded in other socioeconomic dimensions to assess existing and potential socioeconomic pressures that result in habitat destruction in these regions. In the context of this book, we can also look at this with respect to other global pressures, such as global

warming. Sixteen hotspots are in the tropics, while none are in the polar regions. This is the reverse of the potential global warming changes described in Chapter 16 where the magnitude of the temperature change is greatest at high latitudes and least in the tropics. The remaining hotspots are predominantly mediterranean or savanna regions that are still very diverse, but also have the potential to see some significant climate change during this century. The *IPCC Climate Change 2001: Impacts, Adaptation and Vulnerability* report specifically focuses on the Cape Floristic Province and the Succulent Karoo in South Africa as examples of mediterranean and savanna hotspots that, beyond all of the existing threats to biodiversity, are further threatened as these geographic settings limit the ability of species to migrate as climate changes.

While these hotspots represent one attempt to identify biologically diverse regions that are under pressure from human activities, the list does not end here; many more places could be included. This discussion of hotspots neglects a huge area of the planet: the oceans. However, we know much less about life in the oceans than we do about the terrestrial biota. One ocean ecosystem that is receiving considerable attention at present is the coral reef. Despite its appearance, coral is actually an animal. Found in shallow tropical seas, coral reefs are the most productive and diverse of ocean ecosystems. The reef is dynamic, continuously changing, and very fragile. It is easily damaged by shifting sand and storms; left alone, it normally recovers quickly. However, most coral reefs are located close to shore, so any natural damage is now augmented by human activities, notably pollution, mining for coral rock, collection of coral specimens, overfishing, and damage from accidental grounding by ships.

Coral reefs tolerate only a narrow range of ocean temperatures between 21°C and 29°C—growth rates are very slow at temperatures higher and lower than this. They also require sunlight for their symbiotic algae (*zooxanphellae*), so while they have been found at depths of 90 meters (300 feet), they grow better at depths shallower than 18–27 meters (60–90 feet). Consequently, coral

reefs tend to occupy a very narrow range of coastal locations in the tropics (Figure 18-5). It is estimated that coral reefs cover a little over 280,000 km². However, they are found off the coasts of over 100 different countries and they are thought to contain at least 25% of all marine species, including 700 species of coral and over 4,000 different fish species.

While coral reefs all around the world are directly affected by human activities such as those listed above, possibly the greatest impact is likely to be an indirect effect of climate change. Increased ocean temperatures result in episodes of coral bleaching on many reefs. Bleaching is caused by the loss of the symbiotic zooxanthellae (single-celled algae that live in the tissue of the coral animals). This loss is considered to be a general response of the coral to stress. Stress can result from numerous causes, including excessive hot or cold temperatures, chemical pollution, or dilution by freshwater. Even a 1°C temperature rise, if it persists for a few days, can cause the coral to begin to expel its zooxanthellae. In recent years there have been documented occurrences of coral bleaching associated with ENSO events (Chapter 15), and it is thought that an increased incidence of ENSO events may be one

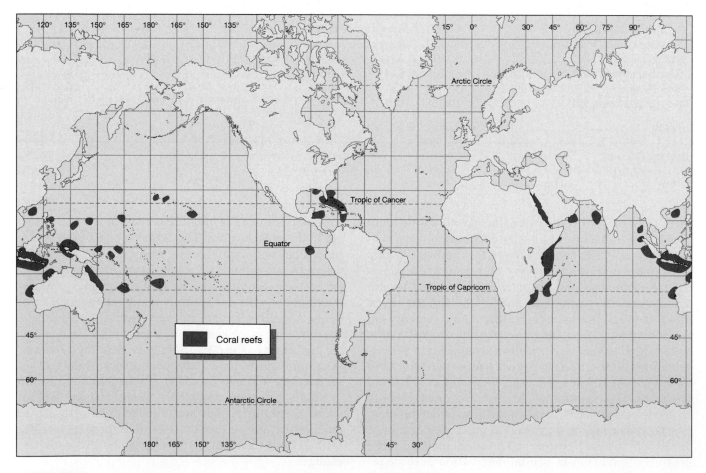

FIGURE 18-5

Distribution of tropical coral reefs.

outcome of global warming. At the same time, one of the more certain impacts of global warming will also be a general increase in surface ocean temperatures. There is a further direct effect of increasing concentrations of atmospheric carbon dioxide. As we saw in Chapter 16, additions of CO_2 to the atmosphere causes a reduction in the carbonate ion concentration of seawater. Carbonate and calcium ions are combined by corals as they build their skeletons. The reduction in carbonate ion concentration is another stress that is applied globally to all coral reefs. Taken together (increasing ocean temperatures, possible increases in ENSO events, and reduced carbonate ion concentrations), there is growing concern that damage to coral reefs may be a significant and early impact of global warming on the world's ecosystems.

Studies in the late 1990s suggested that almost 60% of the world's coral reefs may be threatened by human activity. A recent study led by Callum Roberts of Harvard University identified 10 marine biodiversity hotspots analogous to those described earlier for terrestrial systems. These are all coral reef systems covering almost a quarter of the world's coral reefs and 8 out of 10 of these reefs are adjacent to terrestrial biodiversity hotspots. Many areas have attempted to reverse the damage by establishing marine parks and by controlling the human activities that take place on or near the reefs. These measures help prevent accidental damage to the reef but offer little protection from the more pervasive damage of environmental pollution and global warming.

Why We Should Care about Biodiversity

You probably live in or close to a city; approximately half the world's population lives in urban areas. Although you may have seen one city building knocked down to be replaced by another or an open field turned into a shopping mall, you probably do not have much firsthand experience in large-scale habitat destruction. Unless you are a tropical ecologist or a scuba diver, the closest you may ever get to the backwaters of the Amazon or to a coral reef is a nature program on television. In the 1600s and 1700s, forests were cut down in Europe and in what would become the United States to power steam engines and to expand agriculture to support growing populations. You could say that, in doing so, we were able to support a workforce and develop the technology that eventually enabled us to see the remaining forests or the reefs on those television sets. You might also argue that if the industrializing nations need or want to do the same now, or if Californians prefer suburbs to chaparral, that is their decision and why should we get excited about it? These are valid questions. We are going to try to answer them in

several ways—first by looking at why individual species may be important, and then by stepping back and looking at the importance of biodiversity to the Earth system as a whole.

Instrumental and Intrinsic Values of Species

Ecologists tend to discuss the value of a species in terms of its *instrumental* or *intrinsic value*. A species' **instrumental value** is the degree to which the existence of the species benefits another species in some way. We are normally the other species, and the value or benefit is typically economic. A species' **intrinsic value,** conversely, is its value for its own sake, regardless of whether we benefit by it or not. Some argue, for example, that all species and all individual organisms have the same rights to exist that humans do—that humans have no greater value than any other species. Others argue that humans are unique: We dominate over nature, nature is ours to exploit, and other species have no individual rights *per se*. Still others argue that this very uniqueness in fact imposes a special responsibility. We are capable of moral and ethical judgments, and these should include concern for other species. Assigning an economic value is relatively easy; assigning intrinsic value, however, is a moral and ethical issue that is more subjective. The assignment of intrinsic value varies from society to society and from individual to individual, according to their outlook on nature.

Instrumental Value. We can look at instrumental value from several perspectives. We can view species as being resources:

- *Medicine.* Of all pharmaceuticals produced in the United States, 25% contain ingredients originally derived from native plants. The often-quoted example is a plant called the rosy periwinkle, which grows only on Madagascar and fortunately was not a victim of the large species loss that has occurred on the island. In the 1960s scientists extracted two substances from the plant: vincristine and vinblastine, which became our primary defense against childhood leukemia and Hodgkin's disease. Before the discovery of these drugs, childhood leukemia was almost always fatal; since the 1960s, with the use of vincristine, there is now a 95% chance of remission. These two drugs, developed from a single plant species, now represent a $100 million-a-year industry—just a small part of the $8 billion a year derived from pharmaceuticals based on "natural" products. How many other rosy periwinkles are waiting to be discovered, and how many disappear every year without us even knowing they existed?

- *Scientific value.* Although a considerable amount of discovery takes place in scientific laboratories, much of what we know about evolution and ecology, and our understanding of how natural systems work, has been derived from the study of species and ecosystems in their natural state. However, our understanding of these systems and how they interact, particularly at the global scale, is far from complete. At the present rate of habitat destruction, we are in danger of wiping out whole ecosystems without ever understanding how they operate or what role they may have played in regional or larger-scale ecology.

- *Recreational and aesthetic value.* It is difficult to assign a monetary amount to the scientific value of species and ecosystems. However, it is much easier to assign such a number to their recreational value. Each year, more than 50 million people in the United States spend a total of $38 billion on recreational hunting or fishing. There are also the millions of people who escape the cities to spend their weekends or vacations walking, climbing, and photographing the mountains, forests, and beaches of the more "natural" parts of the world. They may sail on rivers, lakes, or the ocean, sit on shore and admire the view, or simply drive by and look. The continuing popularity of television nature shows also attests to the aesthetic value we place on the natural world, whether we ever get to see it in person or not.

- *Commerce.* There are both direct and indirect commercial benefits to be gained from the natural environment. Direct commercial interests revolve around activities such as logging, commercial fishing, and the collection and sale of exotic plant and animal species. Indirect benefits arise from recreational activities that support commercial interests, for example, in travel, accommodations, restaurants, sporting goods stores, and so on. The business of arranging tours to remote locations, such as trekking in the Himalayas and hunting and photographic safaris in East Africa, is thriving. There is also now a growing business in ecotourism, where tourists visit places specifically to observe rare or endangered species or ecosystems.

- *Agriculture, forestry, aquaculture, and animal husbandry.* Those of us who live in the developed world, in particular, tend to regard food production as simply one component of our industrialized society; we see little connection between miles of uniform crops and mechanized production, and the grasslands or forests that would otherwise be there. In reality, there is a very close connection, and the existence and maintenance of our agricultural food supply is much more dependent on natural species than many people realize. We will return to this connection when we look at the value of species from the perspective of a global system.

The Loss of Biodiversity

Now we will step back from individual species and consider the whole diversity of life. What are the consequences involved in the loss of that biodiversity?

Biodiversity and Food Supply. Biodiversity is of critical importance to the modern agricultural techniques we employ to feed our ever-growing global population. Most people are well aware of many of the local environmental problems that can arise from our attempts to increase productivity: soil degradation and erosion, salinization, groundwater pollution from fertilizer and pesticides, and so on. From our larger-scale or global perspective, however, a more serious problem is developing with respect to the basic ingredients we need for modern agriculture—the seeds themselves.

Early farmers and agronomists noted that some varieties of a crop are more productive, more disease-resistant, or more drought-tolerant than others. The selective use of different strains or varieties resulted in increased productivity. Today biotechnology and gene splicing can produce new crop varieties to order. At first glance it is hard to see the problem: With modern techniques we can produce seeds with high yields and *uniform* characteristics. Crops can be planted, harvested, processed, packaged, and transported all with minimum human intervention—just what we need to keep the world fed. Unfortunately, we never quite manage to feed the world. However, the lack of adequate food supplies is a result of social, political, and economic factors that determine the global distribution of resources, not an inability to produce enough food for all. The potential problem with modern agriculture is not that it is not productive enough but that it is *uniform.*

We have already seen that the ability of an ecosystem to resist change is partly determined by its biodiversity. The same goes for individual species. A food crop such as corn or rice in its *natural* state is likely to consist of many different strains growing together. Some are more productive than others, some more drought-resistant; some would be resistant to one pest, others more resistant to a different pest. In any year, whatever the conditions, whatever pests flourish, there is a good chance that some corn or rice will grow, whichever strain is best adapted to that particular set of conditions.

Over time the environmental conditions change, and the pests and diseases that attack a particular plant evolve. At the same time the crop plant also changes, and new strains evolve that are resistant to the new conditions. Every strain carries its own particular genes. *Genes* are units of heredity; they control the life processes of all cells and contain the information that controls how the genes work—in other words, the information that produces exactly that strain of corn or rice. With thousands of subtly different strains, there is a vast natural reservoir of genetic material available that ensures the long-term health and survival of the species. Vitality depends on diversity.

However, the whole approach to modern agriculture is uniformity. The object is to reduce diversity as much as possible, to make a high-yield variety of a crop that is resistant to drought and a particular assemblage of pests and diseases. This variety flourishes for a few years until a new strain of pest or disease evolves, against which it has no defense. At that point it is back to the drawing board, or the biotechnology lab, where the scientists mix-and-match until they produce a corn variety that is resistant to the new conditions. Everything is fine for a few more years. So, where does the genetic mix-and-match material come from? It comes from very specialized seed banks that have been established around the world. We have not actually destroyed the diversity of our food crops, we have simply concentrated it in a few locations. The International Storage Center for Rice Genes in the Philippines, for example, keeps more than 86,000 varieties of rice. The geneticist needing some new rice strains to work with simply goes to the seed bank and makes a withdrawal.

But the genes in the seed bank are, in a sense, frozen in time. They represent the state of genetic diversity that the rice had achieved when the bank was established. The diseases and pests, conversely, do not live in seed banks—they exist in the wild and mutate and evolve, continuously producing new strains. In some cases they produce something that the seed bank cannot deal with. Then our only recourse is to return to the wild. The geneticist goes back to the "genetic home" of the species and searches for a wild variety that has the traits needed to solve this particular problem. However, if the natural habitat is destroyed, the only repository of genetic diversity is in seed banks, which are more vulnerable to accidents (such as fire), war, or sabotage.

Despite our modern techniques, despite seed banks and biotechnology, ultimately the health of our food crops depends on the continued survival of a diverse population of wild strains. This is where our real problem emerges.

Humans make use of a huge number of different food crops. Yet, the bulk of the world's food supply and animal feed comes from just 130 species of plants. The ultimate storehouses of the genetic material we need to ensure long-term health and high productivity are restricted to only 12 locations around the world. These 12 centers of genetic diversity, called *Vavilovian centers* in honor of Russian geneticist N. I. Vavilov, who first described them, are shown in Figure 18-6. Each of these centers is locat-

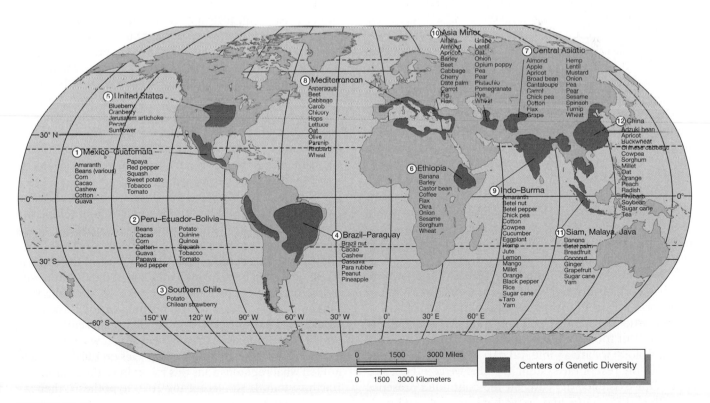

FIGURE 18-6

The 12 centers of genetic diversity. All of the genetic diversity essential for maintaining the world's food supply is limited to just these 12 areas. (After S. C. Witt. 1985. *BriefBook: Biotechnology and Genetic Diversity*. San Francisco: CSI.)

ed in an area where population pressures or development are placing an increasing stress on the existing natural habitat.

Just how vulnerable we might be is illustrated by an episode in the late 1970s that threatened the rice crop of Southeast Asia. The rice, a basic food source for hundreds of millions of people in the region, was threatened by a disease called grassy stunt virus. Scientists searched through the gene banks—through 47,000 varieties of rice—and eventually found the genetic material that would allow the crop to resist the disease, in a single wild species from a valley in India. Soon after the retrieval of the material, the valley was flooded in the construction of a new hydroelectric project. Less fortunate were the Irish in the mid-1800s. It is thought that potatoes were cultivated by the Incas as early as 400 B.C. They were brought to Europe by the Spanish in the mid-16th century. Potatoes eventually became a staple crop for the Irish. The potato blight (a fungus, *Phytophthora infestans*) arrived at the end of the growing season in 1845. It had little effect that year, but the next year it completely devastated the potato crop. Together with other political, economic, and social factors that all played a role in the ensuing famine, the collapse of the farm economy killed over 1 million people and eventually forced the emigration of a further 1.5 million. Ireland lost over a quarter of its 8 million population. The blight is still present in various parts of the world, including the U.S., where, in fact, there have been recent concerns that strains of the fungus have been developing that show increased resistance to the fungicides currently used to protect the crops.

Biodiversity and Stability.   As an extreme case of loss of biodiversity, imagine what would happen if we were suddenly to cut down all the world's forests. Would we suffocate? No, we would not quickly run out of oxygen, even though green plants are responsible for the buildup of oxygen that gave us the atmosphere we breathe today. Even if we removed all the photosynthesizing organisms on the land and the phytoplankton in the sea, it would still take us millions of years to suffocate. (You can calculate just how long in "Critical-Thinking" Problem 3 at the end of this chapter.)

We would not suffocate if we lost all the photosynthesizing organisms, but would we starve? Yes. Without these primary producers at the base of the food chain, all higher-order consumers, including humans, would quickly run out of food. We are not separate from the rest of Earth's biota; we are an integral part of the Earth system. Without the rest of the system in place, we could not survive. Our existence depends on the continued presence of a flourishing biota, and, as we have seen, diversity enhances the health and vitality of the biota. The greater the number of species in an ecosystem, the healthier the ecosystem will be. The more diverse the ecosystem, the

greater chance it can survive disruption: If one species is lost from a diverse ecosystem, some other species is likely to continue in a similar role, performing a similar function in the ecosystem. For the ecosystem to be healthy or productive, how diverse does it have to be?

Studies of productivity in different ecosystems have shown that productivity increases as biodiversity increases. Other studies have shown that the resilience of an ecosystem—how well it is able to withstand different types of stress—also increases as biodiversity increases. Both the productivity and the stability of an ecosystem are strongly affected by biodiversity. However, these studies also suggest that beyond a certain level of biodiversity, there is little further increase in either productivity or stability. Although we can document this threshold for individual ecosystems in particular locations, however, we still do not fully understand what this means at the global scale.

Beyond the moral or ethical issue and beyond the potential medical or economic value of individual species, we can still ask the following question: If we kill off a large number of the species existing on Earth, will it matter?

At present there are two very different views on the importance of biodiversity. One, put forward by Stanford University ecologists Anne and Paul Ehrlich, likens biodiversity to the rivets on an airplane. Each species has a small but important role to play in the overall functioning of the ecosystem, just like the rivets on the airplane. If you remove the rivets (or species) one by one, sooner or later the stress on the system becomes so great that it fails: The airplane falls out of the sky, or the ecosystem collapses. The second hypothesis was proposed by Brian Walker, an ecologist for the Australian Commonwealth Scientific and Industrial Research Organization. He suggested that, in fact, most species are superfluous and that the system is maintained by a few keystone species. In this case, loss of species is not a problem as long as you do not lose the keystone species.

Where does this leave us? If we wiped out all tropical rainforests but replanted an equal area of temperate forest, while the new forest continues to expand it would release oxygen and take up CO_2. If we were to plant a large enough area, we could replace the biomass of the tropical forests and even begin to take up some of the CO_2 released by the burning of fossil fuels. (To take up all of the CO_2 that could be released by the burning of fossil fuels, we would need to more than double the organic carbon content of living biomass and soils.) However, we would have exterminated more than half the species on Earth. We cannot tell what repercussions this might have throughout the Earth system. If we accept the rivet hypothesis, then we might imagine a loss of half the species on Earth as catastrophic. If we assume that only the keystone species are really important or that, for any ecosystem, there will be redundancy in the roles played by different species, then we can accept the loss of some species. Multiple species of insects, for example, might fill similar niches and interact with the rest of the ecosystem in similar ways. Wipe out one, and the rest of the ecosystem might never notice. If we could pick and choose which species should live and which should die, then maybe we could reduce the number of species in the world by half, and the rest of the Earth system would continue along quite happily.

But extinction does not happen that way. As the human population grows, so do our demands on natural resources. The loss of habitat caused by our building of cities, extraction of minerals, logging of forests, and conversion of forests and grasslands to agriculture and grazing lands kills many species and can destroy whole ecosystems. The destruction is not selective; neither are the effects. From a long-term perspective (Chapter 13), we can judge that the Earth system is very resilient. It can recover from very large perturbations, and no doubt Earth will recover from this one, although it may take tens of millions of years. In terms of individual species, it is a very different matter. Although past extinction events were different in cause and structure to the present one, it might be worth considering that whenever such a large extinction event occurred in the past, the species at the top of the food chain before the event were no longer there once it was all over. Dinosaurs ruled the Mesozoic Earth and are now extinct. Humans could be next.

We do not know enough about how most ecosystems operate or about how important different ecosystems are for the overall global system. For purely pragmatic reasons, it makes sense to slow the rate of habitat destruction and the extinction of species. Pragmatism aside, we pose the following ethical question: Does any species have the right to exterminate another?

Chapter Summary

1. A mass extinction episode as large as those that have occurred periodically throughout Earth's history is taking place today. It is a result of the spread of human populations and the destruction of much of the world's most productive natural habitats.

2. Much of this species loss is taking place in the tropical rainforests of the world, which are also the most diverse of the terrestrial ecosystems.
   a. The tropical rainforests are thought to contain three-quarters of all the living things on Earth and two-thirds of all plant and animal species.
   b. Deforestation in the tropics is wiping out species faster than they can be catalogued and much faster than they can be studied. Every day, species are going extinct that we do not even know exist.

3. The instrumental value of a species can be viewed in different ways—for example, their potential value as pharmaceuticals, their recreational value, their scientific value, or their commercial value.
   a. The diversity of species has value in terms of the productivity and stability of ecosystems and plays an important role in the stability, or health, of the Earth system as a whole.
   b. The destruction of tropical forests has other impacts besides the reduction of global biodiversity. It also results in large changes in local climate and the hydrologic cycle, and, through the release of CO_2, it has an impact on the global-scale climate.

4. Biodiversity, or the loss of biodiversity, is of critical importance to modern agricultural production. Current practices, particularly in industrialized nations, are to increase productivity through increased specialization and the use of a limited number of selectively developed varieties of crops.
   a. When only a few varieties of a particular species exist in a location, they become particularly vulnerable to new strains of diseases or pests that may evolve.
   b. Diversity is essential for maintaining the long-term health and survival of a species, but modern agriculture is dependent on reducing that diversity.
   c. In the case of particular food crops, we compromise by building large reservoirs of genetic diversity in specialized seed banks.
   d. As new diseases arise or as new pests evolve, the genetic material contained in these seed banks might not be sufficient to develop a resistant variety of the crop. We then have to turn to the region in the world where the plant first evolved (the genetic home of the plant) to look for a wild strain that might be resistant to that particular threat.
   e. The vast majority of the world's food and animal feed crops come from just 12 regions, each located in an area where population pressures or development are placing an increasing stress on the existing natural habitat.

Key Terms

endemic species
hot spots

instrumental value
intrinsic value

keystone species
speciation

Review Questions

1. Explain why it is difficult to give an accurate count of the number of species that go extinct each year.
2. Use a diagram to help explain the relationships among ecosystems, communities, and species.
3. What is meant by the term *habitat*? Explain its importance to issues of biodiversity and extinction.
4. What is meant by the term *keystone species?* Explain why it is important to issues of biodiversity and extinction.
5. Where does the greatest rate of species loss occur today?
6. If a large area of temperate forest is cleared, it can commonly grow back fairly rapidly to something that closely resembles the original forest. If a large area of tropical forest is cleared, regrowth is slow and the original forest cover may never return. Explain why these forests respond differently to clearing.
7. Describe three consequences of tropical deforestation other than loss of biodiversity.
8. List four social and economic factors that contribute to tropical deforestation.
9. What characterizes the 24 hotspots identified by Norman Myers?
10. What is meant by the instrumental and intrinsic values of species?
11. Explain why lack of diversity in food crops might present problems for the world food supply.
12. Why is there a relationship between biodiversity and the long-term health of an ecosystem?
13. Explain the difference between the Ehrlichs' rivet theory and Walker's theory of the importance of biodiversity.

Critical-Thinking Problems

1. Figure 15-2 presents a systems diagram of the feedbacks involving boreal forest cover, albedo, temperatures, sea ice, and the oceans. We used this diagram to show that it is possible for the northern boreal forest to have a significant impact on the larger-scale climate. Using the information you now have about the possible impacts of anthropogenically induced greenhouse climate change, expand on this diagram and discuss the implications in terms of climate and forest cover.
2. Consider the following set of statements for a situation in the tropical forest where forest is cleared for crops:

   - Forest clearance requires access to the forest (roads).
   - An increasing population requires access to services and commodities (a town or city).
   - Settlements require roads.
   - Construction and forest clearance are easier (may increase) if roads are present.
   - The presence of roads and settlements attracts more people into the region.
   - Forest clearings rapidly lose productivity, requiring more clearing.
   - Clearing results in loss of biodiversity.

   Take these statements as a minimum; you might think of some other conditions or relationships that could also be included. Construct a systems diagram showing how the statements all interact. Describe what the diagram shows. Are any feedback loops present? If so, are they positive or negative? Is this a useful way of looking at some of the forces that promote deforestation? Is anything missing that you think should also be taken into account?

3. What would happen to atmospheric oxygen if photosynthesis were to shut off suddenly? Photosynthesis is the source of most of the O_2 in the atmosphere, so it is reasonable to guess that O_2 would decline. Analyze how it would do so by performing the following calculations:
   a. Forests and soils together contain about 2200 Gton of carbon. The amount of carbon stored in the living marine biomass is negligible in comparison. The atmosphere contains 3.8×10^{19} mol of O_2. By what percentage would atmospheric O_2 decrease if Earth's biota were to die suddenly and if the forest and soil organic carbon were to be oxidized? (Hint: Remember that in the decay of organic matter, 1 mol of carbon reacts with 1 mol of O_2.)
   b. Approximately how long would this process take? Assume the decay proceeded at the modern rate of 30 Gton(C)/yr.
   c. The remaining atmospheric O_2 would have to be removed by the weathering of kerogen and other reduced minerals (iron and sulfides) in rocks. Kerogen is weathered at a rate of 0.05 Gton(C)/yr. Express this rate in terms of moles of carbon per year. If the oxidation of iron and sulfides were to use up oxygen at this same rate, how long would it take for atmospheric O_2 to disappear?

Further Reading

General

Baskin, Y. 1997. *The Work of Nature: How the Diversity of Life Sustains Us.* Washington, D.C.: Island Press.
Eldredge, N. 1998. *Life in the Balance: Humanity and the Biodiversity Crisis.* Princeton, N.J.: Princeton University Press.
Wilson, E. O. 1992. *The Diversity of Life.* New York: W. W. Norton.

Advanced

Balich, M. J., E. Elizabetsky, and S. A. Laird (eds.). 1996. *Medical Resources of the Tropical Forest.* New York: Columbia University Press.

Climate Stability on Earth and Earthlike Planets

Key Questions

- How will Earth's climate evolve in the distant future as the Sun continues to brighten?
- Why are Venus and Mars so different from Earth?
- What determines the width of the habitable zone around our Sun and other stars?
- What are the chances that life exists elsewhere in the universe, and can we detect it if it is there?

a result of the planet's small size, which prevents it from recycling CO_2 back into its atmosphere rather than its distance from the Sun. Habitable, Earthlike planets may exist around other nearby stars. Some of these planets may harbor life; of those, a few may harbor intelligent beings like ourselves. Space-based visible or infrared telescopes and ground-based radio telescopes may eventually tell us whether extraterrestrial life does indeed exist.

Chapter Overview

The Sun will continue to brighten in the future, as it has done throughout Earth's past. As Earth's climate warms, atmospheric CO_2 concentrations should decline in response to the negative feedback provided by the carbonate–silicate cycle. Eventually, this decline may pose a problem for life, because there will not be enough CO_2 in the atmosphere to support photosynthesis; thus, the life-span of Earth's biota is limited.

Although the CO_2–climate feedback may have stabilized Earth's climate over billions of years, it evidently did not operate on our neighboring planets, Venus and Mars. Venus is clearly outside the habitable zone around the Sun—that is, the region in which a planet can support liquid water at its surface. Mars appears to be outside the habitable zone as well, though perhaps as

Introduction

A recurrent theme of ours has been the interaction between atmospheric CO_2 levels and climate. On long time scales, we have argued that these variables are connected by way of a strong negative feedback loop. A warmer climate increases the amount of precipitation, which, in turn, increases the rates at which silicate rocks are weathered and carbonate sediments are formed. As a result, the atmospheric CO_2 concentration drops, thereby decreasing the magnitude of the greenhouse effect and causing the climate to cool. The negative feedback produced by this mechanism was probably a major reason why Earth's climate has remained relatively stable for most of the past 3.5 billion years (the time during which we know that life has existed).

Climate, of course, is not absolutely stable. On shorter time scales, Earth's climate exhibits substantial fluctuations associated with glacial–interglacial cycles, and there do seem to have been relatively brief Snowball Earth episodes when climate stabilization has temporarily failed. But Earth has recovered from these episodes by way of the CO_2–climate feedback, and life has made it through essentially unscathed. The reason is that liquid water has always been present either at, or within a few meters of, the planet's surface. The presence of liquid water is the fundamental criterion for a habitable planet, because all known organisms require liquid water during at least part of their life cycle. And most advanced organisms, such as humans, require liquid water on a day-to-day basis.

Could life exist elsewhere in the universe? To answer this question, we need to determine the width of the liquid water *habitable zone* around the Sun and other stars. We know that within our own solar system only one planet, Earth, has liquid water at its surface. As we pointed out in Chapter 3, Venus is much too hot to have liquid water, and Mars is currently much too cold. Liquid water could be present on Mars several kilometers beneath its surface, where heat from Mars' interior could help keep temperatures above the freezing point. Jupiter's moon Europa may also have liquid water beneath its icy crust. In this case, the heat source is thought to be *tidal flexing* as Europa follows its slightly eccentric orbit around Jupiter. (Jupiter's gravity raises tides on Europa, just as the Moon raises tides on Earth. The same side of Europa always faces Jupiter, so these tides do not move across the surface as they do on Earth. However, because Europa's orbit is noncircular, it wobbles back and forth a little; this wobble causes heat to be released within Europa's crust.) But both of these environments are highly specialized. If life does exist in either place, it will only be detected by going there and drilling down beneath the planet's (or moon's) surface. Furthermore, the chances for finding anything larger than microbes on either Mars or Europa are extremely small. If we hope to find advanced life, we need to look for a planet not too different from Earth. There are no such other planets in our own solar system, but there may be around other stars. If we want to examine such planets remotely, which is all that seems feasible at the current time, we need to look for planets that have liquid water, and life, present at their surfaces.

What went wrong with Venus and Mars in our own solar system? Why were their climates not stabilized by the same silicate-weathering feedback that appears to have operated on Earth? In this chapter, we attempt to answer that question and to determine its implications for the possibility of life on planets around other stars. But first let us consider what the silicate–weathering feedback may mean to Earth's own climatic future.

Climate Evolution in the Distant Future

As we noted in earlier chapters, the Sun has gotten brighter throughout its history as a result of the gradual conversion of hydrogen to helium. This process is ongoing, so the Sun is expected to continue to get brighter in the future. The Sun's luminosity is currently increasing by about 1% every 100 million years.

How long will Earth continue to be habitable? We know that the CO_2–weathering feedback must break down at some point. The Sun is currently a **main-sequence star**—that is, a normal, middle-aged star that shines by "burning" hydrogen. (See the Box "A Closer Look: Main-Sequence Stars and the Hertzsprung–Russell Diagram" in Chapter 10.) About 5 billion years from now, the Sun will reach the end of its main-sequence lifetime. Over the next few hundreds of millions of years, it will expand into a **red giant**, a large, reddish star that is intermediate in the post-main-sequence phase of its evolution. Its luminosity will increase by a factor of more than 1000, and its outer envelope will extend to somewhere between the orbits of Mercury and Venus. All life on Earth will, of course, have been fried to a crisp long before this point is reached. After that event, the Sun will contract and become a **white dwarf**, a small (Earth-sized), hot star that is at the final stage of evolution of a star such as our Sun. Earth itself will then become very cold, but this change will not matter to anyone because all life will have been extinguished during the prior, red-giant phase of the Sun's evolution.

In actuality, Earth's climate regulation system could begin to fail long before the Sun's red-giant stage is reached. Figure 19-1 shows the result of a computer simulation of Earth's climate over the next 1.6 billion years. During this time, solar luminosity should increase by about 16% (Figure 19-1a). Earth's surface temperature, T_S, is predicted to increase slowly at first, then more rapidly starting about 1 billion years away (Figure 19-1b). By 1.6 billion years from now, T_S could exceed 100°C (212°F), and Earth's surface could become uninhabitable for all but the most thermophilic (heat-loving) of microbes. Furthermore, the computer models indicate that another, irreversible change will take place, starting in about 1 billion years when solar luminosity has increased by about 10%: The stratosphere will become wet, and water will be lost rapidly by photodissociation followed by the escape of hydrogen to space. Eventually, Earth should evolve into a planet similar to Venus.

The CO_2 Compensation Point

Earth's biota could run into trouble long before the temperature becomes high and the oceans disappear. The rea-

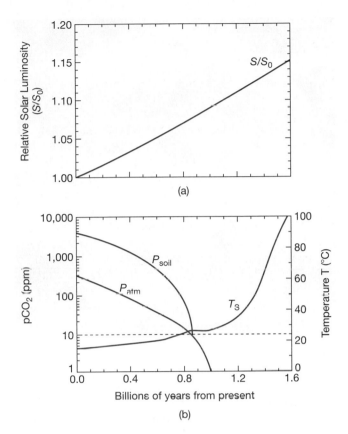

FIGURE 19-1

Long-term projections of (a) solar luminosity and (b) surface temperature (T_s) and CO_2. Here S_0 is the present solar flux and P_{atm} and P_{soil} represent the CO_2 concentrations in the atmosphere and in the soil, respectively. (Reprinted with permission from K. Caldeira and J.F. Kasting. *Nature* 300:721–723, 1992. Copyright 1992 Macmillan Magazines, Ltd.)

son that T_s remains low for the next billion years in this simulation is that atmospheric CO_2 is predicted to decrease rather rapidly (Figure 19-1b). (The current fossil-fuel CO_2 pulse, which will last for only a million years or so, is neglected in this calculation.) The calculation assumes that the silicate weathering rate is controlled by the partial pressure of CO_2 in the soil, P_{soil}, which is higher than that in the atmosphere, P_{atm}, as a result of the microbial decomposition of organic matter in soils (Chapter 8).

This pumping is assumed to be provided by C_4 plants, which are capable of photosynthesizing down to CO_2 levels of about 10 ppm (Chapter 9). The critical concentration of CO_2 required for photosynthesis to occur is called the **CO_2 compensation point**. Below this level, the rate of photosynthesis is slower than the rate of **photorespiration** (respiration induced by the absorption of sunlight), so the plants cannot grow. In the simulation shown in Figure 19-1b, C_4 plants are predicted to become extinct about 900 million years from now, when the atmospheric CO_2 concentration falls below the CO_2 compensation point.

At this time, P_{soil} becomes equal to P_{atm} because the biological pumping action disappears.

The C_3 plants, which fix carbon by means of the Calvin cycle (Chapter 9) and constitute about 95% of the current terrestrial vegetation, could go extinct even earlier. Their CO_2 compensation point is about 150 ppm, approximately half of today's atmospheric concentration. A similar simulation (not shown) in which C_3 plants control the CO_2 partial pressure in soils indicates that C_3 plants should survive for only about another 500 million years. As their CO_2 compensation point is approached, they should gradually be replaced by C_4 plants (which photosynthesize at much lower CO_2 concentrations than do C_3 plants) or by new species of plants that evolve mechanisms for photosynthesizing at low atmospheric CO_2 concentrations.

Although the predicted demise of the biota is not exactly imminent, the results of this simulation are in some sense disturbing. It took almost 4 billion years to evolve multicellular life and another half billion years to produce humans. After all of this evolutionary hard work, less than 1 billion years may remain before the planet becomes uninhabitable. Evidently, even a planet as well-suited for life as Earth is cannot be expected to harbor advanced life for more than a small fraction of its lifetime.

Climate Evolution on Venus and Mars

Why are the climates of Venus and Mars so different from that of Earth? Both these planets probably started out with the necessary ingredients (water, carbon, and silicate rocks) to allow the carbonate–silicate cycle to operate. Thus, the CO_2–climate feedback mechanism could in principle have operated on these planets as well. But this mechanism was clearly incapable of keeping their surface temperatures within the liquid-water regime.

The Runaway Greenhouse on Venus

The answer to our question about the climates of Venus and Mars is that even systems containing negative feedback loops commonly have limits beyond which they become unstable (see Chapter 2). In the case of Venus, the negative feedback loop between the surface temperature and the outgoing infrared flux (see Figure 3-22) breaks down when the solar heating is too strong. Venus orbits the Sun at about 0.72 AU, so the inverse-square law (see Chapter 3) shows that the solar flux at the surface of Venus is about 1.9 ($=1/0.72$)^2 times that at Earth's surface. Even early in the solar system's history when the Sun was 25–30% dimmer, the solar flux at Venus' orbit would still have been some 40% higher than the present flux at Earth's orbit.

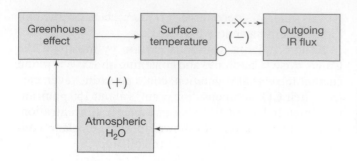

FIGURE 19-2

Systems diagram illustrating the runaway greenhouse on Venus.

FIGURE 19-3

[See color section] The surface of Venus, as observed by radar from the Magellan spacecraft. (From Jet Propulsion Laboratory, NASA Headquarters.)

Under these circumstances, climate models predict that the positive coupling between surface temperature and the outgoing IR flux disappears (Figure 19-2). Recall from Chapter 3 that this coupling is part of the fundamental feedback loop that keeps Earth's climate stable. On Venus, the atmosphere became so warm and so full of water vapor that virtually no infrared radiation from the surface was able to escape to space. Only the radiation from the cold upper troposphere and stratosphere was able to escape, and the flux of this radiation did not change as the surface became warmer and warmer. The resulting out-of-control climate system is called a **runaway greenhouse**. With the stabilizing feedback loop gone, the surface temperature "ran away" to values as high as 1500 K, the oceans evaporated, and all the water on the planet's surface turned to vapor. Even the stratosphere became filled with water vapor.

Once the stratosphere of Venus became wet, water would have been lost by photodissociation, followed by the escape of hydrogen to space. The oxygen left behind would have reacted with reduced materials, such as ferrous iron-bearing minerals, present at the planet's surface and with any reduced gases remaining in its atmosphere. With no liquid water remaining to facilitate silicate weathering, volcanic CO_2 would have accumulated, forming the dense atmosphere that exists today. The loss of water would also have removed the sink for sulfur-containing volcanic gases. These gases would likewise have accumulated in the atmosphere, giving rise to the dense sulfuric acid clouds that presently obscure the Venusian surface. The end result of this process is the hot, dry planet that we see today (Figure 19-3).

How do we know that this is actually what happened on Venus? Could Venus have simply started out with less water than Earth did as a consequence of being closer to the Sun? We cannot be sure because we cannot look back in time and because Venus is so hot that it would be extremely difficult to go there and look for the oxygen left behind from the loss of water. An important clue has been left behind in the Venusian atmosphere, however: The ratio of deuterium to hydrogen in the little water vapor

that remains is more than 150 times higher than that ratio in Earth's oceans. Recall from Chapter 14 that deuterium, D, is an isotope of hydrogen that has an extra neutron in its nucleus. If Venus initially had a significant amount of water and if this water was lost by photodissociation followed by escape of hydrogen to space, we would expect that the lighter hydrogen isotope, H, would escape more quickly and that the heavier one, D, would become enriched in the water vapor that was left behind. The fact that this is exactly what we observe supports the idea that Venus was once a water-rich planet and that the runaway greenhouse hypothesis is correct.

Martian Climate Evolution

Mars, in contrast, faced just the opposite of Venus' climate problem. The solar flux reaching Mars is only 43% of that reaching Earth and would have been even lower early in the solar system's history. The solar heating was so low that CO_2 itself should have condensed, forming clouds and polar ice caps. We know CO_2 ice by the name *dry ice,* and it has many familiar uses, such as keeping ice cream cold. The present Martian polar caps (see Figure 3-1) contain a mixture of CO_2 ice and water ice.

The fact that CO_2 can condense on Mars makes that planet's climate evolution very different from that of Earth. On Mars, volcanic CO_2 would not simply have accumulated in the atmosphere if the surface temperature fell below freezing. Thus, the stabilizing feedback provided by the carbonate–silicate cycle would not have operated in the same way as on Earth.

The surprising thing about Mars is that, despite this problem of CO_2 condensation, the Martian surface was once warm enough to allow liquid water to flow on its surface (Figure 19-4). Most of the Martian channels, or *valleys*, appear to have formed prior to about 3.8 b.y ago. (On Earth, a valley is the relatively wide region carved by a stream and a channel is the narrow stream that runs through it.) We infer this because the terrain on which most of the valleys are located is covered with lots of impact craters. By analogy with the cratering record of the Moon, this cratering occurred during the first 700 million years of the solar system's history (see Chapter 10). Evidently, Mars had a denser atmosphere at that time with enough greenhouse effect to keep its surface warm despite the low solar luminosity. Climate modelers are still not sure how to explain this warm climate, although greenhouse warming by CO_2 was almost certainly involved. High-altitude CO_2 clouds, which act somewhat like cirrus clouds on Earth, may also have played a role. (Recall from Chapter 3 that high-altitude clouds warm a planet's surface whereas low-altitude clouds tend to cool it.)

Why, then, did Mars become so cold? The answer may lie in the planet's small size. Mars has about half Earth's diameter and only about one-tenth its mass. A planet that size should have cooled off relatively quickly after its formation and would have a smaller radioactive heat source. (A planet's internal heat energy is proportional to its volume, which decreases more rapidly with radius than does the planet's surface area.) With less geothermal energy available to drive plate tectonics, the Martian crust might have solidified into a single unit early in the planet's history. This solidification would have shut down the carbonate–silicate cycle, because carbonate

rocks would no longer have undergone metamorphism. With most of the planet's CO_2 trapped in carbonates or stored as dry ice in the Martian soil, the greenhouse effect would have diminished and Mars would have approached its present, frozen state.

Habitable Planets around Other Stars

The question of what happened to Venus and Mars bears on another, even more interesting question: What is the chance that habitable planets exist elsewhere in our galaxy? We could broaden this to include the rest of the universe, but the question would then become entirely philosophical because we would have no way of verifying our predictions. As we will soon see, within the next 20 to 30 years we may be able to determine whether habitable (or even inhabited) planets exist within at least part of our own galaxy. Thus, it is not too soon to start thinking about whether they should be expected.

We know now that planets exist around other stars. At the time of this writing, at least 100 stars other than the Sun have been shown to have planetary companions. All these planets are large, gas-giant planets like Jupiter (see Chapter 10). (At least we assume they are gas giants because they are so massive.) Large planets are easier to detect than small ones, because their gravitational pull on their parent star is strong; it is the motion of the stars themselves that has been observed. But gas-giant planets are not likely abodes for life. They have no solid surface, and their atmospheres are highly convective. Thus, even though organisms could theoretically exist at some level in their atmospheres where liquid water is stable, the organisms would periodically be lofted to great heights, where the temperature is very cold, or carried to great depths, where they would be incinerated. Life *as we know it* requires a much more stable environment that is best provided by small, rocky planets like our own.

Formation of Earthlike Planets

What are the chances that we will find Earthlike planets around other stars? The answer to this question depends partly on how planets form and partly on what their climates are like later on. How a planet's climate evolves depends on a number of factors that we have discussed in previous chapters.

The formation of terrestrial-type planets is probably a common process. The rock-forming elements (silicon, oxygen, iron, and magnesium) that make up the bulk of Earth have been identified in the spectra of other stars and are thought to be abundant throughout the universe. The volatile elements on which life depends—carbon, nitrogen, hydrogen (as water), phosphorus, and sulfur—are

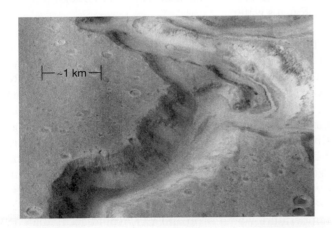

FIGURE 19-4

A Martian valley (Nanedi Vallis), seen from the Mars Global Surveyor spacecraft. (Courtesy of NASA.)

seen in interstellar clouds and are likewise thought to be abundant elsewhere. Disks of dust and gas are known to exist around many young stars, so the general conditions that are believed to lead to planet formation are commonplace. If the formation of other planetary systems has followed the same general pattern as that of our solar system, it is likely that planets formed in their inner parts would have been chemically much like Earth. In particular, they are likely to have had enough water to form an ocean and enough carbon to have a carbon cycle. A planet that is similar in size to Earth might also be expected to have enough internal heat to power Earthlike plate tectonics.

The Habitable Zone around the Sun

Whether another planet could have an Earthlike climate depends on how far from its star the planet forms and on how fast the parent star evolves in luminosity. As mentioned earlier, a fundamental requirement for life as we know it is that the planet's surface temperature remain in the range at which liquid water can exist. The region in space around a star where this condition is satisfied is called the **habitable zone** (HZ), or **ecosphere**. Because stars increase in luminosity as they age, their habitable zone moves outward with time. The region where a planet can remain habitable for some finite time interval is called the **continuously habitable zone**, or CHZ. The CHZ represents the overlap between the HZ at some initial time, t_0, and at some later time, t_1 (Figure 19-5). It is defined for some specific interval, $\Delta t = t_1 - t_0$. For our

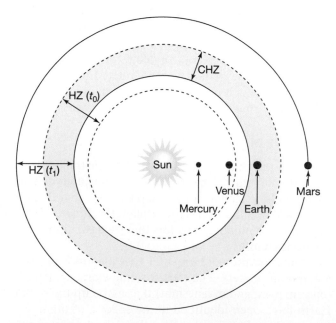

FIGURE 19-5

The relationship between the habitable zone (HZ) and the continuously habitable zone (CHZ). The CHZ is defined for some particular time interval $\Delta t = t_1 - t_0$, taken here to be 4.6 b.y.

Sun, we usually take t_0 to be the time the Sun formed (4.6 b.y. ago) and t_1 to be the present, so $\Delta t = 4.6$ b.y.

A point of clarification should be added here. Liquid water may actually exist on a planet at almost any distance from its parent star. Jupiter's moon Europa is a prime example. Europa is thought to have a subsurface ocean that is kept warm by tidal heating. Jupiter and Europa are well outside the conventional habitable zone in our own solar system, and yet liquid water and life may exist there. However, if life does exist on Europa, it is buried beneath the icy surface and is not detectable remotely. If we are interested in finding life on planets around other stars, we must concentrate on those on which organisms can modify the planet's atmosphere. So, the habitable zone of interest is that for which liquid water exists *at a planet's surface*.

How wide is the current habitable zone around our own Sun? Climate models can be used to help answer this question. So far, only one-dimensional, radiative-convective models have been used, but ultimately this question could be investigated with three-dimensional general circulation models. The one-dimensional models predict that the inner edge of the HZ around our own Sun is at about 0.95 AU. Inside this distance, a planet's stratosphere is predicted to become wet, and water is lost rapidly by photodissociation followed by the escape of hydrogen to space, as is thought to have occurred on Venus. Note that this result is consistent with the predictions for Earth's future climate discussed earlier in this chapter. Figure 19-1 indicates that Earth's surface temperature should start to increase rapidly beginning about 1 billion years in the future, when solar luminosity has increased by about 10%. If Earth were located at 0.95 AU rather than at 1 AU, the inverse-square law (Chapter 3) predicts that the solar flux would be higher by a factor of $1/(0.95)^2$, or 1.11. In either case, an increase of approximately 10% in the solar flux would be enough to drive off Earth's water.

Identifying the outer edge of the habitable zone is a little trickier. Early calculations by some climate modelers suggested that the outer edge is very close to Earth's current orbital distance of 1 AU. Moving Earth out to even 1.01 AU is enough to produce **runaway glaciation** in some models, in which the planet's surface becomes entirely covered by snow and ice. The reason was the positive feedback provided by snow and ice (see Figure 3-21). But we now believe that this positive feedback loop is countered by a very strong negative feedback loop involving the carbonate-silicate cycle. If Earth were somehow to be moved farther from the Sun, Earth's surface temperature would drop and CO_2 would accumulate in its atmosphere, just as it is thought to have done in the distant past when solar luminosity was lower (and especially during the possible "Snowball Earth" episodes in the Paleo- and Neoproterozoic). When this positive feedback loop is included, the climate models predict that the outer edge of the current HZ is somewhere near the orbit of Mars, around 1.5 AU. The exact distance is difficult to determine, because the effect of CO_2 clouds cannot be estimated reliably with one-dimensional climate models and because other greenhouse gases, such as methane, might extend the HZ even farther out if they were present in appreciable concentrations in a planet's atmosphere. As we mentioned earlier, Mars had liquid water on its surface early in its history even though it appears to have been outside the HZ at that time, according to Figure 19-5. Perhaps the HZ is wider than we think.

Because the CHZ represents the overlap between two HZs at different times in the Sun's history, it is narrower than the HZ. A planet near the outer edge of the current HZ would have been outside the HZ 4.6 b.y. ago when the Sun was only 70% as bright as it is today. When we take into account the change in solar luminosity, the outer edge of the 4.6-b.y. CHZ is estimated to lie in the vicinity of 1.25 AU. (Note again that the CHZ is defined only for some specific time interval.) This estimate is conservative because it does not take into account the possible warming caused by CO_2 clouds on planets near the outer edge of the HZ. So, the actual width of the CHZ may be even greater. Evidently, the habitable region around our own Sun is much wider than early predictions had suggested.

Habitable Zones around Other Stars

The same type of climate stability calculations described above can be performed for stars other than our Sun. The results are summarized in Figure 19-6, which shows the in-

stantaneous HZ midway through a star's main-sequence lifetime. (It is difficult to show CHZs on such a plot, because stars of different masses evolve at different rates.) The vertical scale shows stellar mass relative to the Sun's mass; the horizontal scale shows distance in astronomical units. The basic result is very simple: To have a surface temperature similar to Earth's, a planet orbiting a dim, red star would have to be closer than 1 AU to that star, whereas a planet orbiting a bright, blue star would have to be farther than 1 AU from that star.

Problems occur, however, for stars much earlier or later in the main sequence than our own Sun. For dim, red stars, the habitable zone falls within the **tidal locking radius** of the star, that is, the distance from a star within which a planet is likely to rotate such that it always faces the star. The Moon shows only one side to Earth; the reason is that it is close enough to Earth for its rotation to have been damped by tides. An astronomer would say that the Moon exhibits **captured rotation**. Close to the star, the tidal forces increase faster than the amount of available starlight. Thus, a planet that is close enough to a dim, red star to be within its habitable zone is likely to develop a captured rotation. (Note from Figure 19-6 that Mercury is within our Sun's tidal locking radius. However, it avoids this fate because it is caught in a resonance that causes it to spin 3 times for every 2 orbits around the Sun.) The planet would then have one hot, permanently sunlit side and one cold, perpetually dark side. Its atmosphere might freeze out on the dark side, making the entire planet uninhabitable.

Bright, bluish stars pose other problems for planetary habitability. The most serious is that they have relatively short main-sequence lifetimes. The rate of energy

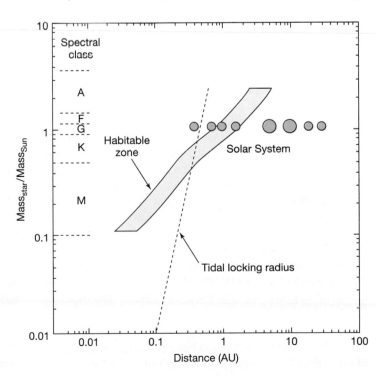

FIGURE 19-6

Habitable zones for other main-sequence stars. The HZ is plotted midway through the star's main-sequence lifetime. For a discussion of spectral classes of stars, see the Box "A Closer Look: Main-Sequence Stars and the Hertzsprung–Russell Diagram" in Chapter 10. (After J.F. Kasting et al. *Icarus*, 101:108–128, 1993.)

generation in a star increases with a star's mass much faster than the mass itself does. The result is that hot, massive stars exhaust the hydrogen in their cores much faster than does a star like the Sun. The hottest, most massive type of star stays on the main sequence for only a few tens of millions of years, compared with 10 billion years for our Sun. Thus, if life were to originate on a planet circling a bright star, it would have much less time to evolve than it did here on Earth.

Hot, blue stars pose another potential problem for life in that they emit a significant fraction of their radiation at ultraviolet wavelengths. According to Wien's law (Chapter 3), as the temperature of the star's surface increases, the peak wavelength of its emitted radiation decreases. The Sun's effective surface temperature is 5780 K; hence, its radiation peaks at about 0.5 μm, near the middle of the visible part of the electromagnetic spectrum. Hot, blue stars can have surface temperatures of up to 25,000 K, so their radiation peaks well into the ultraviolet. The relative amount of UVB and UVC radiation emitted by such a star is orders of magnitude greater than that emitted by our own Sun. Hence, a planet orbiting a hot, blue star would need to have an extremely well-developed ozone screen in order for its surface to be habitable by advanced life. The good news is that the shorter UV wavelengths that split O_2 and make ozone are even more strongly enhanced in these stars, so such "super" ozone screens might actually exist.

We conclude that stars not too different from our Sun are the ones most likely to harbor habitable planets. Among such stars, the chances of there being an Earthlike planet appear to be pretty good.

The Drake Equation

The last chapter of many introductory astronomy textbooks is devoted to speculation about the possibility of life elsewhere in the universe. The discussion is generally framed in terms of the **Drake equation**, a relatively innocuous-looking formula used to estimate the number of other intelligent civilizations in our galaxy with which we might one day establish radio communication. The equation was proposed by Frank Drake, an astronomer now at the University of Santa Cruz. The late astronomer Carl Sagan of Cornell University also had a hand in crafting the equation and wrote about it extensively in his books *Cosmos* and *Intelligent Life in the Universe* (with I. S. Shklovskii).

The equation can be written as follows: *N,* the number of advanced, communicating civilizations in the galaxy, is expressed as the product of seven terms:

$$N = N_g f_p n_e f_l f_i f_c f_L$$

where N_g = number of stars in our galaxy; f_p = the fraction of stars that have planets; n_e = the number of Earthlike

planets per planetary system; f_l = the fraction of habitable planets on which life evolves; f_i = the probability that life will evolve to an intelligent state; f_c = the probability that intelligent life will develop the capacity to communicate over long distances (e.g., by radio telescope); and f_L = the fraction of a planet's lifetime during which it supports a technological civilization.

An alternative form of the Drake equation replaces the factors N_g and f_L with R (the average rate of star formation in the galaxy) and L (the lifetime of a technological civilization). If the number of stars in the galaxy is in steady state, which is probably approximately true, then R is equal to the number of stars in the galaxy, N_g, divided by the average lifetime of a star, L_s. The product RL is thus $(N_g/L_s)L = N_g(L/L_s)$. But L/L_s is essentially the same as f_L, so the two forms of the equation are equivalent.

Despite its apparent simplicity, the Drake equation cannot be solved, because many of the individual terms are difficult to estimate. It is nonetheless useful, because it helps us identify the factors that are important to biological evolution, and to human evolution in particular. It is also an appropriate way to conclude our study of Earth system science, because many of the issues that we have addressed bear on terms in the equation. Hence, we offer the following thoughts as a way of summing up our thoughts about the Earth system and its degree of stability.

N_g—The Number of Stars in a Galaxy

We can estimate the number of stars in a galaxy by looking at small but representative patches of sky and counting the individual stars. The current estimate for N_g is about 4×10^{11}, or 400 billion, stars. Other galaxies, of which there are billions, have comparable numbers of stars, so the total number of stars in the universe is very much larger than this. But the distances separating galaxies are so large (2 million light years or greater) that the possibility of radio communication with another civilization is too remote to consider.

f_p—The Fraction of Stars That Have Planets

In the previous section, we discussed briefly the fraction of stars that have planets. Astronomers believe that planetary formation is a natural accompaniment to the formation of stars. One problem that we did not mention is that as many as 90% of the stars in our galaxy are members of binary (two-star) systems or other multiple-star systems. Such stars may not form planets, but theoreticians are not certain. Even if they do, it seems unlikely that many of these planets would be located within the habitable zone of one of the stars. So, let us assume that only single stars have planets; hence we assign f_p a value of about 0.1.

For now, this value for f_p is only a guess. Within the next few years, however, it should be possible to measure this number directly. As mentioned earlier, planets have now been discovered around 10 main-sequence stars. So far, only the largest and closest-in planets have been discovered, because these are the easiest to detect. As observational technology improves, however, and as astronomers accumulate data on stellar motions over longer periods, it should be possible to determine whether planetary systems like our own are commonplace.

n_e—The Number of Earthlike Planets per Planetary System

We have already addressed the number of Earthlike planets per planetary system in our earlier discussion of habitable zones around stars. On the basis of our relatively optimistic estimate of the width of the CHZ around our own Sun, we predict that solar-type stars are very likely to harbor habitable planets. Our estimate of the 4.6-b.y. CHZ width, about 0.3 AU, is roughly equal to the mean spacing between the four innermost planets in our own solar system, which lie between 0.4 AU and 1.5 AU. Thus, if planets are spaced similarly around other solar-type stars, there is a very good chance that one of them will be within the CHZ. To be conservative, we will estimate this probability at 50%. Only stars of spectral type close to our own sun look to be good candidates, however. This cuts down the number of available stars by a factor of about 5. So, a preliminary estimate of n_e would be 0.5 × 0.2, or 0.1 Earthlike planets per star.

f_l—The Fraction of Habitable Planets on Which Life Evolves

The fraction of habitable planets on which life evolves is one of the most poorly understood terms in the Drake equation. We discussed the question of life's origin in Chapter 10. Chemists and biologists are uncertain as to how it occurred, so it is impossible to say whether it was a likely or an unlikely event here on Earth.

Two experimental approaches could eventually shed light on the origin-of-life question. One would be to create life in a test tube. The other would be to detect life elsewhere in the universe. Of these, the approach that appears most feasible at present is the detection of life elsewhere. We have already sent two *Viking* spacecraft out from Earth in the late 1970s with the express intent of searching for life on Mars. Neither spacecraft found life on the martian surface; indeed, they all but proved that it could not exist. (The martian soil is so highly oxidized that organic material cannot survive very long.) It now seems unlikely, though not impossible, that life will be found on Mars or anywhere else in our own solar system.

Other planetary systems are a different matter, however. Life around other stars may be detectable by at least two methods. One method is radio communication. The giant radio telescope at Arecibo, Puerto Rico, is capable

FIGURE 19-7

The Arecibo radio telescope. (From David Parker/Science Photo Library, Photo Researchers, Inc.)

of communicating with another, similarly equipped civilization anywhere in our galaxy (Figure 19-7). If we manage to communicate, however, we will no longer need the Drake equation. The second method of detecting life is to build large telescopes in space and use them to view planets around other stars directly. If connected electronically, these telescopes could act as an **interferometer**, an instrument consisting of two or more telescopes with their images combined. They act effectively as one big telescope. In principle, such an instrument could be used to cancel out the light from the star and to let through only the light from the planets around it. NASA has long-range plans for a four-telescope interferometer that could image Earth-sized planets out to a distance of about 50 light years. This instrument would allow us to look for planets around more than 100 single, solar-type stars.

Such an instrument may be able to perform **spectroscopy** on a planet's atmosphere—a technique in which the composition of a material, such as a gas, is analyzed according to the details of its electromagnetic spectrum. By doing so, an interferometer may also be able to look for the presence of CO_2, H_2O, and O_3. The first two compounds are required for life as we know it. The third, ozone, is formed photochemically from O_2, as we discussed in Chapter 17. Almost all of Earth's O_2 is generated by photosynthesis. The atmospheres of Venus and Mars have only small amounts of O_2 and, hence, do not show evidence of O_3 (Figure 19-8). So, the detection of ozone in the atmosphere of an extrasolar planet would be

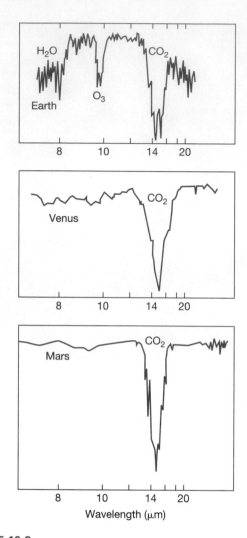

FIGURE 19-8

Infrared spectra of the atmosphere of Earth, Venus, and Mars. Only Earth shows the presence of H_2O and O_3. (After R. Hanel, Goddard Space Flight Center.)

fairly strong evidence that the planet was inhabited. Note that this method of detecting extraterrestrial life is much more general than radio communication, because it requires evolution only to the bacterial level of complexity.

In the meantime, as we do not know how to estimate f_l, we shall for the moment leave its value unassigned.

f_i—The Probability That Life Will Evolve to an Intelligent State

The probability that life will evolve to an intelligent state is another poorly understood term in the Drake equation, in part because it is not entirely obvious what intelligence actually is. Are humans the only intelligent species on this planet? Or should we include dolphins and perhaps a few other species that are not that far behind in terms of mental capabilities?

If we confine our discussion of intelligence to the Drake equation, the appropriate definition becomes clear: To be detectable by radio communication, a species must have the type of intelligence to develop electronic equipment and other aspects of a technological civilization. For this ability, they probably have to have fingers instead of fins. So, in terms of the Drake equation, an intelligent species needs to be fairly similar to our own.

We will not attempt to elucidate here the many factors involved in the evolution of human intelligence. We can, however, point out two topics that we have discussed in previous chapters that bear on this question. The first is the rise of atmospheric oxygen about 2.3 billion years ago. The rise of oxygen was an absolute necessity not only for human evolution, but also for the evolution of all multicellular life. Indeed, its importance is even more fundamental: Even single-celled eukaryotes (cells with nuclei) require oxygen for their metabolism. Although we cannot be certain that biological evolution would have proceeded elsewhere as it did on Earth, chances are that oxygen is a metabolic requirement for all advanced organisms.

Was the rise of oxygen a likely event? Now that the cyanobacteria-like microfossils found in the Apex Chert Formation of Australia have been called into question (Chapter 10), we are no longer certain that oxygenic photosynthesis was an early biological invention. Evolving this capability could have taken as long as 2 billion years. Thus, it is difficult to say how long it might take elsewhere. Once photosynthesis had been invented on Earth, the atmosphere was almost bound to become oxidizing, sooner or later. If the hypothesis outlined in Chapter 11 is correct, the key factor was the gradual oxidation of Earth's mantle and crust caused by the recycling of seawater through the mantle and the escape of hydrogen to space. A similar type of process would presumably operate on any planet that had a surface gravity similar to Earth's (so hydrogen could escape) and that had enough internal heat to support plate tectonics. Thus, the rise of oxygen in an Earthlike planet's atmosphere may be a relatively commonplace event *if* photosynthesis itself has been invented elsewhere.

A second important factor that must influence biological evolution is the frequency of large impacts. Energetic collisions with asteroids or comets have evidently caused mass extinctions of organisms on our planet and have paved the way for the evolution of new life forms (see Chapter 13). Humans might not have evolved if the Cretaceous-Tertiary impact event had not eliminated the dinosaurs. Impacts that were either too large or too frequent would have a detrimental effect on biological evolution, however. The lunar cratering record indicates that large (100-km diameter) impactors probably hit Earth fairly often until about 3.8 b.y. ago. These events would have sterilized most of Earth's surface. In our own solar system, the flux of these large impactors decreased to close to zero after the first few hundred million years. The same, however, would not necessarily be true of other planetary systems.

The key factor in our solar system that controls the impact flux that hits Earth is the planet Jupiter. Jupiter is directly responsible for at least some of the impactors that have hit Earth over the past few billion years, because that giant planet perturbs asteroids from certain regions in the asteroid belt into Earth-crossing orbits. But Jupiter may also protect us from cometary impacts by deflecting incoming comets before they can enter Earth-crossing orbits. If other planetary systems have similar cometary fluxes, the presence of a giant planet like Jupiter may be essential to ensuring environmental stability on the inner planets.

Like f_l, we do not know what f_i should be, so we will leave it undefined as well.

f_c—The Probability That Intelligent Life Will Develop the Capacity to Communicate over Long Distances

We cannot assess with confidence the numerical probability that an extraterrestrial civilization will develop radio communication. But we may be able to say something about how the planetary environment influenced the emergence of a technologically advanced civilization on Earth. Yale anthropologist Elizabeth Vrba has suggested that our prehuman ancestors were forced out of their forested environments in Africa around 2.5 m.y. ago by ecological changes associated with the Pleistocene glacial cooling. According to Vrba, these changes led to the replacement of forests by vast areas of savannah (grassland). Our ancestors adapted to their changed environment by developing a more upright manner of walking and by developing weapons with which to hunt the large herbivores that grazed on the savannah.

These changes in lifestyle may also have accelerated intellectual development in humans. Whether the same evolutionary changes would have taken place in the absence of climate change is not clear. But we can see how mildly destabilizing events such as climate change may have helped guide, and perhaps accelerate, both the evolution of humans and the gradual emergence of civilization.

A second, somewhat better documented correlation between climate change and the development of civilization is seen at the end of the last ice age, about 10,000 years ago. At about this time, agricultural societies emerged spontaneously in several different regions of the world, including Mesopotamia, China, and the Indus River valley in India. Very likely, the switch from hunter-gatherer societies to agricultural-based societies was facilitated by the improved climatic conditions. Again, this switch could have occurred without any forcing from climate. But, Milankovitch cycles do seem to

have played some role in the development of our current civilization.

Imagine what might happen, though, on an Earthlike planet without a large moon. The Milankovitch climate cycles might then be driven by a chaotically varying planetary obliquity (see Chapter 14). They would probably appear to be regular and to have relatively small amplitudes on a time scale of tens to hundreds of thousands of years, but on longer time scales the planetary obliquity might reach extreme values. What this would do to the emergence of agriculture (and to the soil itself) is unclear. We suspect, however, that it would make the emergence of civilization much more difficult.

Collecting Terms to This Point

Let us stop for a moment to take stock of what we have learned so far. If we combine what we know with what we do not know about the terms in the Drake equation, we are left with

$$
\begin{aligned}
N &= N_g f_p n_e f_l f_i f_c f_L, \\
&\approx (4 \times 10^{11}) \times (0.1) \times (0.1)\, f_e f_i f_c f_L \\
&= 4 \times 10^9\, f_l f_i f_c f_L,
\end{aligned}
$$

where we have inserted approximate values for the three leading terms, N_g, f_p, and n_e, and where the symbol $\approx$ means "approximately equals." Our point in writing the equation in this manner is the following: Only the physical science terms in the Drake equation can be assigned numerical values with any degree of certainty. (Even these terms might be subject to considerable debate.) The last four terms deal with biological and sociological evolution and are much more speculative.

Suppose, however, that we are optimistic about the chances of life originating elsewhere and of the evolution of intelligence and civilization. Carl Sagan, for example, was always an optimist, and he estimated (in *Cosmos*) that the product $f_l f_i f_c$ is equal to about 1/300. If we accept his estimate and round off our numbers, the Drake equation boils down to

$$
N = 10^7 f_L
$$

Under these assumptions, whether N is large or small depends exclusively on f_L, the fraction of a planet's lifetime during which it supports a technological civilization. So, let us speculate for a moment about the magnitude of that factor.

f_L—The Fraction of a Planet's Lifetime during which It Supports a Technological Civilization

The remaining uncertainty in the Drake equation is the fraction of a planet's lifetime during which it supports a technological civilization. Let us assume that the average lifetime of planets themselves is about 10 b.y., the approximate main-sequence lifetime of the Sun. The question then becomes: How long does an average technological civilization last? How long will ours last? The answer clearly depends on what eventually causes our demise. If we do ourselves in by a nuclear war, the answer could be as short as 100 years. If other civilizations annihilated themselves in the same way, the factor f_L would be equal to 10^2 years/10^{10} years = 1×10^{-8}, and N would be a paltry 0.1. A value of N below one implies that we might well be the only technological civilization in the entire galaxy.

If left to natural causes, however, the lifetime of our civilization could be much longer. The average longevity of a species in the geologic record is about 1 million years (Chapter 13). If we use this value for civilized species, we get $f_L = 10^6/10^{10} = 1 \times 10^{-4}$ and $N = 1000$. But civilized species are much smarter than average and, thus, might survive much longer. If we are able to deal successfully with all our other problems, humans should be able to survive for at least another 500 m.y., the time scale for the loss of C_3 photosynthesis. If technical civilizations in general were this long-lived, f_L could be as high as $5 \times 10^8/10^{10} = 0.05$ and N would be 500,000.

Future technological innovations, such as interstellar space flight or planet-sized solar shields, could allow us to outlive even this disaster. But the basic point has already been made: If technological civilizations can survive for an appreciable fraction of a planetary lifetime, then the number of such civilizations in the galaxy could be quite large. For $N = 500,000$, the average distance between civilizations should be a few hundred light years. We can imagine that a slow dialogue might take place by radio communication were two such civilizations to discover each other's existence.

Ensuring Our Long-Term Survival

What are the chances that we can overcome all our potential problems and survive for hundreds of millions years? We have discussed a number of current environmental problems in previous chapters. Could any of them trigger our downfall?

Global warming might lead to drastic environmental changes and even to the relocation of large numbers of people, but it is not likely to result in the demise of all of civilization. Ozone depletion is potentially more dangerous, because UV radiation is directly harmful to almost all forms of life. However, we are already dealing with it in what looks to be an effective manner. Loss of biodiversity might pose a significant threat to global food production (see Chapter 18), but it does not seem likely by itself

to wipe out the human population. Human population growth itself might be the most significant problem, but even this will ultimately be self-limiting. Either the availability of food or competition for other resources will eventually cause the population to stabilize, hopefully at a comfortable and sustainable level.

If we avoid nuclear self-destruction, the most significant threat to our continued existence over very long time scales might come from large impacts. As we pointed out in Chapter 13, the expected frequency of impact events the size of that at the Cretaceous–Tertiary is about one per 10^8 years. Would a 10-km-diameter impactor destroy civilization? It is difficult to say. On the basis of the amount of energy released, the effects would be far more devastating than those of an all-out nuclear war (see Chapter 1). The initial firestorm ignited by the reentry of impact ejecta into the atmosphere could destroy agricultural crops around the world within hours. The dust generated by the impact would reach the stratosphere and would linger there for 6 months to a year. Photosynthesis might be shut off during much of this time, and continental interiors could become quite cold. (Coastal areas would remain relatively warm because of the large heat capacity of the oceans.) Some of us might be able to survive such a catastrophe by taking advantage of our already fairly advanced technology, but the number of people killed would still likely be in the billions.

Fortunately, there is no reason why such a catastrophe need ever happen. We are already aware of the existence and approximate orbital parameters of many of the Earth-crossing asteroids, and we have the technology to improve considerably on this database. Within a few years, a space-based telescope dedicated to searching for asteroids and comets could map out all of the potentially dangerous objects in the solar system. Indeed, the construction of such a telescope has already been suggested and seems likely to happen in the wake of the Comet Shoemaker–Levy impact with Jupiter, which helped alert lawmakers to the potential danger of such impacts. With sufficient advance warning of an impact, we may already be capable of avoiding it. A spacecraft, for example, could be sent to rendezvous with the object and then to detonate a small nuclear device on one side of the object to deflect its orbit. (Blowing it to pieces as done in the movie "Deep Impact" would be energetically difficult or impossible for large bodies and would create showers of smaller objects that might still cause significant damage.) As time passes and our space technology improves, there is little doubt that we will be able to construct an asteroid defense system, if we decide to do so. So, there is no reason why we need go the way of the dinosaurs. We know enough about the dangers of the Earth system to survive for longer than they did, if we can survive the danger that we pose to ourselves.

Chapter Summary

1. The Sun continues to brighten at a rate of about 1% per hundred million years.
   a. The resulting increase in Earth's surface temperature could lead to rapid loss of water in about a billion and a half years.
   b. Long before this occurs, first C_3 and then C_4 plants might die off as atmospheric CO_2 concentrations fall below the level required to sustain photosynthesis. The CO_2 decrease is predicted to result from the same feedback of surface temperature–silicate weathering rate that is thought to have stabilized Earth's climate throughout the planet's history.
2. Venus and Mars may have been more Earthlike in the distant past, but they evolved along very different climatic paths.
   a. Venus lost its water because it was too close to the Sun. After that, CO_2 and sulfur gases accumulated, forming the hot, dense atmosphere that exists today.
   b. Mars had liquid water flowing on its surface early in its history, despite the lower solar luminosity at that time. A dense CO_2 atmosphere filled with clouds of CO_2 ice may have provided the necessary greenhouse warming. Mars may have cooled down because it was too small to maintain plate tectonics and to keep recycling CO_2 back into its atmosphere.
3. Jupiter-sized planets have been detected around at least 100 other main-sequence stars. Earth-sized planets have not yet been observed, but it should eventually be possible to look for them. Climate calculations predict that at least some of these planets should lie within the habitable zone around their parent star. Stars not too different from our Sun appear to be the best candidates for harboring habitable planets.
4. The information gleaned about climatic stability on Earth may be fed into the Drake equation to estimate the probability that advanced, technological civilizations exist elsewhere in the galaxy.
   a. Some of the factors in this equation are already known, and other factors may be determined observationally in the foreseeable future. In particular, we may be able to look for the presence of photosynthetic life on planets in other solar systems by analyzing their atmospheres spectroscopically and by looking for the presence of ozone.
   b. The largest single uncertainty in the Drake equation is the lifetime of a technological civilization, a factor that is largely under that civilization's control. If technological civilizations are able to persist for an appreciable fraction of their planet's lifetime, then the chances that other civilizations exist in our galaxy appear to be very good.

Key Terms

captured rotation
CO_2 compensation point
continuously habitable zone
Drake equation
ecosphere

habitable zone
interferometer
main-sequence star
photorespiration
red giant

runaway glaciation
runaway greenhouse
spectroscopy
tidal locking radius
white dwarf

Review Questions

1. How should future solar evolution affect climate and life here on Earth?
2. What is the evidence that Venus once possessed more water than it does today? How did it lose its water, and how did its atmosphere evolve afterward?
3. What is the evidence that Mars was warmer and wetter in the past? Why did Mars cool off over time, even though the Sun has become brighter?
4. How is the habitable zone around a star defined?
5. What is the relationship between the instantaneous habitable zone and the continuously habitable zone around the Sun? How wide are these zones thought to be?
6. Why are bright, blue stars and dim, red stars not good candidates for supporting habitable planets?
7. What factors in the Drake equation are we capable of estimating today?
8. How do we know that planets exist around other stars? How many such planets have been detected to date?
9. How might the presence of life on another planet be inferred from the composition of that planet's atmosphere?
10. What factors might limit the lifetime of a technological civilization? Could any of today's global environmental problems (global warming, ozone depletion, and loss of biodiversity) destroy our technological society?

Critical-Thinking Problems

1. The inner edge of the HZ is currently estimated to be at a distance of 0.95 AU from the Sun. Where would it have been 4.6 b.y. ago when solar luminosity was about 70% of its present value? Would Venus, which orbits the Sun at a distance of 0.72 AU, have been inside or outside this critical distance? (Hint: Remember the inverse-square law from Chapter 3.)

2. How long do you think our present, technological civilization will last, and what factor (or combination of factors) do you think will bring it to an end? Write a 1- to 2-page essay defending your opinion.

Further Reading

General

Darling, D. 2001. *Life Everywhere: The Maverick Science of Astrobiology*. New York: Basic Books.

Dorminey, B. 2002. *Distant Wanderers: The Search for Planets beyond the Solar System*. New York: Copernicus Books.

Kasting, J. F. 1988. How climate evolved on the terrestrial planets. *Scientific American*, 256:90–97.

Sagan, C. 1980. *Cosmos*. New York: Random House.

Shklovskii, I. S., and C. Sagan. 1966. *Intelligent Life in the Universe*. San Francisco: Holden-Day.

Ward, P., and D. Brownlee. 2000. *Rare Earth: Why Complex Life is Uncommon in the Universe*. New York: Copernicus Books.

Ward, P., and D. Brownlee. 2003. *The Life and Death of Planet Earth*. New York: Times Books.

Advanced

Caldeira, K., and J. F. Kasting. 1992. The life span of the biosphere revisited. *Nature*, 360:721–723.

APPENDIX *A*

Units and Unit Conversions

Common Unit Equivalences

1 kilometer (km)	= 1000 meters (m)
1 meter (m)	= 100 centimeters (cm)
1 centimeter (cm)	= 0.39 inches (in.)
1 mile (mi)	= 5280 feet (ft)
1 foot (ft)	= 12 inches (in.)
1 inch (in.)	= 2.54 centimeters (cm)
1 square mile (mi^2)	= 640 acres (a)
1 kilogram (kg)	= 1000 grams (g)
1 pound (lb)	= 16 ounces (oz)
1 fathom	= 6 feet (ft)

Unit Conversions

To convert:	Multiply by:	To find:
Length		
inches	2.54	centimeters
centimeters	0.39	inches
feet	0.30	meters
meters	3.28	feet
yards	0.91	meters
meters	1.09	yards
miles	1.61	kilometers
kilometers	0.62	miles
Area		
square inches	6.45	square centimeters
square centimeters	0.15	square inches
square feet	0.09	square meters
square meters	10.76	square feet
square miles	2.59	square kilometers
square kilometers	0.39	square miles
hectares	10,000	square kilometers
Volume		
cubic inches	16.38	cubic centimeters
cubic centimeters	0.06	cubic inches
cubic feet	0.028	cubic meters
cubic meters	35.3	cubic feet
cubic miles	4.17	cubic kilometers
cubic kilometers	0.24	cubic miles
liters	1.06	quarts
liters	0.26	gallons
gallons	3.78	liters
Masses and Weights		
ounces	20.33	grams
grams	0.035	ounces
pounds	0.45	kilograms
kilograms	2.205	pounds

The International System of Units (SI)

Basic Units

Quantity	Unit	SI Symbol
Length	meter	m
Mass	kilogram	kg
Time	second	s
Thermodynamic temperature	kelvin	K
Amount of substance	mole	mol

Prefixes

Prefix	Multiply Unit by:	Symbol
tera	10^{12}	T
giga	10^9	G
mega	10^6	M
kilo	10^3	k
hecto	10^2	h
deka	10	da
deci	10^{-1}	d
centi	10^{-2}	c
milli	10^{-3}	m
micro	10^{-6}	μ
nano	10^{-9}	n
pico	10^{-12}	p
femto	10^{-15}	τ
atto	10^{-18}	a

Derived Units

Quantity	Unit	Expression
Area	square meter	m^2
Volume	cubic meter	m^3
Frequency	hertz (Hz)	s^{-1}
Density	kilogram per cubic meter	kg/m^3
Velocity	meter per second	m/s
Acceleration	meter per second squared	m/s^2
Force	newton (N)	$kg \cdot m/s^2$
Pressure	newton per square meter	N/m^2
Work, energy, quantity of heat	Joule (J)	$N \cdot m$
Power	watt (W)	J/s

APPENDIX *B*
Temperature Conversions

$T(°F) = T(°C) \times 1.8 + 32$

$T(°C) = \dfrac{T(°F) - 32}{1.8}$

$T(°C) = T(K) - 273.15$

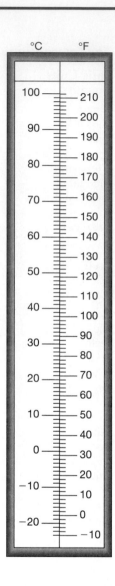

APPENDIX C

Periodic Table

Main groups

Transition metals

Main groups

1 / 1A	2 / 2A	3 / 3B	4 / 4B	5 / 5B	6 / 6B	7 / 7B	8 / 8B	9 / 8B	10 / 8B	11 / 1B	12 / 2B	13 / 3A	14 / 4A	15 / 5A	16 / 6A	17 / 7A	18 / 8A
1 **H** 1.00794																	2 **He** 4.002602
3 **Li** 6.941	4 **Be** 9.012182											5 **B** 10.811	6 **C** 12.0107	7 **N** 14.0067	8 **O** 15.9994	9 **F** 18.998403	10 **Ne** 20.1797
11 **Na** 22.989770	12 **Mg** 24.3050											13 **Al** 26.981538	14 **Si** 28.0855	15 **P** 30.973761	16 **S** 32.065	17 **Cl** 35.453	18 **Ar** 39.948
19 **K** 39.0983	20 **Ca** 40.078	21 **Sc** 44.955910	22 **Ti** 47.867	23 **V** 50.9415	24 **Cr** 51.9961	25 **Mn** 54.938049	26 **Fe** 55.845	27 **Co** 58.933200	28 **Ni** 58.6934	29 **Cu** 63.546	30 **Zn** 65.39	31 **Ga** 69.723	32 **Ge** 72.64	33 **As** 74.92160	34 **Se** 78.96	35 **Br** 79.904	36 **Kr** 83.798
37 **Rb** 85.4678	38 **Sr** 87.62	39 **Y** 88.90585	40 **Zr** 91.224	41 **Nb** 92.90638	42 **Mo** 95.94	43 **Tc** (98)	44 **Ru** 101.07	45 **Rh** 102.90550	46 **Pd** 106.42	47 **Ag** 107.8682	48 **Cd** 112.411	49 **In** 114.818	50 **Sn** 118.710	51 **Sb** 121.760	52 **Te** 127.60	53 **I** 126.90447	54 **Xe** 131.293
55 **Cs** 132.90545	56 **Ba** 137.327	57 ***La** 138.9055	72 **Hf** 178.49	73 **Ta** 180.9479	74 **W** 183.84	75 **Re** 186.207	76 **Os** 190.23	77 **Ir** 192.217	78 **Pt** 195.078	79 **Au** 196.96655	80 **Hg** 200.59	81 **Tl** 204.3833	82 **Pb** 207.2	83 **Bi** 208.98038	84 **Po** (209)	85 **At** (210)	86 **Rn** (222)
87 **Fr** (223)	88 **Ra** (225)	89 **†Ac** (227)	104 **Rf** (261)	105 **Db** (262)	106 **Sg** (266)	107 **Bh** (264)	108 **Hs** (269)	109 **Mt** (268)	110 †† (271)	111 (272)	112 (285)		114 (289)		116 (292)		

*Lanthanide series (rare-earth elements)	58 **Ce** 140.116	59 **Pr** 140.90765	60 **Nd** 144.24	61 **Pm** (145)	62 **Sm** 150.35	63 **Eu** 151.964	64 **Gd** 157.25	65 **Tb** 158.92534	66 **Dy** 162.50	67 **Ho** 164.93032	68 **Er** 167.259	69 **Tm** 168.93421	70 **Yb** 173.04	71 **Lu** 174.967
†Actinide series	90 **Th** 232.0381	91 **Pa** 231.03588	92 **U** 238.02891	93 **Np** (237)	94 **Pu** (244)	95 **Am** (243)	96 **Cm** (247)	97 **Bk** (247)	98 **Cf** (251)	99 **Es** (252)	100 **Fm** (257)	101 **Md** (258)	102 **No** (259)	103 **Lr** (262)

APPENDIX *D*

Useful Facts

Fundamental Physical Constants

Quantity	Symbol	Value
Speed of light in a vacuum	C	2.998×10^8 m/s
Universal gas constant	R	8.314 J/K/mol
Stefan-Boltzmann constant	σ	5.670×10^{-8} W/m^2/s
Planck's constant	h	6.626×10^{-20} J $\cdot$ s
Avogadro's number	N_a	6.023×10^{23} /mol
Mass of a hydrogen atom	m_H	1.67×10^{-27} kg
Boltzmann's constant	k	1.38×10^{-23} J/K

Physical Properties of Earth

Quantity	Value
Mass	5.974×10^{24} kg
Equatorial radius	6378 km
Polar radius	6357 km
Surface area	5.1006×10^8 km^2
Average density	5142 kg/m^3
Average Earth–Sun distance	1.496×10^8 km
Average Earth–Moon distance	3.844×10^5 km
Volume of oceans	1.4×10^{18} m^3
Surface area of oceans	3.6×10^8 km^2 (71% of surface)
Average depth of oceans	3800 m
Mass of atmosphere	5.0×10^{18} kg
Mean sea level atmospheric pressure	1.013×10^5 N/m^2 (= 1013 mbar)
Density of air at sea level	1.225 kg/m^3

Glossary

15-μm CO_2 band An absorption band that is a result of a vibration frequency of the CO_2 molecule that allows the molecule to absorb infrared radiation at a wavelength of about 15μm, near the peak of Earth's outgoing radiation. Very little of this radiation is allowed to escape directly to space, and so CO_2 is an important contributor to the greenhouse effect.

Absolute vorticity The sum of the planetary and relative vorticities experienced by a moving fluid.

Absorption coefficient The efficiency with which a molecule absorbs electromagnetic radiation of a given wavelength.

Accretion A process by which small bodies orbiting the Sun (or a planet) collide to form larger ones.

Acid A solution with a high concentration of hydrogen ions, that is, with a pH less than 7.

Acid rain Acidic rainwater produced when various acids, including sulfuric acid produced from SO_2 oxidation, combine with natural rainwater.

Action spectrum The relative efficiency with which UV photons at different wavelengths contribute to a specific biological response in a specific organism.

Adaptation A characteristic that enhances an organism's survival or reproductive success.

Albedo The reflectivity of a surface, usually expressed as a decimal fraction of the total incident sunlight reflected from the surface.

Amino acids Organic compounds containing an amino (NH_2) group and a carboxyl (COOH) group. They are the basic building blocks for proteins.

Anion A negatively charged ion.

Anoxic Totally devoid of oxygen.

Anoxic basin A region of the ocean in which deep water is completely devoid of oxygen; rare in today's ocean.

Anoxygenic photosynthesis A type of photosynthesis, carried out by certain types of bacteria, in which no O_2 is released.

Antarctic Bottom Water (AABW) Bottom water that forms in the Weddell Sea off Antarctica. The AABW circles Antarctica and flows northward as the deepest layer in the Atlantic, Pacific, and Indian Ocean basins.

Anthropogenic Induced by humans.

Anti-greenhouse effect Absorption of sunlight (and radiation of infrared energy) by particles or gases high up in the stratosphere. This prevents sunlight from reaching a planet's surface and, hence, causes surface cooling.

Aphelion The position in a planet's orbit that is farthest from the Sun.

Archaea One of the three primary domains of life, as determined by sequencing of ribosomal RNA. The Archaea (like the *Bacteria*) are all single-celled, prokaryotic organisms. Includes the methanogens.

Asteroid belt A region between Mars and Jupiter that has a high concentration of asteroids. The asteroid belt represents the remains of a tenth, inner planet that failed to form or was destroyed by a large meteorite impact early in the history of the solar system.

Asteroids Pieces of rocky material composed of minerals and metallic elements; they range in size from dust size to 1000 km in diameter and most reside in the *asteroid belt,* but some are diverted to planet-crossing orbits by gravitational disturbance.

Asthenosphere The zone of most ductile upper mantle just below the lithosphere; may contain small amounts of molten rock.

Atmosphere The thin envelope of gases that surrounds most planets; one of the four major components of the Earth system.

Atomic number The number of protons in the nucleus of an atom. All atoms of a given element have the same atomic number.

Autotroph An organism (*producer*) that can derive its energy for growth and reproduction from either solar or chemical energy.

Bacteria One of the three primary domains of life, as determined by sequencing of ribosomal RNA. The Bacteria (like the *Archaea*) are all single-celled, prokaryotic organisms.

Banded iron-formations Laminated rocks consisting of alternating layers of iron minerals and chert that were formed almost exclusively during the first half of Earth's history.

Barometric law A relationship stating that atmospheric pressure decreases by about a factor of 10 for each 16-km increase in altitude. More technically, atmospheric pressure decreases exponentially with altitude.

Basalt An igneous rock that reaches Earth's surface as lava and cools rapidly; the major rock type in the oceanic crust.

Base A solution with a low concentration of hydrogen ions, that is, with a pH greater than 7.

Biodiversity The variety of life forms, for example, the number of species in an area.

Biological pump A transfer of CO_2 and nutrients from the surface waters to the deep ocean as a result of photosynthesis in shallow waters, of the settling of organic matter, and of decomposition in deep waters. This process operates on a much faster time scale (days to months) than the thermohaline circulation of the oceans.

Biological sulfate reduction A metabolic pathway in which sulfate is combined with organic matter to produce energy. The sulfate is reduced and the organic matter is oxidized during the reaction.

Biomass The total combined weight of organic material in each trophic level. In terrestrial ecosystems the biomass is reduced by 90 to 99% at each higher trophic level.

Biomass-based fuels Liquid fuels, such as methanol and ethanol, that are produced from fast-growing plants. Unlike fossil fuels, burning these substances does not contribute to atmospheric CO_2 buildup.

Biome A region with a characteristic plant community.

Biosphere The part of the Earth system that directly supports life, including the oceans, atmosphere, land surface, and soils.

Biota All living organisms.

Bioturbation The stirring of sediments by worms and other burrowing organisms that live at the sea floor.

Black smoker A mid-ocean ridge hydrothermal vent from which clouds of black particles of iron sulfide (FeS) are precipitating.

Blackbody radiation Radiation given off by a blackbody, or a body that emits equally well at all wavelengths. This radiation is characterized by the body's absolute temperature.

Body waves Seismic waves that travel through Earth's interior as they spread outward from the earthquake's focus; categorized as either P waves or S waves.

Bottom water Very dense, cold water that forms along the edges of sea ice in certain areas near the poles and subsides, making up the bottom layer of ocean water as it circulates throughout the world's oceans.

Boundary conditions The physical conditions (such as land/sea distribution, surface elevation, solar constant) that remained fixed for the duration of a model simulation.

Boyle's law A relationship stating that, with temperature held constant, the pressure and volume of a gas are inversely proportional. In other words, the product of pressure and volume is a constant.

Buffer A dissolved substance that helps maintain a stable pH. Carbonate ion is an important pH buffer in seawater.

Buoyancy A tendency of an object to float, rise, or sink when submerged in a fluid.

C_3 plants Plants that fix carbon into a three-carbon chain during photosynthesis. These plants (which include about 95% of all terrestrial plants) should be positively affected by CO_2 fertilization.

C_4 plants Plants that fix carbon into a four-carbon chain during the initial step of photosynthesis. These plants (which include about 5% of all terrestrial plants, such as corn and sugar cane) experience little CO_2 fertilization at CO_2 levels near 350 ppm.

Calvin cycle The biochemical pathway by which green plants convert CO_2 into organic carbon during photosynthesis, a method of carbon fixation.

Cap carbonates Thick layers of carbonate rocks overlying glacial deposits. They are thought by some geologists to be formed from rapid weathering in the aftermath of Snowball Earth episodes.

Captured rotation A rotation in which a planet (or moon) always shows the same side to a star (or planet). Earth's Moon is a good example.

Carbon dioxide (CO_2) A gas containing one carbon atom and two oxygen atoms that is one of two primary greenhouse gases in Earth's atmosphere (the other one being water vapor).

Carbon fixation The biochemical process that occurs during photosynthesis by which atmospheric CO_2 is converted to organic carbon.

Carbonate metamorphism A chemical reaction occurring at high temperatures and pressures between sedimentary carbonate minerals and silica-rich sediments that forms calcium silicate minerals and releases CO_2.

Catalyst A compound that increases the rate of a chemical reaction and is itself unchanged by the reaction.

Catalytic cycle A set of chemical reactions facilitated by the presence of a catalyst. Ozone can be destroyed by catalytic cycles involving nitrogen oxides and chlorine.

Cation A positively charged ion.

Chaotic orbit An orbit that is particularly sensitive to slight perturbations that lead to large, unpredictable changes in an orbiting object's position at a later time.

Chapman mechanism A set of four chemical reactions that describe the production and loss of ozone in a simplified (pure N_2–O_2) atmosphere.

Characteristic response time How long a reservoir takes to respond measurably to large imbalances in inflow or outflow, calculated in the same way as residence time.

Charles's law A relationship stating that, with pressure held constant, the volume and temperature of a gas are directly proportional. In other words, the quotient of volume and temperature is a constant.

Chemical evolution The sequence of chemical reactions leading up to the origin of life.

Chemosynthesis The biological production of organic matter by bacteria that utilize energy stored in chemical compounds (inorganic matter) rather than solar energy.

Chlorofluorocarbons (CFCs) Synthetic compounds containing chlorine, fluorine, and carbon. These gases, also called freons, contribute to the greenhouse effect and are harmful to the ozone layer.

Chloroplasts Inclusions (organelles) within cells that are responsible for photosynthesis. They contain DNA that is remarkably similar to cyanobacterial DNA.

Chondrites "Primitive" meteorites that have never been melted. They are thought to be remnants of the original material from which the planets formed.

Cloud condensation nuclei (CCN) Small particles, often consisting of sulfate aerosols, that catalyze the condensation of water vapor into cloud droplets.

CO_2 compensation point The critical CO_2 concentration below which plants are unable to photosynthesize and hence, grow. Technically, it is the CO_2 level below which the rate of photorespiration exceeds that of photosynthesis.

CO_2 fertilization An increase in the growth rate of plants by the addition of CO_2 to the atmosphere.

CO_2 tax A proposed tax on CO_2 emissions; one mechanism that has been suggested to reduce the rate at which fossil fuels are consumed.

Coal A hydrogen- and carbon-bearing compound produced under high-pressure and high-temperature conditions of burial deep within the solid Earth; a fossil fuel formed from high concentrations of organic matter in terrestrial sediments.

Column depth The total amount of ozone per unit area above Earth's surface. Usually reported in Dobson units.

Comets Balls of frozen gases (water, ammonia, methane, and carbon dioxide) containing rocky and metallic debris that are mostly in distant orbits around the Sun, beyond Pluto (see *Oort cloud*).

Community A characteristic assemblage of two or more groups of interacting species.

Component An individual part of a system. A component may be a reservoir of matter or energy, a system attribute, or a subsystem.

Condensation The process by which a gas becomes a liquid.

Conduction Transfer of heat energy by direct contact between individual molecules.

Consumers Organisms, such as animals, that are incapable of utilizing solar energy directly but must instead consume plants or other photosynthesizers to utilize the chemical energy stored in plant tissues.

Continental drift Theory stating that the supercontinent Pangea began to break apart at the beginning of the Mesozoic era (200 million years ago) and that the continents then slowly drifted into their current positions; a concept proposed by Alfred Wegener in the early 20th century that has been supplanted by the (related) theory of plate tectonics.

Continental shelves The shallow, submerged part of continental margins, bounded on the landward side by the coastline and on the seaward side by steep slopes falling to great depths.

Continuously habitable zone The region around a star where a planet can remain habitable for some finite time interval.

Convection Transfer of heat energy by the circulating motions of a fluid that is heated from below; one of the three primary mechanisms of heat transfer.

Convergence As used in this book, the inward movement of air or water to a region in the atmosphere or ocean.

Core The central part of a planet or of the Sun. Earth's core—one of the three components of the solid Earth—is dense, is composed mostly of metallic iron and nickel and has a solid inner and a liquid outer part.

Coriolis effect The apparent tendency for a fluid (air or water) moving across Earth's surface to be deflected from its straight-line path. Fluids are deflected to the right of their initial path in the Northern Hemisphere and to the left in the Southern Hemisphere.

Corona The extremely hot (1–2 million K) outer layer of the Sun from which the solar wind arises. The corona becomes visible during a total solar eclipse.

Cost–benefit analysis An economic analysis in which the costs of a certain action are weighed against its benefits.

Coupling The links between any two components of a system. Couplings can be positive or negative.

Craton The old, tectonically dormant regions of the continents that have served as the nuclei for continental accretion over the past 3 to 4 billion years.

Crust The thin, outer layer of the solid Earth; consists of light, rocky matter that is in contact with the atmosphere, hydrosphere, and biota.

Cyanobacteria Bacteria capable of both oxygenic and anoxygenic photosynthesis; formerly known as blue-green algae.

Decomposition The breakdown of organic matter, a process typically carried out by certain bacteria and fungi.

Deforestation The clearing of all the trees off an area of land.

Dendrochronology A method of dating trees by counting their annual growth rings. The ring widths can be used to infer environmental conditions during the growing season.

Denitrification The conversion of fixed nitrogen into N_2 or N_2O by bacteria.

Deoxyribonucleic acid (DNA) A double-stranded molecule that is the basis for information storage by all living organisms.

Deserts Areas of low rainfall, generally less than 250 mm per year. Deserts may be warm (such as those located in the subtropics) or cold (as in the central plateau of Antarctica).

Detrital mineral A mineral grain found in a sediment that was never completely dissolved during the weathering process.

Deuterium A stable isotope of hydrogen that has one neutron in its nucleus; denoted as ^2H or D.

Discount rate A factor by which future costs or benefits are reduced in each succeeding year to account for economic growth and/or the preference of people for having money sooner rather than later.

Dissociate To come apart. Acids dissociate in solution, forming anions plus hydrogen ions (protons).

Divergence As used in this book, the outward movement of air or water from a region in the atmosphere or ocean.

Dobson unit (DU) A unit for measuring the column depth of ozone. One DU is equivalent to a layer of pure ozone 0.001 cm thick at 1 atmosphere pressure.

Domains The three primary divisions of life, as determined from sequencing of ribosomal RNA. Includes the *Archaea,* the *Bacteria,* and the *Eukarya.*

Dose rate The number of UV photons per unit time that lead to a specific biological response, such as sunburn or skin cancer.

Double helix The spiral pattern formed by two interlinked strands of DNA.

Downwelling The sinking of surface water caused by convergence and water accumulation at the surface.

Drake equation A formula used to estimate the number of other intelligent civilizations in our galaxy with which we might one day establish radio communication.

Dropstones "Misplaced" chunks of rock dropped into marine sediments by melting icebergs.

Earthquake The sudden release of stored energy as the result of rapid movement between two lithospheric blocks.

Eccentricity The degree to which a rotating object's orbit is elliptical. Eccentricity is defined for Earth's orbit as the distance from the center of the orbit to its foci (one of which is occupied by the Sun), divided by the average distance between Earth and the Sun.

Ecosphere Synonym for *habitable zone.*

Ecosystem A subset of the biota, consisting of assemblages of plant, animal, and microbial species that interact with each other and with their surrounding environment.

Ecotone The diffuse boundary between two ecosystems.

Effective radiating temperature The temperature a planet such as Earth would have if the planet radiated as a blackbody (or if it had no atmosphere).

Ejecta The debris thrown away from an impact event.

Ekman transport The net direction of transport in the water column as a result of the Ekman spiral. The movement is at 90° to the wind direction (to the right in the Northern Hemisphere and to the left in the Southern Hemisphere).

El Niño Originally, a warm ocean current that appears off the coast of Peru and Ecuador shortly after Christmas and flows for only a few weeks; the name now describes a major shift in the oceanic circulation that occurs in this region every 2 to 10 years.

El Niño–Southern Oscillation (ENSO) A climatic event in the tropical Pacific Ocean in which the main area of surface convection moves from the western to the central Pacific. This event is associated with large-scale changes in the ocean circulation, the atmospheric circulation, and tropical precipitation patterns. The effects of an ENSO event may also spread beyond the tropics, causing anomalous weather conditions in many mid-latitude locations.

Electromagnetic radiation A self-propagating electric and magnetic wave such as visible light, ultraviolet, or infrared radiation.

Electromagnetic spectrum The full range of different forms of electromagnetic radiation, which differ by wavelength (or, conversely, by frequency).

Endemic species Species that are indigenous, or native, to a particular location.

Endosymbiosis A biological system in which one organism lives inside another in a mutually interdependent relationship.

Energy-balance climate model Two-dimensional numerical climate model that computes vertical and latitudinal exchanges of energy and allows for some latitudinal transport of energy and moisture.

Enzymes Proteins that catalyze various biochemical reactions.

Eon One of the four major subdivisions of geologic time.

Equilibrium climate experiment Experiment in which the model is driven by a fixed set of boundary conditions and the model is integrated forward in time until it reaches equilibrium. There is year-to-year variability but little further change in the trend over time. The usual practice is then to change one of the boundary conditions (e.g., atmospheric greenhouse gas composition) and run the model again. The difference in the two equilibrium states is then analyzed.

Equilibrium state A state in which the system is in equilibrium, that is, the state in which the system will remain unless something disturbs it. An equilibrium state can be stable or unstable.

Equilibrium surface warming The surface warming that will occur when Earth's surface achieves full thermal balance.

Erosion The transport of the products of weathering to basins where sediment accumulates; the process whereby crustal materials, decomposed and loosened by weathering, are transported by winds, landslides, and streams.

Erythemal action spectrum The action spectrum for sunburn in humans.

Eukaryotes (Eukarya) One of the three primary domains of life, as determined by sequencing of ribosomal RNA. The Eukarya (also known as eukaryotes) are organisms that have cell nuclei. They include all higher plants and animals.

Evaporation The process by which a liquid is converted to a gas.

Evaporite deposit Mineral deposit formed by the evaporation of seawater from shallow seas. The remaining salts are concentrated and precipitate from solution.

Evolution The descent, with modification, of preexisting life forms. The *natural selection* of favorable genetic mutations is the dominant mechanism of evolution.

Exploitation efficiency The proportion of the available biomass that is transferred from one trophic level to the next.

Extinction The loss of all individuals within a species.

Feedback factor The ratio of the equilibrium response to forcing (the response with feedback) to the response without feedback. Feedback factors less than 1 are indicative of negative feedback: The equilibrium response (with feedback) is smaller than the response to forcing without feedback. Feedback factors greater than 1 indicate positive feedback: The equilibrium response is larger than the response to the forcing itself.

Feedback loop A linkage of two or more system components that forms a round-trip flow of information. Feedback loops can be positive or negative.

Fermentation An anaerobic form of metabolism in which both oxidized carbon and reduced carbon are produced (often generating methane); a decomposition process typically performed by anaerobic bacteria.

Ferric iron Iron in its highest (3+) oxidation state; insoluble in water.

Ferrous iron Iron in its intermediate (2+) oxidation state; soluble in seawater.

Finite difference model Model in which the domain is divided into grid boxes. The calculations of energy, mass, and momentum transfer are computed as some function of the difference in various quantities (e.g., temperature) between vertically and horizontally adjacent grid boxes.

Fixed nitrogen Compounds containing a nitrogen atom bonded to something other than another nitrogen atom.

Flux The amount of energy (or number of photons) in an electromagnetic wave that passes perpendicularly through a unit surface area per unit time.

Food chain A progression of organisms each dependent on the one before for food.

Food web An intricate interlacing of food chains, more typical of natural communities.

Forcing A persistent disturbance of a system; a longer-term disturbance than a perturbation.

Formaldehyde (H_2CO) A chemical that could have formed in the early atmosphere and that can be used to synthesize sugars and amino acids.

Fossil fuels Fuels such as coal, oil, and natural gas that are formed from the nondecomposed organic remains of organisms, concentrated in sedimentary rocks.

Fractionated Separated according to some parameter. When applied to different isotopes, the term implies separation by mass.

Frequency The number of wave crests that pass a fixed point in 1 second.

Gaia hypothesis A theory suggesting that Earth is a self-regulating system in which the biota play an integral role.

General circulation model (GCM) A three-dimensional computer model of the global atmosphere (or ocean) that simulates winds (currents), moisture transport, and energy balance; also called a global climate model.

General gas law The relationship that combines Boyle's law and Charles's law and states that the product of the pressure and volume of a gas is directly proportional to temperature. In other words, the quotient of the product (pressure) × (volume) and temperature is a constant.

Genetic code The key that determines how sequences of three different nucleotides represent different amino acids. Using this key, cells are able to manufacture various proteins on the basis of instructions stored in the cell DNA.

Geostrophic current A current that flows around oceanic gyres; produced where the Coriolis effect that deflects the flow into the center of a gyre is balanced by the downslope flow from the higher sea-surface elevations in the gyre center. The resulting flow is clockwise in the Northern Hemisphere (counterclockwise in the Southern Hemisphere), approximately parallel to the ocean slope, and in the same direction as the wind-driven flow.

Geothermal heat Heat flowing from Earth's interior up to its surface.

Geothermal power The production of electricity by using temperature gradients within the solid Earth as the energy source.

Giant impact hypothesis A theory of lunar formation in which the Moon forms as the result of a glancing collision between Earth and a Mars-sized body.

Glacial interval An interval of time during the Pleistocene when ice sheets covered much of northern North America and Scandinavia and other parts of northern Europe and Asia (as well as Greenland and Antarctica). Globally averaged surface temperatures during glacial intervals were about 10°C, and atmospheric CO_2 concentrations were about 200 ppm.

Glacial striations Grooves carved into bedrock by rocks frozen to the base of a moving glacier.

Glacial surge The sudden, rapid movement of a glacier.

Global climate model Three-dimensional numerical model that includes many components of the global climate system. These models now incorporate atmosphere, ocean, land, and sea ice components, as well as various aerosol and carbon cycle models.

Global warming A warming of Earth's atmosphere due to an anthropogenic enhancement of the greenhouse effect.

Granite A common igneous silicate rock that solidifies below Earth's surface and forms the cores of many mountain ranges. Granite is less dense than basalt, so when a continental plate collides with an oceanic plate, the oceanic plate is subducted beneath the continental plate.

Greenhouse effect The natural mechanism by which a planet's surface is warmed by infrared-absorbing gases in its atmosphere.

Greenhouse gases Gases such as carbon dioxide, methane, nitrous oxide, and water vapor that warm a planet's surface by absorbing infrared radiation and reradiating some of it back toward the surface. Greenhouse gases, whether natural or anthropogenic, contribute to the atmospheric greenhouse effect.

Groundwater Water that penetrates through soil and rock and collects below the surface.

Gyre A large, circular circulation pattern in the ocean. Gyres in the Northern Hemisphere circulate clockwise, whereas those in the Southern Hemisphere circulate counterclockwise.

H_2O rotation band A strong absorption band that is a result of rotational frequencies of the water molecule in vapor form. The H_2O molecule absorbs infrared radiation of wavelengths about 12 μm and longer.

Habitable zone The region around a star in which a life-supporting planet might be found; generally, the region where liquid water could exist on a planet's surface.

Habitat The plant community and physical environment that supports a single species.

Hadley circulation The process by which an air mass undergoes convergence at the tropics and divergence at about 30° N or 30° S latitude in one large convection cell.

Half-life The time it takes for half the initial quantity of radioactive isotope to decay.

Halocline A steep salinity gradient in the pycnocline zone that marks the transition between the surface zone and the deep ocean. Salinity rises rapidly with increasing depth in the halocline.

Halons Artificial chemicals containing the elements bromine, chlorine, fluorine, and carbon, used mostly in fire extinguishers.

Heavy bombardment period The time interval between 4.6 Ga and 3.8 Ga when Earth and the other terrestrial planets were being regularly bombarded by large planetesimals.

Hematite (Fe_2O_3) A mineral composed of fully oxidized iron. It has a reddish color when present as small, dispersed grains.

Hertzsprung-Russell diagram (H-R diagram) A type of graph in which stars are displayed in terms of their absolute luminosities (vertical scale) and surface temperatures (horizontal scale).

Heterocysts Specialized cells within some filamentous bacteria that are responsible for nitrogen fixation.

Heterogeneous reaction A chemical reaction that occurs on a solid surface.

Heterotroph An organism (*consumer*) that depends on other organisms (*autotrophs* or *producers*) to produce its food.

Holocene The geological epoch extending from 10,000 years to the present; an interglacial interval.

Holocene Climatic Optimum A warm period that occurred during the mid-Holocene.

Homogeneous reaction A chemical reaction between molecules that are in the gas phase.

Hydrogen cyanide (HCN) A molecule believed to be an essential building block for life.

Hydrologic cycle The major reservoirs of water in the Earth system and the pattern of water storage and movement throughout that system.

Hydrosphere The component of Earth system that includes the various reservoirs of water and ice on Earth's surface.

Hydrothermal vents Cracks in the sea floor, especially around mid-ocean spreading ridges, through which seawater circulates.

Hydroxyl radical A highly reactive molecule that plays a variety of important roles in atmospheric chemistry; chemical formula OH.

Hyperthermophilic bacteria Organisms that have optimal growth temperatures above 80°C (176°F).

Ice shelves Floating sea ice formed at the margins of continents.

Ice-rafting A process by which chunks of rock are carried to sea by icebergs. The icebergs eventually melt, and the rocks may be preserved as dropstones in sediments.

Igneous rock Rock formed by the cooling and solidification of magma. If the magma solidifies beneath Earth's surface, the rocks are intrusive; if the magma erupts as lava at a volcano, the magma cools rapidly into extrusive rocks.

Impact degassing The venting of water and other volatile compounds directly into a planet's atmosphere during impacts of comets or asteroids onto its surface.

Infrared (IR) radiation Electromagnetic radiation of fairly low energy and wavelengths longer than those of visible light from 0.7 μm to 1000 μm.

Inorganic carbon Carbon not associated with compounds that are typically formed by living organisms and that do not contain carbon–carbon or carbon–hydrogen bonds.

Instrumental value The degree to which a species' existence benefits another species (normally, humans) in some way.

Interferometer An instrument consisting of two or more telescopes connected electronically so that their images can be combined.

Intergenerational equity Treating different generations of people equally (in economic terms). Multigenerational problems such as global warming bring up considerations of intergenerational equity.

Interglacial period An interval during the Pleistocene, such as the Holocene, when continental ice sheets were restricted to Greenland and Antarctica. Globally averaged surface temperatures during interglacial intervals were about 15°C, and atmospheric CO_2 concentrations were about 280 ppm.

Interplanetary dust particles (IDPs) Small particles collected in Earth's stratosphere that originated in space.

Interstellar clouds Clouds of dust and gas from which stars occasionally form.

Intertropical convergence zone (ITCZ) A region of the tropics where surface heating causes uplift in the atmosphere, allowing subtropical air to flow inward to produce a convergence zone. This zone moves north and south of the equator as the seasons change.

Intrinsic value A species' value for its own sake, regardless of whether it benefits humans.

Inverse-square law A relationship describing the rate at which the solar flux decreases with increasing distance.

Iridium An element that is rare at Earth's surface but is concentrated in Earth's core and in extraterrestrial materials.

Isochron diagram A diagram in which one plots the abundance of two different radiogenic isotopes (e.g., ^{206}Pb and ^{207}Pb) measured in a number of different rock samples against each other. If the rocks all have the same age, then the data should fall on a straight line. The slope of the line gives the age of the rocks.

Isotopes Atoms of a given element that have different numbers of neutrons in their nuclei.

K-T boundary The boundary between the Cretaceous (K) and Tertiary (T) periods, about 65 million years ago, when the dinosaurs and many other species went extinct.

Kelvin (absolute) temperature scale A metric temperature scale in which the degree has the same size as a Celsius degree, but in which the zero point is moved downward by 273.15°, to absolute zero.

Kerogen Dispersed organic matter (hydrocarbons) in rocks.

Keystone species A species that plays a vital function in the operation of an ecosystem.

La Niña The opposite phase of the Southern Oscillation from El Niño conditions. It represents a stronger or more extreme version of the "normal" circulation in the tropical Pacific.

Latent heat The heat energy released or absorbed during the transition from one phase to another, such as when water evaporates.

Latent heat of fusion The energy required to effect a change of phase between a solid and a liquid. Converting a solid to a liquid requires an addition of energy; converting a liquid to a solid releases energy to the environment.

Latent heat of vaporization The energy required to effect a change of phase between a liquid and a gas. Converting a liquid to a gas requires an addition of energy; converting a gas to a liquid releases energy to the environment.

Lead–lead dating A radiometric age dating technique that relies on measurements of two different isotopes of lead produced from two uranium isotopes that have different half-lives.

Limestone A sedimentary rock composed largely of calcium carbonate minerals, mainly calcite.

Lithification The transformation of sediments into sedimentary rocks; typically involves compaction and the precipitation of mineral cements between sediment grains.

Lithosphere The uppermost mantle and rigid crust, above the asthenosphere. The lithosphere is divided into plates that move relative to one another in the process of plate tectonics.

Little Ice Age An interval of colder temperatures from the 15th to 19th centuries, interrupted by a warmer interval in the 17th century.

Loess Wind-blown glacial silt.

Logistic growth Growth of a population that increases in size exponentially when the population is small, but then approaches a constant size when the population is large. For logistic growth to occur, as the population size increases, either the population growth rate decreases or the death rate increases.

Luminosity The brightness of a star such as our Sun.

Macrofossils Fossils of multicellular organisms that are large enough to see easily with the naked eye.

Magma Molten, or liquid, rock that forms igneous rock when cooled.

Magma ocean A layer of molten rock covering the entire surface of a planet. The Moon is thought to have had one early in its history. Earth may have had one as well.

Magnetic dynamo The mechanism whereby convection of the liquid-iron outer core generates Earth's magnetic field.

Main-sequence star A middle-aged star that lies on a band of stars running from the upper left to the lower right of a plot of luminosity versus effective radiating temperature (H-R diagram).

Mantle One of the three layers of the solid Earth; a thick, rocky layer between the core and crust. Composed primarily of silicate minerals.

Mass extinction An extraordinary extinction event in which more than 25% of all extant families are lost.

Mass number The combined number of protons and neutrons in the nucleus of an atom.

Maunder Minimum The interval from A.D. 1645 to 1715, when very few sunspots were recorded.

Medieval warm period A period of mild conditions in Northern Europe and North Atlantic that reached a maximum at about A.D. 1100.

Mesosphere An atmospheric layer that extends from about 50 km to 90 km above the surface; temperature decreases with altitude there.

Metamorphic rocks Rocks formed from exposure to high temperatures, high pressures, chemically active fluids, or any combination of these agents. The rocks must remain in the solid phase to be classified as metamorphic; if melting occurs, they are called igneous.

Meteorite A comet or asteroid that strikes Earth. Strictly speaking, the material that remains after such an impact.

Methane clathrate hydrate An ice-like combination of methane (CH_4) and water (H_2O) in which one methane molecule is trapped in a "cage" of 5 or 6 water molecules.

Methane sulfonic acid (MSA) An acid that is produced from biogenic dimethyl sulfide and forms cloud condensation nuclei in the atmosphere.

Methanogenesis Methane production carried out by certain bacteria.

Methanogenic bacteria Anaerobic bacteria that convert carbon dioxide and hydrogen (or other substances) into methane.

Methanogens Methanogenic bacteria.

Methanotropic bacteria Aerobic bacteria that consume methane and incorporate the carbon from that methane into their body tissues.

Microfossils The fossilized remains of single-celled organisms.

Mid-ocean ridge A linear chain of subsea volcanic mountains on the sea floor that is the site of formation of new oceanic lithosphere.

Mie scattering Scattering of radiation by particles that are of approximately the same size as the wavelength of radiation being scattered. Reduces to Rayleigh scattering (or geometric optics, i.e., ray tracing) in the small (large) particle limit.

Mixed layer The surface layer of the ocean that is mixed by wind action.

Moho The crust/mantle boundary, marked by a sharp increase in seismic wave speeds.

Monsoon A seasonal reversal in the surface winds caused by large-scale differential heating of land and ocean surfaces. Monsoon circulation is defined by the windfields but usually also has a direct impact on rainfall.

Montreal Protocol An international treaty signed in 1987 that limited freon and halon emissions.

Moraine A ridge of sediment deposited by a glacier.

Mountain glaciers Ice fields formed on the cold, upper reaches of mountains.

Mutation A random change in the DNA of an organism. This can lead to evolution if the mutation is advantageous to the survival of the organism.

Natural selection The unequal survival and reproduction of organisms, owing to environmental pressures that result in the preservation of favorable adaptations. The relatively nonselective nature of mass extinctions suggests that natural selection may not play an important role in these large events.

Negative coupling A link indicating that a change (increase or decrease) in one component leads to a change of the opposite direction (decrease or increase, respectively) in the linked component.

Negative feedback loop A feedback loop with an odd number of negative couplings. Negative feedback loops tend to diminish the effects of disturbances.

Neutrinos Massless particles given off in nuclear reactions.

Nitrogen fertilization The stimulation of plant growth caused by the addition of anthropogenic fixed nitrogen, commonly from automobile exhaust.

Nitrogen fixation A process by which some organisms (primarily prokaryotes) convert N_2 into fixed nitrogen.

North Atlantic Deep Water (NADW) Cold, dense water that forms in the northernmost Atlantic Ocean, sinks, and flows southward at depth into the Atlantic Ocean.

Nuclear fission The splitting of a heavy atomic nucleus into two fragments, accompanied by the release of energy.

Nuclear fusion The combining of lightweight atomic nuclei into a heavier nucleus accompanied by the release of energy.

Nucleotides Organic compounds consisting of a sugar, a base, and a phosphate group. Nucleotides are the basic building blocks of DNA and RNA.

Number density A measure of the concentration of a gas. The units of number density are molecules per cubic centimeter.

Nutrient Substances normally obtained in the diet that are essential to organisms.

Obliquity The angle of a planet's spin axis relative to a line drawn perpendicular to the plane of the planet's orbit around the Sun; also called *tilt*.

Odd nitrogen Any pure nitrogen-containing compound that has an odd number of nitrogen atoms; NO and NO_2. Odd nitrogen is similar to fixed nitrogen in that the nitrogen atom is coupled to atoms other than another N atom.

Odd oxygen Any pure oxygen-containing atom or molecule that has an odd number of oxygen atoms; O and O_3.

Oort cloud A distant region of the solar system, beyond Pluto, from which long-period comets originate.

Organic carbon Carbon associated with compounds that are typically formed by living organisms and that contain carbon–carbon or carbon–hydrogen bonds.

Organism A living system.

Oxidation state The degree of oxidation of an atom, molecule, or compound. Substances with a low oxidation state have a large number of available electrons; substances with a high oxidation state do not.

Oxidative weathering Reactions between the atmosphere and surface rocks in which minerals react with O_2.

Oxidized carbon Carbon that, in compounds, is combined with oxygen. The carbon atoms in skeletons composed of $CaCO_3$ and in atmospheric CO_2 are oxidized carbon.

Oxygen minimum zone A zone at intermediate depths in the ocean—about 1 km from the surface—where dissolved oxygen concentrations reach a minimum (and nutrient concentrations reach a maximum), as a result of high oxygen demand by aerobic decomposers and low oxygen supply from the surface ocean or from below.

Oxygenic photosynthesis The normal, O_2-generating type of photosynthesis carried out by plants.

Ozone (O_3) A form of oxygen that is much less abundant than, and chemically unlike, the oxygen that we breathe. The ozone that is dispersed in the stratosphere blocks the Sun's harmful ultraviolet radiation.

Ozone hole A patch of extremely low ozone concentration in the ozone layer. This hole has appeared near the South Pole each October since about 1976.

Ozone layer A chemically distinct region of the atmosphere (specifically, the stratosphere) that protects Earth's surface from the Sun's harmful ultraviolet radiation.

Ozonesondes A balloon-borne instrument that measures the concentration of stratospheric ozone.

P wave Primary wave; a seismic body wave transmitted as a series of compressions and expansions in the overall direction of wave movement through Earth's interior. P waves can travel through fluids or solids.

Paleoclimate Past climate.

Paleoclimatology The study of past climates.

Paleontologist A scientist who studies the history of life using the fossil record.

Paleosols Ancient soils that can be used as an indicator of past atmospheric oxygen levels.

Palynology The study of pollen and organic microfossils.

Pangea A supercontinent consisting of all the land masses of Earth that formed about 300 million years ago and broke apart about 200 million years ago.

Parameterized Simulated on the basis of experiments or observations, as opposed to being calculated directly. For example, clouds generally need to be parameterized to be included in atmospheric general circulation models.

Partial pressure In a mixture of gases, the pressure a gas would exert if it were the only gas present (i.e., the contribution of each individual gas to the total pressure exerted by the mixture).

Perihelion The position in a planet's orbit that is nearest to the Sun.

Period A unit of geologic time, shorter than an era and longer than an epoch. Also, the time required for a planet to go around the sun.

Periodicity A time interval of regular recurrence of a phenomenon.

Perturbation A temporary disturbance of a system; a shorter-term disturbance than a forcing.

Petroleum A hydrogen- and carbon-bearing compound produced under high-temperature conditions of burial deep (3–4 km) within the crust; a fossil fuel formed from high concentrations of organic matter in marine sediments.

pH A measure of acidity, defined as the negative logarithm (to the base 10) of the hydrogen ion concentration (in moles per liter). Acids have pH values less than 7, and bases have pH values greater than 7. A pH of 7 is defined as neutral.

Photic zone The portion of the oceans where there is sufficient sunlight for photosynthesis; about the upper 100m of the water column.

Photochemical models Computer models used to simulate atmospheric chemistry.

Photochemical reactions Chemical reactions initiated by the absorption of a photon.

Photodissociation The splitting of a molecule by the absorption of light or by UV radiation; also called *photolysis.*

Photolysis The splitting of a molecule by the absorption of light or by UV radiation; also called *photodissociation.*

Photolyze To split a molecule apart with visible or ultraviolet radiation.

Photon A single, discrete particle, or pulse, of electromagnetic radiation.

Photorespiration Respiration induced by the absorption of sunlight. (The combining of organic matter with oxygen to yield CO_2, H_2O, and energy.)

Photosphere The surface layer of the Sun from which most of its energy, including visible radiation, is emitted.

Photosynthesis The process by which an organism such as a green plant uses sunlight, carbon dioxide, and water to produce organic matter and oxygen.

Phototrophic Attracted to light. Most phototrophic organisms use sunlight for some metabolic purpose, such as (but not necessarily limited to) photosynthesis.

Photovoltaic cells Specially designed panels that directly convert sunlight into electricity.

Plages Bright (higher temperature) areas that surround sunspots on the surface of the Sun.

Planetary vorticity The angular rotation about a vertical axis at Earth's surface brought about because of Earth's rotation.

Planetesimals Small proto-planets formed during planetary accretion.

Plate tectonics Theory by which the Earth's surface is divided into rigid plates of sea floor and continent that move relative to one another through time.

Pleistocene Epoch The geological epoch, extending from 1.8 million years ago to 10,000 years ago, that is characterized by oscillations in and out of the glacial state.

Polar front zone A zone of steep temperature gradients formed at approximately 60° N and 60° S latitude, where cold, polar air meets the warm air moving poleward from the subtropics.

Polar stratospheric clouds (PSCs) Collections of droplets, consisting of a mixture of water and nitric acid, that are involved in the formation of the Antarctic ozone hole.

Polar vortex The circular, downward-sinking whirlpool of stratospheric air over the poles in winter. The polar vortex is more pronounced in the Southern Hemisphere than in the Northern Hemisphere.

Polarity The direction of orientation of a magnetic field. Earth's polarity (the geographic location of the North and South Poles) flips irregularly on geological time scales.

Polymerase chain reaction (PCR) A method of amplifying small strands of DNA. This allows organisms (including humans) to be unambiguously identified from extremely small tissue samples.

Polymerize To join together in a long, repeating chain. Photolysis of methane leads to the formation of long hydrocarbon polymers.

Population All of the members of a single species that live in a given area.

Positive coupling A link indicating that a change (increase or decrease) in one component leads to a change of the same direction (increase or decrease, respectively) in the linked component.

Positive feedback loop A feedback loop with an even number of, or zero, negative couplings. Positive feedback loops tend to amplify the effects of disturbances.

Prebiotic synthesis The formation of complex, biologically important molecules from compounds existing in the natural environment.

Precession The rotation of a planet's spin axis around a line drawn perpendicular to its orbital plane.

Primary producer A plant (or other type of photosynthesizer or chemosynthesizer) that provides energy used by consumers.

Primary productivity The amount of organic matter produced by photosynthesis in a unit time over a unit area of Earth's surface; also called productivity.

Prokaryotes Single-celled organisms that lack a cell nucleus.

Proteins Biochemical compounds, composed of amino acids, that perform a wide variety of functions within cells, including assisting with DNA replication.

Proto-Sun The central bulge in the solar nebula that later formed the Sun.

Proxy data Data that cannot be obtained by direct measurement but can be inferred from other evidence.

Pycnocline A steep density gradient (caused by changing temperature, salinity, or both) that marks the transition between the surface zone and the deep ocean. On the order of a kilometer in thickness, it is characterized by a rapid downward increase in density. The steep density gradient in the pycnocline zone makes this layer very stable.

Pyrite (FeS_2) An iron-bearing mineral that becomes oxidized during weathering in today's oxygen-rich atmosphere.

Radiation (of heat energy) The outward transfer of heat energy in the form of electromagnetic rays emitted by a body; one of the three primary mechanisms of heat transfer.

Radiative-convective model (RCM) A one-dimensional computer model of the atmosphere that can be used to simulate the greenhouse effect. In an RCM, the climate system is greatly simplified by averaging the incoming solar and outgoing infrared radiation over Earth's entire surface.

Radiative forcing A term used by climatologists to describe the change in the net downward infrared flux at the tropopause caused by a given concentration of greenhouse gases, clouds, or aerosols.

Radical A molecule that is highly reactive because it has an unpaired electron in its outer shell.

Radioactive decay Radioactivity; the spontaneous disintegration of an unstable nucleus of one element, creating a different nucleus of a different element and releasing particles and radiation.

Radiometric dating A method of calculating the age of a sample of material by knowing the half-life of a radioisotope within it. This method has provided absolute dates for the geologic time scale and for other specific events in Earth's history.

Rare earth elements Elements with atomic numbers 57 to 71. The pattern of rare earth elements in banded iron-formations resembles that in modern mid-ocean ridge hydrothermal fluids.

Rayleigh scattering Scattering of radiation by particles that are small compared to the wavelength of radiation being scattered. Applies to scattering of sunlight by air molecules.

Reactive chlorine Chlorine compounds that either destroy ozone directly, such as Cl and ClO, or are readily converted into these compounds by photolysis.

Red giant A large, reddish star that is in the immediate post-main-sequence phase of its evolution.

Redbeds Sedimentary deposits in which the individual grains are coated with the mineral hematite (Fe_2O_3); thought to have formed under conditions of relatively high atmospheric O_2.

Redfield ratio The ratios of the nutrient elements that marine organisms incorporate into their tissues; these ratios appear to be nearly identical in all species of photoplankton. For these organisms, and for seawater, the atomic ratio of carbon:nitrogen:phosphorus is very nearly 106:16:1.

Reduced carbon Carbon that, in compounds, is combined mainly with other carbon atoms, hydrogen, or nitrogen; organic carbon is a form of reduced carbon. Reduced carbon tends to be reactive in the presence of oxygen gas.

Reduced gases Gases, such as H_2 or CH_4, that can react with oxygen. Such gases typically contain the element hydrogen.

Relative humidity The amount of water vapor contained by a unit volume of air divided by the amount of water vapor that volume would contain if the air were saturated.

Relative vorticity Vorticity produced by processes (other than Earth's rotation) that induce a rotary motion in a fluid (such as clockwise or counterclockwise surface wind patterns acting on the ocean surface, or current shear in the oceans).

Replication The process by which an organism produces a new version of itself, generally modified by mutation. One of the three critical steps of Darwinian evolution.

Residence time The average length of time a substance spends in a given reservoir that is at a steady state with respect to the processes that add and remove the substance to and from the reservoir. Residence time is calculated as the ratio of the reservoir size to the rate of inflow or outflow (which are equal at steady state).

Resolution In this context, resolution refers to the effective grid size of the model (spatial resolution) and the model time step (temporal resolution). It represents the spatial and temporal frequency with which the model carries out its calculations.

Ribonucleic acid (RNA) A single-stranded molecule involved in protein synthesis. RNA is capable of copying the genetic information stored in double-stranded DNA.

Ribosomes Small inclusions (organelles) within cells that are responsible for protein synthesis.

RNA world A hypothetical period of evolutionary history in which organisms were based entirely on RNA, unlike the modern world, in which organisms depend on RNA, DNA, and proteins.

Rock cycle The cyclical process of creation and destruction of rocks through tectonic, weathering, metamorphic, and igneous processes.

Rocks Consolidated mixtures of crystalline materials called minerals.

Runaway glaciation An out-of-control climate in which a planet's surface becomes entirely covered by snow and ice.

Runaway greenhouse An out-of-control climate in which all water on a planet's surface is present as vapor. The oceans are completely evaporated.

S wave Secondary or shear wave; a seismic body wave transmitted as displacements perpendicular to the overall direction of wave travel through Earth's interior. S waves can travel only through solids.

Salinity The salt content of a water mass; often expressed in parts per (modern usage requires that salinity be expressed without units).

Scattering Redirection of the path of an incident electromagnetic wave by particles (including air molecules) in its path.

Sea-floor spreading The formation of hot, new oceanic crust from magma extruded at mid-ocean ridges. Once it forms, the new sea floor spreads to the sides of the ridges and is replaced at the ridge axis by an even younger new sea floor.

Seasonal temperature contrast The difference in average temperatures between summer and winter.

Sedimentary rock Rock formed by the compaction and lithification of sediments or by the chemical or biochemical precipitation of minerals.

Sediments Layers of unconsolidated material that is transported by water or by air.

Seismic wave An earthquake-produced wave that ripples through Earth's interior, away from the earthquake's focus, as a result of the elastic deformation of the solid Earth. Two types of seismic waves are generated: body waves and surface waves.

Serpentinization Production of serpentine minerals via reaction of water with ultramafic (iron- and magnesium-rich) rocks. Produces either hydrogen or methane.

Shelly fossils The fossilized remains of shelled organisms. The first such fossils occur just before the Cambrian period at the beginning of the Phanerozoic eon.

Siderophile Literally "iron-loving," refers to elements that tend to dissolve readily into molten iron, and thus are concentrated in Earth's core.

Silicate mineral A mineral rich in silicon and oxygen. A major class of rock-forming minerals, the silicates make up most of Earth's crust.

Slab Downgoing plate; the portion of oceanic lithosphere that is subducted into the mantle at a subduction zone.

Solar maximum A time of high sunspot activity and corresponding high solar UV flux.

Solar minimum A time of low sunspot activity and corresponding low solar UV flux.

Solar nebula The cloud of gas and dust surrounding the Sun shortly after it formed.

Solar-power satellites Large satellites whose proposed function is to harvest sunlight in space and beam the energy to Earth.

Solar thermal power A method of conversion of sunlight into electricity in which the sunlight is first used to heat a fluid that, in turn, drives a turbine.

Solar wind A stream of charged particles, mostly hydrogen and helium ions, emanating from the Sun's corona.

Solid Earth The component of the Earth system that includes all rocks and all unconsolidated rock fragments. The core, mantle, and crust make up the solid Earth.

Southern Oscillation An oscillation in sea-level pressure between the western and central/eastern portions of the tropical Pacific Ocean.

Southern Oscillation Index A measure of the pressure difference between the western and central eastern parts of the tropical Pacific Ocean. Strong negative values indicate ENSO conditions, whereas strong positive values indicate La Niña (non-ENSO) conditions.

Speciation The origination of new species.

Spectral Model Three-dimensional atmospheric general circulation model in which the transfer equations are expressed as wave functions. The number of waves resolved determines the spatial resolution of the model—equivalent to the grid size in the finite difference models. Most current general circulation models are spectral models.

Spectroscopy A technique in which the composition of a material, such as a gas, is analyzed according to the details of its electromagnetic spectrum.

Spörer Minimum An interval of low sunspot activity between A.D. 1450 and 1534.

Stable equilibrium A state in which the system will remain if left undisturbed and to which the system will return when disturbed.

Stable isotope An isotope that does not spontaneously change into another isotope or into an atom of another element by radioactive decay.

State The set of important attributes of a system that characterize the system at a particular time.

Steady state A condition in which the state of a system component is unchanging in time. A reservoir is in steady state when the rates of inflow and outflow are equal.

Stefan-Boltzmann law A relationship stating that the flux of radiation emitted by a blackbody is related to the fourth power of the body's absolute temperature; derived from the Planck function.

Stratosphere The stable atmospheric layer between 10–15 km and 50 km above the surface; temperature increases with altitude there. The stratosphere contains most of Earth's ozone.

Stromatolites Laminated or domed structures in rocks produced by layers of single-celled organisms.

Strongly reduced atmosphere An atmosphere composed mostly of compounds rich in hydrogen (e.g., CH_4 and NH_3) and containing no molecular oxygen (O_2).

Subduction The process whereby one plate of oceanic lithosphere, and at times its sediment cover, is carried underneath another plate (either oceanic or continental) in a convergent margin.

Sub-grid-scale process Physical process that takes place on spatial scales that are smaller than the grid-scale of the model. These processes cannot be modeled directly using the primitive equations of energy, mass and momentum, but instead are represented by empirical functions based on the observed climate system.

Subsidence The sinking of air from higher levels in the atmosphere down toward the surface. Also the vertical movement of Earth's crust toward the mantle.

Succession A progressive change in species composition of a community, often in response to a disturbance, but sometimes resulting from the colonization of previously uninhabited areas (e.g., bare rock). Fast growing, rapidly reproducing species usually colonize first, but then are replaced by slower-growing but competitively advantaged species.

Sulfate-reducing bacteria Bacteria that make a metabolic living by combining sulfate with organic matter.

Sunspot Dark areas of lower-than-normal temperatures on the surface of the sun.

Surface waves Seismic waves that travel only across Earth's surface as they spread outward from the earthquake's surface.

Symbiosis A relationship between two species in which the two are dependent on each other to the degree that neither can live alone.

System An entity comprised of diverse but interrelated parts (components) that function as a complex whole.

Taxon An individual taxonomic group (species, genus, and so on).

Taxonomy The systematic organization of living or fossil organisms into a hierarchy.

Terrestrial planets The four rocky, inner planets of the solar system: Mercury, Venus, Earth, and Mars.

Thermal drag The ocean's ability to delay climate warming (or cooling) as a result of its large heat capacity.

Thermal expansion The tendency of a substance (such as water above 4°C) to expand when heated.

Thermocline A steep temperature gradient in the pycnocline zone that marks the transition between the surface zone and the deep ocean. Temperature drops rapidly with increasing depth in the thermocline.

Thermohaline circulation The circulation of the deep oceans; driven by density differences that result from variations in temperature and salinity.

Thermophiles Shorthand for *thermophilic bacteria.*

Thermophilic bacteria Bacteria that have optimal growth temperatures between 40°C and 80°C.

Thermosphere The atmospheric layer higher than about 90 km above the surface; temperature increases with altitude there.

Tidal locking radius The distance from a star within which a planet is likely to develop a captured rotation, that is, the same side of the planet always faces the star.

Tidal power The production of electricity by using long, floating booms to harness the energy of ocean tides.

Till Sediment transported by glaciers and ice sheets, comprising materials of a large range of grain sizes and compositions.

Tillite Rock composed of lithified till. One of several pieces of evidence used to identify past glaciations.

Trace gases Gases such as methane, nitrous oxide, and freons that are present in Earth's atmosphere in very low concentrations.

Tracers Atoms, ions, or chemical compounds whose sources and sinks are easily measured and are relatively well understood, so they can be used to determine oceanic (or atmospheric) circulation.

Transform fault A fracture in the lithosphere between two lithospheric plates that, relative to each other, are sliding past one another (parallel to the boundary) in opposite directions.

Transient model experiment A climate model experiment, different from an equilibrium experiment in that the forcing function (e.g., atmospheric greenhouse gas composition) is allowed to vary with time. The model then simulates the time-dependent response to the forcing—rather than the equilibrium response.

Transient response The time-dependent (nonequilibrium) change in a quantity, such as temperature, in response to a perturbation.

Tritium An unstable isotope of hydrogen in which two neutrons are present in each atomic nucleus; denoted as ^3H or T.

Troposphere The lowermost, convective layer of Earth's atmosphere between the surface and 10 to 15 km above it; temperature decreases rapidly with altitude there. Weather is confined to the troposphere.

T-Tauri wind An enhanced stellar wind that emanates from certain young stars. The T-Tauri wind is thought to be driven by accretion of material onto the star from the surrounding nebula. Hence, it is gone well before planets are fully formed.

Ultramafic rocks Igneous rocks that are rich in iron and magnesium and low in quartz. Peridotite is a common example.

Ultraviolet (UV) radiation Electromagnetic radiation of fairly high energy and wavelengths from 400 nm to about 10 nm, shorter than those of visible radiation.

Uniformitarianism The principle that processes operating today also operated in the past, and that past geologic events can be explained in terms of processes active today.

Unstable equilibrium A state in which the system will remain if left undisturbed, but even slight disturbances will carry the system to some other (stable) equilibrium state.

Unstable isotope An isotope that spontaneously changes into another isotope or into an atom of another element by radioactive decay.

Uplift Any process by which air is forced to rise upward in the atmosphere. Also an upward vertical tectonic movement of Earth's crust.

Upwelling The rising of cooler, nutrient-rich ocean water to the surface to replace warm, divergent surface water.

Uraninite (UO$_2$) A uranium-containing mineral that is normally destroyed during weathering in an O_2-rich atmosphere.

Vascular plants Plants that have a well-developed stem or trunk for transporting water and nutrients from the ground up to their leaves. They also tend to have well-developed root systems, which take up water and nutrients and help support the plant.

Visible radiation Visible light; electromagnetic radiation of moderate energy and a relatively narrow range of wavelengths, from about 400 nm to 700 nm. Within this range, the color of the light depends on its wavelength.

Visible spectrum The range of component wavelengths of visible light; the colors of the rainbow.

Volatile compounds Chemical compounds that vaporize (turn into gases) at relatively low temperatures.

Volcanic aerosol forcing A perturbation in the radiation budget caused by the presence of volcanic aerosols.

Vorticity The tendency of a fluid to undergo rotary motion. A tendency for counterclockwise motion is referred to as positive vorticity; a tendency for clockwise motion is negative vorticity.

Wavelength The distance between two adjacent wave crests.

Weakly reduced atmosphere An atmosphere composed mainly of N_2 and CO_2, plus small amounts of reduced (II-rich) gases and virtually no free O_2.

Weathering The physical or chemical break-up of rocks exposed at the Earth surface.

White dwarf A small, compact star that represents the final stage in the evolution of a star such as our Sun.

Wien's law A relationship stating that the flux of radiation emitted by a blackbody reaches its peak value at a wavelength that depends inversely on the body's absolute temperature; derived from the Planck function.

Wilson cycle A plate tectonic cycle of supercontinent assembly and destruction; each cycle lasts about 500 million years.

Wind power The production of electricity from windmills, utilizing Earth's solar-energy-driven winds.

Wolf Minimum An interval of low sunspot activity between A.D. 1282 and 1342.

Younger Dryas A climatic reversal to glacial conditions that occurred 10,500 years ago after the end of the last glaciation. Evidence of climate change at this time is seen in various parts of the world, but the main effects of the Younger Dryas event appear to be centered on the North Atlantic region.

Zircon A mineral composed of zirconium silicate that has been used to date some of the oldest rocks on Earth.

Zooplankton Free-floating, marine consumers, including small invertebrates and microorganisms that cannot photosynthesize and therefore feed on phytoplankton.

Index

Photo Credits

Chapter 1 p. 8 PH ESM, GEOSYSTEMS, 3/e, by Christopherson, (c)1997, Adapted by permission of Pearson Education, Inc., Upper Saddle River, N.J.

Chapter 3 p. 35 Getty Images, Inc. Photodisc. **p. 36** PH ESM, 1999 Prentice Hall. **p. 49** Corbis/Stock Market, (c) Claudia Parks/CORBIS.

Chapter 4 p. 58 PH ESM, GEOSYSTEMS, 3/e, by Christopherson, (c)1997, Adapted by permission of Pearson Education, Inc.,Upper Saddle River, N.J. **p. 61** National Climatic Data Center Federal Building, courtesy of NOAA/National Climatic Data Center. **p. 67** PH ESM, PHYSICAL GEOGRAPHY: A LANDSCAPE APPRECIATION 6/e by McKnight/Hess, (c)2000 Adapted by permission of Pearson Education, Inc., Upper Saddle River, NJ. **p.68** PH ESM, EARTH SYSTEMS, by Kump/Kasting/Crane, (c)1999, Reprinted by permission of Pearson Education, Inc., Upper Saddle River, NJ. **p.69** PH ESM, PHYSICAL GEOGRAPHY: A LANDSCAPE APPRECIATION 6/e by McKnight/Hess, (c)2000 Adapted by permission of Pearson Education, Inc., Upper Saddle River, NJ. **p.80** PH ESM, GEOSYSTEMS, 3/e, by Christopherson, (c)1997, Adapted by permission of Pearson Education, Inc., Upper Saddle River, NJ.

Chapter 7 p.120 PH ESM, EARTH: AN INTRODUCTION TO GEOLOGIC CHANGE by Judson/Richardson, (c)1995, Adapted by permission of Pearson Education, Inc., Upper Saddle River, NJ

Chapter 10 p.201 Photo Researchers Inc.

Chapter 11 p.209 Peter Arnold, Inc. **p.211** Calvin Hamilton. **p.212** Photo Researches, Inc. **p.213** Visuals Unlimited. Dr. Gernot Arp/Gernot Arp, University of Gottingen, Gottingen, Germany, and Christian Boker, Carl Zeiss Jena, Germany, Susan Barns/Susan Barns and Norman R. Pace, University of Colorado. **p.217** Estate of Preston Cloud, from P. Cloud. "Oasis in Space: Earth History from the Beginning." (New York: Norton and Company, 1988). **p.222** Visuals Unlimited. **p.225** The Field Museum (C) The Field Museum, Neg #GEO85637C, Chicago. Photographer: John Weinstein.

Chapter 12 p.231 Photo Researchers, Inc. **p.237** J. William Schopt Courtesy of William Schopf, UCLA.

Chapter 14 p.271 PH ESM, EARTH'S DYNAMIC SYSTEMS 8/e by Hamblin/Christiansen, (c)1998, Adapted by permission of Pearson Education, Inc., Upper Saddle River, NJ. **p. 272** Peter Arnold, Inc., TLM Photo by Tom L. McKnight., **p.274** PH ESM, EXPLORING EARTH: AN INTRODUCTION TO PHYSICAL GEOLOGY by Davidson/Reed/Davis, (c)1997 Adapted by permission of Pearson Education, Inc., Upper Saddle River, N.J., **p.284** Photo Researchers, Inc.

Chapter 15 p.301 PH ESM, 1999 Prentice Hall. After Bluth, G. J. S., et al., 1992, Global Tracking of the SO2 clouds from the June 1991 Mount Pinatubo eruption. Geophysical Research Letters, 19:151-154. **p.310** PH ESM, EARTH SYSTEMS, by Kump/Kasting/Crane, (c)1999 Reprinted by permission of Pearson Education, Inc., Upper Saddle River, NJ.

Color Insert p.1 National Oceanic and Atmospheric Administration/Seattle. **p.11** Peter Arnold, Inc., Calvin Hamilton. **p.12** Photo Researchers, Inc., Visuals Unlimited, Gernot Arp/Gernot Arp, University of Gottingen, Gottingen, Germany, and Christian Boker, Carl Zeiss Jena, Germany, Susan Barns/Susan Barns and Norman R. Pace, University of Colorado. **p.14** The Field Museum (c) The Field Museum, Neg #GEO85637C, Chicago. Photographer: John Weinstein. **p.15** Photo Researchers, Inc.